SEMITOPOLOGICAL VECTOR SPACES

Hypernorms, Hyperseminorms, and Operators

SEMITOPOLOGICAL VECTOR SPACES

Hypernorms, Hyperseminorms, and Operators

Mark Burgin, PhD

AAP | APPLE ACADEMIC PRESS

Apple Academic Press Inc.
3333 Mistwell Crescent
Oakville, ON L6L 0A2 Canada

Apple Academic Press Inc.
9 Spinnaker Way
Waretown, NJ 08758 USA

First issued in paperback 2021

Exclusive worldwide distribution by CRC Press, a member of Taylor & Francis Group

ISBN-13: 978-1-77463-666-4 (pbk)
ISBN-13: 978-1-77188-534-8 (hbk)

Library and Archives Canada Cataloguing in Publication

Burgin, M. S. (Mark Semenovich), author
Semitopological vector spaces : hypernorms, hyperseminorms, and operators / Mark Burgin, PhD.

Includes bibliographical references and index.
Issued in print and electronic formats.
ISBN 978-1-77188-534-8 (hardcover).--ISBN 978-1-315-20747-6 (PDF)
1. Vector spaces. 2. Linear topological spaces. 3. Operator theory. I. Title.

QA186.B87 2017 515'.73 C2016-908260-1 C2016-908261-X

Library of Congress Cataloging-in-Publication Data

Names: Burgin, M. S. (Mark Semenovich)
Title: Semitopological vector spaces : hypernorms, hyperseminorms, and operators / Mark Burgin, PhD.
Description: Toronto : Apple Academic Press, 2017. | Includes bibliographical references and index.
Identifiers: LCCN 2016057228 (print) | LCCN 2016057711 (ebook) | ISBN 9781771885348 (hardcover : alk. paper) | ISBN 9781315207476 (ebook)
Subjects: LCSH: Vector spaces. | Linear topological spaces. | Operator theory.
Classification: LCC QA186 .B87 2017 (print) | LCC QA186 (ebook) | DDC 515/.73--dc23
LC record available at https://lccn.loc.gov/2016057228

Apple Academic Press also publishes its books in a variety of electronic formats. Some content that appears in print may not be available in electronic format. For information about Apple Academic Press products, visit our website at **www.appleacademicpress.com** and the CRC Press website at **www.crcpress.com**

CONTENTS

PREFACE

Mathematical structures that combine topological and algebraic configurations, such as normed vector spaces defined in an axiomatic form by Banach, topological vector spaces introduced by Kolmogorov and locally convex spaces brought in by von Neumann, which form an important type of topological vector spaces, play an important role in functional analysis and are studied in many books and papers. These spaces are related to such areas as distributions, integration and measure theory. A new direction in functional analysis, called quantum functional analysis, has been developed based on polynormed and multinormed vector spaces and linear algebras. In addition, normed vector spaces and topological vector spaces play an important role in physics and in control theory.

However, there are useful structures in functional analysis that are neither normed nor topological vector spaces. To include these structures in the context of functional analysis, semitopological vector spaces, hypernormed vector spaces and hyperseminormed vector spaces have been introduced and studied. Although they satisfy fewer conditions than topological vector spaces while encompassing a larger domain of mathematical structures, semitopological vector spaces preserve many useful properties of topological vector spaces, providing a solid base for the further development of functional analysis and its applications. Exploration of semitopological vector spaces, hypernormed vector spaces, hyperseminormed vector spaces and hypermetric vector spaces is the main topic of this book, which provides an introduction to the theory of such spaces. The emphasis is made on the operators in these spaces because operators in general and linear operators, in particular, occupy a central place in the contemporary theory of information, in control theory, in contemporary physics, especially, in quantum physics, in chemistry and in functional analysis.

This book belongs to physical mathematics, which is a recently delineated area in mathematics inherently related to and induced by theoretical physics.

In this book, the study of operators and their spaces goes in three directions. One of them is related to the extension of operators to hyperoperators and functionals to hyperfunctionals. The second direction analyzes properties of operators in the environment of vector spaces with a multiplicity of hypernorms and hyperseminorms. To reflect the situation, the concepts of relative continuity and relative boundedness are introduced and studied. The third direction employs methods of neoclassical analysis where techniques and procedures of the classical mathematical analysis are broadened by utilization of fuzzy properties of conventional mathematical structures. In our case, the system of continuous operators is enlarged to the arrangement of fuzzy continuous operators, while linear and additive operators are treated as subclasses of approximately linear and approximately additive operators, correspondingly.

To make this book easier for the reader and more adjusted to students with some basic knowledge in mathematics, denotations and definitions of the main mathematical concepts and structures used in the book are included in Appendix.

As various novel mathematical structures are studied in this book, it is possible to use this book for enhancing traditional courses of calculus for undergraduates, as well as for teaching a separate course for graduate students because the material of this book is closely related to what is taught at colleges and universities. To achieve this goal, we discuss standard topics and structures from functional analysis before introducing related new structures, which elucidate power of the conventional functional analysis and expand its applications. For instance, we define norms, metrics and seminorms describing their characteristics before we introduce hypernorms, hypermetrics and hyperseminorms exploring their properties. In addition, it is possible to use a definite number of statements from the book as exercises for students because their proofs are not given in the book but left for the reader. Besides, it is possible to suggest problems listed in Chapter 11 to PhD students for individual research.

ABOUT THE AUTHOR

Mark Burgin, PhD
University of California, Los Angeles, USA;
formerly, Head, Assessment Laboratory, Research Center of Science,
National Academy of Sciences of Ukraine

Dr. Mark Burgin received his MA and PhD in mathematics from Moscow State University, which was one of the best universities in the world at that time, and Doctor of Science in logic and philosophy from the National Academy of Sciences of Ukraine. He was a Professor at several institutions, including the Institute of Education, Kiev; International Solomon University, Kiev; Kiev State University, Ukraine; as well as Head of the Assessment Laboratory in the Research Center of Science at the National Academy of Sciences of Ukraine. Currently he is working at the University of California, Los Angeles, USA.

Dr. Burgin is a member of the New York Academy of Sciences and an Honorary Professor of the Aerospace Academy of Ukraine. Dr. Burgin is also a member of the Science Advisory Committee at Science of Information Institute, Washington, DC, USA.

Dr. Burgin is doing research, has publications, and taught courses in various areas of mathematics, artificial intelligence, computer science, information sciences, system theory, logic, psychology, social sciences, and methodology of science. He originated such theories as system theory of time, general information theory, theory of named sets, hyperprobability theory, and neoclassical analysis (in mathematics) and has made essential contributions to such fields as foundations of mathematics, theory of algorithms and computation, theory of knowledge, theory of intellectual activity, and complexity studies. He was the first to discover non-diophantine arithmetics; the first to axiomatize and build mathematical foundations for negative probability used in physics, finance, and economics; and the first to explicitly overcome the barrier posed by the Church-Turing Thesis.

Dr. Burgin has authored and co-authored more than 500 papers and 21 books, including *Structural Reality* (2012), *Theory of Named Sets* (2011),

Theory of Information (2010), *Neoclassical Analysis: Calculus Closer to the Real World* (2008), *Super-Recursive Algorithms* (2005), *On the Nature and Essence of Mathematics* (1998), *Intellectual Components of Creativity* (1998), *Fundamental Structures of Knowledge and Information* (1997), *The World of Theories and Power of Mind* (1992), and *Axiological Aspects of Scientific Theories* (1991). Dr. Burgin has also edited eight books.

CHAPTER 1

INTRODUCTION

Normed vector spaces axiomatically defined by Riesz (1918) and in a more complete form by Banach (1922), topological vector spaces introduced by Kolmogorov (1934) and the important type of topological vector spaces, locally convex spaces, brought in by von Neumann (1935) play an important role in functional analysis [cf., e.g., (Bourbaki, 1953–1955; Dunford and Schwartz, 1958; Robertson and Robertson, 1964; Choquet, 1969; Smolyanov and Fomin, 1976; Rudin, 1991; Grothendieck, 1992; Kurzweil, 2000)]. These spaces are related to such mathematical areas as distributions (Horváth, 1966; Treves, 1995), integration (Masani, 1947; Edwards and Wayment, 1970; Shuchat, 1972; Kurzweil, 2000) and measure theory (Smolyanov and Fomin, 1976; Pantsulaia, 2007). In addition, locally convex topological vector spaces offer a convenient structure for studies of summation, which is integration of functions on natural numbers (Pietsch, 1965). As an extension of these mathematical fields, a new direction in functional analysis, called quantum functional analysis, has been developed based on polinormed and multinormed vector spaces and linear algebras [cf., e.g., (Effros and Ruan, 2000; Dosiev, 2008; Helemski, 2010; Dosi, 2011)].

As a result, normed vector spaces and topological vector spaces have become an intrinsic tool for theoretical physics (von Neumann, 1927, 1929, 1932, 1932a; von Neumann and Halmos, 1942; von Neumann, Jordan and Wigner, 1934; von Neumann and Murray, 1936, 1937; Geroch, 1985; Exner and Havlíček, 2008; Eschrig, 2011), as well as for control theory (Partington, 2004).

At the same time, there are useful structures in functional analysis that are not topological vector spaces but are in some aspects similar to them. To include such structures in the context of functional analysis and find their properties, semitopological vector spaces are introduced and studied.

Although they satisfy fewer conditions than topological vector spaces, semitopological vector spaces preserve many useful properties of topological vector spaces, providing a solid base for the further development of functional analysis and its applications. In addition to semitopological vector spaces, we introduce and study such new mathematical structures as hypernormed vector spaces, hyperseminormed vector spaces and hypermetric vector spaces, making emphasis on operators in these spaces. We explore various properties of these operators such as different types of continuity, linearity, boundedness and relations between studied properties.

Operators in general and linear operators, in particular, occupy a central place in contemporary physics – in classical physics [cf., e.g., (Heaviside, 1893, 1894; von Neumann, 1932a; von Neumann and Halmos, 1942; Lurie, 1950; Ditkin and Prudnikov, 1966)], including optics and acoustics [cf., e.g., (Maslov, 1976, 1976a, 1983)], and especially, in quantum mechanics and quantum field theory [cf., e.g., (von Neumann, 1932; Davies and Lewis, 1970; Kato, 1980; Byron and Fuller, 1992; Huang, 1998; Carfi, 2004; Dosi, 2011)]. In physics, an *operator* is a dynamic object acting on the space of physical states when transformation of one physical state to another physical state is obtained while very often generating some extra relevant information.

Besides, operators are resourcefully used in quantum chemistry [cf., e.g., (Hui-Yun and Zu Sen, 1989)]; in economics [cf., e.g., (Carfi, 2004; Schwartz, 1961)]; in the complexity analysis of programs and algorithms [cf., e.g., (García-Raffi et al., 2002)]; in psychology [cf., e.g., (Newell and Simon, 1972; Dominowski and Dallob, 1995; Goldstein, 2005)]; in control theory [cf., e.g., (Lee and Markus, 1989)]; in optimization [cf., e.g., (Attouch et al., 1994; Hauser, 2002; Vlach, 1981)]; in signal processing [cf., e.g., (Combettes and Pesquet, 2009)]; in the calculus of variations [cf., e.g., (Dacorogna, 1989)]; as well as in functional analysis and its applications [cf., e.g., (Stone, 1930, 1932; Dunford and Schwartz, 1958; Rudin, 1991; Kolmogorov and Fomin, 1999)].

One of the basic concepts of physics is *observable*. Physicists interpret observables as physical objects or properties of such objects, which some system (in a more restricted sense, people) can observe and/or measure in a physical experiment. To make this concept exact, it was necessary to build a corresponding mathematical model, and in the later development of quantum theory, an observable was associated with a self-adjoint linear operator. For instance, in contemporary physics, time is also considered as an operator (Prigogine, 1980; Nicolis and Prigogine, 1977).

Later several authors suggested the generalized representation of observables as positive operator measures (Ludwig, 1964; Davies and Lewis, 1970; Kraus, 1983). This concept advanced the mathematical coherence and conceptual clarity of the quantum theory. From physics, mathematical operators were extended to quantum chemistry where they also have been successfully used to model different processes.

At the same time, operators play an important role in the contemporary theory of information (Burgin, 2010, 2010b, 2011a, 2001b, 2014; Brenner and Burgin, 2011), while many researchers treat the theory of information as the foundation for physics [cf., e.g., (Frieden, 1998)].

In addition, operators are frequently employed in programming languages [cf., e.g., (Burgin, 1976; Louden, 2003)], models of computing systems and computations [cf., e.g., (Baeck et al., 2000; Burgin, 1980; Burgin and Karasik, 1976)], databases [cf., e.g., (Codd, 1970; Date et al., 2002] and system theory [cf., e.g., (Burgin, 1982)].

An essential property of operators is continuity. One of the central results of functional analysis is the theorem that establishes equivalence between continuity and boundedness for linear operators. Thus, presented in this book study of approximately linear operators and their fuzzy properties, including fuzzy continuity, approximate linearity and multiboundedness, provides potent tools for applications in theoretical physics, as well as in information theory and practice.

One more field where linear operators play a key role is control theory and mathematical systems theory (Partington, 2004). Consequently, a study of approximately linear operators and their fuzzy properties provides means for making control theory and mathematical systems theory more exact and better connected to applications.

Other fields where linear operators are efficiently applied are economics, time-frequency analysis and classification, numerical analysis, mathematical physics and mathematical methods in electrical engineering. Utilization of approximately linear operators studied in this book allows extending applications of functional analysis in these areas. This book consists of three parts.

In the first part (Chapters 2 and 4), extensions of number systems to hypernumbers and function spaces to extrafunction spaces are described and studied. This study has the three-fold aim. The first goal is the further development of functional analysis. The second objective is formation of foundations and necessary means for the theory of hypernormed vector spaces, hyperseminormed vector spaces and semitopological vector spaces. The third purpose is the extension of the scope of applications of functional analysis to science in general and to physics, in particular. One of the main problems of contemporary physics is, as Close calls it, the "infinity puzzle" (Close, 2011). Its essence is that physicists have been trying for decades to eliminate infinite values that have emerged in physical theories but such quantities appeared again and again. As Salam wrote, "Field-theoretic infinities first encountered in Lorentz's computation of electron have persisted in classical electrodynamics for 70 and in quantum electrodynamics for some 35 years" (Isham et al., 1972). Namely, from the very beginning, quantum physics has been haunted by divergences, while physicists and mathematicians fought to exterminate these divergences one after another. However, chased from the door, divergences returned from the window.

The new approach, the theory of hypernumbers and extrafunctions, suggests not to straightforwardly eliminate all infinities but to learn how to work with them in a mathematically rigorous and physically meaningful way. To achieve this goal, the theory of hypernumbers and extrafunctions provides novel mathematical structures and techniques for physicists. The concept of a real or complex extrafunction essentially extends the concept of a real or complex function, encompassing, in particular, the concept of a distribution, i.e., distributions are a kind of extrafunctions (Burgin, 2012). Extrafunctions have many advantages in comparison with functions and distributions. For instance, integration of extrafunctions is more powerful than integration of functions allowing integration of a much larger range of functions as it is demonstrated in Burgin (2012).

Hypernumbers essentially increased the capacity of the calculus allowing researchers to perform summation of any series of real or complex numbers (Burgin, 2008a), to integrate any real function (Burgin, 2012), to build a rigorous construction of the Feynman path integral (Burgin, 2008b, 2012b), to solve operator equations (Burgin and Dantsker, 1995) and to solve some problems in probability theory (Burgin and Krinik, 2009, 2010, 2012; Luu, 2011). For instance, mathematicians have been searching for various conditions of series summability in the classical sense [cf., e.g., (Jolley, 1961; Davis, 1962; Sofo, 2003)], while hypernumbers allow one to find the sum of any series of real or complex numbers (Burgin, 2008a, 2012).

It is useful to note that in this book, hypernumbers are constructed and studied in the framework of normed fields. This approach allows unification of real and complex hypernumbers in a more general schema.

It is necessary to understand that, as it is demonstrated in Burgin (2012) and in Chapter 2, real hypernumbers differ from other generalizations of real numbers such as hyperreal numbers from nonstandard analysis introduced by Robinson (1961, 1966), number systems with numerical sequences as infinitesimals introduced by Chwistek (1935) and studied by Schmieden and Laugwitz (1958); Laugwitz (1960, 1961); Henle and Kleinberg (1979); Henle (1999), transfinite numbers introduced by Cantor (1883, 1886), surreal numbers introduced by Conway (1976) and described by Knuth (1974) in popular form, superreal numbers introduced by Tall (1980), and generalized numbers introduced by Egorov (1989).

In particular, the difference between nonstandard analysis and the theory of hypernumbers and extrafunctions stems from the indispensable difference between the construction principles and methods of both theories: nonstandard analysis is based on set-theoretical principles and methods inherent for classical mathematics (Robinson, 1961, 1966), while the theory of hypernumbers and extrafunctions is based on topological principles and methods inherent for modern physics (Burgin, 2012). Besides, nonstandard analysis maintains both infinitely big and infinitely small numbers, or infinitesimals, which were called "the ghosts of departed quantities" by George Berkeley. In contrast to this, the theory of hypernumbers and extrafunctions includes only infinitely big numbers.

In addition, as it is demonstrated in Chapter 2, real hypernumbers studied in this book are not included in the scope of surreal numbers unlike hyperreal numbers from nonstandard analysis.

Other basic constructions of the new theory are extrafunctions, hyperfunctionals and hyperoperators, which constitute the further development of distribution theory being constructed and exposed in Chapter 4. Alike hypernumbers, they provide extremely powerful means for differentiation and integration. For instance, while distributions allow one to differentiate any continuous real function (Schwartz, 1950, 1951; Antosik et al., 1973; Liverman, 1964), extrafunctions permit one to differentiate any real function (Burgin, 2012). In addition, extrafunctions make available integration of any real function, as well as summation of any functional series (Burgin, 2012, 2008a). This peculiarity allows to be exceedingly useful for differential equations allowing solving such equations that are not solvable even in distributions (Burgin and Ralston, 2004; Burgin, 2010). In addition, functional hyperintegrals, which are a kind of hyperfunctionals, provide a rigorous mathematical foundation for the Feynman path integral (Burgin,

2004, 2015), which has become one of the primary key tools in contemporary physics (Feynman and Hibbs, 1965; Collins, 1984) finding applications in information theory (Lerner, 2010), stochastic analysis (Albeverio et al., 2008), control theory (Lerner, 2012), chemistry, statistics and financial mathematics (Kleinert, 2004), stochastic differential equations (Chow and Buice, 2012) and biology (Jarvis et al., 2004).

At the same time, the differential calculus of extraderivatives includes as particular cases, or as subcalculi, and extends different calculi of finite differences (Boole, 2003; Milne-Thomson, 1951; Richardson, 1954; Jordan, 1965; Spiegel, 1971), as well as the quantum calculus developed by Katz and Cheung (2002).

Hyperfunctionals and hyperoperators constructed in Chapter 4 include and extend extrafunctions as their particular case. In particular, the spaces of generalized distributions are also enlarged demonstrating that they include conventional distributions of Schwartz, tempered distributions, ultradistributions and many other generalized functions.

In the second part of the book (Chapters 3, 5 and 6), new mathematical structures, such as hypernormed vector spaces, hyperseminormed vector spaces, polyhypernormed vector spaces, polyhyperseminormed vector spaces, and semitopological vector spaces, as well as operators in these spaces and their mappings are constructed and studied. In Chapter 3, we start with a study of norms, seminorms, metrics and their generalizations, such as quasi-norms, pseudonorms, ultranorms, ultraseminorms, quasi-seminorms, quasi-norms, semimetrics, pseudometrics and quasi-metrics. Then we introduce and explore hyperseminorms, hypernorms, hypermetrics and hyperpseudometrics. It is demonstrated that hyperseminorms induce hyperpseudometrics, while hypernorms induce hypermetrics in vector spaces. Such classical mathematical structures as norms are special cases of hypernorms, while seminorms are special cases of hyperseminorms. Sufficient and necessary conditions for a hyperpseudometric (hypermetric) to be induced by a hyperseminorm (hypernorm) are found. This provides tools for finding properties of extrafunction spaces in the abstract setting of algebraic systems and topological spaces.

Obtained results for hyperseminorms, hypernorms and hypermetrics have various corollaries for norms, seminorms and metrics. The majority of these corollaries is new containing some classical results from functional analysis as special cases.

In such a way, we come (in Chapter 5) to a new topological-algebraic structure, which generalizes topological vector spaces and is called a semitopological vector space. We use hyperseminorms and hypernorms to study semitopological vector spaces, in particular, demonstrating that the structure of semitopological vector space is closely related to systems of hyperseminorms. It is useful to note that obtained results for hyperseminorms, hypernorms, hypermetrics and hyperpseudometrics have various corollaries for norms, seminorms, metrics and pseudometrics. The majority of these corollaries are new containing some classical results from functional analysis as special cases.

In addition (Chapters 6), we study relations between relative continuity and relative boundedness as essential properties of operators in polyhyperseminormed and polyhypernormed vector spaces. For instance, continuity of operators plays a central role in mathematics and physics for various models of studied phenomena [cf. (Dunford and Schwartz, 1958; Rudin, 1991; Kolmogorov and Fomin, 1999)]. When we have seminormed or normed vector spaces, continuity and boundedness are determined by the unique norm or seminorm in the considered space. However, when there are many norms, seminorms, hyperseminorms and hypernorms in a vector space, it is possible to define continuity and boundedness based on a separate norm, seminorm, hyperseminorm or hypernorm or based on a system of norms, seminorm, hyperseminorms and hypernorms. This gives us the concepts of relative continuity and relative boundedness. Results obtained in this chapter contain some classical results from functional analysis as special cases extending possibilities for their applications.

In the third part of the book (Chapters 7–10), conventional constructions of mathematics in general and functional analysis, in particular, are extended at first, from normed and seminormed vector spaces to polyhyperseminormed and polyhypernormed vector spaces and then to the corresponding structures of neoclassical analysis (Burgin, 2008) in particular, and of fuzzy mathematics (Zimmermann, 2001) in general. Neoclassical analysis extends methods of the classical calculus to reflect vagueness, imprecision and uncertainties that arise in computations and measurements. In it, ordinary structures of analysis, that is, functions, sequences, series, and operators, are studied by means of fuzzy concepts: fuzzy limits, fuzzy continuity, fuzzy derivatives and fuzzy gradients. For instance, neoclassical analysis studies functions that may be continuous only to some extent or in some sense. They are called fuzzy continuous functions providing better tools for exact modeling of natural and social processes. Continuous functions form a subclass of the class of fuzzy continuous functions.

It is persuasively demonstrated that fuzzy continuous functions better represent and model reality than continuous functions (Burgin, 2008). In particular, fuzzy continuity allows elaboration of the natural concept of continuity for functions in discrete spaces (Burgin, 2012c). As a result, fuzzy continuous functions emerge and are used in diverse practical applications, especially in those that are related to computations. For instance, Steimann (2001) analyzes an interesting interplay between discrete structures in Artificial Intelligence and continuous processes and characteristics in medicine. Let us consider some of his arguments.

Firstly, even when medical actions are inherently discrete (e.g., a surgery is either performed or not), their grounds usually involve some continuity and gradations.

Secondly, discretization can lead to the undue amplifications of differences. Would it not seem natural that patients with comparable symptoms be given comparable diagnoses?

Analogously and thirdly, discreteness may cause erratic behavior in the context of change. Would it not seem likely that a slight alteration over time in the vital parameters of a patient changes the diagnosis only slightly? Instead, however, discrete dynamic systems usually respond to the continuous change in the patient condition with an abrupt change in state. It is necessary to materialize an informal concept of continuity: a system is continuous (or more exactly, behaves continuously) if it maps close inputs to close outputs.

These arguments show that continuity in medicine is not the classical continuity with its absolute precision. Only fuzzy continuity studied for functions in Burgin (1993a, 1995a, 2008, 2012c) and Burgin and Glushchenko (1997, 1998) and for functionals and operators in this book gives an adequate model for incessant systems and processes in medicine.

Another system of concepts from neoclassical analysis, which are introduced and explored in this book (Chapter 8), consists of the concepts of approximately additive, approximately homogeneous and approximately linear operators in vector spaces. This direction in neoclassical analysis is related to Ulam's problem (Ulam, 1964), which has been extensively studied by different authors [cf., e.g., (Rassias, 1978; Forti, 1987, 1995; Gajda, 1991; Isac and Rassias, 1993, 2007; Găvruta, 1994; Borelli and Forti, 1995; Jung, 1996, 1997, 1998, 1998a; Kanovei and Reeken, 2000; Rassias and Rassias,

2003, 2005)]. Different approaches to approximate additivity, homogeneity and linearity of operators in hyperseminormed, hypernormed, seminormed and normed vector spaces are developed and investigated. An important property of introduced concepts is that all of them include conventional additivity, homogeneity and linearity of operators as their limit case.

Note that in the context of neoclassical analysis, being approximate is a kind of being fuzzy as it is possible to discern three kinds of *fuzziness as a relation* between objects, systems or concepts:

- An object (system or concept) A is *approximately* another object (system or concept) B if it is possible to estimate the difference between these objects or in other words, how far are they from one another, and this difference (distance) is sufficiently small.
- An object (system or concept) A is *vaguely* another object (system or concept) B if A and B are similar but it is impossible to estimate the difference between these objects.
- An object (system or concept) A is *roughly* another object (system or concept) B if it is possible to estimate the difference between these objects or in other words, how far are they from one another, and this difference (distance) is not small.

One of the primary goals in this area exposed in this book is to explore relations between fuzzy continuity and boundedness of approximately linear operators in the context of neoclassical analysis. It is demonstrated, for example, that for approximately linear operators, fuzzy continuity is equivalent to boundedness when the continuity defect (or measure of discontinuity) is sufficiently small. The classical result that describes continuity of linear operators in terms of boundedness becomes a direct corollary of this theorem. We also demonstrate that for linear operators in normed vector spaces, fuzzy continuity coincides with continuity when the continuity defect is sufficiently small, i.e., when it is less than one. The obtained results are oriented at applications in physics, the theory of information, control theory and other fields where operator equations play an important role.

It is useful to understand that concepts and constructions from neoclassical analysis studied in this book and in other works, such as fuzzy continuity, approximate linearity and relative boundedness, contain their classical counterparts as limit cases. For instance, fuzzy continuity becomes classical continuity when the continuity defect turns into zero.

As the materials presented here form only the beginning of the theory of semitopological vector spaces, hypernormed vector spaces and hyperseminormed vector spaces, several open problems and directions for future research in the fields, which are developed and explored in this book, are considered in the concluding chapter. These problems and directions are oriented both at the current situation in functional analysis and the needs of extending applications of mathematical methods in various scientific and technological areas.

To make this book easier for the reader and more adjusted to students with some basic knowledge in mathematics, denotations and definitions of the main mathematical concepts and structures used in the book are included in Appendix.

As new mathematical structures are studied in this book, it is possible to use this book for enhancing traditional courses of calculus for undergraduates, as well as for teaching a separate course for graduate students because the material of this book is closely related to what is taught at colleges and universities. To achieve this goal, we discuss standard topics and structures from functional analysis before introducing related new structures, which elucidate power of the conventional functional analysis. For instance, we define norms, metrics and seminorms describing their characteristics before we introduce hypernorms, hypermetrics and hyperseminorms exploring their properties. In addition, it is possible to use a definite number of statements from the book as exercises for students because their proofs are not given in the book but left for the reader. Besides, it is possible to suggest problems listed in Chapter 11 to PhD students for individual research.

ACKNOWLEDGEMENTS

I would like to thank all my teachers and especially, my thesis advisor, Alexander Gennadievich Kurosh, who helped shaping my scientific viewpoint and research style. I also appreciate the advice and help of Andrei Nikolayevich Kolmogorov from Moscow State University in the development of the holistic view on mathematics and its connections with physics. In developing ideas in the theory of hypernormed vector spaces, I have benefited from conversations with many friends and colleagues. In particular, discussions on extrafunctions with Yury Daletsky from Kiev Polytechnic University and participants of the Kiev Functional Analysis Seminar directed by Daletsky were useful at the early stages of the theory development. My collaboration with James Ralston from UCLA and Alan Krinik from California State Polytechnic University, Pomona gave much to the development of ideas and concepts of the theory of hypernormed vector spaces and especially, to applications of this theory. Collaboration with Oktay Duman, Vitalii Glushchenko, Martin Kalina and Alexander Ŝostak gave much to the development of ideas and concepts of fuzzy continuity and approximate linearity of operators in hypernormed vector spaces. Credit for my desire to write this book must go to my academic colleagues. Their questions and queries made significant contribution to my understanding of functional analysis and its role in contemporary physics. I would particularly like to thank many fine participants of the Applied Mathematics Colloquium, Analysis and PDE Seminar, and Topology Seminar at UCLA, Mathematics and Statistics Department Colloquium at California State Polytechnic University, Pomona, International Conference "The Feynman Integral and Related Topics in Mathematics and Physics," University of Nebraska, and Meetings of the American Mathematical Society for extensive and helpful discussions on hypernumbers and extrafunctions that gave me much encouragement for further work in this direction. I would also like to thank the Departments of Mathematics and Computer Science in the School of Engineering at UCLA for providing space, equipment, and helpful discussions.

KEYWORDS

- **continuity**
- **functional analysis**
- **fuzzy continuity**
- **hypernormed vector space**
- **locally convex space**
- **normed vector space**
- **operator**
- **semitopological vector space**
- **theoretical physics**
- **topological vector space**

CHAPTER 2

HYPERNUMBERS: CONSTRUCTIONS AND OPERATIONS

In this chapter, we construct spaces of hypernumbers and study their properties as the basic tool for the theory of hypernorms, hyperseminorms and semitopological vector spaces.

Let us consider a partially ordered set P with the order relation $\leq$ and a set X. For instance, the set $\boldsymbol{N}$ of all natural numbers, the set $\boldsymbol{R}$ of all real numbers and the set $\boldsymbol{Z}$ of all integer numbers are partially ordered sets.

Definition 2.1. A *sequence* $\boldsymbol{q}$ in P is a one-to-one ordered mapping (monomorphism) of the ordered set $\omega = \{1, 2, 3, \ldots\}$ of natural numbers into P, i.e., $\boldsymbol{q}: \omega \to P$.

By definition, $p_n \leq p_{n+1}$ for all $n = 1, 2, 3, \ldots$

The sequence $\boldsymbol{q}$ is usually represented either by its graph $\boldsymbol{q} = \{p_1, p_2, p_3, \ldots, p_n, \ldots; p_n \in P, n = 1, 2, 3, \ldots\}$ or by its graph $\boldsymbol{q} = \{p_n; p_n \in P, n = 1, 2, 3, \ldots\}$ or by its graph $\boldsymbol{q} = (p_i)_{i\in\omega}$. So, there is a sequence as a mapping and sequence as a graph in the form of an indexed set. The latter is the most popular representation of sequences in mathematical literature and in what follows, we use this representation denoting the graph and the mapping by the same letter.

Example 2.1. $\boldsymbol{q} = \omega = \{1, 2, 3, \ldots\}$ is a sequence in $\boldsymbol{R}$.

Example 2.2. $\boldsymbol{q} = \{-i, -2i, -3i, \ldots$ where $i = \sqrt{-1}\}$ is a sequence in the set $\boldsymbol{C}$ of all complex numbers with the inverse order.

Example 2.3. $\boldsymbol{q} = \{3i + 1, 3i + 2, 3i + 3, \ldots$ where $i = \sqrt{-1}\}$ is a sequence in $\boldsymbol{C}$.

Definition 2.2. A sequence $\boldsymbol{q} = (a_i)_{i\in\omega}$ is a *subsequence* of a sequence $\boldsymbol{p} = (b_i)_{i\in\omega}$ if there is a strictly increasing function $f_{ab}: \omega \to \omega$ such that for any $i \in \omega$, we have $a_i = b_{f_{ab}(i)}$, i.e., the following diagram is commutative

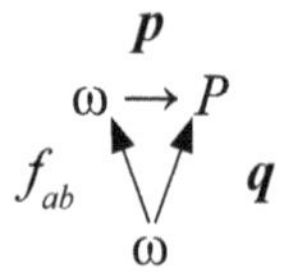

For instance, the sequence $\boldsymbol{q} = \{2, 4, 6, \ldots, 2n, \ldots\}$ is a subsequence of the sequence $\boldsymbol{p} = \{1, 2, 3, \ldots, n, \ldots\}$.

In what follows, there is, at least, one sequence $\boldsymbol{q}$ in P.

Definition 2.3. A sequence $\boldsymbol{q}$ in P is *maximal* if there is no sequence $\boldsymbol{p}$ in P such that $\boldsymbol{q}$ is a subsequence of $\boldsymbol{p}$.

Example 2.4. The set of natural numbers $\boldsymbol{N}$ has one maximal sequence $\omega = \{1, 2, 3, \ldots\}$.

Example 2.5. The set of natural numbers $P = \{1, 2, 3, \ldots; i, 2i, 3i, \ldots$ where $i = \sqrt{-1}\ \}$ has two maximal sequences $\omega = \{1, 2, 3, \ldots\}$ and $i\omega = \{i, 2i, 3i, \ldots\}$.

Example 2.6. The set of integer numbers $\boldsymbol{Z}$ does not have maximal sequences.

Definition 2.4. a) A *P-sequence* $\boldsymbol{l}$ with values in X is a mapping of a sequence $\boldsymbol{q}$ in P into X, i.e., $\boldsymbol{l}$:$\boldsymbol{q}$→X.

b) The P-sequence $\boldsymbol{l}$ with values in X is *defined* by the sequence $\boldsymbol{q}$ in P.

The P-sequence $\boldsymbol{l}$ is also denoted either by $\boldsymbol{l} = \{l_1, l_2, l_3, \ldots, l_n, \ldots; l_n \in X, n = 1, 2, 3, \ldots\}$ or by $\boldsymbol{l} = \{l_n; l_n \in X, n = 1, 2, 3, \ldots\}$ or by $\boldsymbol{l} = (l_p)_{p \in q}$. Elements l_n are called *coordinates* of the P-sequence $\boldsymbol{l}$.

Definition 2.5. A *P-set* $\boldsymbol{a}$ with values in X is a mapping of the set P into X, i.e., $\boldsymbol{a}$: P→X.

The P-set $\boldsymbol{a}$ is also denoted either by $\boldsymbol{a} = \{a_i; a_i \in X, i \in P\ \}$ or by $\boldsymbol{a} = (a_i)_{i \in P}$. Elements a_i are called *coordinates* of the P-set $\boldsymbol{a}$.

$X^P = \{(a_i)_{i \in P}; a_i \in X\}$ denotes the set of all P-sets with values in X.

We remind that a mapping (a nonnegative function) $\|.\|$: $L \to \boldsymbol{R}^+$ is called a *norm* in a real vector space L if it satisfies the following conditions:

N1. For any x from L, $\|x\| = 0$ if and only if $x = \mathbf{0}$.

N2. $\|ax\| = |a| \cdot \|x\|$ for any x from L and any number a from $\boldsymbol{R}$.

N3 (the triangle inequality or subadditivity).

$$\|x + y\| \leq \|x\| + \|y\| \text{ for any } x \text{ and } y \text{ from } L$$

Let us consider a normed vector space X with a norm $\|.\|$. Then we can build generalized P-hypernumbers over X. They are equivalent classes of P-sets.

Definition 2.6. For arbitrary P-sets $\boldsymbol{a} = (a_i)_{i \in P}$ and $\boldsymbol{b} = (b_i)_{i \in P}$ from X^P,

$$\boldsymbol{a} \approx \boldsymbol{b} \text{ means that } \lim_{i \to \infty} \|a_{q(i)} - b_{q(i)}\| = 0 \text{ for any sequence } \boldsymbol{q} \text{ in } P$$

Lemma 2.1. $\boldsymbol{a} \approx \boldsymbol{b}$ if and only if $\lim_{p \in l} \|a_p - b_p\| = 0$ for any P-sequence $\boldsymbol{l}$.
Proof is left as an exercise.

Lemma 2.2. The relation $\approx$ is an equivalence relation in X^P.

Indeed, by definition, this relation is reflexive ($\boldsymbol{a} \approx \boldsymbol{a}$) and symmetric ($\boldsymbol{a} \approx \boldsymbol{b} \Rightarrow \boldsymbol{b} \approx \boldsymbol{a}$). Thus, we need only to show that it is transitive. Let us take three P-sets $\boldsymbol{a} = (a_i)_{i \in \omega}$, $\boldsymbol{b} = (b_i)_{i \in \omega}$ and $\boldsymbol{c} = (c_i)_{i \in \omega}$ from X^P such that $\boldsymbol{a} \approx \boldsymbol{b}$ and $\boldsymbol{b} \approx \boldsymbol{c}$, we have

$$\lim_{i \to \infty} \|a_{q(i)} - b_{q(i)}\| = 0 \text{ for any sequence } \boldsymbol{q} \text{ in } P$$

and

$$\lim_{i \to \infty} \|b_{q(i)} - c_{q(i)}\| = 0 \text{ for any sequence } \boldsymbol{q} \text{ in } P$$

By properties of norms and limits, we have

$$0 \leq \lim_{i \to \infty} \|a_{q(i)} - c_{q(i)}\| = \lim_{i \to \infty} \|a_{q(i)} - b_{q(i)} + b_{q(i)} - c_{q(i)}\| \leq$$

$$\lim_{i \to \infty} \|a_{q(i)} - b_{q(i)}\| + \lim_{i \to \infty} \|b_{q(i)} - c_{q(i)}\| = 0 + 0 = 0$$

Consequently,

$$\lim_{i \to \infty} \|a_{q(i)} - c_{q(i)}\| = 0 \text{ for any sequence } \boldsymbol{q} \text{ in } P$$

i.e., $\boldsymbol{a} \approx \boldsymbol{c}$.

Definition 2.7. Classes of the equivalence $\approx$ are called *generalized P-hypernumbers in X* and their set is denoted by X_P.

Any P-set $\boldsymbol{a} = (a_i)_{i \in P}$ determines (and represents) a generalized P-hypernumber $\alpha = \text{Hn}(a_i)_{i \in P}$ in X and such a P-set $\boldsymbol{a}$ is called a *defining* P-set or *representing* P-set or *representation* of the generalized P-hypernumber $\alpha = \text{Hn}(a_i)_{i \in P}$. Definition 2.7 means that any two generalized P-hypernumbers $\text{Hn}(a_i)_{i \in P}$ and $\text{Hn}(b_i)_{i \in P}$ are equal if their defining P-sets $(a_i)_{i \in P}$ and $(b_i)_{i \in P}$ are equivalent.

Note that any generalized P-hypernumber α does not depend on a finite number of elements from its representation, i.e., if $\alpha = \text{Hn}(a_i)_{i \in P}$ and almost all elements in P-sets $(a_i)_{i \in P}$ and $(b_i)_{i \in P}$ are the same (equal), then $\alpha = \text{Hn}\,(b_i)_{i \in P}$.

It is also possible to define generalized P-hypernumbers in an arbitrary metric space M.

The most interesting for us and the most important for applications is the case when L is a set of numbers and in particular, the sets $\mathbf{R}$ of all real numbers and $\mathbf{C}$ of all complex numbers with absolute values as the standard norms.

Definition 2.8. a) P-hypernumbers in $\mathbf{R}$ are called *real P-hypernumbers.*

b) P-hypernumbers in $\mathbf{C}$ are called *complex P-hypernumbers.*

Such general classes of hypernumbers are introduced and explored in Burgin (2001, 2005b).

Example 2.7. Let $\mathbf{R}^{\omega} = \{(a_i)_{i\in\omega};\ a_i \in \mathbf{R}\}$ be the set of all sequences of real numbers. We remind that a sequence $\mathbf{a} = (a_i)_{i\in\omega}$ of real numbers is a mapping $f_a: \omega \to \mathbf{R}$.

For arbitrary sequences $\mathbf{a} = (a_i)_{i\in\omega}$ and $\mathbf{b} = (b_i)_{i\in\omega}$ from $\mathbf{R}^{\omega}$,

$$\mathbf{a} \sim \mathbf{b} \text{ means that } \lim_{i\to\infty} |a_i - b_i| = 0$$

The introduced relation ~ is an equivalence relation (Burgin, 2002).

The set of all classes of the equivalence ~ is denoted by $\mathbf{R}_{\omega}$.

Elements of the set $\mathbf{R}_{\omega}$ are called real hypernumbers (Burgin, 1990, 2002, 2004, 2005). Any sequence $\mathbf{a} = (a_i)_{i\in\omega}$ of real numbers determines (and represents) a real hypernumber $\alpha = \mathrm{Hn}(a_i)_{i\in\omega}$ and such a sequence is called a *defining sequence* or *representing sequence* or *representation* of the hypernumber $\alpha = \mathrm{Hn}(a_i)_{i\in\omega}$.

In a similar way, it is possible to build complex hypernumbers and study their properties (Burgin, 2002, 2004, 2010).

Definition 2.9. A real P-hypernumber α is *represented* in a set $A \subseteq \mathbf{R}$ if there is a P-set $\mathbf{a} = (a_i)_{i\in P}$ of real numbers such that $a_i \in A$ for all $i \in P$ and $\alpha = \mathrm{Hn}(a_i)_{i\in P}$.

For instance, by Corollary 2.1, all real hypernumbers are represented in the set $\mathbf{Q}$ of all rational numbers or in any set of numbers that contains $\mathbf{Q}$.

Lemma 2.3. The set $\mathbf{R}$ is isomorphic to the subset $\underline{\mathbf{R}} = \{\alpha = \mathrm{Hn}(a_i)_{i\in\omega};\ a_i = a \in \mathbf{R} \text{ for all } i \in \omega\}$ of the set $\mathbf{R}_{\omega}$.

Proof is left as an exercise.

When it does not cause confusion, we call elements from $\underline{\mathbf{R}}$ real numbers and denote this set by $\mathbf{R}$. In what follows, we will identify hypernumbers from $\underline{\mathbf{R}}$ and corresponding real numbers, assuming that $\mathbf{R} = \underline{\mathbf{R}}$ is a subset of $\mathbf{R}_{\omega}$. For instance, $\mathrm{Hn}(10, 10, 10, \ldots, 10, \ldots) = 10$ in $\mathbf{R}_{\omega}$.

Lemma 2.4. If $\alpha = \text{Hn}(a_i)_{i\in P}$, $\beta = \text{Hn}(b_i)_{i\in P}$, and in some P-sequence, all elements a_i and b_i are integer numbers such that $a_i \neq b_i$ for an infinite sequence in P, then $\alpha\neq\beta$.

Proof is left as an exercise.

Let us consider the following example.

Example 2.8. For real numbers, we have $1 = 0.99999...9...$ The same is true for hypernumbers, i.e., $1 = \text{Hn}(a_i)_{i\in\omega}$, where $a_i = 0.999...9$ where 9 is repeated i times, i.e., the sequence 0.9, 0.99, 0.999, … represents the real hypernumber 1, which is also a real number.

Thus, by the described construction, we obtain both all real numbers and many real hypernumbers that are different from real numbers, for example, infinite hypernumbers, such as $\alpha = \text{Hn}(i)_{i\in\omega}$.

As we know, not all real numbers are also rational numbers and consequently, not all sequences of real numbers are also sequences of rational numbers. There are much more real numbers, which form a continuum, than rational numbers, which form a countable set. However, there is a one-to-one correspondence between all introduced equivalence classes of real sequences and all introduced equivalence classes of rational sequences. This follows from the following result.

Let X be a dense subset in $\boldsymbol{R}$.

Theorem 2.1. There is a one-to-one correspondence between sets X_ω and $\boldsymbol{R}_\omega$, i.e., these sets are equipotent.

Proof. At first, we show that any class of equivalent real sequences contains a class of a sequence represented in X. Indeed, let us take a class α of equivalent real sequences and a real sequence $(a_i)_{i\in\omega}$ that belongs to this class. As the set X is dense in the set $\boldsymbol{R}$ of all real numbers, for each number $a_i (i\in\omega)$, there is a number b_i from X, such that

$$| a_i - b_i | < 1/i$$

By Definition 2.7, the sequence $(b_i)_{i\in\omega}$ represented in X belongs to the class α and thus, the whole class of sequences represented in X and equivalent to the sequence $(b_i)_{i\in\omega}$ is a subset of the class α because the norm in X is induced by the conventional norm, i.e., absolute value, in $\boldsymbol{R}$.

This gives us a mapping f of the set $\boldsymbol{R}_\omega$ onto the set X_ω. Besides, two different classes from the set X_ω cannot be subsets of the same class in the set $\boldsymbol{R}_\omega$ because the equivalence $(c_i)_{i\in\omega} \sim (d_i)_{i\in\omega}$ in X_ω implies the same equivalence in $\boldsymbol{R}_\omega$ for any sequences $(c_i)_{i\in\omega}$ and $(d_i)_{i\in\omega}$ represented in X.

Consequently, f is a one-to-one mapping, which means equipotence of the sets X_ω and $\boldsymbol{R}_\omega$.

Theorem is proved.

As we know not all real numbers are also rational numbers and consequently, not all sequences of real numbers are also sequences of rational numbers. There are much more real numbers, which form a continuum, than rational numbers, which form a countable set. However, the set $\boldsymbol{Q}$ of all rational numbers is dense in the set $\boldsymbol{R}$ of all real numbers. Thus, Theorem 2.1 implies the following result.

Corollary 2.1. There is a one-to-one correspondence between sets $\boldsymbol{Q}_\omega$ and $\boldsymbol{R}_\omega$, i.e., these sets are equipotent.

Even more, as it is demonstrated by Burgin (2004), the spaces $\boldsymbol{Q}_\omega$ and $\boldsymbol{R}_\omega$ are isomorphic as topological spaces and as linear spaces.

So, it is natural to assume that both sets $\boldsymbol{R}_\omega$ and $\boldsymbol{Q}_\omega$ represent the same object, which is called the set of all real hypernumbers. It means that rational numbers and real numbers as building blocks for real hypernumbers give us the same construction. Consequently, any sequence $\boldsymbol{a} = (a_i)_{i\in\omega}$ of real numbers determines (and represents) a real hypernumber $\alpha = \mathrm{Hn}(a_i)_{i\in\omega}$ and such a sequence is called a *defining sequence* or *representing sequence* or *representation* of the hypernumber $\alpha = \mathrm{Hn}(a_i)_{i\in\omega}$.

Corollary 2.2. If $\boldsymbol{A}$ is the set of all algebraic numbers, then there is a one-to-one correspondence between sets $\boldsymbol{A}_\omega$ and $\boldsymbol{R}_\omega$, i.e., these sets are equipotent.

So, it is possible to use only algebraic numbers for building all real numbers and all hypernumbers.

As the $\boldsymbol{Ir}$ of all irrational numbers is dense in the set $\boldsymbol{R}$ of all real numbers, Theorem 2.1 implies the following results.

Corollary 2.3. There is a one-to-one correspondence between sets $\boldsymbol{Ir}_\omega$ and $\boldsymbol{R}_\omega$, i.e., these sets are equipotent.

So, it is possible to use irrational numbers for building all real numbers and all hypernumbers.

However, each of the approaches to the hypernumber construction has its advantages. When we build hypernumbers from rational numbers, we use finite constructive elements, while this is not true for the second construction. Indeed, in contrast to rational numbers, real numbers are inherently infinite because they emerge as a result of an infinite process or as a relation in infinite sets. In addition, when the first construction is used, it is not necessary to construct separately real numbers – they automatically emerge as a subclass of real hypernumbers. At the same time, when we construct real hypernumbers from real numbers, it makes the construction more transparent and helps researchers to better understand properties and behavior of hypernumbers. That is why here we assume that hypernumbers are generated by sequences of real numbers and any sequence $\boldsymbol{a} = (a_i)_{i\in\omega}$ of real numbers determines/represents a hypernumber $\alpha = \mathrm{Hn}(a_i)_{i\in\omega}$.

The construction of real hypernumbers using sequences of rational numbers is similar to (in essence the same as) the construction of real numbers by taking equivalence classes of Cauchy sequences of rational numbers. The only but essential difference is that here we take all sequences of rational numbers.

As the equivalence relation is the same in a set and in any of its subsets, we have the following result.

Lemma 2.5. If Z is a dense subset of a set X, then any P-hypernumber $\alpha = \mathrm{Hn}(a_i)_{i\in\omega}$ in X is represented in Z.

A proof that is similar to the proof of Theorem 2.1 and based on Lemma 2.5 gives us the following result.

Theorem 2.2. There is a one-to-one correspondence between sets Z_P and X_P of P-hypernumbers, i.e., these sets are equipotent. For instance, as in the case $P = \omega$, it is natural to assume that both sets $\boldsymbol{R}_P$ and $\boldsymbol{Q}_P$ represent the same object, which is called the set of all real P-hypernumbers. It means that rational numbers and real numbers as building blocks for real P-hypernumbers give us the same construction. Consequently, any P-set $\boldsymbol{a} = (a_i)_{i\in P}$ of real numbers determines (and represents) a real P-hypernumber $\alpha = \mathrm{Hn}(a_i)_{i\in P}$ and such a P-set $\boldsymbol{a} = (a_i)_{i\in P}$ is called a *defining P-set* or *representing P-set* or *representation* of the P-hypernumber $\alpha = \mathrm{Hn}(a_i)_{i\in P}$.

Corollary 2.4. There is a one-to-one correspondence between sets $\boldsymbol{Ir}_P$ and $\boldsymbol{R}_P$, i.e., these sets are equipotent.

Note that a real P-hypernumber $\alpha = \mathrm{Hn}(a_i)_{i\in P}$ or a real hypernumber $\beta = \mathrm{Hn}(b_i)_{i\in\omega}$ is infinite if the set of all numbers a_i (of all numbers b_i) in its representation is unbounded.

As many scientists assume, there is no infinity in the real world concluding that there is no need in mathematics of infinite, we explain, as it is done in Burgin (2012), that infinity is already inherent to the majority of real numbers, such as π or e, while these numbers are very useful to people. Indeed, an explicit digital representation of irrational numbers, such as π or e, demands infinite number of digits, while there are much more irrational numbers (continuum) than rational numbers (a countable set). This situation worried mathematicians who were suspicious of infinity as an actual object. For instance, Emile Borel (1952) devoted a whole book to discuss these problems of the existence of real numbers, in particular, the problem of "inaccessible numbers." As Borel defines, an accessible number is a number that can be described as a mathematical object. The problem is that it is permissible only to use a finite process or finite description of a process to describe a real number. Thus, it is possible to describe integer numbers easily enough by an algorithm that generates the decimal representation of an arbitrary integer number. An algorithmic description of integer numbers allows us to do the same with rational numbers. For instance, it is possible to represent a rational number either as a pair of integer numbers or by specifying the repeating decimal expansion. Hence, integer and rational numbers are accessible. We can do the same with some real numbers but not with all. For example, if we take the, so-called, Liouville transcendental number, then it can be described (built) by an algorithm that puts 1 in the place $n!$ and 0 elsewhere. A finite way of specifying the n-th term in a Cauchy sequence of rational numbers gives us a finite description of the resulting real number. However, as Borel pointed out, there are a countable number of such descriptions or algorithms. At the same time, there are uncountably many (the continuum) of real numbers and the continuum is much bigger than any countable number. So, the natural conclusion is that the majority of real numbers are inaccessible and thus, as Borel claimed, it is impossible to operate with them.

Besides, Borel (1927) pointed out that when we consider a real number as an infinite sequence of digits, then we could put an infinite amount of information into a single number, building the "know-it-all" number. First, it is possible to build a real number q such that q contains every English sentence in a coded form. Each sentence is coded by a block of decimal digits. All English sentences are ordered and these blocks go after the decimal point in q one after another in the same order. In particular, q contains every possible true/false question that can be asked in English. Then we construct a real number r as follows. If the n-th block of q translates into a true/false question, then we set the n-th digit of q after the decimal point equal to 1 if the answer to the question is true and equal to 2 if the answer is false. If the n-th block of q does not translate into a true/false question, then we set the n-th digit of r after the decimal point equal to 3. Thus, using r, it is possible to answer every possible question that has ever been asked, or ever will be asked, in English. Borel calls this number r an unnatural real number, or an "unreal" real. This argument was further developed in Burgin (2005a). It was demonstrated that a possibility to operate with arbitrary real numbers makes it possible to compute any function defined for words in a finite alphabet.

Thus, there are problems with inherent infinity in transcendental real numbers, but in spite of this these numbers are efficiently used in many areas. In the same way, hypernumbers could be useful in many areas and specifically in physics.

Besides, in the decimal representation many rational numbers, e.g., 1/3 or 1/9, also have infinite representations. However, taking the base three for positional representation of numbers using four digits {0, 1, 2}, we see that 1/3 has a finite representation 0.01. Thus, in some cases, infinity in a number depends on its representation.

Real numbers contain different subclasses, such as rational numbers, irrational numbers, transcendental numbers, integer numbers, etc. In the universe of P-hypernumbers $\boldsymbol{R}_P$, there are even more distinct subclasses. As $\boldsymbol{R} \subseteq \boldsymbol{R}_P$, the set $\boldsymbol{R}_P$ also contains all subclasses of $\boldsymbol{R}$. However, there are also many new subclasses. For instance, it is possible to introduce three types of hypernumbers: stable, infinite, and oscillating hypernumbers.

For the theory of hypernormed vector spaces and semitopological vector spaces, we need only real hypernumbers, i.e., only the case where $P = \omega$ and $X = \boldsymbol{R}$. That is why we build our classification only for real hypernumbers.

Important classes of hypernumber are defined by the types of coordinates in their representation. Accordingly, we have:

- Whole hypernumbers with representations, coordinates of which are whole numbers.
- Integer hypernumbers with representations, coordinates of which are integer numbers.
- Positive hypernumbers with representations, coordinates of which are positive numbers.
- Negative hypernumbers with representations, coordinates of which are negative numbers.
- Prime hypernumbers with representations, coordinates of which are prime numbers.
- Compound hypernumbers with representations, coordinates of which are whole numbers.
- Odd hypernumbers with representations, coordinates of which are odd numbers.
- Even hypernumbers with representations, coordinates of which are even numbers.

There are specific relations between these classes.

Proposition 2.1.

a) The class of all integer hypernumbers contains the class of all whole hypernumbers.
b) The class of all whole hypernumbers contains the classes of all prime hypernumbers, all compound hypernumbers, all odd hypernumbers and all eve hypernumbers.
c) The classes of all prime hypernumbers and all compound hypernumbers do not intersect.
d) The classes of all odd hypernumbers and all even hypernumbers do not intersect.

Proof is left as an exercise.

There are also other important classes of hypernumbers. Let us consider some examples.

Example 2.9. An infinite increasing hypernumber: $\alpha = \mathrm{Hn}(a_i)_{i\in\omega}$ where $a_i = i$, $i = 1, 2, 3, \ldots$.

Example 2.10. An infinite increasing hypernumber: $\beta = \mathrm{Hn}(b_i)_{i\in\omega}$ where $b_i = 2^i$, $i = 1, 2, 3, \ldots$.

Example 2.11. A finite (bounded) oscillating hypernumber: $\gamma = \mathrm{Hn}(a_i)_{i\in\omega}$ where $a_i = (-1)^i$, $i = 1, 2, 3, \ldots$.

Example 2.12. An infinite oscillating hypernumber: $\delta = \mathrm{Hn}(a_i)_{i\in\omega}$ where $a_i = (-1)^i \cdot i$, $i = 1, 2, 3, \ldots$.

Example 2.13. An infinite decreasing hypernumber: $\nu = \mathrm{Hn}(a_i)_{i\in\omega}$ where $a_i = -5i$, $i = 1, 2, 3, \ldots$

Example 2.14. An infinite oscillating hypernumber: $\theta = \mathrm{Hn}(a_i)_{i\in\omega}$ where $a_i = 2^i + (-1)^i \cdot i$, $i = 1, 2, 3, \ldots$.

Hypernumbers from Examples 2.9–2.11 and 2.14 are whole hypernumbers, while the hypernumbers from Examples 2.12 and 2.13 are integer hypernumbers.

Let us give exact definitions for these classes of hypernumbers obtaining a classification of hypernumbers based on their size and form. This classification is based on the corresponding classification of hypernumber representations.

Definition 2.10. A sequence $(a_i)_{i\in\omega}$ is called *bounded* if there is a real number b and a natural number n such that $|a_i| < b$ for all $i > n$.

As any finite set of real numbers is bounded, we have the following result.

Lemma 2.6. A sequence $\boldsymbol{a} = (a_i)_{i\in\omega}$ is bounded if and only if there is a real number q such that $|a_i| < q$ for all i.

Proof is left as an exercise.

Lemma 2.7. If a sequence $\boldsymbol{a} = (a_i)_{i\in\omega}$ is bounded and $\boldsymbol{a} \approx \boldsymbol{b} = (b_i)_{i\in\omega}$, then the sequence $\boldsymbol{b}$ is also bounded.

Proof. By Lemma 2.6, there is a real number q such that $|a_i| < q$ for all i. At the same time, $\lim_{i\to\infty} |a_i - b_i| = 0$. Thus, there is a number m such that $|a_i - b_i| < m$ for all i. Then by the properties of absolute value, we have

$$|b_i| = |b_i - a_i + a_i| \leq |a_i - b_i| + |a_i| < m + q$$

It means that the sequence $\boldsymbol{b}$ is bounded.

Lemma is proved.

Remark 2.1. It is possible to define bounded P-sets and prove Lemma 2.7 when almost all elements in P belong to some sequence in P.

Bounded sequences from $\boldsymbol{R}^\omega$ are mapped onto finite hypernumbers from $\boldsymbol{R}_\omega$.

Definition 2.11. A real hypernumber α is called *finite* or *bounded* if there is a sequence $(a_i)_{i\in\omega}$ such that $\alpha = \text{Hn}(a_i)_{i\in\omega}$ and for some positive real number a, we have $|a_i| < a$ for almost all $i\in\omega$. The set of all finite/bounded real hypernumbers is denoted by $\mathbf{F}\boldsymbol{R}_\omega$.

Proposition 2.2. The following conditions are equivalent:

a) α is a finite real hypernumber;
b) there is a sequence $(a_i)_{i\in\omega}$ such that $\alpha = \text{Hn}(a_i)_{i\in\omega}$ and for some real number a, $|a_i| < a$ for all $i\in\omega$;
c) for any sequence $(a_i)_{i\in\omega}$ such that $\alpha = \text{Hn}(a_i)_{i\in\omega}$, there is a real number a such that $|a_i| < a$ for almost all $i\in\omega$;
d) α is represented in some interval from the real line $\boldsymbol{R}$.

Proof. **(a)⇔(b).** If $\alpha = \text{Hn}(a_i)_{i\in\omega}$ and $|a_i| < a$ for almost all $i\in\omega$. It means that in the sequence $(a_i)_{i\in\omega}$, there is only a finite number of elements a_i such that the absolute value of each of them is larger than a. Then taking b equal to $c + 1$ where c is the maximum the absolute values of all elements from the sequence $(a_i)_{i\in\omega}$, we have $|a_i| < b$ for all $i\in\omega$. Thus, (a) implies (b) and naturally, b) implies a) as b) is a stronger condition.

(a)⇔(c). Let us assume that $\alpha = \text{Hn}(a_i)_{i\in\omega} = \text{Hn}(b_i)_{i\in\omega}$ and for some positive real number a, we have $|a_i| < a$ for almost all $i\in\omega$. By Definition 2.6, the sequences $(a_i)_{i\in\omega}$ and $(b_i)_{i\in\omega}$ are equivalent. It means that $\lim_{i\to\infty}|a_i - b_i| = 0$. Thus, if the absolute values of almost all a_i are bounded, then the absolute values of almost all b_i are also bounded. Consequently, (a) implies (c) and naturally, (c) implies (a) as (c) is a stronger condition.

Equivalence of conditions (b) and (c) is also true as any equivalence relation is transitive, e.g., (b)⇔(a) and (a)⇔(c) imply (b)⇔(c).

In addition, conditions (b) and (d) are also equivalent because if $| a_i | < a$ for all $i \in \omega$, then α is represented in the interval $[-a, a]$, and if α is represented in the interval $[d, a]$, then $| a_i | < \max \{|a|, |d|\} + 1$ for all $i \in \omega$.

Proposition is proved.

Definition 2.12. A real hypernumber $\alpha = \text{Hn}(a_i)_{i\in\omega}$ is called *proper* if $\alpha \notin \boldsymbol{R}$.

Improper real hypernumbers are exactly real numbers.

Note that a subhypernumber of a proper hypernumber is always a proper hypernumber, while subhypernumber of an improper hypernumber may be a proper hypernumber.

Definition 2.13. A real hypernumber $\alpha = \text{Hn}(a_i)_{i\in\omega}$ is called *stable* if there is a real number b such that $\alpha = \text{Hn}(b_i)_{i\in\omega}$ and $b_i = b \in \boldsymbol{R}$ for almost all $i \in \omega$.

For such a hypernumber, we have $\alpha = b \in \boldsymbol{R}$.

Definitions 2.12 and 2.13 imply the following result.

Lemma 2.8. Any stable hypernumber $\alpha = \text{Hn}(a_i)_{i\in\omega}$ is finite.

Proof is left as an exercise.

Proposition 2.2. The following conditions are equivalent:

a) α is a stable real hypernumber;
b) there is a sequence $(b_i)_{i\in\omega}$ such that $\alpha = \text{Hn}(b_i)_{i\in\omega}$ and for some real number b, we have $b_i = b$ for all $i \in \omega$;
c) there is a real number a such that for any sequence $(c_i)_{i\in\omega}$ such that $\alpha = \text{Hn}(c_i)_{i\in\omega}$, we have $\lim_{i\to\infty} c_i = a$;
d) $\alpha \in \boldsymbol{R}$;
e) for some sequence $(c_i)_{i\in\omega}$ such that $\alpha = \text{Hn}(c_i)_{i\in\omega}$, the limit $\lim_{i\to\infty} c_i$ exists and is equal to α.

Proof. If $\alpha = \text{Hn}(b_i)_{i\in\omega}$ and $b_i = b$ for almost all $i \in \omega$. Then by Definition 2.6, $\alpha = \text{Hn}(a_i)_{i\in\omega}$, where $a_i = b$ for all $i \in \omega$. Thus, Condition (a) implies Condition (b) and naturally, Condition (b) implies Condition (a) as (b) is a stronger condition.

If α is a stable real hypernumber, then by Definition 2.11, there is a real number a and a sequence $(b_i)_{i\in\omega}$ such that $\alpha = \text{Hn}(b_i)_{i\in\omega}$ and $b_i = a$ for almost all $i \in \omega$. At the same time, as $\text{Hn}(b_i)_{i\in\omega} = \text{Hn}(c_i)_{i\in\omega}$, we have by Definitions 2.6 and 2.7, that $\lim_{i\to\infty} |c_i - b_i| = 0$. Thus, $\lim_{i\to\infty} c_i = a$, and Condition (a) implies Condition (c).

If $\lim_{i\to\infty} c_i = a$ and in a sequence $(b_i)_{i\in\omega}$ almost all $b_i = b$, then $\text{Hn}(b_i)_{i\in\omega} = \text{Hn}(c_i)_{i\in\omega}$ is a stable hypernumber. Thus, Condition (c) implies Condition (a).

If $\alpha = \text{Hn}(b_i)_{i\in\omega} = \text{Hn}(c_i)_{i\in\omega}$ and the limit $\lim_{i\to\infty} a_i$ exists, then by Definitions 2.6 and 2.7, $\lim_{i\to\infty}|c_i - a_i| = 0$. Thus, $\lim_{i\to\infty} c_i = \lim_{i\to\infty} a_i$, and Condition (e) implies Condition (c). At the same time, Condition (c) implies Condition (e) as (c) is a stronger condition.

In addition, Conditions (a) and (d) are equivalent by definition.

Proposition is proved.

Thus, it is possible to assume that stable real hypernumbers are real numbers, which belong to $\boldsymbol{R}_\omega$. In what follows, we will identify stable real hypernumbers and corresponding real numbers. For instance, Hn(10, 5, 1, 1, 1, 8, 8, …, 8, 8, 8, …) = 8 in $\boldsymbol{R}_\omega$.

Definition 2.14. A real hypernumber $\alpha = \text{Hn}(a_i)_{i\in\omega}$ is called *oscillating* if there is a positive real number k such that there are two infinite sequences of natural numbers $m(i)$ and $n(i)$ with $i = 1, 2, 3, \ldots$ such that $m(i) < n(i) < m(i + 1)$, $a_{m(i)} - a_{n(i)} > k$ and $a_{m(i+1)} - a_{n(i)} > k$ for all $i = 1, 2, 3, \ldots$

Remark 2.2. Oscillating hypernumbers may be bounded/finite (cf. Example 2.11) and unbounded or infinite (cf. Examples 2.12 and 2.14). In Examples 2.11, 2.12 and 2.14, $m(i) = 2i$ and $m(i) = 2i + 1$ ($i = 1, 2, 3, \ldots$).

Note that oscillation in a hypernumber can be very fast as in the hypernumbers from Examples 2.11 and 2.12 or it can be very slow as in the hypernumber $\alpha = \text{Hn}(a_i)_{i\in\omega}$ where $a_i = 10^{2i}$ when $10^{2i} \le i < 10^{2i+1}$ and $a_i = -10^{2i+1}$ when $10^{2i+1} \le i < 10^{2(i+1)}$.

Definition 2.15. A real hypernumber α is called:

1) *increasing* if $\alpha = \text{Hn}(a_i)_{i\in\omega}$ for some sequence $(a_i)_{i\in\omega}$ such that

$$\exists j \in \omega \forall i > j\ (a_{i+1} - a_i \ge 0)$$

2) *decreasing* if $\alpha = \text{Hn}(a_i)_{i\in\omega}$ for some sequence $(a_i)_{i\in\omega}$ such that

$$\exists j \in \omega \forall i > j\ (a_{i+1} - a_i \le 0)$$

3) *strictly increasing* if $\alpha = \text{Hn}(a_i)_{i\in\omega}$ for some sequence $(a_i)_{i\in\omega}$ such that

$$\exists j \in \omega \forall i > j\ (a_{i+1} - a_i > 0)$$

4) *strictly decreasing* if $\alpha = \text{Hn}(a_i)_{i\in\omega}$ for some sequence $(a_i)_{i\in\omega}$ such that

$$\exists j \in \omega \forall i > j\ (a_{i+1} - a_i < 0)$$

5) *infinite increasing* if $\alpha = \text{Hn}(a_i)_{i\in\omega}$ for some sequence $(a_i)_{i\in\omega}$ such that

$$\exists j \in \omega \forall i > j\,(a_{i+1} - a_i \geq 0))\&(\forall p \in \boldsymbol{R}^+ \exists i \in \omega\,(a_i > p));$$

6) *infinite decreasing* if $\alpha = \text{Hn}(a_i)_{i\in\omega}$ for some sequence $(a_i)_{i\in\omega}$ such that

$$\exists j \in \omega \forall i > j(a_{i+1} - a_i \leq 0))\&(\forall p \in \boldsymbol{R}^+ \exists i \in \omega(a_i < -p));$$

7) *infinite strictly increasing* if $\alpha = \text{Hn}(a_i)_{i\in\omega}$ for some sequence $(a_i)_{i\in\omega}$ such that

$$\exists j \in \omega \forall i > j\,(a_{i+1} - a_i > 0))\&(\forall p \in \boldsymbol{R}^+ \exists i \in \omega\,(a_i > p));$$

8) *infinite strictly decreasing* if $\alpha = \text{Hn}(a_i)_{i\in\omega}$ for some sequence $(a_i)_{i\in\omega}$ such that

$$\exists j \in \omega \forall i > j\,(a_{i+1} - a_i < 0))\&(\forall p \in \boldsymbol{R}^+ \exists i \in \omega\,(a_i < -p));$$

9) *infinite expanding* if $\alpha = \text{Hn}(a_i)_{i\in\omega}$ for some sequence $(a_i)_{i\in\omega}$ such that there are subsequences $(b_i)_{i\in\omega}$ and $(c_i)_{i\in\omega}$ of the sequence $(a_i)_{i\in\omega}$ such that $\text{Hn}(b_i)_{i\in\omega}$ is an infinite increasing hypernumber and $\text{Hn}(c_i)_{i\in\omega}$ is an infinite decreasing hypernumber;
10) *faithfully infinite expanding* if it is infinite expanding and the sequence $(a_i)_{i\in\omega}$ is the union of the sequences $(b_i)_{i\in\omega}$ and $(c_i)_{i\in\omega}$ such that $\text{Hn}(b_i)_{i\in\omega}$ is an infinite increasing hypernumber and $\text{Hn}(c_i)_{i\in\omega}$ is an infinite decreasing hypernumber;
11) *infinite monotone* if $\alpha = \text{Hn}(a_i)_{i\in\omega}$ for some monotone sequence $(a_i)_{i\in\omega}$, i.e., $a_{i+1} \geq a_i$ for almost all i or $a_{i+1} \leq a_i$ for almost all i, that tends to infinity.

It is possible to show that definitions of hypernumber classes are invariant with respect to the choice of sequences representing hypernumbers, i.e., a hypernumber stays in the same class when the representing sequence is changed.

Definition 2.15 directly implies the following result.

Lemma 2.9. Infinite monotone hypernumbers are exactly infinite increasing hypernumbers or infinite decreasing hypernumbers.

Infinite monotone hypernumbers form two classes:

- Positive infinite monotone hypernumbers, in which almost all elements of each representation are positive.

- Negative infinite monotone hypernumbers, in which almost all elements of each representation negative.

Lemma 2.10.

a) Positive infinite monotone hypernumbers are exactly infinite increasing hypernumbers.

b) Negative infinite monotone hypernumbers are exactly infinite decreasing hypernumbers.

Indeed, a representative of positive hypernumber cannot have elements that tend to negative infinity, while a representative of negative hypernumber cannot have elements that tend to positive infinity.

Lemma 2.11. Any infinite expanding hypernumber is an unbounded oscillating hypernumber and vice versa.

Proof. 1. Let us take an infinite expanding hypernumber $\alpha = \mathrm{Hn}(a_i)_{i\in\omega}$. Then there are subsequences $(b_i)_{i\in\omega}$ and $(c_i)_{i\in\omega}$ of the sequence $(a_i)_{i\in\omega}$ such that $\mathrm{Hn}(b_i)_{i\in\omega}$ is an infinite increasing hypernumber and $\mathrm{Hn}(c_i)_{i\in\omega}$ is an infinite decreasing hypernumber. Because the distance between elements b_i and c_i grows, there is a number $k \in \boldsymbol{R}^{++}$ such that $b_i - c_i > k$ and $b_{i+1} - c_i > k$ for all $i = 1, 2, 3, \ldots$. Indeed, as the sequence $(b_i)_{i\in\omega}$ is infinitely increasing, while the sequence $(c_i)_{i\in\omega}$ is decreasing, there is an element b_i larger than any element from the sequence $(c_i)_{i\in\omega}$ plus k. If $b_i = a_j$, we take $m(1) = j$. As the sequence $(c_i)_{i\in\omega}$ is infinite, we can find an element c_k such that $c_k = a_t$ and $t > j$. Then we take $n(1) = t$. As the sequence $(b_i)_{i\in\omega}$ is infinite, we can find an element b_l such that $b_l = a_q$ and $q > t$. Then we take $m(2) = q$. By construction, we have $a_{m(1)} - a_{n(1)} > k$ and $a_{m(2)} - a_{n(1)} > a_{m(1)} - a_{n(1)} > k$.

As the sequence $(c_i)_{i\in\omega}$ is infinite, we can find an element c_h such that $c_h = a_r$ and $r > q$. Then we take $n(2) = r$ and continue this process. It gives us two infinite sequences of natural numbers $m(i)$ and $n(i)$ with $i = 1, 2, 3, \ldots$ such that $m(i) < n(i) < m(i + 1)$, $a_{m(i)} - a_{n(i)} > k$ and $a_{m(i+1)} - a_{n(i)} > k$ for all $i = 1, 2, 3, \ldots$. Consequently, by Definition 2.14, α is an unbounded oscillating hypernumber.

2. Let us take an unbounded oscillating hypernumber $\alpha = \mathrm{Hn}(a_i)_{i\in\omega}$. As α is not bounded from above, the sequence $(a_i)_{i\in\omega}$ has a subsequence $(b_i)_{i\in\omega}$ that infinitely increases. As α is not bounded from below, the sequence $(a_i)_{i\in\omega}$ has a subsequence $(c_i)_{i\in\omega}$ that infinitely decreases. Consequently, α is an infinite expanding hypernumber.

Lemma is proved.

The following lemma gives characteristic properties of real hypernumbers that are outside the real line.

Lemma 2.12. A real hypernumber $\alpha = \text{Hn}(a_i)_{i\in\omega}$ is proper, that is, $\alpha \notin \boldsymbol{R}$, if and only if either the sequence $\boldsymbol{a} = (a_i)_{i\in\omega}$ is unbounded or (*condition* D) there is an interval $(d, b) \subset \boldsymbol{R}$ such that there are infinitely many elements a_i larger than b and infinitely many elements a_i smaller than d.

Proof. Sufficiency. Let $\alpha \in \boldsymbol{R}$. Then by Lemma 2.10, $\lim_{i\to\infty} a_i = a \in \boldsymbol{R}$. Consequently, the sequence $\boldsymbol{a} = (a_i)_{i\in\omega}$ is bounded (cf. (Ross, 1996)). Thus, if the sequence $\boldsymbol{a} = (a_i)_{i\in\omega}$ is unbounded, $\alpha \notin \boldsymbol{R}$.

Let us assume that there is an interval $(d, b) \subseteq \boldsymbol{R}$ such that there are infinitely many elements a_i larger than b and infinitely many elements a_i smaller than d. Then taking some element a_i larger than b, we can find an elements a_j smaller than d with $j > i$. Then we can find an elements a_t larger than b with $t > j$. Then we can find an elements a_p smaller than d with $p > t$. Taking $k = b - d$, we see that this process will give us two infinite sequences of natural numbers $m(i)$ and $n(i)$ with $i = 1, 2, 3, \dots$ such that $a_{m(i)} - a_{n(i)} > k$ and $a_{n(i)} - a_{m(i+1)} > k$ for all $i = 1, 2, 3, \dots$ By Definition 2.14, α is an oscillating hypernumber. It means that Condition D implies that α is a proper hypernumber.

Necessity. For any real hypernumber $\alpha = \text{Hn}(a_i)_{i\in\omega}$, we have two cases: the sequence $\boldsymbol{a} = (a_i)_{i\in\omega}$ is bounded or unbounded. If the sequence $\boldsymbol{a}$ is unbounded, then the first condition of the lemma is satisfied. So, to prove the lemma, we need to consider only the case when this sequence is bounded. It means that all elements a_i belong to some interval $[u, v]$.

Let us divide $[u, v]$ into three equal parts: $[u, u_1]$, $[u_1, u_2]$, and $[u_2, v]$. Then either (case 1) only one of these intervals (say, $[u, u_1]$) contains infinite number of elements a_i or (case 2) only two adjacent intervals (say, $[u, u_1]$ and $[u_1, u_2]$) contain infinite number of elements a_i or (case 3) both non-adjacent intervals $[u, u_1]$ and $[u_2, v]$ contain infinite number of elements a_i. In the latter case, everything is proved because it shows that α is an oscillating hypernumber with $k = b - d$ as we can take (u_1, u_2) as (d, b), validating Condition D. Otherwise we continue decomposition of intervals: in the first case, of the interval $[u, u_1]$ and in the second case, of the interval $[u, u_2]$. Note that the length of the new interval obtained in the decomposition is equal, at most, to two thirds of the length of the interval that has been decomposed. We continue this process of decomposition.

If at some step of this process, we get the third case considered above, then we have an interval (d, b) that is necessary for validity of the lemma because it indicates that α is an oscillating hypernumber. If we always get cases one or two, then it gives us a system of nested closed intervals. The lengths of these intervals converge to 0 because the length of each of them is less then or equal to two thirds of the length of the previous interval and each of them contains almost all elements a_i. By the standard argument, this implies that the sequence $\boldsymbol{a} = (a_i)_{i\in\omega}$ has the limit. Consequently, $\alpha \in \boldsymbol{R}$.

Thus, if $\alpha \notin \boldsymbol{R}$, then the condition of the lemma must be satisfied.

Lemma is proved.

Proposition 2.3. Any finite real hypernumber is either a real number or an oscillating real hypernumber.

Proof. If a hypernumber α is defined by a bounded sequence $\boldsymbol{l}$, then $\boldsymbol{l}$ either converges or has, at least, two subsequences that converge to two different points. In the first case, α is a real number. In the second case, α is an oscillating real hypernumber because if a subsequence $\boldsymbol{h}$ of $\boldsymbol{l}$ converges to the point a, while a subsequence $\boldsymbol{g}$ of $\boldsymbol{l}$ converges to the point b and $a < b$, then the number $\frac{1}{2}(b - a)$ satisfies the condition for the number k from Definition 2.14.

Proposition is proved.

Corollary 2.5. Any finite increasing (decreasing) real hypernumber is a real number.

Definitions 2.14 and 2.15 imply the following result.

Proposition 2.4. Any infinite real hypernumber is either an infinite increasing hypernumber or an infinite decreasing hypernumber or an oscillating real hypernumber.

Proof. If a hypernumber α is defined by a sequence $\boldsymbol{l}$, then being unbounded, $\boldsymbol{l}$ is either unbounded only above (case 1) or unbounded only below (case 2) or unbounded below and above (case 3). In the first case, there are two options for the hypernumber α: the sequence $\boldsymbol{l}$ either converges to ∞ or has two subsequences one of which converges to ∞, while the another converges to some real number a. In the first case, α is an infinite increasing hypernumber. The second option is true when in $\boldsymbol{l}$, there are infinitely many elements less than some real number c and it shows that α is an unbounded above oscillating real hypernumber because in this case, any positive real number satisfies the condition for the number k from Definition 2.14.

In the second case, which is symmetric to the first case, there are two options for α: the sequence $\boldsymbol{l}$either converges to $-\infty$ or has a subsequence that either converges to $-\infty$, while another subsequence converges to some real number a. In the first case, α is an infinite decreasing hypernumber. The second option is true when in $\boldsymbol{l}$, there are infinitely many elements larger than some real number c and it shows that α is an unbounded below oscillating real hypernumber because any positive real number satisfies the condition for the number k from Definition 2.15.

In the third case, α is an oscillating infinite expanding real hypernumber because in this case, any positive real number satisfies the condition for the number k from Definition 2.15.

Proposition is proved.

Lemma 2.8 and Propositions 2.3 and 2.4 are used to give a complete characterization of real hypernumbers.

Theorem 2.3. There are four disjoint classes of real hypernumbers and each hypernumber belongs to one of them:

- Stable hypernumbers.
- Bounded/finite oscillating hypernumbers.
- Unbounded oscillating hypernumbers.
- Infinite monotone hypernumbers, which form two groups: increasing and decreasing hypernumbers.

In turn, unbounded oscillating hypernumbers form three groups:

- Bounded above oscillating hypernumbers.
- Bounded below oscillating hypernumbers.
- Unbounded below and above (infinite expanding) oscillating hypernumbers.

Proof. Definitions 2.12 and 2.13 imply that any hypernumber α is either finite (bounded) or infinite. Indeed, if $\alpha = \text{Hn}(a_i)_{i\in\omega} = \text{Hn}(b_i)_{i\in\omega}$, then either both sequences are bounded or both sequences are unbounded. In the first case, by Proposition 2.3, α is either a real number or an oscillating real hypernumber. In the second case, by Proposition 2.4, α is either an infinite increasing hypernumber or an infinite decreasing hypernumber or an unbounded oscillating real hypernumber.

In addition, Propositions 2.3 and 2.4 show that an oscillating hypernumber is either bounded or unbounded above or unbounded below or unbounded below and above (infinite expanding) oscillating hypernumber.

Theorem is proved.

Now let us consider some relational and algebraic properties of P-sets with values in $\boldsymbol{R}$ and P-hypernumbers.

Relations on $\boldsymbol{R}$ induce corresponding relations on the set $\boldsymbol{R}^P$ of all P-sets in $\boldsymbol{R}$ through pairwise relations between members of representing sequences with the same index number. The most important are order relations.

Definition 2.16. For any *P*-sets $\boldsymbol{a} = (a_i)_{i\in P}$, $\boldsymbol{b} = (b_i)_{i\in P} \in \boldsymbol{R}^P$, we define:

1. *Partial order*:

$$\boldsymbol{a} \leq \boldsymbol{b}$$

if for any *P*-sequences $\boldsymbol{l} = (a_i)_{i\in\omega}$ and $\boldsymbol{h} = (b_i)_{i\in\omega}$, defined in $\boldsymbol{a}$ and $\boldsymbol{b}$ by the same sequence $\boldsymbol{q}$ in *P*, we have

$$\exists n\in\omega \forall i \geq n(a_i \leq b_i)$$

2. *Strict partial order*:

$$\boldsymbol{a} < \boldsymbol{b}$$

if for any *P*-sequences $\boldsymbol{l} = (a_i)_{i\in\omega}$ and $\boldsymbol{h} = (b_i)_{i\in\omega}$, defined in $\boldsymbol{a}$ and $\boldsymbol{b}$ by the same sequence $\boldsymbol{q}$ in *P*, we have

$$\exists n\in\omega \forall i \geq n(a_i \leq b_i)$$

and there are, at least, two *P*-sequences $\boldsymbol{l} = (a_i)_{i\in\omega}$ and $\boldsymbol{h} = (b_i)_{i\in\omega}$ defined in $\boldsymbol{a}$ and $\boldsymbol{b}$ by the same sequence $\boldsymbol{q}$ in *P* such that

$$a_i < b_i$$

for infinitely many elements *i* from $\boldsymbol{q}$

3. *Accurate partial order*:

$$\boldsymbol{a} \ll \boldsymbol{b}$$

if for any *P*-sequences $\boldsymbol{l} = (a_i)_{i\in\omega}$ and $\boldsymbol{h} = (b_i)_{i\in\omega}$, defined in $\boldsymbol{a}$ and $\boldsymbol{b}$ by the same sequence $\boldsymbol{q}$ in *P*, we have

$$\exists n\in\omega \forall i \geq n(a_i \leq b_i)$$

and there are, at least, two *P*-sequences $\boldsymbol{l} = (a_i)_{i\in\omega}$ and $\boldsymbol{h} = (b_i)_{i\in\omega}$ defined in $\boldsymbol{a}$ and $\boldsymbol{b}$ by the same sequence $\boldsymbol{q}$ in *P* such that there is a number $k \in \boldsymbol{R}^{++}$ for which

$$b_i - a_i > k$$

for infinitely many elements *i* from $\boldsymbol{q}$.

For instance,

$(1, 2, 3, \ldots, n, \ldots) \leq (10, 20, 3, \ldots, n, \ldots)$

$(1, 2, 3, \ldots, n, \ldots) < (1, 2 + ½, 3 + 1/3, \ldots, n + 1/n, \ldots)$

$(1, 2, 3, \ldots, n, \ldots) \ll (1 + 1/n, 2 + 1/n, 3 + 1/n, \ldots, n + 1/n, \ldots)$

As always, the inequality $\boldsymbol{b} \geq \boldsymbol{a}$ means $\boldsymbol{a} \leq \boldsymbol{b}$, the inequality $\boldsymbol{b} > \boldsymbol{a}$ means $\boldsymbol{a} < \boldsymbol{b}$ and the inequality $\boldsymbol{b} \gg \boldsymbol{a}$ means $\boldsymbol{a} \ll \boldsymbol{b}$.

Lemma 2.13. The relation $\ll$ implies the relation $<$, while the relation $<$ implies the relation $\leq$.

Proof is left as an exercise.

It is possible that neither $\boldsymbol{a} \leq \boldsymbol{b}$ nor $\boldsymbol{b} \leq \boldsymbol{a}$. It means that these P-sets are incomparable and this relation is denoted by $\boldsymbol{a} \perp \boldsymbol{b}$.

Lemma 2.14. $\boldsymbol{a} < \boldsymbol{b}$ if and only if $\boldsymbol{a} \leq \boldsymbol{b}$ and there are, at least, two P-sequences $\boldsymbol{l} = (a_i)_{i\in\omega}$ and $\boldsymbol{h} = (b_i)_{i\in\omega}$ defined in $\boldsymbol{a}$ and $\boldsymbol{b}$ by the same sequence $\boldsymbol{r}$ in P such that $b_i > a_i$ for all elements i from $\boldsymbol{r}$.

Indeed, we can take the subsequence of $\boldsymbol{q}$ where $b_i - a_i > k$ as $\boldsymbol{r}$.

Lemma 2.15. $\boldsymbol{a} \ll \boldsymbol{b}$ if and only if $\boldsymbol{a} \leq \boldsymbol{b}$ and there are, at least, two P-sequences $\boldsymbol{l} = (a_i)_{i\in\omega}$ and $\boldsymbol{h} = (b_i)_{i\in\omega}$ defined in $\boldsymbol{a}$ and $\boldsymbol{b}$ by the same sequence $\boldsymbol{r}$ in P such that there is a number $k \in \boldsymbol{R}^{++}$ for which $b_i - a_i > k$ for all elements i from $\boldsymbol{r}$.

Indeed, we can take the subsequence of $\boldsymbol{q}$ where $a_i < b_i$ as $\boldsymbol{r}$.

We remind that a *preorder*, also called *quasiorder*, on a set (class) X is a binary relation Q in X that satisfies the following axioms:

O1. Q is *reflexive*, i.e. $x\mathrm{Q}x$ for all x from X.

O2. Q is *transitive*, i.e., $x\mathrm{Q}y$ and $y\mathrm{Q}z$ imply $x\mathrm{Q}z$ for all $x, y, z \in X$.

A *partial order* is a preorder that satisfies the following additional axiom:

O3. Q is *antisymmetric*, i.e., $x\mathrm{Q}y$ and $y\mathrm{Q}x$ imply $x = y$ for all $x, y \in X$.

A *strict* also called *sharp partial order* is a preorder that is not reflexive, is transitive and satisfies the following additional axiom:

O4. Q is *asymmetric*, i.e., only one relation $x\mathrm{Q}y$ or $y\mathrm{Q}x$ is true for all x, $y \in X$.

Proposition 2.5. Relation $\leq$ is a partial preorder, while relations $<$ and $\ll$ in the set $\boldsymbol{R}^P$ of all P-sets in $\boldsymbol{R}$ are strict partial orders.

Proof. We begin with the relation $\leq$. By definition, a partial preorder is a reflexive transitive relation. Thus, we have to test these properties for $\leq$ in $\boldsymbol{R}^P$.

1. Reflexivity of $\leq$.

Indeed, for any P-set $\boldsymbol{a} = (a_i)_{i\in P}$ from $\boldsymbol{R}^P$, we have $\boldsymbol{a} \leq \boldsymbol{a}$ because by the definition of a partial order, $a_i \leq a_i$ for all $i \in P$.

2. Transitivity of ≤.

Let us consider three P-sets $\boldsymbol{a} = (a_i)_{i\in P}$, $\boldsymbol{b} = (b_i)_{i\in P}$ and $\boldsymbol{c} = (c_i)_{i\in P}$ from $\boldsymbol{R}^P$ such that $\boldsymbol{a} \le \boldsymbol{b}$ and $\boldsymbol{b} \le \boldsymbol{c}$. By definition, for any P-sequences $\boldsymbol{l} = (a_i)_{i\in\omega}$ and $\boldsymbol{h} = (b_i)_{i\in\omega}$, defined in $\boldsymbol{a}$ and $\boldsymbol{b}$ by the same sequence $\boldsymbol{q}$ in P, there is an element n from $\boldsymbol{q}$ such that if i belongs to $\boldsymbol{q}$ and $i > n$, then $a_i \le b_i$, and there is an element m from $\boldsymbol{q}$ such that if j belongs to $\boldsymbol{q}$ and $j > m$, then $b_j \le c_j$. Consequently, $a_i \le b_i$ and $b_j \le c_j$ when $i > t$ and $j > t$, where $t = \max\{n, m\}$. Thus, $a_i \le c_j$ when $i > t$ because ≤ is a partial order in X. It means that $\boldsymbol{a} \le \boldsymbol{c}$. Hence, in the set $\boldsymbol{R}^P$, the relation ≤ is transitive.

Now we treat the relation <. By definition, a strict partial order is a transitive antisymmetric relation. Thus, we have to test these properties for < in $\boldsymbol{R}^P$.

3. Transitivity of <.

Let us consider three P-sets $\boldsymbol{a} = (a_i)_{i\in P}$, $\boldsymbol{b} = (b_i)_{i\in P}$ and $\boldsymbol{c} = (c_i)_{i\in P}$ from $\boldsymbol{R}^P$ such that $\boldsymbol{a} < \boldsymbol{b}$ and $\boldsymbol{b} < \boldsymbol{c}$. By definition, for any P-sequences $\boldsymbol{l} = (a_i)_{i\in\omega}$ and $\boldsymbol{h} = (b_i)_{i\in\omega}$, defined in $\boldsymbol{a}$ and $\boldsymbol{b}$ by the same sequence $\boldsymbol{q}$ in P, there is an element n from $\boldsymbol{q}$ such that if i belongs to $\boldsymbol{q}$ and $i > n$, then $a_i \le b_i$, and for any P-sequences $\boldsymbol{g} = (c_i)_{i\in\omega}$ and $\boldsymbol{h} = (b_i)_{i\in\omega}$, defined in $\boldsymbol{c}$ and $\boldsymbol{b}$ by the same sequence $\boldsymbol{q}$ in P, there is an element m from $\boldsymbol{q}$ such that if j belongs to $\boldsymbol{q}$ and $j > m$, then $b_j \le c_j$. Consequently, $a_i \le b_i$ and $b_j \le c_j$ when $i > t$ and $j > t$, where $t = \max\{n, m\}$. Thus, $a_i \le c_j$ when $i > t$ because ≤ is a partial order in $\boldsymbol{R}$.

In addition, by Lemma 2.13, there are two P-sequences $\boldsymbol{l} = (a_i)_{i\in\omega}$ and $\boldsymbol{h} = (b_i)_{i\in\omega}$ defined in $\boldsymbol{a}$ and $\boldsymbol{b}$ by the same sequence $\boldsymbol{r}$ in P such that $a_i < b_i$ for all elements i from $\boldsymbol{r}$ and there are two P-sequences $\boldsymbol{p} = (c_i)_{i\in\omega}$ and $\boldsymbol{u} = (b_i)_{i\in\omega}$ defined in $\boldsymbol{c}$ and $\boldsymbol{b}$ by the same sequence $\boldsymbol{t}$ in P such that $b_i < c_i$ for all elements i from $\boldsymbol{t}$.

By transitivity of ≤, for any P-sequences $\boldsymbol{l} = (a_i)_{i\in\omega}$ and $\boldsymbol{h} = (c_i)_{i\in\omega}$, defined in $\boldsymbol{a}$ and $\boldsymbol{c}$ by the same sequence $\boldsymbol{q}$ in P, there is an element n from $\boldsymbol{q}$ such that if i belongs to $\boldsymbol{q}$ and $i > n$, then $a_i \le c_i$. In addition, for the P-sequences $\boldsymbol{l} = (a_i)_{i\in\omega}$ and $\boldsymbol{k} = (c_i)_{i\in\omega}$ defined in $\boldsymbol{a}$ and $\boldsymbol{c}$ by the sequence $\boldsymbol{r}$ in P, we have $a_i < c_i$ for all elements i from $\boldsymbol{r}$ because $a_i < b_i$ and $b_i \le c_i$ for all elements i from $\boldsymbol{r}$.

It means that $\boldsymbol{a} < \boldsymbol{c}$. Hence, in the set $\boldsymbol{R}^P$, the relation < is transitive.

4. Asymmetry of <.

Let us consider two P-sets $\boldsymbol{a} = (a_i)_{i\in P}$ and $\boldsymbol{b} = (b_i)_{i\in P}$ from $\boldsymbol{R}^P$ such that $\boldsymbol{a} < \boldsymbol{b}$. By definition, for any P-sequences $\boldsymbol{l} = (a_i)_{i\in\omega}$ and $\boldsymbol{h} = (b_i)_{i\in\omega}$, defined in $\boldsymbol{a}$ and $\boldsymbol{b}$ by the same sequence $\boldsymbol{q}$ in P, there is an element n from $\boldsymbol{q}$ such that if i belongs to $\boldsymbol{q}$ and $i > n$, then $a_i \leq b_i$, and byLemma 2.13, there are two P-sequences $\boldsymbol{l} = (a_i)_{i\in\omega}$ and $\boldsymbol{h} = (b_i)_{i\in\omega}$ defined in $\boldsymbol{a}$ and $\boldsymbol{b}$ by the same sequence $\boldsymbol{r}$ in P such that $a_i < b_i$ for all elements i from $\boldsymbol{r}$.

Then the inequality $\boldsymbol{a} > \boldsymbol{b}$ and even the inequality $\boldsymbol{a} \geq \boldsymbol{b}$ are impossible because the inequality $a_i < b_i$ prohibits the inequality $a_i \geq b_i$. Hence, in the set $\boldsymbol{R}^P$, the relation $<$ is antisymmetric.

Thus, $<$ is a partial order.

Now we treat the relation $\ll$. By definition, a strict partial order is a transitive antisymmetric relation. Thus, we have to test these properties for $\ll$ in $\boldsymbol{R}^P$.

5. Transitivity of $\ll$.

Let us consider three P-sets $\boldsymbol{a} = (a_i)_{i\in P}$, $\boldsymbol{b} = (b_i)_{i\in P}$ and $\boldsymbol{c} = (c_i)_{i\in P}$ from $\boldsymbol{R}^P$ such that $\boldsymbol{a} \ll \boldsymbol{b}$ and $\boldsymbol{b} \ll \boldsymbol{c}$. By definition, for any P-sequences $\boldsymbol{l} = (a_i)_{i\in\omega}$ and $\boldsymbol{h} = (b_i)_{i\in\omega}$, defined in $\boldsymbol{a}$ and $\boldsymbol{b}$ by the same sequence $\boldsymbol{q}$ in P, there is an element n from $\boldsymbol{q}$ such that if i belongs to $\boldsymbol{q}$ and $i > n$, then $a_i \leq b_i$, and for any P-sequences $\boldsymbol{g} = (c_i)_{i\in\omega}$ and $\boldsymbol{h} = (b_i)_{i\in\omega}$, defined in $\boldsymbol{c}$ and $\boldsymbol{b}$ by the same sequence $\boldsymbol{q}$ in P, there is an element m from $\boldsymbol{q}$ such that if j belongs to $\boldsymbol{q}$ and $j > m$, then $b_j \leq c_j$. Consequently, $a_i \leq b_i$ and $b_j \leq c_j$ when $i > t$ and $j > t$, where $t = \max\{n, m\}$. Thus, $a_i \leq c_j$ when $i > t$ because $\leq$ is a partial order in X.

In addition, by Lemma 2.14, there are two P-sequences $\boldsymbol{l} = (a_i)_{i\in\omega}$ and $\boldsymbol{h} = (b_i)_{i\in\omega}$ defined in $\boldsymbol{a}$ and $\boldsymbol{b}$ by the same sequence $\boldsymbol{r}$ in P such that there is a number $k \in \boldsymbol{R}^{++}$ for which $\|b_i - a_i\| > k$ for all elements i from $\boldsymbol{r}$ and there are two P-sequences $\boldsymbol{p} = (c_i)_{i\in\omega}$ and $\boldsymbol{u} = (b_i)_{i\in\omega}$ defined in $\boldsymbol{c}$ and $\boldsymbol{b}$ by the same sequence $\boldsymbol{t}$ in P such that there is a number $h \in \boldsymbol{R}^{++}$ for which $\| c_i - b_i \| > h$ for all elements i from $\boldsymbol{t}$.

By transitivity of $\leq$, for any P-sequences $\boldsymbol{l} = (a_i)_{i\in\omega}$ and $\boldsymbol{h} = (c_i)_{i\in\omega}$, defined in $\boldsymbol{a}$ and $\boldsymbol{c}$ by the same sequence $\boldsymbol{q}$ in P, there is such an element n from $\boldsymbol{q}$ that if i belongs to $\boldsymbol{q}$ and $i > n$, then $a_i \leq c_i$. In addition, for the P-sequences $\boldsymbol{l} = (a_i)_{i\in\omega}$ and $\boldsymbol{k} = (c_i)_{i\in\omega}$ defined in $\boldsymbol{a}$ and $\boldsymbol{c}$ by the sequence $\boldsymbol{r}$ in P, we have $\| c_i - a_i \| \geq \|b_i - a_i\| > k$ for all elements i from $\boldsymbol{r}$ because $b_i - a_i > k$ and $b_i \leq c_i$ for all elements i from $\boldsymbol{r}$.

It means that $\boldsymbol{a} \ll \boldsymbol{c}$. Hence, in the set $\boldsymbol{R}^P$, the relation $<$ is transitive.

6. Asymmetry of$\ll$.

Let us consider two P-sets $\boldsymbol{a} = (a_i)_{i\in P}$ and $\boldsymbol{b} = (b_i)_{i\in P}$ from $\boldsymbol{R}^P$ such that $\boldsymbol{a} \ll \boldsymbol{b}$. By Lemma 2.13, the relation $\ll$implies the relation $<$ and thus, $\boldsymbol{a} < \boldsymbol{b}$. By antisymmetry of the relation $<$, the inequality $\boldsymbol{a} > \boldsymbol{b}$ is impossible.

Consequently, the inequality $\boldsymbol{a} \gg \boldsymbol{b}$ is impossible because by Lemma 2.13, the relation $\ll$ implies the relation <.

Thus, in the set $\boldsymbol{R}^P$, $\ll$ is a partial order.

Proposition is proved.

Remark 2.3. The relation $\leq$ is not a partial order because for any P-sets $\boldsymbol{a} = (a_i)_{i\in P}$ and $\boldsymbol{b} = (b_i)_{i\in P}$, we have $\boldsymbol{a} \leq \boldsymbol{b}$ and $\boldsymbol{b} \leq \boldsymbol{a}$ if only one a_i is not equal to the corresponding b_i.

For instance, $\boldsymbol{a} = \{1, 2, 3, \ldots, n, \ldots\}$ is not equal to $\boldsymbol{b} = \{10, 2, 3, 4, 5, \ldots, n, \ldots\}$ but the inequalities $\boldsymbol{a} \leq \boldsymbol{b}$ and $\boldsymbol{b} \leq \boldsymbol{a}$ are true.

These relations in $\boldsymbol{R}^P$ induce similar relations on $\boldsymbol{R}_P$ of all real P-hypernumbers.

Definition 2.17. If α and β are elements from $\boldsymbol{R}_P$, then

$\alpha \leq \beta$ means that $\exists \boldsymbol{a} \in \alpha \exists \boldsymbol{b} \in \beta(\boldsymbol{a} \leq \boldsymbol{b})$;

$\alpha < \beta$ means that $\exists \boldsymbol{a} \in \alpha \exists \boldsymbol{b} \in \beta(\boldsymbol{a} << \boldsymbol{b})$

When it is neither $\alpha \leq \beta$ nor $\beta \leq \alpha$, we denote this relation by $\alpha \perp \beta$.

If we take hypernumbers α, β, γ, δ, θ, and ν from the Examples 2.9–2.14, then we have the following relations:

$\alpha < \beta$, $\alpha > \nu$, $\gamma < \alpha$, $\gamma < \beta$, $\delta < \theta$, $\alpha < \theta$, $\beta > \delta$, $\alpha \geq \delta$, and $\nu < \gamma$, while hypernumbers γ and δ as well as hypernumbers β and θ, are incomparable with respect to the order relations. This shows that real hypernumbers may be incomparable in contrast to real numbers, where for any two numbers a and b, we have either $a = b$ or $a > b$ or $a < b$.

Lemma 2.16. The relation < implies the relation ≤.

Proof is left as an exercise.

Lemma 2.17. If $\alpha < \beta$, then $\alpha \neq \beta$.

Proof. If $\alpha < \beta$, then by definition, there are a P-set $\boldsymbol{a} \in \alpha$ and a P-set $\boldsymbol{b} \in \beta$ such that $\boldsymbol{a} \ll \boldsymbol{b}$. It means that there are two P-sequences $\boldsymbol{l} = (a_i)_{i\in\omega}$ and $\boldsymbol{h} = (b_i)_{i\in\omega}$ defined in $\boldsymbol{a}$ and $\boldsymbol{b}$ by the same sequence $\boldsymbol{r}$ in P such that there is a number $k \in \boldsymbol{R}^{++}$ for which $b_i - a_i > k$ for all elements i from $\boldsymbol{r}$.

At the same time, the equality $\alpha = \beta$ implies $\boldsymbol{a} \approx \boldsymbol{b}$. Consequently,

$$\lim_{i\to\infty} |a_{r(i)} - b_{r(i)}| = 0 \text{ for the sequence } \boldsymbol{r} \text{ in } P$$

This is impossible because $|b_i - a_i| = b_i - a_i > k$ for all elements i from $\boldsymbol{r}$.

Lemma is proved.

Proposition 2.6. Relations ≤ and < in $\boldsymbol{R}_P$ are a partial order and a strict partial order, respectively.

Proof. We begin with the relation $\leq$. By definition, a partial preorder is a reflexive transitive relation. Thus, we have to test these properties for < on $\boldsymbol{R}^P$.

1. Reflexivity of $\leq$.

Indeed, if $\alpha = \text{Hn}(a_i)_{i\in P}$ and $\boldsymbol{a} = (a_i)_{i\in P}$, then $\alpha \leq \alpha$because $\boldsymbol{a} \leq \boldsymbol{a}$.

2. Transitivity of $\leq$.

Let us consider three P-hypernumbers $\alpha, \beta, \gamma \in \boldsymbol{R}_P$ such that $\alpha \leq \beta$ and $\beta \leq \gamma$. By definition, there are four P-sets $\boldsymbol{a} = (a_i)_{i\in P}$, $\boldsymbol{b} = (b_i)_{i\in P}$, $\boldsymbol{l} = (l_i)_{i\in P}$ and $\boldsymbol{c} = (c_i)_{i\in P}$ from $\boldsymbol{R}^P$ such that $\alpha = \text{Hn}(a_i)_{i\in P}$, $\beta = \text{Hn}(b_i)_{i\in P} = \text{Hn}(l_i)_{i\in P}$, $\gamma = \text{Hn}(c_i)_{i\in P}$, $\boldsymbol{a} \leq \boldsymbol{b}$ and $\boldsymbol{l} \leq \boldsymbol{c}$. Thus, for any P-sequences $\boldsymbol{k} = (a_i)_{i\in P}$ and $\boldsymbol{h} = (b_i)_{i\in P}$, defined in $\boldsymbol{a}$ and $\boldsymbol{b}$ by the same sequence $\boldsymbol{q}$ in P, there is an element n from $\boldsymbol{q}$ such that if i belongs to $\boldsymbol{q}$ and $i > n$, then $a_i \leq b_i$. Besides, for any P-sequences $\boldsymbol{u} = (b_i)_{i\in P}$ and $\boldsymbol{v} = (c_i)_{i\in P}$, defined in $\boldsymbol{l}$ and $\boldsymbol{c}$ by the same sequence $\boldsymbol{q}$ in P, there is an element m from $\boldsymbol{r}$ such that if j belongs to $\boldsymbol{q}$ and $j > m$, then $l_j \leq c_j$. It means that $b_i - a_i \geq 0$ when $i > n$, and $c_i - l_i \geq 0$ when $i > m$. Consequently, taking $t = \max \{n, m\}$, we have $b_i \geq a_i$ and $c_i \geq l_i$ when $i > t$.

For any $i > t$, there are two cases for b_i and l_i – either $l_i \geq b_i$ or $b_i > l_i$. In the first case, we put $d_i = c_i \geq l_i \geq b_i \geq a_i$. In the second case, we take $d_i = c_i + (b_i - l_i)$. As a result, we have

$$\lim_{i\to\infty} \| d_i - c_i \| = 0$$

because either $\| d_i - c_i \| = 0$ or $\| d_i - c_i \| = \| b_i - l_i \|$ and $\lim_{i\to\infty} \| b_i - l_i \| = 0$.

Thus, $\gamma = \text{Hn}(c_i)_{i\in P} = \text{Hn}(d_i)_{i\in P}$ and when $i > t$, we have $d_i \geq b_i \geq a_i$ as $c_i \geq l_i$. Consequently, $\alpha \leq \gamma$.

Hence, in $\boldsymbol{R}_P$, the relation $\leq$ is transitive.

3. Antisymmetry of $\leq$.

Let us consider two P-hypernumbers $\alpha, \beta \in \boldsymbol{R}_P$ such that $\alpha \leq \beta$ and $\beta \leq \alpha$. By definition, there are four P-sets $\boldsymbol{a} = (a_i)_{i\in P}$, $\boldsymbol{b} = (b_i)_{i\in P}$, $\boldsymbol{d} = (l_i)_{i\in P}$ and $\boldsymbol{c} = (c_i)_{i\in P}$ from $\boldsymbol{R}^P$ such that $\alpha = \text{Hn}(a_i)_{i\in P} = \text{Hn}(c_i)_{i\in P}$, $\beta = \text{Hn}(b_i)_{i\in P} = \text{Hn}(d_i)_{i\in P}$, $\boldsymbol{a} \leq \boldsymbol{b}$ and $\boldsymbol{d} \leq \boldsymbol{c}$. Thus, for any P-sequences $\boldsymbol{k} = (a_i)_{i\in P}$ and $\boldsymbol{h} = (b_i)_{i\in P}$, defined in $\boldsymbol{a}$ and $\boldsymbol{b}$ by the same sequence $\boldsymbol{q}$ in P, there is an element n from $\boldsymbol{q}$ such that if i belongs to $\boldsymbol{q}$ and $i > n$, then $a_i \leq b_i$. Besides, for any P-sequences $\boldsymbol{u} = (b_i)_{i\in P}$ and $\boldsymbol{v} = (c_i)_{i\in P}$, defined in $\boldsymbol{l}$ and $\boldsymbol{c}$ by the same sequence $\boldsymbol{q}$ in P, there is an element m from $\boldsymbol{r}$ such that if j belongs to $\boldsymbol{q}$ and $j > m$, then $d_j \leq c_j$. It means that $b_i - a_i \geq 0$ when $i > n$, and $c_i - d_i \geq 0$ when $i > m$. Consequently, taking $t = \max \{n, m\}$, we have $b_i \geq a_i$ and $c_i \geq d_i$ when $i > t$.

Let us consider the difference $b_i - a_i$. If $\lim_{i\to\infty}|a_i - b_i| = 0$ for any sequence $\boldsymbol{q}$ in P, then $\alpha = \beta$. If $\lim_{i\to\infty}|a_i - b_i| \neq 0$ for some sequence $\boldsymbol{q}$ in P, then there is a positive real number k such that $b_i - a_i > k$ for infinitely many indices i.

At the same time, we have

$$b_i - a_i = b_i - d_i + d_i - c_i + c_i - a_i$$

Thus,

$$|b_i - a_i| \leq |b_i - d_i| + |d_i - c_i| + |c_i - a_i|$$

As $\lim_{i\to\infty}|a_i - c_i| = 0$ and $\lim_{i\to\infty}|d_i - b_i| = 0$, we can take an index i such that $b_i - a_i > k$, $b_i - d_i < \frac{1}{4}k$, and $c_i - a_i < \frac{1}{4}k$. This implies $|b_i - a_i| = b_i - a_i < \frac{1}{2}k$ because $d_i - c_i \leq 0$.

Obtained contradiction shows that $\lim_{i\to\infty}|a_i - b_i| = 0$ for any sequence $\boldsymbol{q}$ in P and consequently, $\alpha = \beta$.

4. Transitivity of <.

Let us consider three P-hypernumbers $\alpha, \beta, \gamma \in \boldsymbol{R}_P$ such that $\alpha \leq \beta$ and $\beta \leq \gamma$. By definition, there are four P-sets $\boldsymbol{a} = (a_i)_{i\in P}$, $\boldsymbol{b} = (b_i)_{i\in P}$, $\boldsymbol{d} = (d_i)_{i\in P}$ and $\boldsymbol{c} = (c_i)_{i\in P}$ from $\boldsymbol{R}^P$ such that $\alpha = \mathrm{Hn}(a_i)_{i\in P}$, $\beta = \mathrm{Hn}(b_i)_{i\in P} = \mathrm{Hn}(d_i)_{i\in P}$, $\gamma = \mathrm{Hn}(c_i)_{i\in P}$, $\boldsymbol{a} \leq \boldsymbol{b}$ and $\boldsymbol{l} \leq \boldsymbol{c}$. Thus, for any P-sequences $\boldsymbol{k} = (a_i)_{i\in P}$ and $\boldsymbol{h} = (b_i)_{i\in P}$, defined in $\boldsymbol{a}$ and $\boldsymbol{b}$ by the same sequence $\boldsymbol{q}$ in P, there is an element n from $\boldsymbol{q}$ such that if i belongs to $\boldsymbol{q}$ and $i > n$, then $a_i \leq b_i$. Besides, for any P-sequences $\boldsymbol{u} = (d_i)_{i\in P}$ and $\boldsymbol{v} = (c_i)_{i\in P}$, defined in $\boldsymbol{l}$ and $\boldsymbol{c}$ by the same sequence $\boldsymbol{q}$ in P, there is an element m from $\boldsymbol{r}$ such that if j belongs to $\boldsymbol{q}$ and $j > m$, then $d_j \leq c_j$. It means that $b_i - a_i \geq 0$ when $i > n$, and $c_i - d_i \geq 0$ when $i > m$. Consequently, taking $t = \max \{n, m\}$, we have $b_i \geq a_i$ and $c_i \geq d_i$ when $i > t$.

In addition, by Lemma 2.11, there are two P-sequences $\boldsymbol{l} = (a_i)_{i\in P}$ and $\boldsymbol{h} = (b_i)_{i\in P}$ defined in $\boldsymbol{a}$ and $\boldsymbol{b}$ by the same sequence $\boldsymbol{r}$ in P and a real number $k \in \boldsymbol{R}^{++}$ such that $b_i - a_i > k$ for all elements i from $\boldsymbol{r}$ and there are two P-sequences $\boldsymbol{v} = (c_i)_{i\in P}$ and $\boldsymbol{u} = (d_i)_{i\in P}$ defined in $\boldsymbol{c}$ and $\boldsymbol{b}$ by the same sequence $\boldsymbol{p}$ in P and a real number $h \in \boldsymbol{R}^{++}$ such that $c_i - d_i > h$ for all elements i from $\boldsymbol{p}$.

Taking P-sets $\boldsymbol{b} = (b_i)_{i\in P}$ and $\boldsymbol{d} = (d_i)_{i\in P}$, we construct a new P-set $\boldsymbol{l} = (l_i)_{i\in P}$ by the rule $l_i = \max \{b_i, d_i\}$. Then

$$\lim_{i\to\infty}|l_{q(i)} - b_{q(i)}| = 0 \text{ for any sequence } \boldsymbol{q} \text{ in } P$$

because $\boldsymbol{d} \approx \boldsymbol{b}$.

Taking P-sets $\boldsymbol{c} = (c_i)_{i \in P}$ and $\boldsymbol{l} = (d_i)_{i \in P}$, we construct a new P-set $\boldsymbol{g} = (g_i)_{i \in P}$ by the rule $g_i = c_i + (l_i - d_i)$. Then we have

$$g_i = c_i + (l_i - d_i) \geq d_i + (l_i - d_i) = l_i \geq b_i$$

for all i because $l_i \geq b_i$ for all i.

Besides, $\gamma = \mathrm{Hn}(l_i)_{i \in P}$ because $\lim_{i \to \infty} |l_{q(i)} - d_{q(i)}| = 0$ for any sequence $\boldsymbol{q}$ in P and for any P-sequences $\boldsymbol{k} = (a_i)_{i \in P}$ and $\boldsymbol{h} = (b_i)_{i \in P}$, defined in $\boldsymbol{a}$ and $\boldsymbol{g}$ by the same sequence $\boldsymbol{q}$ in P, there is an element n from $\boldsymbol{q}$ such that if i belongs to $\boldsymbol{q}$ and $i > n$, then $a_i \leq g_i$ because $g_i \geq b_i \geq a_i$.

Thus, we have

$$l_i - a_i \geq b_i - a_i > k$$

for all elements i from $\boldsymbol{r}$.

It means that $\boldsymbol{a} \ll \boldsymbol{g}$ and therefore, $\alpha < \gamma$. Hence, in $\boldsymbol{R}_P$, the relation $<$ is transitive.

5. Asymmetry of $<$.

Let us consider two real P-hypernumbers $\alpha, \beta \in \boldsymbol{R}_P$ such that $\alpha < \beta$. If at the same time, $\beta < \alpha$, we have the equality $\alpha = \beta$ because $<$ implies $\leq$. However by Lemma 2.17, $\alpha < \beta$ implies $\alpha \neq \beta$.

Thus, $<$ is asymmetric.

Proposition is proved.

This allows us to define:

- the set $\boldsymbol{R}_P^{++}$ of all positive real P-hypernumbers, i.e., $\alpha > 0$.
- the set $\boldsymbol{R}_P^{+}$ of all non-negative real P-hypernumbers, i.e., $\alpha \geq 0$.
- the set $\boldsymbol{R}_P^{-}$ of all non-positive real P-hypernumbers, i.e., $\alpha \leq 0$.
- the set $\boldsymbol{R}_P^{--}$ of all negative real P-hypernumbers, i.e., $\alpha < 0$.

Corollary 2.6. Relations $\leq$ and $<$ in $\boldsymbol{R}_\omega$ are a partial order and a strict partial order, respectively.

This allows us to define:

- the set $\boldsymbol{R}_\omega^{++}$ of all positive real hypernumbers.
- the set $\boldsymbol{R}_\omega^{+}$ of all non-negative real hypernumbers.
- the set $\boldsymbol{R}_\omega^{-}$ of all non-positive real hypernumbers.
- the set $\boldsymbol{R}_\omega^{-}$ of all negative real hypernumbers.

It is interesting that although real numbers are isomorphically included into the set of all real hypernumbers, some concepts change their meaning. One of them is the concept of an interval. As we know an interval in $\boldsymbol{R}$ is uniquely determined by its endpoints. The same is true for an interval in $\boldsymbol{R}_\omega$. However, the same endpoints determine different intervals in $\boldsymbol{R}$ and in $\boldsymbol{R}_\omega$. For instance, the interval $[-1, 1]$ in $\boldsymbol{R}$ contains only real numbers, while the interval $[-1, 1]$ in $\boldsymbol{R}_\omega$ also contains proper real hypernumbers, such as $\gamma = \mathrm{Hn}(a_i)_{i\in\omega}$ where $a_i = (-1)^i$, $i = 1, 2, 3, \ldots$ or $\beta = \mathrm{Hn}(b_i)_{i\in\omega}$ where $b_i = ½$ for $i = 2, 4, \ldots, 2n, \ldots$ and $b_i = ¼$ for $i = 1, 3, \ldots, 2n - 1, \ldots$.

Let us assume that almost all elements from P belong to some sequence in P.

Lemma 2.18. (a) $\alpha \leq \beta$ if and only if there are a P-sets $\boldsymbol{a} = (a_i)_{i\in P}$ that represents α and a sequence $\boldsymbol{b} = (b_i)_{i\in P}$ that represents β such that $a_i \leq b_i$ for all $i\in P$.

(b) $\alpha < \beta$ if and only if there are a sequence $\boldsymbol{a} = (a_i)_{i\in\omega}$ that represents α and a sequence $\boldsymbol{b} = (b_i)_{i\in\omega}$ that represents β such that $k < b_i - a_i$ for some positive real number k and all $i = 1, 2, 3, \ldots$

Proof is left as an exercise.

It is also possible to transfer other relations from the set X(or $\boldsymbol{R}$) to the sets X^P(or $\boldsymbol{R}^P$) as well as the sets X_p(or $\boldsymbol{R}_p$). The same is true for operations.

Operations in $\boldsymbol{R}$ induce corresponding operations in $\boldsymbol{R}^P$ as it is possible to apply operations from $\boldsymbol{R}$ to the corresponding coordinates of P-hypernumbers.

Let $\boldsymbol{a} = (a_i)_{i\in P}$ and $\boldsymbol{b} = (b_i)_{i\in P}$ are elements from $\boldsymbol{R}^P$ and a is a real number. Then we have the following constructions.

Definition 2.18.

a) Operation of *addition* in $\boldsymbol{R}^P$ is defined as $\boldsymbol{a} + \boldsymbol{b} = (c_i)_{i\in P}$ where $c_i = a_i + b_i$ for all $i\in P$.
b) Operation of *subtraction* in $\boldsymbol{R}^P$ is defined as $\boldsymbol{a} - \boldsymbol{b} = (c_i)_{i\in P}$ where $c_i = a_i - b_i$ for all $i\in P$.
c) Operation of *scalar multiplication* in $\boldsymbol{R}^P$ is defined as $a \cdot \boldsymbol{b} = (c_i)_{i\in P}$ where $c_i = a\cdot b_i$ for all $i\in P$.
d) Operation of *multiplication* in $\boldsymbol{R}^P$ is defined as $\boldsymbol{a}\cdot\boldsymbol{b} = (c_i)_{i\in P}$ where $c_i = a_i\cdot b_i$ for all $i\in P$.
e) If $b_i \neq 0$ for all $i\in P$, then operation of *division* in $\boldsymbol{R}^P$ is defined as $\boldsymbol{a}/\boldsymbol{b} = (c_i)_{i\in P}$ where $c_i = a_i/b_i$ for all $i\in P$.
f) Operation of taking the *maximum* in $\boldsymbol{R}^P$ is defined as $max\{\boldsymbol{a}, \boldsymbol{b}\} = (c_i)_{i\in P}$ where $c_i = max\ \{a_i, b_i\}$ for all $i\in P$.

g) Operation of taking the *minimum* in $\mathbf{R}^P$ is defined as $min\{\mathbf{a}, \mathbf{b}\} = (c_i)_{i\in P}$ where $c_i = min\{a_i, b_i\}$ for all i P.

Remark 2.4. By the definition of addition and multiplication in the set $\mathbf{R}^P$, all laws of operations with real numbers (commutativity of addition, associativity of addition, commutativity of multiplication, associativity of multiplication, and distributivity) are valid for corresponding operations with sequences of real numbers.

In particular, we have the following result.

Theorem 2.4. The set $\mathbf{R}^P$ of all P-sets in $\mathbf{R}$ is an ordered algebra over the field $\mathbf{R}$ of real numbers with the binary operations *max* and *min*.

Corollary 2.7 (Any course of functional analysis). The set $\mathbf{R}^\omega$ of all sequences of real numbers is an ordered algebra over the field $\mathbf{R}$ of real numbers with the binary operations *max* and *min*.

It is possible to transfer some operations defined by operations on coordinates of P-sets from the set $\mathbf{R}^P$ to the set $\mathbf{R}_P$.

Definition 2.19. Operations on P-hypernumbers defined by operations with coordinates of these P-hypernumbers are called *regular*.

Operations defined above are regular as the following results show.

Proposition 2.7. Operations of *min* and *max* in $\mathbf{R}^P$ induce similar regular operations in $\mathbf{R}_P$.

Proof is left as an exercise.

Proposition 2.8. Operations of addition and subtraction in $\mathbf{R}^P$ induce similar regular operations in $\mathbf{R}_P$.

Proof. Let us take some P-hypernumbers $\alpha = \text{Hn}(a_i)_{i\in P}$ and $\beta = \text{Hn}(b_i)_{i\in P}$ from $\mathbf{R}_P$ We define $\alpha + \beta = \gamma$, where $\gamma = \text{Hn}(a_i + b_i)_{i\in P}$. To show that this is an operation in $\mathbf{R}_P$, it is necessary to prove that γ does not depend on the choice of the representing P-sets $\mathbf{a} = (a_i)_{i\in P}$ and $\mathbf{b} = (b_i)_{i\in P}$. To do this, let us take another P-set $\mathbf{d} = (d_i)_{i\in P}$ in β and show that if the P-hypernumber δ is equal to $\text{Hn}(a_i + d_i)_{i\in P}$, then $\delta = \gamma$.

The equality $\beta = \text{Hn}(b_i)_{i\in P} = \text{Hn}(d_i)_{i\in P}$ means that for any P-sequences $\mathbf{l} = (a_i)_{i\in P}$ and $\mathbf{h} = (b_i)_{i\in P}$, defined in $\mathbf{d}$ and $\mathbf{b}$ by the same arbitrary sequence $\mathbf{q}$ in P, we have

$$\lim_{i\to\infty}|b_i - d_i| = 0$$

Consequently,

$$\lim_{i\to\infty}|\,(a_i + b_i) - (a_i + l_i)\,| = \lim_{i\to\infty}|b_i - l_i| = 0$$

Then by Definitions 2.6 and 2.7, $\delta = \gamma$.

If we take another P-set that also represents the P-hypernumber α, a similar proof shows that the result of addition $\alpha + \beta$ will be the same.

The proof for the difference $\alpha - \beta = \mathrm{Hn}(a_i - b_i)_{i\in P}$ of two P-hypernumbers α and β is similar to the proof for the sum of two P-hypernumbers.

Proposition is proved.

Corollary 2.8 (Burgin, 2012). Operations of addition and subtraction in $\boldsymbol{R}^{\omega}$ induce similar regular operations in $\boldsymbol{R}_{\omega}$.

Proposition 2.9. a) If $\alpha \le \beta$ and $\gamma \in \boldsymbol{R}_P$, then $\alpha + \gamma \le \beta + \gamma$, i.e., operation of addition is monotone in $\boldsymbol{R}_P$.

b) If $\alpha < \beta$ and γ and $\gamma \in \boldsymbol{R}_P$, then $\alpha + \gamma < \beta + \gamma$,

Proof. a) The inequality $\alpha \le \beta$ of P-hypernumbers implies existence of P-sets $\boldsymbol{a} = (a_i)_{i\in P}$ and $\boldsymbol{b} = (b_i)_{i\in P}$ such that $\alpha = \mathrm{Hn}(a_i)_{i\in P}$, $\beta = \mathrm{Hn}(b_i)_{i\in P}$ and for any P-sequences $\boldsymbol{l} = (a_i)_{i\in\omega}$ and $\boldsymbol{h} = (b_i)_{i\in\omega}$, defined in $\boldsymbol{a}$ and $\boldsymbol{b}$ by the same sequence $\boldsymbol{q}$ in P, we have

$$\exists n\in\omega\forall i \ge n\ (a_i \le b_i)$$

Thus, by the properties of the relation $\le$ in $\boldsymbol{R}$, for any P-sequences $\boldsymbol{l} = (a_i)_{i\in\omega}$ and $\boldsymbol{h} = (b_i)_{i\in\omega}$, defined in $\boldsymbol{a}$ and $\boldsymbol{b}$ by the same sequence $\boldsymbol{q}$ in P and $\gamma = \mathrm{Hn}(c_i)_{i\in\omega}$, we have

$$\exists n\in\omega\forall i \ge n\ (a_i + c_i \le b_i + c_i)$$

By Definition 2.18, it means $\alpha + \gamma \le \beta + \gamma$.

b) The inequality $\alpha < \beta$ of P-hypernumbers implies existence of P-sets $\boldsymbol{a} = (a_i)_{i\in P}$ and $\boldsymbol{b} = (b_i)_{i\in P}$ such that $\alpha = \mathrm{Hn}(a_i)_{i\in P}$, $\beta = \mathrm{Hn}(b_i)_{i\in P}$ and for any P-sequences $\boldsymbol{l} = (a_i)_{i\in\omega}$ and $\boldsymbol{h} = (b_i)_{i\in\omega}$, defined in $\boldsymbol{a}$ and $\boldsymbol{b}$ by the same sequence $\boldsymbol{q}$ in P, we have

$$\exists n\in\omega\forall i \ge n\ (a_i \le b_i)$$

Thus, by the properties of the relation $\le$ in $\boldsymbol{R}$, for any P-sequences $\boldsymbol{l} = (a_i)_{i\in\omega}$ and $\boldsymbol{h} = (b_i)_{i\in\omega}$, defined in $\boldsymbol{a}$ and $\boldsymbol{b}$ by the same sequence $\boldsymbol{q}$ in P and $\gamma = \mathrm{Hn}(c_i)_{i\in\omega}$, we have

$$\exists n\in\omega\forall i \ge n\ (a_i + c_i \le b_i + c_i)$$

In addition, by Lemma 2.11, there are two P-sequences $\boldsymbol{l} = (a_i)_{i\in\omega}$ and $\boldsymbol{h} = (b_i)_{i\in\omega}$ defined in $\boldsymbol{a}$ and $\boldsymbol{b}$ by the same sequence $\boldsymbol{r}$ in P such that $b_i - a_i > k$ for all elements i from $\boldsymbol{r}$. Then we have

$$(a_i + c_i) - (b_i + c_i) = b_i - a_i > k$$

for all elements i from $\boldsymbol{r}$. By Definition 2.18, it means $\alpha + \gamma < \beta + \gamma$.

Proposition is proved.

Corollary 2.9. a) If $\alpha \leq \beta$ and $\gamma \in \boldsymbol{R}_\omega$, then $\alpha + \gamma \leq \beta + \gamma$, i.e., operation of addition is monotone in $\boldsymbol{R}_\omega$.

b) If $\alpha < \beta$ and γ and $\gamma \in \boldsymbol{R}_P$, then $\alpha + \gamma < \beta + \gamma$.

Proposition 2.10. The operation of scalar multiplication in $\boldsymbol{R}^P$ induces a similar operation in $\boldsymbol{R}_P$.

Proof. Let us take a P-hypernumber $\alpha = \mathrm{Hn}(a_i)_{i\in P}$ from $\boldsymbol{R}_P$ and a real number a. We define $a\alpha = \gamma$, where $\gamma = \mathrm{Hn}(aa_i)_{i\in P}$. To prove that this is an operation in $\boldsymbol{R}_P$, it is necessary to show that γ does not depend on the choice of the P-set $\boldsymbol{a} = (a_i)_{i\in P}$ that defines α. To do this, let us take another P-set $\boldsymbol{d} = (d_i)_{i\in\omega}$ in α and demonstrate that if the P-hypernumber δ is equal to $\mathrm{Hn}(ad_i)_{i\in P}$, then $\delta = \gamma$.

The equality $\alpha = \mathrm{Hn}(a_i)_{i\in P} = \mathrm{Hn}(d_i)_{i\in P}$ means that for any P-sequences $\boldsymbol{l} = (a_i)_{i\in\omega}$ and $\boldsymbol{h} = (d_i)_{i\in\omega}$, defined in $\boldsymbol{d}$ and $\boldsymbol{a}$ by the same arbitrary sequence $\boldsymbol{q}$ in P, we have

$$\lim_{i\to\infty} |a_i - d_i| = 0$$

Consequently,

$$\lim_{i\to\infty} |\, aa_i - ad_i \,| = \lim_{i\to\infty} |a(a_i - d_i)| = \lim_{i\to\infty} (|a| \cdot |a_i - d_i|) = |a| \cdot \lim_{i\to\infty} |a_i - d_i| = 0$$

Then by Definitions 2.6 and 2.7, $\delta = \gamma$.

Proposition is proved.

Corollary 2.10. The operation of scalar multiplication in $\boldsymbol{R}^\omega$ induces a similar operation in $\boldsymbol{R}_\omega$.

Proposition 2.11.

a) If α and β are P-hypernumbers, $\alpha \leq \beta$ and $a \in \boldsymbol{R}^+$, then $a\alpha \leq a\beta$, i.e., operation of scalar multiplication by positive numbers is monotone in $\boldsymbol{R}_P$.
b) If α and β are P-hypernumbers, $\alpha \leq \beta$ and $a \in \boldsymbol{R}^-$, then $a\alpha \geq a\beta$, i.e., operation of scalar multiplication by negative numbers is antitone in $\boldsymbol{R}_P$.

c) If α and β are P-hypernumbers, $\alpha < \beta$ and $a \in \boldsymbol{R}^{++}$, then $a\alpha < a\beta$.

d) If α and β are P-hypernumbers, $\alpha < \beta$ and $a \in \boldsymbol{R}^{--}$, then $a\alpha > a\beta$.

Proof. a) The inequality $\alpha \leq \beta$ of P-hypernumbers implies existence of P-sets $\boldsymbol{a} = (a_i)_{i\in P}$ and $\boldsymbol{b} = (b_i)_{i\in P}$ such that $\alpha = \mathrm{Hn}(a_i)_{i\in P}$, $\beta = \mathrm{Hn}(b_i)_{i\in P}$ and for any P-sequences $\boldsymbol{l} = (a_i)_{i\in\omega}$ and $\boldsymbol{h} = (b_i)_{i\in\omega}$, defined in $\boldsymbol{a}$ and $\boldsymbol{b}$ by the same sequence $\boldsymbol{q}$ in P, we have

$$\exists n\in\omega\forall i \geq n\ (a_i \leq b_i)$$

Thus, by the properties of the relation $\leq$ in $\boldsymbol{R}$, for any P-sequences $\boldsymbol{l} = (a_i)_{i\in\omega}$ and $\boldsymbol{h} = (b_i)_{i\in\omega}$, defined in $\boldsymbol{a}$ and $\boldsymbol{b}$ by the same sequence $\boldsymbol{q}$ in P and a non-negative real number a, we have

$$\exists n\in\omega\forall i \geq n\ (aa_i \leq ab_i)$$

By Definition 2.18, it means $a\alpha \leq a\beta$ because $a\alpha = \mathrm{Hn}(aa_i)_{i\in P}$ and $a\beta = \mathrm{Hn}(ab_i)_{i\in P}$.

b) The inequality $\alpha \leq \beta$ of P-hypernumbers implies existence of P-sets $\boldsymbol{a} = (a_i)_{i\in P}$ and $\boldsymbol{b} = (b_i)_{i\in P}$ such that $\alpha = \mathrm{Hn}(a_i)_{i\in P}$, $\beta = \mathrm{Hn}(b_i)_{i\in P}$ and for any P-sequences $\boldsymbol{l} = (a_i)_{i\in\omega}$ and $\boldsymbol{h} = (b_i)_{i\in\omega}$, defined in $\boldsymbol{a}$ and $\boldsymbol{b}$ by the same sequence $\boldsymbol{q}$ in P, we have

$$\exists n\in\omega\forall i \geq n\ (a_i \leq b_i)$$

Thus, by the properties of the relation $\leq$ in $\boldsymbol{R}$, for any P-sequences $\boldsymbol{l} = (a_i)_{i\in\omega}$ and $\boldsymbol{h} = (b_i)_{i\in\omega}$, defined in $\boldsymbol{a}$ and $\boldsymbol{b}$ by the same sequence $\boldsymbol{q}$ in P and a non-positive real number a, we have

$$\exists n\in\omega\forall i \geq n\ (aa_i \geq ab_i)$$

By Definition 2.18, it means $a\alpha \geq a\beta$ because $a\alpha = \mathrm{Hn}(aa_i)_{i\in P}$ and $a\beta = \mathrm{Hn}(ab_i)_{i\in P}$.

c) The inequality $\alpha < \beta$ of P-hypernumbers implies existence of P-sets $\boldsymbol{a} = (a_i)_{i\in P}$ and $\boldsymbol{b} = (b_i)_{i\in P}$ such that $\alpha = \mathrm{Hn}(a_i)_{i\in P}$, $\beta = \mathrm{Hn}(b_i)_{i\in P}$ and for any P-sequences $\boldsymbol{l} = (a_i)_{i\in\omega}$ and $\boldsymbol{h} = (b_i)_{i\in\omega}$, defined in $\boldsymbol{a}$ and $\boldsymbol{b}$ by the same sequence $\boldsymbol{q}$ in P, we have

$$\exists n\in\omega\forall i \geq n\ (a_i \leq b_i)$$

Thus, by the properties of the relation $\leq$ in $\boldsymbol{R}$, for any P-sequences $\boldsymbol{l} = (a_i)_{i\in\omega}$ and $\boldsymbol{h} = (b_i)_{i\in\omega}$, defined in $\boldsymbol{a}$ and $\boldsymbol{b}$ by the same sequence $\boldsymbol{q}$ in P and $\gamma = \mathrm{Hn}(c_i)_{i\in\omega}$, we have

$$\exists n\in\omega\forall i\geq n\ (aa_i \leq ab_i)$$

In addition, by Lemma 2.11, there are two P-sequences $\boldsymbol{l} = (a_i)_{i\in\omega}$ and $\boldsymbol{h} = (b_i)_{i\in\omega}$ defined in $\boldsymbol{a}$ and $\boldsymbol{b}$ by the same sequence $\boldsymbol{r}$ in P such that $b_i - a_i > k$ for all elements i from $\boldsymbol{r}$. Then we have

$$ab_i - aa_i = a(b_i - a_i) > ak$$

for all elements i from $\boldsymbol{r}$. By Definition 2.18, it means $a\alpha < a\beta$ as ak is a positive real number.

d) The inequality $\alpha < \beta$ of P-hypernumbers implies existence of P-sets $\boldsymbol{a} = (a_i)_{i\in P}$ and $\boldsymbol{b} = (b_i)_{i\in P}$ such that $\alpha = \mathrm{Hn}(a_i)_{i\in P}$, $\beta = \mathrm{Hn}(b_i)_{i\in P}$ and for any P-sequences $\boldsymbol{l} = (a_i)_{i\in\omega}$ and $\boldsymbol{h} = (b_i)_{i\in\omega}$, defined in $\boldsymbol{a}$ and $\boldsymbol{b}$ by the same sequence $\boldsymbol{q}$ in P, we have

$$\exists n\in\omega\forall i\geq n\ (a_i \leq b_i)$$

Thus, by the properties of the relation $\leq$ in $\boldsymbol{R}$, for any P-sequences $\boldsymbol{l} = (a_i)_{i\in\omega}$ and $\boldsymbol{h} = (b_i)_{i\in\omega}$, defined in $\boldsymbol{a}$ and $\boldsymbol{b}$ by the same sequence $\boldsymbol{q}$ in P and $\gamma = \mathrm{Hn}(c_i)_{i\in\omega}$, we have

$$\exists n\in\omega\forall i\geq n\ (aa_i \geq ab_i)$$

In addition, by Lemma 2.11, there are two P-sequences $\boldsymbol{l} = (a_i)_{i\in\omega}$ and $\boldsymbol{h} = (b_i)_{i\in\omega}$ defined in $\boldsymbol{a}$ and $\boldsymbol{b}$ by the same sequence $\boldsymbol{r}$ in P such that $b_i - a_i > k$ for all elements i from $\boldsymbol{r}$. Then we have

$$aa_i - ab_i = (-a)(b_i - a_i) > -ak$$

for all elements i from $\boldsymbol{r}$. By Definition 2.18, it means $a\alpha > a\beta$ as $-ak$ is a positive real number.

Proposition is proved.

Corollary 2.11.

a) If α and β are hypernumbers, $\alpha \leq \beta$ and $a \in \boldsymbol{R}^+$, then $a\alpha \leq a\beta$, i.e., operation of scalar multiplication by positive numbers is monotone in $\boldsymbol{R}_P$.

b) If α and β are hypernumbers, $\alpha \leq \beta$ and $a \in \boldsymbol{R}^{-}$, then $a\alpha \geq a\beta$, i.e., operation of scalar multiplication by negative numbers is antitone in $\boldsymbol{R}_P$.

c) If α and β are hypernumbers, $\alpha < \beta$ and $a \in \boldsymbol{R}^{++}$, then $a\alpha < a\beta$.

d) If α and β are hypernumbers, $\alpha < \beta$ and $a \in \boldsymbol{R}^{--}$, then $a\alpha > a\beta$.

Propositions 2.8–2.11 give us the following result.

Theorem 2.5. The set $\boldsymbol{R}_P$ of all *P*-hypernumbers in $\boldsymbol{R}$ is an ordered infinite-dimensional vector space over the field $\boldsymbol{R}$ of real numbers with the binary operations *max* and *min.*

Proof. By Propositions 2.8 and 2.10, the set $\boldsymbol{R}_P$ is a vector space over the field $\boldsymbol{R}$. Let us consider *P*-hypernumbers $\alpha^t = \mathrm{Hn}(a_i^t)_{i\in\omega}$ with $t \in \omega$ and $a_i^t = (i)^t$ for all $t, i \in \omega$. We can see that these elements are linearly independent in $\boldsymbol{R}_P$. Consequently, the vector space $\boldsymbol{R}_P$ is infinite-dimensional.

The set $\boldsymbol{R}_P$ is partially ordered. So, to prove the theorem, we have only to demonstrate that this order is compatible with the operations in $\boldsymbol{R}_P$ (Fuchs, 1963). In the case of vector spaces, it means that we have to show that the inequality $\alpha \leq \beta$ implies the inequality $a\alpha \leq a\beta$ for any $\alpha, \beta \in \boldsymbol{R}_{\backslash P}$ and $a \in \boldsymbol{R}^{+}$, and it implies $\alpha + \gamma \leq \beta + \gamma$ for any $\gamma \in \boldsymbol{R}_P$. However, this is already demonstrated in Propositions 2.9 and 2.11.

Theorem is proved.

Corollary 2.12 (Burgin, 2012)**.** The set $\boldsymbol{R}_\omega$ of all real hypernumbers is an ordered infinite dimensional vector space over the field $\boldsymbol{R}$ of real numbers with the binary operations *max* and *min.*

Remark 2.5. Not all operations in $\boldsymbol{R}_P$ and in $\boldsymbol{R}_\omega$ are regular. For instance, it is impossible to determine regular multiplication (Burgin, 2005a) or regular integration in the space $\boldsymbol{R}_\omega$ (Burgin, 2010a).

Definition 2.20. A *P*-set $(a_i)_{i\in P}$ in $\boldsymbol{R}$ is called *bounded* if there is a real number b such that $|a_i| < b$ for almost all i from P.

For instance, the ω-set $(1, -1, 1, -1, 1, \ldots)$ is bounded, while the ω-set $(n)_{n\in\omega}$ is unbounded.

Lemma 2.19. a) If $\boldsymbol{a} = (a_i)_{i\in P}$ and $\boldsymbol{b} = (b_i)_{i\in P}$ are bounded *P*-sets with values in $\boldsymbol{R}$, then the *P*-set $\boldsymbol{a} + \boldsymbol{b}$ is also bounded.

b) If $\boldsymbol{a} = (a_i)_{i\in P}$ is a bounded *P*-set with values in $\boldsymbol{R}$ and c is a positive (negative) real number, then the *P*-set $c\boldsymbol{a}$ is also bounded.

Proof is left as an exercise.

Corollary 2.13. a) If $\boldsymbol{a} = (a_i)_{i\in P}$ and $\boldsymbol{b} = (b_i)_{i\in P}$ are bounded sequences with values in $\boldsymbol{R}$, then the sequence $\boldsymbol{a} + \boldsymbol{b}$ is also bounded.

b) If $\boldsymbol{a} = (a_i)_{i\in P}$ is a bounded sequence with values in $\boldsymbol{R}$ and c is a positive (negative) real number, then the sequence $c\boldsymbol{a}$ is also bounded.

Bounded P-sets from $\boldsymbol{R}^P$ are mapped onto finite real P-hypernumbers from $\boldsymbol{R}_P$.

Definition 2.21. A P-hypernumber α is called *finite* or *bounded* if there is a bounded representation $(a_i)_{i\in P}$ of α, i.e., $\alpha = \mathrm{Hn}(a_i)_{i\in P}$.

The set of all finite/bounded P-hypernumbers is denoted by $\mathbf{FP}\boldsymbol{R}_P$.

Theorem 2.6. The operation of regular multiplication is defined in $\mathbf{FP}\boldsymbol{R}_P$.

Proof. Let us take arbitrary finite real P-hypernumbers $\alpha = \mathrm{Hn}(a_i)_{i\in P}$ and $\beta = \mathrm{Hn}(b_i)_{i\in P}$ from $\mathbf{FP}\boldsymbol{R}_P$. Then we define the product of P-sets as $\boldsymbol{a}\cdot\boldsymbol{b} = (c_i)_{i\in\omega}$ where $\boldsymbol{a} = (a_i)_{i\in\omega}$, $\boldsymbol{b} = (b_i)_{i\in\omega}$, $c_i = a_i\cdot b_i$ for all $i\in P$ and the product of real P-hypernumbersα and β as $\alpha\cdot\beta = \gamma = \mathrm{Hn}(a_i\cdot b_i)_{i\in\omega}$.

To show that this is an operation in $\boldsymbol{R}_P$, it is necessary to prove that γ does not depend on the choice of the sequences $\boldsymbol{a}$ and $\boldsymbol{b}$. To do this, let us, at first, take another P-set $\boldsymbol{d} = (d_i)_{i\in P}$ such that that represents β, i.e., $\beta = \mathrm{Hn}(d_i)_{i\in P}$, and show that if the real P-hypernumber δ is equal to $\mathrm{Hn}(a_i\cdot d_i)_{i\in P}$, then $\delta = \gamma$.

By definition,

$$\lim_{i\to\infty} \|d_{q(i)} - b_{q(i)}\| = 0 \text{ for any sequence } \boldsymbol{q} \text{ in } P$$

Besides, there is a number $c \in \boldsymbol{R}^+$ such that $|a_i| \le c$ for all $i\in P$. Consequently, for any sequence $\boldsymbol{q}$ in P, we have

$$\lim_{i\to\infty} |\,(b_{q(i)}\cdot a_{q(i)}) - (d_{q(i)}\cdot a_{q(i)})\,| = \lim_{i\to\infty} (|b_i - d_i| \cdot |\,a_i\,|) \le$$

$$\lim_{i\to\infty} (|\mathrm{b}_{q(i)} - d_{q(i)}| \cdot c) = (\lim_{i\to\infty} |b_{q(i)} - d_{q(i)}|)\cdot c = 0$$

Then by Definitions 2.6 and 2.7, $\beta/\alpha = \delta = \mathrm{Hn}(d_i\cdot a_i)_{i\in P}$.

It means that the definition of the product $\alpha\cdot\beta$ does not depend on the choice of a sequence from β.

The same is true for the real P-hypernumber α.

Thus, multiplication is correctly defined in $\mathbf{FP}\boldsymbol{R}_P$

Theorem is proved.

Corollary 2. 14. The operation of regular multiplication is defined in $\mathbf{F}\boldsymbol{R}_\omega$.

As any stable sequence is bounded, we have the following result.

Corollary 2.15. The operation of multiplication by stable real sequences induces the operation of multiplication of real hypernumbers by elements of $\boldsymbol{R}$, i.e., by real numbers.

Remark 2.6. By the definition of addition and multiplication in the set $\boldsymbol{R}_P$, all laws of operations with real numbers (commutativity of addition, associativity of addition, commutativity of multiplication, associativity of multiplication, and distributivity) are valid for corresponding operations with bounded real hypernumbers.

In particular, Propositions 2.8–2.10 and Theorem 2.6 imply the following result.

Theorem 2.7. The set $\mathbf{FP}\boldsymbol{R}_P$ of all finite P-hypernumbers in $\boldsymbol{R}$ is an ordered linear algebra over the field $\boldsymbol{R}$ of real numbers with the binary operations *max* and *min*.

Corollary 2.16 (Burgin, 2012). The set $\mathbf{FP}\boldsymbol{R}_\omega$ of all finite real hypernumbers is an ordered linear algebra over the field $\boldsymbol{R}$ of real numbers with the binary operations *max* and *min*.

Let us explore division of P-hypernumbers.

Definition 2.22. A P-set $\boldsymbol{a} = (a_i)_{i\in P}$ of real numbers is *separated from* a real number a if $|\, a - a_i| > k > 0$ for some number k and for almost all $i \in P$.

For instance, a sequence $(a_i)_{i\in\omega}$ of real numbers is *separated from* 0 if there is a natural number n such that $|\, a_i| > k$ for some number $k > 0$ and all $i > n$.

Lemma 2.20. A P-set $\boldsymbol{a} = (a_i)_{i\in P}$ with values in $\boldsymbol{R}$ is separated from a real number a if and only if the P-set $(a - a_i)_{i\in P}$ is separated from 0.

Proof is left as an exercise.

Corollary 2.17. A sequence $\boldsymbol{a} = (a_i)_{i\in\omega}$ with values in $\boldsymbol{R}$ is separated from a real number a if and only if the sequence $(a - a_i)_{i\in\omega}$ is separated from 0.

Lemma 2.21. a) If $\boldsymbol{a} = (a_i)_{i\in P}$ and $\boldsymbol{b} = (b_i)_{i\in P}$ are separated from a real number a positive (negative) P-sets with values in $\boldsymbol{R}$, then the P-set $\boldsymbol{a} + \boldsymbol{b}$ is also separated from a.

b) If $\boldsymbol{a} = (a_i)_{i\in P}$ is a separated from 0 positive (negative) P-set with values in $\boldsymbol{R}$ and c is a positive (negative) real number, then the P-set $c\boldsymbol{a}$ is also separated from 0.

Proof is left as an exercise.

Corollary 2.18. a) If $\boldsymbol{a} = (a_i)_{i\in P}$ and $\boldsymbol{b} = (b_i)_{i\in P}$ are separated from a real number a positive (negative) sequences with values in $\boldsymbol{R}$, then the sequence $\boldsymbol{a} + \boldsymbol{b}$ is also separated from a.

b) If $\boldsymbol{a} = (a_i)_{i\in P}$ is a separated from 0 positive (negative) sequence with values in $\boldsymbol{R}$ and c is a positive (negative) real number, then the sequence $c\boldsymbol{a}$ is also separated from 0.

Lemma 2.22. If $\boldsymbol{a} = (a_i)_{i\in P}$ and $\boldsymbol{b} = (b_i)_{i\in P}$ are separated from a real number a positive (negative) bounded P-sets with values in $\boldsymbol{R}$, then the P-set $\boldsymbol{a} \cdot \boldsymbol{b}$ is also separated from a.

Proof is left as an exercise.

Corollary 2.19. a) If $\boldsymbol{a} = (a_i)_{i\in P}$ and $\boldsymbol{b} = (b_i)_{i\in P}$ are separated from a real number a positive (negative) sequences with values in $\boldsymbol{R}$, then the sequence $\boldsymbol{a} \cdot \boldsymbol{b}$ is also separated from a.

Lemmas 2.20–2.22 imply the following result.

Proposition 2.12. a) The set $\mathbf{SPS}\boldsymbol{R}_P^+$ ($\mathbf{SPS}\boldsymbol{R}_P^-$) of all separated from 0 positive (negative) P-sets in $\boldsymbol{R}$ is a commutative semigroup.

b) The set $\mathbf{SBPS}\boldsymbol{R}_P^+$ ($\mathbf{SBPS}\boldsymbol{R}_P^-$) of all bounded separated from 0 positive (negative) P-sets in $\boldsymbol{R}$ is a commutative semiring.

Corollary 2.20. a) The set $\mathbf{SPS}\boldsymbol{R}_\omega^+$ ($\mathbf{SPS}\boldsymbol{R}_P^-$) of all separated from 0 positive (negative) sequences of real numbers is a commutative semigroup.

b) The set $\mathbf{SBPS}\boldsymbol{R}_\omega^+$ ($\mathbf{SPSS}\boldsymbol{R}_P^-$) of all bounded separated from 0 positive (negative) sequences of real numbers is a commutative semiring.

Definition 2.23. A P-hypernumber α is *separated from* a real number a if there is a P-set $(a_i)_{i\in P}$ with values in $\boldsymbol{R}$, which is separated from a and for which $\alpha = \mathrm{Hn}(a_i)_{i\in P}$.

For instance, a hypernumber α is *separated from* 0 if there is a separated from 0 sequence $(a_i)_{i\in\omega}$ of real numbers such that $\alpha = \mathrm{Hn}(a_i)_{i\in\omega}$.

We denote by $\mathbf{SP}\boldsymbol{R}_P$ the set of all P-hypernumbers separated from 0, by $\mathbf{S}\boldsymbol{R}_\omega$ the set of all real hypernumbers separated from 0, by $\mathbf{SB}\boldsymbol{R}_\omega$ the set that consists of all bounded real hypernumbers that are separated from 0 and by $\mathbf{SBP}\boldsymbol{R}_P$ the set that consists of all bounded P-hypernumbers that are separated from 0.

Lemma 2.23. A P-hypernumber α is separated from 0 if and only if there is a P-set $(a_i)_{i\in P}$ with values in $\boldsymbol{R}$ such that $|a_i| > k > 0$ for some number k and all $i\in P$.

Proof is left as an exercise.

Lemma 2.20 implies the following result.

Lemma 2.24. A P-hypernumber α is separated from a real number a if and only if the P-hypernumber $\alpha - a$ is separated from 0.

Corollary 2.21. A real hypernumber α is separated from a real number a if and only if the real hypernumber $\alpha - a$ is separated from 0.

Lemma 2.25. a) If α and β are separated from a real number a positive (negative) P-hypernumbers, then the P-hypernumber $\alpha + \beta$ is also separated from a.

b) If α is a separated from 0 positive (negative) P-hypernumber and c is a positive (negative) real number, then the P-hypernumber $c\alpha$ is also separated from 0.

Proof is left as an exercise.

Corollary 2.22. a) If α and β are separated from a real number a positive (negative) real hypernumber, then the hypernumber $\alpha + \beta$ is also separated from a.

b) If α is a separated from 0 positive (negative) real hypernumber and c is a positive (negative) real number, then the hypernumber $c\alpha$ is also separated from 0.

Lemma 2.26. If α and β are separated from a real number a positive (negative) bounded P-hypernumbers with values in $\boldsymbol{R}$, then the P-hypernumber $\alpha \cdot \beta$ is also separated from a.

Proof is left as an exercise.

Corollary 2.23. a) If α and β are separated from a real number a positive (negative) real hypernumbers, then the hypernumber $\alpha \cdot \beta$ is also separated from a.

Lemmas 2.24–2.26 imply the following result.

Theorem 2.8. a) The set $\mathbf{S}\mathbf{P}\boldsymbol{R}_P^{\,+}$ ($\mathbf{S}\mathbf{P}\boldsymbol{R}_P^{\,-}$) of all separated from 0 positive (negative) P-hypernumbers is a commutative semigroup.

b) The set $\mathbf{SBP}\boldsymbol{R}_P^{\,+}$ ($\mathbf{SBP}\boldsymbol{R}_P^{\,-}$) of all finite separated from 0 positive (negative) P-hypernumbers is a commutative semiring.

Corollary 2.24. a) The set $\mathbf{S}\mathbf{P}\boldsymbol{R}_P^{\,+}$ ($\mathbf{S}\mathbf{P}\boldsymbol{R}_P^{\,-}$) of all separated from 0 positive (negative) hypernumbers is a commutative semigroup.

b) The set $\mathbf{SBP}\boldsymbol{R}_P^{\,+}$ ($\mathbf{SBP}\boldsymbol{R}_P^{\,-}$) of all finite separated from 0 positive (negative) hypernumbers is a commutative semiring.

Definition 2.24. If all $b_i \neq 0$, then the sequence $(a_i/b_i)_{i\in\omega}$ of real numbers defines the *sequential division* of sequences $\boldsymbol{a} = (a_i)_{i\in\omega}$ and $\boldsymbol{b} = (b_i)_{i\in\omega}$ of real numbers.

Sequential division of real sequences allows us to define sequential division of real hypernumbers.

Theorem 2.9. The sequential division of sequences correctly defines division of bounded P-hypernumbers by P-hypernumbers separated from 0.

Proof. Let us take an arbitrary bounded P-hypernumber $\beta = \mathrm{Hn}(b_i)_{i\in P} \in \mathbf{F}\boldsymbol{R}_P$ and an arbitrary P-hypernumber $\alpha \in \boldsymbol{R}_P$, which is separated from 0. By Definition 2.23, there is a P-set $\boldsymbol{a} = (a_i)_{i\in\omega}$ of real numbers such that $\alpha = \mathrm{Hn}(a_i)_{i\in P}$ and $|\, a_i \,| > k$ for some number $k > 0$ and all $i \in P$.

Then we define the P-set $\boldsymbol{b}/\boldsymbol{a} = (c_i)_{i\in\omega}$ where $c_i = b_i / a_i$ for all $i\in\omega$ and $\beta/\alpha = \gamma$ from $\boldsymbol{R}_\omega$, where $\gamma = \text{Hn}(b_i / a_i)_{i\in P}$. The P-set $(b_i / a_i)_{i\in P}$ is correctly defined because all numbers a_i are not equal to 0. Thus, P-hypernumber is also correctly defined.

To show that this is an operation in $\boldsymbol{R}_\omega$, it is necessary to prove that γ does not depend on the choice of the P-sets $\boldsymbol{a}$ and $\boldsymbol{b}$ from P-hypernumbers α and β. To do this, let us, at first, take another P-set $\boldsymbol{d} = (d_i)_{i\in P}$ such that that represents β, i.e., $\beta = \text{Hn}(d_i)_{i\in P}$, and show that if the real P-hypernumber δ is equal to $\text{Hn}(d_i / a_i)_{i\in P}$, then $\delta = \gamma$.

By definition,

$$\lim_{i\to\infty} ||d_{q(i)} - b_{q(i)}|| = 0 \text{ for any sequence } \boldsymbol{q} \text{ in } P$$

Consequently, for any sequence $\boldsymbol{q}$ in P, we have

$$\lim_{i\to\infty} | (b_{q(i)}/a_{q(i)}) - (d_{q(i)}/a_{q(i)}) | = \lim_{i\to\infty} |b_i - l_i|/| a_i | \leq$$

$$\lim_{i\to\infty} (|\mathrm{b}_{q(i)} - d_{q(i)}| /|k|) = (\lim_{i\to\infty} |b_{q(i)} - d_{q(i)}|) /|k| = 0$$

Then by Definitions 2.6 and 2.7, $\beta/\alpha = \delta = \text{Hn}(d_i / a_i)_{i\in\omega}$. It means that the definition of the ratio β/α does not depend on the choice of a P-set that represents β.

Now let us, at first, take another separated from 0 P-set $\boldsymbol{t} = (t_i)_{i\in P}$ in α and show that if the P-hypernumber η is equal to $\text{Hn}(b_i / t_i)_{i\in P}$, then $\eta = \gamma$.

By definition,

$$\lim_{i\to\infty} ||a_{q(i)} - t_{q(i)}|| = 0 \text{ for any sequence } \boldsymbol{q} \text{ in } P$$

Besides, as the hypernumber β is bounded, there is such $c \in \boldsymbol{R}^+$ that $| b_i | \leq c$ for all $i\in P$. Consequently, for any sequence $\boldsymbol{q}$ in P, we have

$$\lim_{i\to\infty} | (b_{q(i)}/a_{q(i)}) - (b_{q(i)}/t_{q(i)}) | = \lim_{i\to\infty} (| b_{q(i)} |\cdot|(t_{q(i)} - a_{q(i)})/(a_{q(i)}t_{q(i)}) |) =$$

$$\lim_{i\to\infty} (| b_{q(i)} |/|a_{q(i)}t_{q(i)} |)\cdot|t_{q(i)} - a_{q(i)}| \leq \lim_{i\to\infty} (| b_{q(i)} |/ k^2)\cdot |t_{q(i)} - a_{q(i)} | \leq$$

$$\lim_{i\to\infty} (c/ k^2)\cdot |t_{q(i)} - a_{q(i)} | = (c/ k^2) \lim_{i\to\infty} |t_{q(i)} - a_{q(i)} | = 0$$

Then by Definitions 2.6 and 2.7, $\beta/\alpha = \eta = \text{Hn}(b_i / t_i)_{i\in P}$. It means that the definition of the ratio β/α does not depend on the choice of a P-set that represents the P-hypernumber α.

Consequently, division of bounded P-hypernumbers by P-hypernumbers separated from 0 is correctly defined.

Theorem is proved.

Lemma 2.27. If β is a bounded P-hypernumber and α is a P-hypernumber separated from 0, then β/α is a bounded P-hypernumber.

Proof is left as an exercise.

Corollary 2.25. If β is a bounded real hypernumber and α is a real hypernumber separated from 0, then β/α is a bounded real hypernumber.

Corollary 2.26. Any element from $\boldsymbol{SR}_\omega$ has an inverse element in $\boldsymbol{R}_\omega$.

Indeed, if α is a P-hypernumber separated from 0, then the P-hypernumber 1/α is inverse to α.

However, such an inverse element does not necessarily belong to $\boldsymbol{SR}_P$.

Thus, we have demonstrated that operations in $\boldsymbol{R}^\omega$ induce similar regular operations in $\boldsymbol{R}_\omega$. Namely, if $\alpha = \text{Hn}(a_i)_{i\in\omega}$ and $\beta = \text{Hn}(b_i)_{i\in\omega}$, then

$$\alpha + \beta = \text{Hn}(a_i + b_i)_{i\in\omega}$$

$$\alpha - \beta = \text{Hn}(a_i - b_i)_{i\in\omega}$$

$\alpha\cdot\beta = \text{Hn}(a_i\cdot b_i)_{i\in\omega}$ when the set of all numbers a_i or all numbers b_i is bounded

$\alpha/\beta = \text{Hn}(a_i/b_i)_{i\in\omega}$ when there is $k > 0$, such that $|b_i| > k$ for all $i\in\omega$

$$max\{\alpha = \text{Hn}(a_i)_{i\in\omega}, \beta = \text{Hn}(b_i)_{i\in\omega}\} = \text{Hn}(max\{a_i, b_i\})_{i\in\omega}$$

$$min\{\alpha = \text{Hn}(a_i)_{i\in\omega}, \beta = \text{Hn}(b_i)_{i\in\omega}\} = \{\text{Hn}(min\{a_i, b_i\})_{i\in\omega}.$$

The concept of a Q-subset of a P-set brings us to the concept of a Q-subhypernumber of a P-hypernumber and the concept of a subsequence brings us to the concept of a subhypernumber of a hypernumber. For simplicity, we consider here only the case of subhypernumbers of real hypernumbers.

Definition 2.25. a) A real hypernumber α is called a *subhypernumber* of a real hypernumber β if there are sequences $\boldsymbol{a} = (a_i)_{i\in\omega}$ and $\boldsymbol{b} = (b_i)_{i\in\omega}$ such that $\alpha = \text{Hn}(a_i)_{i\in\omega}$, $\beta = \text{Hn}(b_i)_{i\in\omega}$, and $(a_i)_{i\in\omega}$ is a subsequence of the sequence $(b_i)_{i\in\omega}$. b) If α is a subhypernumber of β and α≠β, then α is called a *proper subhypernumber* of β.

We denote this relation by $\alpha \Subset \beta$ and $\alpha \subset \beta$, correspondingly. Sub $\alpha = \{\gamma; \gamma \Subset \alpha\}$ denotes the set of all subhypernumbers of the real hypernumber α. If $\alpha \Subset \beta$, then β is called a *superhypernumber* of α.

Note that real numbers do not have proper subhypernumbers.

Lemma 2.28. If $\alpha \Subset \beta$ and $\beta \in \boldsymbol{R}$, then $\alpha \in \boldsymbol{R}$ and $\alpha = \beta$.

Proof is left as an exercise.

The concept of a subhypernumber is defined using representatives of hypernumbers. However, it does not depend on the choice of a representing sequence of its superhypernumber as is implied by the following result.

Lemma 2.29. If $\beta = \text{Hn}(b_i)_{i\in\omega} = \text{Hn}(c_i)_{i\in\omega}$ and $\alpha = \text{Hn}(a_j)_{j\in\omega}$ where $(a_j)_{j\in\omega}$ is a subsequence of $(b_i)_{i\in\omega}$, then there is a subsequence $(d_j)_{j\in\omega}$ of $(c_i)_{i\in\omega}$ such that $\alpha = \text{Hn}(d_j)_{j\in\omega}$.

Indeed, to build the necessary subsequence of the sequence $(c_i)_{i\in\omega}$, we can take elements d_j from the sequence $(c_i)_{i\in\omega}$ that that have the same indices as elements a_j in the sequence $(b_i)_{i\in\omega}$. As $\text{Hn}(b_i)_{i\in\omega} = \text{Hn}(c_i)_{i\in\omega}$, we have $\lim_{i\to\infty} |b_i - c_i| = 0$. Consequently, $\lim_{i\to\infty} |a_i - d_i| = 0$. Thus, $\text{Hn}(a_j)_{j\in\omega} = \text{Hn}(d_j)_{j\in\omega}$.

Proposition 2.13. If $\alpha \Subset \beta$ and α is an oscillating finite real hypernumber, then β is an oscillating real hypernumber.

Indeed, if $\alpha = \text{Hn}(a_i)_{i\in\omega}$, $\beta = \text{Hn}(b_i)_{i\in\omega}$, $(a_i)_{i\in\omega}$ is a subsequence of $(b_i)_{i\in\omega}$ and there is a number $k \in \boldsymbol{R}^{++}$ such that there are two infinite sequences of natural numbers $m(i)$ and $n(i)$ with $i = 1, 2, 3, \ldots$ such that $a_{m(i)} - a_{n(i)} > k$ and $a_{m(i+1)} - a_{n(i)} > k$ for all $i = 1, 2, 3, \ldots$, then the same is true for the elements b_i because $(a_i)_{i\in\omega}$ is a subsequence of $(b_i)_{i\in\omega}$.

Proposition 2.14. If $\alpha \Subset \beta$ and β is an increasing (decreasing) real hypernumber, then α is an increasing (decreasing) real hypernumber.

Indeed, if we have an increasing (decreasing) sequence of real numbers, then any its subsequence is also increasing (decreasing).

Proposition 2.15. If β is an increasing (decreasing) infinite real hypernumber, then for any hypernumber $\alpha \in \boldsymbol{R}_\omega$, there is a subhypernumber γ of hypernumber β such that $\gamma \geq \alpha$ ($\gamma \leq \alpha$).

Proof is left as an exercise.

Proposition 2.16. A hypernumber α is separated from 0 if and only if all of its subhypernumbers are separated from 0.

Indeed, ifα is separated from 0, then there is a sequence $(a_i)_{i\in\omega}$ of real numbers such that $\alpha = \text{Hn}(a_i)_{i\in\omega}$ and $|\, a_i| > k > 0$ for some number k and all $i \in \omega$. However, the latter condition is true for any subsequence of the sequence $(a_i)_{i\in\omega}$. So, any subhypernumber of α is separated from 0. At the same time, the condition of the proposition is sufficient because α is its own subhypernumber.

Lemma 2.30. If $\alpha \Subset \beta$ and $\gamma \Subset \alpha$, then $\gamma \Subset \beta$, i.e., relation $\Subset$ is transitive.

Proof is left as an exercise.

As relation $\Subset$ is reflexive, we have the following result.

Proposition 2.17. Relation $\Subset$ is a preorder.

In a general case, relation $\Subset$ is not a partial order (cf. Appendix). To show that relations $\alpha \Subset \beta$ and $\beta \Subset \alpha$, do not imply the relation $\alpha = \beta$ in a general case, we consider the following example.

Example 2.15. Let us consider two real sequences $\boldsymbol{a} = (1, 0, 1, 0, 1, 0, 1, \ldots, 1, 0, 1, 0, \ldots)$ and $\boldsymbol{b} = (1,0, 0, 0, 1, 0, 0, 0, 1, 0, 0, 0, 1, \ldots, 1, 0, 0, 0, 1, 0, 0, 0, \ldots)$. Then $\boldsymbol{a}$ is a subsequence of $\boldsymbol{b}$ where each pair 1,0 from $\boldsymbol{a}$ is mapped into the corresponding pair 1,0 from $\boldsymbol{b}$. Consequently, the hypernumber α defined by $\boldsymbol{a}$ is a subhypernumber of the hypernumber β defined by $\boldsymbol{b}$, i.e., $\alpha \Subset \beta$. At the same time, there is a mapping f of the sequence $\boldsymbol{b}$ into the sequence $\boldsymbol{a}$ such that numbers 0 from $\boldsymbol{b}$ are mapped into numbers 0 from $\boldsymbol{a}$ that have the same number in $\boldsymbol{b}$, that is, the first number 0 from $\boldsymbol{b}$ is mapped into the first number 0 from $\boldsymbol{a}$, the second number 0 from $\boldsymbol{b}$ is mapped into the second number 0 from $\boldsymbol{a}$ and so on. At the same time, numbers 1 from $\boldsymbol{b}$ are mapped into numbers 1 from $\boldsymbol{a}$ so that when a number 1 from $\boldsymbol{b}$ is mapped, then the next numbers 1 from $\boldsymbol{b}$ is mapped into the number 1 from $\boldsymbol{a}$ that goes after three zeroes and two ones after the previous image, i.e., the first number 1 from $\boldsymbol{b}$ is mapped into the first number 1 from $\boldsymbol{a}$, the second number 1 from $\boldsymbol{b}$ is mapped into the fourth number 1 from $\boldsymbol{a}$ and so on. Consequently, the hypernumber β defined by $\boldsymbol{b}$ is a subhypernumber of the hypernumberα defined by $\boldsymbol{a}$, i.e., $\beta \Subset \alpha$. However, $\alpha \neq \beta$.

We know that if $\alpha \leq \beta$ and $\gamma \geq 0$, then $\alpha + \gamma \leq \beta + \gamma$. A similar property is not true for the relation $\Subset$ as the following example demonstrates.

Example 2.16. Let us consider two real sequences $\boldsymbol{a} = (2, 4, 6, 8, 10, \ldots)$ and $\boldsymbol{b} = (1, 2, 3, \ldots)$. Then $\boldsymbol{a}$ is a subsequence of $\boldsymbol{b}$. Consequently, the hypernumber $\alpha = \mathrm{Hn}\, \boldsymbol{a}$ is a subhypernumber of the hypernumber $\beta = \mathrm{Hn}\, \boldsymbol{b}$, i.e., $\alpha \Subset \beta$. Taking $\gamma = \mathrm{Hn}\, \boldsymbol{b}$, we see that $\gamma \geq 0$. At the same time, $\alpha + \gamma = \mathrm{Hn}(3, 6, 9, 12, 15, \ldots)$, while $\beta + \gamma = \mathrm{Hn}(2, 4, 6, 8, 10, \ldots)$. Then any subhypernumber of the hypernumber $\beta +\gamma$is defined by a sequence of even numbers, while a representative of $\alpha + \gamma$ contains infinitely many odd numbers. So, by Lemma 2.30, the hypernumber $\alpha + \gamma$ cannot be a subhypernumber of the hypernumber $\beta + \gamma$.

Theorem 2.3 implies the following result.

Proposition 2.18. $\alpha = \mathrm{Hn}(a_i)_{i\in\omega}$ is an oscillating real *P*-hypernumber if and only if it has, at least, two stable real *P*-subhypernumbers or one infinite monotone real *P*-hypernumber and, at least, one stable real *P*-subhypernumber or one infinite increasing and one infinite decreasing real *P*-hypernumbers.

It is also possible to define new operations for real hypernumbers.

Definition 2.26. A real *P*-hypernumber α is called a *disjunctive union* of real hypernumbers β and γ if there are sequences $\boldsymbol{a} = (a_i)_{i\in\omega}$, $\boldsymbol{b} = (b_i)_{i\in\omega}$ and $\boldsymbol{c} = (c_i)_{i\in\omega}$ such that $\alpha = \mathrm{Hn}(a_i)_{i\in\omega}$, $\beta = \mathrm{Hn}(b_i)_{i\in\omega}$, $\gamma = \mathrm{Hn}(c_i)_{i\in\omega}$, $\boldsymbol{b} = (b_i)_{i\in\omega}$ and $\boldsymbol{c} = (c_i)_{i\in\omega}$ are subsequences of the sequence $\boldsymbol{a} = (a_i)_{i\in\omega}$ such that images of the injections $f_{ba}: \omega \to \omega$ and $f_{ca}: \omega \to \omega$ do not have common elements and each element from $(a_i)_{i\in\omega}$ belongs either to $(b_i)_{i\in\omega}$ or to $(c_i)_{i\in\omega}$.

The disjunctive union of real *P*-hypernumbers β and γ is denoted by $\beta\bigcup\gamma$.

Proposition 2.19. For any real *P*-hypernumbers α, β, δ and γ, if $\gamma\Subset\alpha$, and $\delta\Subset\beta$, then $\gamma\bigcup\delta\Subset\alpha\bigcup\beta$.

Proof is left as an exercise.

Proposition 2.20. $a(\beta\bigcup\gamma) = a\beta\bigcup a\gamma$ for any real number a and any real *P*-hypernumbers β and γ.

Proof is left as an exercise.

Note that the set $\boldsymbol{C}_\omega$ of all complex hypernumbers has similar algebraic properties.

Some researchers assume that the system of surreal numbers contains "all numbers great and small" [cf., e.g., (Ehrlich, 1994, 2001)]. So, it is possible to ask a question whether hypernumbers are some kinds of surreal numbers. Thus, to conclude this chapter, we show that surreal numbers cannot contain hypernumbers that are not separated from zero. Indeed, let us take two hypernumbers $\alpha = \mathrm{Hn}(a_i)_{i\in\omega}$ and $\beta = \mathrm{Hn}(b_i)_{i\in\omega}$ in which $a_{2i} = 1$, $a_{2i-1} = 0$, $b_{2i} = 0$, and $b_{2i-1} = 1$, for all $i = 1, 2, 3, \ldots$ The system **No** of all surreal numbers is a field (Gonshor, 1986). It means that any element δ from **No** has the inverse, i.e., an element γ such that $\delta\gamma = 1$. Thus, if $\boldsymbol{R}_\omega$ is homomorphically included in **No**, then there is an element η such that $\beta\eta = 1$. Then $\alpha(\beta\eta) = \alpha\cdot 1 = \alpha$. At the same time, $(\alpha\beta)\eta = 0\cdot\eta = 0$ because by the definition of multiplication of bounded hypernumbers, we have $\alpha\beta = 0$. Because multiplication in a field is associative (Kurosh, 1963) and $\alpha \neq 0$, we come to a contradictory equality $\alpha = \alpha(\beta\eta) = (\alpha\beta)\eta = 0$. This contradiction shows that surreal numbers do not include real hypernumbers.

This also shows the essential difference between real hypernumbers and hyperreal numbers from nonstandard analysis (Robinson, 1961, 1966) because hyperreal numbers form a subfield of the field of surreal numbers while as we demonstrated above, there are real hypernumbers that do not belong to surreal numbers.

Similar reasoning explains that complex hypernumbers and hypercomplex numbers from nonstandard analysis are fundamentally different. The basic difference between nonstandard analysis and the theory of hypernumbers and extrafunctions stems from the indispensable difference between the construction principles and methods of both theories: nonstandard analysis is based on set-theoretical principles and methods inherent for classical mathematics, while the theory of hypernumbers and extrafunctions is based on topological principles and methods inherent for modern physics (Burgin, 2012).

Besides, nonstandard analysis has both infinitely big and infinitely small numbers, or infinitesimals, while the theory of hypernumbers and extrafunctions has only infinitely big numbers. This allows researchers to utilize hypernumbers for measuring infinite distances, lengths of infinite lines, areas of infinite shapes and volumes bodies although the result of their measurement, as in physics, depends on the measurement procedure. For instance, if we apply the same measurement procedure to finding the length of the intervals $[0, \infty)$ and $(-\infty, \infty)$, we will find that the length of the first interval is equal to one half of the length of the second one. Note that in the conventional approach to measures, either the length of both intervals is undefined or it is equal to ∞. We see that employment of hypernumbers for measurement of distances and lengths gives results, which are in much better correlation with our intuition than the standard measure theory.

In addition, taking the most popular mathematical theories, which provide mathematical models of infinity allowing operation with infinite quantities, nonstandard analysis and set theory, we see that nonstandard analysis is oriented on problems of calculus foundations and originated by Georg Cantor set theory is oriented on problems of the foundations of mathematics. In contrast to this, the theory of hypernumbers and extrafunctions is aimed at providing an efficient technology to physicists for solving problems of physics. Indeed, quantum theory regularly encounters infinitely big values in the form of divergent series and integrals but never meets infinitely small numbers or infinitesimals, which were called "the ghosts of departed quantities" by George Berkeley.

However, it is necessary to admit that each of the three theories – set theory, nonstandard analysis and the theory of hypernumbers and extrafunctions – is important in its own area and cannot substitute any of the other two theories.

It is also necessary to remark that hypernumbers constructed in this chapter are essentially different from hypernumbers of Mikusinski. Indeed, a hypernumber in the sense of (Mikusinski, 1983) is a pair of elements (c, f), in which c is a complex number and f is a vector.

KEYWORDS

- **addition**
- **bounded hypernumber**
- **finite hypernumber**
- **hypernumber**
- **infinite hypernumber**
- **oscillating hypernumber**
- **subtraction**

CHAPTER 3

FROM NORMS, METRICS, AND SEMINORMS TO HYPERNORMS, HYPERMETRICS, AND HYPERSEMINORMS

In this chapter, our main interest is in mathematical structures that combine algebraic and topological properties. Examples of such structures are normed vector spaces, normed rings, seminormed vector spaces and topological vector spaces. Here, we extend the conventional and very useful in mathematics and its applications concepts of norm, seminorm, semimetric, pseudometric and metric utilizing hypernumbers. This allows introduction of extended metric and norm structures into a much larger class of spaces, extending the scope of applications and achieving higher precision in measuring distances, areas and volumes.

At first, we summarize and study conventional concepts of norms, seminorms, metrics and some of their generalizations.

Definition 3.1. a) A ring K with a mapping q: $K \to \boldsymbol{R}$ is called a (*submultiplicative*) *seminormed ring*, while q is called a *submultiplicative seminorm* in K, if the following axioms are satisfied:

SN1. $q(x) = 0$ if $x = 0$.

N3 (*the triangle inequality*).

$$q(x + y) \leq q(x) + q(y) \text{ for any } x \text{ and } y \text{ from } K.$$

SN4. $q(x \cdot y) \leq q(x) \cdot q(y)$ for any x and y from K.

b) A ring K with a mapping q: $K \to \boldsymbol{R}$ is called a *multiplicative seminormed ring*, or *composition ring*, while q is called a *multiplicative seminorm* in K, if the following axioms are satisfied:

SN1. $q(x) = 0$ if $x = 0$.

N3 (*the triangle inequality*).

$$q(x + y) \leq q(x) + q(y) \text{ for any } x \text{ and } y \text{ from } K.$$

N4. $q(x\cdot y) = q(x)\cdot q(y)$ for any x and y from K.

Note that any multiplicative seminormed ring is a submultiplicative seminormed ring.

Definition 3.2. a) A ring K with a submultiplicative seminorm q is called a (*submultiplicative*) *normed ring* and q is called a norm in K if q also satisfies one more axiom:

N1. $q(x) = 0$ if and only if $x = 0$ for any x from L

b) A ring K with a multiplicative seminorm q is called a *multiplicative normed ring*, or *valuated ring*, and q is called a norm in K if q also satisfies Axiom N1.

Note that any multiplicative normed ring is a submultiplicative normed ring.

It is possible to introduce other algebraic operations in a ring. With additional operations, an associative and commutative ring becomes a field (cf. Appendix).

Definition 3.3. a) If a seminormed ring K is a field, it is called a *seminormed field*.

b) If a normed ring K is a field, it is called a *normed field* or a *valuated field*. In this case, the norm in K is also called *absolute value*.

Lemma 3.1. If q: $K \to \boldsymbol{R}$ is a seminorm in a seminormed (normed) ring (field) K, then $q(x) \geq 0$ for any $u \in K$.

Proof. By Axiom N3, we have

$$q(u) + q(-u) \geq q(u + (-u)) = q(\mathbf{0})$$

At the same time, by N2, we have $q(\mathbf{0}) = 0\cdot q(\mathbf{0}) = 0$ and $q(-u) = q(u)$. This gives us $q(u) + q(-u) = q(u) + q(u) = 2q(u) \geq q(u + (-u)) = q(\mathbf{0}) = 0$ and thus, $q(u) \geq 0$.

Lemma is proved.

It is necessary to remark that in algebra, seminormed and normed rings and fields are sometimes defined and studied with seminorms (norms) that take values in a more general structure than the real numbers. This more general structure is called an ordered ring (Kurosh, 1963). However, in this book, we utilize norms and seminorms that take values in a field $\boldsymbol{F}$ or mostly, in the field $\boldsymbol{R}$ of all real numbers as this is the most prevalent case in mathematics and its applications.

Let us consider an abelian group G.

Definition 3.4. A mapping q: $G \to \boldsymbol{R}^+$ is called an *ultraseminorm* if it satisfies the following conditions:

SN1. $q(x) = 0$ if $x = \mathbf{0}$.

UN2 (symmetry). $q(x) = q(-x)$ for any x from G.

UN3 (*the ultranorm inequality*).

$$q(x + y) \leq \max\,(q(x), q(y)) \text{ for any } x \text{ and } y \text{ from } L$$

Lemma 3.2. Any ultraseminorm is a seminorm.

Indeed,

$$q(x + y) \leq \max\,(q(x), q(y)) \leq q(x) + q(y)$$

Definition 3.5. An ultraseminorm q is called an *ultranorm* (Warner, 1993; Hasler, 2006) if it satisfies Axioms UN2, UN3 and axiom:

N1. $q(x) = 0$ if and only if $x = 0$ for any x from L

Definitions imply the following result.

Lemma 3.3. Any ultranorm is a norm.

Modules over a ring R are more complex structures than abelian groups because they are commutative groups with multiplication by elements from the ring R (cf. Appendix).

Let us assume that L is a module over a seminormed (normed) ring K with a seminorm (norm) p.

Definition 3.6. The module L with a mapping q: $L \to \boldsymbol{R}^+$ is called a *seminormed module*, while q is called a seminorm in L, if q satisfies Axiom N3 and one more axiom:

N5. $q(u \cdot y) \leq p(u) \cdot q(y)$ for any y from L and any u from K

For instance, the set of all polynomials of the first degree with integer coefficients form a seminormed module over the ring $\boldsymbol{Z}$.

As any vector space is a module over a field, Definition 3.6 also introduces the well-known concept of a seminormed vector space.

Definition 3.7. The module L with a norm q is called a *normed module* if q satisfies Axioms N1, N3 and one more axiom:

N5a. $q(u \cdot y) = p(u) \cdot q(y)$ for any y from L and any u from K

For instance, the set of all polynomials of the first degree with integer coefficients defined in the interval [0, 1] form a normed module over the ring $\boldsymbol{Z}$.

As any vector space is a module over a field, Definition 3.7 also introduces the well-known concept of a normed vector space.

Definition 3.8. The module (ring or linear/vector space) L with a set Q of seminorms [norms] is called a *Q-seminormed* [*Q-normed*] *module* (*ring* or *linear/vector space*) if it is a seminormed [normed] module (ring or linear/vector space, respectively) for each $q \in Q$.

Let L be a vector (linear) space over a normed (valuated) field $\boldsymbol{F}$, e.g., a real vector (linear) space over $\boldsymbol{R}$.

Definition 3.9. A mapping q: $L \to \boldsymbol{R}$ is called a *seminorm* if it satisfies the following conditions:

SN1. $q(x) = 0$ if $x = \mathbf{0}$.

N2. $q(ax) = |a| \cdot q(x)$ for any x from L and any element a from $\boldsymbol{F}$.

N3 (the triangle inequality).

$$q(x + y) \leq q(x) + q(y) \text{ for any } x \text{ and } y \text{ from } L$$

We can see that norms and seminorms are non-negative subadditive functionals with additional properties.

Note that for some non-zero elements from L, their seminorm q can be equal to 0. The set of all such elements is called the *kernel* Ker q of the seminorm q, i.e., Ker $q = \{x;\ q(x) = 0\ \}$.

Lemma 3.4. If q: $L \to \boldsymbol{R}$ is a seminorm, then $q(x) \geq 0$ for all $x \in L$.

Proof. By N3, we have

$$q(x) + q(-x) \geq q(x + (-x)) = q(\mathbf{0})$$

At the same time, by N2, we have $q(\mathbf{0}) = 0 \cdot q(\mathbf{0}) = 0$ and $q(-x) = q(x)$. Thus, $q(x) + q(-x) = q(x) + q(x) = 2q(x) \geq 0$ and $q(x) \geq 0$.

Lemma is proved.

So, it is possible to define seminorms as mappings q: $L \to \boldsymbol{R}^+$.

Example 3.1. Let us take the space $C(\boldsymbol{R}, \boldsymbol{R})$ of all continuous functions on the real line $\boldsymbol{R}$. Taking any natural number n, we obtain the following seminorm

$$\|f\|_n = \max \{|f(x)|;\ x \in [-n, n]\}$$

Example 3.2. Let us take the space $L^1(\boldsymbol{R}, \boldsymbol{R})$ of all locally absolutely Riemann integrable real functions on the real line $\boldsymbol{R}$. We remind that a function is locally Riemann integrable if it is Riemann integrable on an arbitrary interval $[a, b]$. Taking any natural number n, we obtain the following seminorm

$$||f||_{L^1} = \int_{-n}^{n} |f(x)|\mathrm{d}x.$$

Example 3.3. Let us take the space $L^2(\boldsymbol{R}, \boldsymbol{R})$ of all locally Riemann square-integrable real functions on the real line $\boldsymbol{R}$. The seminorm in this space is defined by the following formula

$$||f||_{L^2} = [\int_{-n}^{n} |f(x)|^2\mathrm{d}x]^{1/2}$$

Example 3.4. Let us consider the space $L^p(\boldsymbol{R}, \boldsymbol{R})$ of all locally Riemann p-integrable ($p \geq 1$) functions on the real line $\boldsymbol{R}$. The seminorm in this space is defined by the following formula

$$||f||_{L^p} = \mathrm{Hn}(d_n)_{n\in\omega}$$

where $d_n = [\int_{-n}^{n} |f(x)|^p\mathrm{d}x]^{1/p}$.

Example 3.5. Let us take $B(\boldsymbol{R}, \boldsymbol{R})$ is the space of all locally bounded functions on the real line $\boldsymbol{R}$. We remind that a function is locally bounded if it is bounded on an arbitrary interval $[a, b]$. The seminorm in this space is defined by the following formula

$$||f||_{\infty} = \mathrm{Hn}(a_n)_{n\in\omega}$$

where $a_n = \sup \{|f(x)|; x \in [-n, n]\}$

Example 3.6. Let us consider the generalized Sobolev space $H^1(\boldsymbol{R}, \boldsymbol{R})$ of all differentiable functions on the real line $\boldsymbol{R}$ such that they are locally square integrable and their derivatives are locally square integrable. The seminorm in this space is defined by the following formula

$$||f||_{H^1} = ||f||_{L^2} + ||f'||_{L^2} = [\int_{-n}^{n} |f(x)|^2\mathrm{d}x]^{1/2} + [\int_{-n}^{n} |f'(x)|^2\mathrm{d}x]^{1/2}$$

Lemma 3.5. Axiom SN1 follows from Axiom N2.

Indeed, by Axiom N2, we have

$$q(\mathbf{0}) = q(0x) = 0q(x) = 0$$

The triangle inequality implies the following result.

Lemma 3.6. For any x and y from L, the triangle inequality implies:

a) $|q(x) - q(y)| \leq q(x + y)$
b) $|q(x) - q(y)| \leq q(x - y)$

There are different conditions defining seminorms.

Proposition 3.1. A mapping q: $L \to \boldsymbol{R}^+$ is a seminorm if and only if one of the following conditions (a) or (b) is true:

(a) $q(ax + by) \leq |a|q(x) + |b|q(y)$ for any x and y from L and any numbers a and b from $\boldsymbol{R}$.
(b) $q(tx + (1 - t)y) \leq tq(x) + (1 - t)q(y)$ and $q(ax) = |a|\cdot q(x)$ for any x and y from L and any numbers t from the interval $[0, 1]$ and a from $\boldsymbol{R}$.

Proof is left as an exercise.

We remind that a mapping f: $L \to \boldsymbol{R}$ is called convex if its graph never lies above the set $[f(x), f(y)] = \{ tf(x) + (1 - t)f(y); t \in [0, 1]\}$ for any elements x and y from L (Magaril-Il'yaev and Tikhomirov, 2003). Proposition 3.1 shows that any seminorm is a non-negative convex function.

Definition 3.10. A seminorm q is called a *norm* if it satisfies Axioms N2, N3 and one more axiom:

N1. $q(x) = 0$ if and only if $x = 0$ for any x from L

Usually a norm or a seminorm is denoted by $\|.\|$, e.g., the norm (seminorm) of an element v is $\|v\|$.

Example 3.7. Let us take the n-dimensional real vector space $\boldsymbol{R}^n$ as the space L. The *standard norm* in this space is defined by the conventional formula

If $x = (a_1, a_2, a_3, \ldots, a_n)$ with all $a_i \in \boldsymbol{R}$, then $\|x\| = \sqrt{a_1^{\,2} + a_2^{\,2} + x_3^{\,2} + \ldots + a_n^{\,2}}$

Definition 3.11. A vector/linear space L with a norm (seminorm) is called a *normed (seminormed) vector/linear space* or simply, a *normed (seminormed) space*.

The standard notation for the norm of a function f is $\|f\|$.

The most familiar example of a normed space is the set of real numbers $\boldsymbol{R}$ with the absolute value as a norm.

Definition 3.12. A mapping p: $L \to \boldsymbol{R}$ is called a *pseudonorm* (Schaefer and Wolff, 1999) if it satisfies Axioms N1, N3 and the following condition:

PN2. $q(ax) \leq q(x)$ for any x from L and any number a from $\boldsymbol{R}$ with $\|a\| \leq 1$.

Lemma 3.7. If q: $L \to \boldsymbol{R}$ is a pseudonorm, then $q(x) \geq 0$ and $q(-x) = q(x)$ for all $x \in L$.

Proof. By PN2, we have $q(-x) = q(x)$. At the same time, by N3, we have

$$q(x) + q(-x) \geq q(x + (-x)) = q(\mathbf{0})$$

and by N1, we have

$$q(\mathbf{0}) = 0$$

Thus, $q(x) + q(-x) = q(x) + q(x) = 2q(x) \geq 0$ and $q(x) \geq 0$.

Lemma is proved.

Operations in a vector space L allow researchers to define similar operations with subsets of L.

Addition: If $A, B \subseteq L$, then $A + B = \{x + y;\ x \in A, y \in B\}$.

Multiplication by real numbers (*scalar multiplication*): If $B \subseteq L$ and r is a real number, then $rB = \{rx;\ x \in B\}$.

Properties of operations with elements in the vector space L (cf. Appendix) imply similar properties of operations with subsets of L.

Lemma 3.8.

(a) Addition of subsets is associative.

(b) Addition of subsets is commutative.

(c) Addition of subsets has an identity element $\{\mathbf{0}\}$.

(d) Scalar multiplication of subsets by elements from $\boldsymbol{R}$ is distributive over addition of subsets in L.

(e) Scalar multiplication of subsets by elements from $\boldsymbol{R}$ is distributive over addition in $\boldsymbol{R}$.

(f) Scalar multiplication of subsets by elements from $\boldsymbol{R}$ is compatible with multiplication in $\boldsymbol{R}$.

(g) The identity element 1 from $\boldsymbol{R}$ also is an identity element for scalar multiplication of subsets by elements from $\boldsymbol{R}$:

Proof is left as an exercise.

However, not all properties of operations with elements in the vector space are preserved by similar operations with subsets. For instance, not every subset has the inverse element with respect to addition of subsets.

Let us assume that L is a vector space over $\boldsymbol{R}$.

Proposition 3.2. (a) If A and B are vector subspaces of L, then $A + B$ is also a vector subspace of the space L.

(b) If A is a vector subspace of L, then rA is also a vector subspace of L for any real number r.

Proof is left as an exercise.

Let us assume that L is a linear algebra over $\boldsymbol{R}$.

Definition 3.13. a) The linear algebra L with a seminorm q is called a (*submultiplicative*) *seminormed algebra* if it satisfies Axioms SN1, N2, N3 and SN4.

b) The linear algebra L with a seminorm q is called a *multiplicative seminormed algebra*, or *semicomposition algebra* if it satisfies Axioms SN1, N2, N3 and N4.

For instance, the set of all polynomials with real coefficients form a seminormed algebra over the field $\boldsymbol{R}$.

Seminormed algebras were introduced by Arens (1946, 1952).

Note that any multiplicative seminormed algebra is a submultiplicative seminormed algebra.

Definition 3.14. a) The linear algebra L with a norm q is called a (*submultiplicative*) *normed algebra* if it satisfies Axioms N1, N2, N3 and SN4.

b) The linear algebra L with a norm q is called a *multiplicative normed algebra*, as well as *real-valued algebra* or *composition algebra*, if it satisfies Axioms N1, N2, N3 and N4.

For instance, algebras $\boldsymbol{R}$ and $\boldsymbol{C}$ are multiplicative normed (composition) algebras over the field $\boldsymbol{R}$.

Note that any multiplicative normed algebra is a submultiplicative normed algebra.

We can see that basic (axiomatic) properties of norms and seminorms depend in what structures these norms and seminorms are defined because norms and seminorms have to be consistent with other configurations, e.g., algebraic operations, in these structures.

There is a rich theory of normed rings [cf., e.g., (Gelfand, 1941; Gelfand and Naimark, 1943; Naimark, 1959)] and normed, seminormed, composition and real-valuedalgebras [cf., e.g., (Arens, 1946, 1952; Albert, 1947, 1949; Kaplansky, 1949; Wright, 1953; Rickart, 1960; Naimark, 1972; Bonsal and Duncan, 1973; Helemski, 1989; Dosi, 2011)]. These theories originated in the works of Nagumo (1936), Yosida, (1936) and Mazur (1938).

Let us introduce and study operations with norms and seminorms.

Taking a vector space L over a topological (normed) field $\boldsymbol{F}$ with a system Q of norms (seminorms), we introduce the following operations:

Addition:

if $||.||_1$ and $||.||_2$ belong to Q, then for any element v from L, we define their sum $||.|| = ||.||_1 + ||.||_2$ by the following rule

$$||v|| = ||v||_1 + ||v||_2$$

Scalar multiplication by elements from $\boldsymbol{F}^{+}$:

if $||.||_1$ belongs to Q and a is a nonnegative element from $\boldsymbol{F}$, then for any element v from L, we define their scalar multiplication $a||.|| = ||.||_1$ by the following rule

$$||v||_1 = a||v||$$

Maximum:

if $||.||_1$ and $||.||_2$ belong to Q, then for any element v from L, we define their maximum $||.|| = \max(||.||_1, ||.||_2)$ by the following rule

$$||v|| = \max(||v||_1, ||v||_2)$$

Minimum:

if $||.||_1$ and $||.||_2$ belong to Q, then for any element v from L, we define their minimum $||.|| = \min(||.||_1, ||.||_2)$ by the following rule

$$||v|| = \min(||v||_1, ||v||_2)$$

Proposition 3.3. a) If $||.||_1$ and $||.||_2$ are norms (seminorms), then their sum $||.|| = ||.||_1 + ||.||_2$ is a norm (seminorm).

b) If $||.||_1$ is a norm (seminorm) and a is a nonnegative element from $\boldsymbol{F}$, then their scalar multiplication $a||.|| = ||.||_1$ is a norm (seminorm).

c) If $||.||_1$ and $||.||_2$ are ultranorms (ultraseminorms), then their maximum $||.|| = \max(||.||_1, ||.||_2)$ is a ultranorm (ultraseminorm).

Proof. We need to check axioms SN1 (N1), N2, UN3 and N3 for new functions.

(a) SN1: $||\mathbf{0}|| = ||\mathbf{0}||_1 + ||\mathbf{0}||_2 = 0 + 0 = 0$ because $||.||_1$ and $||.||_2$ are seminorms.

N2: $||av|| = ||av||_1 + ||av||_2 = a||v||_1 + a||v||_2 = a(||v||_1 + ||v||_2) = a||v||$ for any v from L and any nonnegative element a from $\boldsymbol{F}$ because $||.||_1$ and $||.||_2$ are seminorms.

N3: $||z + v|| = ||z + v||_1 + ||z + v||_2 \leq (||z||_1 + ||v||_1) + (||z||_2 + ||v||_2) = (||z||_1 + ||z||_2) + (||v||_1 + ||v||_2) = ||z|| + ||v||$ for any z and v from L because $||.||_1$ and $||.||_2$ are seminorms.

N1: Let us assume that $||.||$ is a norm and $0 = ||v|| = ||v||_1 + ||v||_2$, then $||v||_1 = 0$ and $||v||_2 = 0$ because both values $||v||_1$ and $||v||_2$ are nonnegative. Consequently, $v = \mathbf{0}$ because both $||.||_1$ and $||.||_2$ are norms.

(b) SN1: $||\mathbf{0}||_1 = a||\mathbf{0}|| = a0 = 0$ because $||.||$ is a seminorm.

N2: $||bv||_1 = a||bv|| = ab||v|| = b(a(||v||) = b||v||_1$ for any v from L and any nonnegative element b from $\boldsymbol{F}$ because $||.||$ is a seminorm.

N3: $||z + v||_1 = a|| z + v || = || az + av || \leq ||az|| + ||av|| = a||z|| + a||v|| = ||z||_1 + ||v||_1$ for any z and v from L because $||.||$ is a seminorm.

N1: Let us assume that $||.||$ is a norm and $0 = ||v||_1 = a||v|| = a \cdot 0 = 0$, then $v = \mathbf{0}$ because $||.||$ is a norm.

(c) SN1: $||\mathbf{0}|| = \max(||\mathbf{0}||_1, ||\mathbf{0}||_2) = \max(0, 0) = 0$ because $||.||_1$ and $||.||_2$ are ultraseminorms.

N2: $||av|| = \max(||av||_1, ||av||_2) = \max(a||v||_1, a||v||_2 = \max(a(||v||_1, ||v||_2) = a\cdot \max(||v||_1, ||v||_2) = a||v||$ for any v from L and any nonnegative element a from $\boldsymbol{F}$ because $||.||_1$ and $||.||_2$ are ultraseminorms.

UN3: $||z + v|| = = \max(|| z + v ||_1, || z + v ||_2) \leq \max(\max(||z||_1, ||v||_1), \max(||z||_2, ||v||_2)) = \max(||z||_1, ||v||_1, ||z||_2, ||v||_2) = \max(\max(||z||_1, ||z||_2), \max(||v||_1, ||v||_2)) = \max(||z||, ||v||)$ for any z and v from L because $||.||_1$ and $||.||_2$ are ultraseminorms.

N1: Let us assume that $||.||$ is a norm and $0 = ||v|| = ||v||_1 + ||v||_2$, then $||v||_1 = 0$ and $||v||_2 = 0$ because both values $||v||_1$ and $||v||_2$ are nonnegative. Consequently, $v = \mathbf{0}$ because both $||.||_1$ and $||.||_2$ are ultranorms.

Proposition is proved.

The proof of Proposition 3.3 implies the following result.

Corollary 3.1. If $||.||_1$ is a norm and $||.||_2$ is a seminorm, then their sum $||.|| = ||.||_1 + ||.||_2$ is a norm.

Corollary 3.2. Any linear combination with positive coefficients of seminorms (norms) in L is a seminorm (norm).

Corollary 3.3. a) All seminorms in L form a semimodule over the semiring $\boldsymbol{R}^+$.

b) All norms in L form an S-module over the S-ring $\boldsymbol{R}^{++}$.

For norms and seminorms in other mathematical structures, it is possible to define different operations. For instance, it is possible to define multiplication of seminorms for seminorms in seminormed rings or seminormed fields.

Let us consider the system Q of all norms in a ring R.

If $||.||_1$ and $||.||_2$ belong to Q, then for any element v from R, we define their product $||.|| = ||.||_1 \cdot ||.||_2$ by the following rule

$$||v|| = ||v||_1 \cdot ||v||_2,$$

Proposition 3.4. a) If $||.||_1$ and $||.||_2$ are norms (seminorms), then their product $||.|| = ||.||_1 \cdot ||.||_2$ is a norm (seminorm).

b) If $||.||_1$ is a norm (seminorm) and a is a nonnegative element from $\boldsymbol{F}$, then their scalar multiplication $a||.|| = ||.||_1$ is a norm (seminorm).

Proof. We need to check axioms N1 (SN1), N4 (SN4) and N3 for new functions.

(a) SN1: $||\mathbf{0}|| = ||\mathbf{0}||_1 \cdot ||\mathbf{0}||_2 = 0 \cdot 0 = 0$ because $||.||_1$ and $||.||_2$ are seminorms in R.

N3: $||z + v|| = ||\ z + v\ ||_1 + ||\ z + v\ ||_2 \leq (||z||_1 + ||v||_1) + (||z||_2 + ||v||_2) = (||z||_1 + ||z||_2) + (||v||_1 + ||v||_2) = ||z|| + ||v||$ for any z and v from R because $||.||_1$ and $||.||_2$ are seminorms.

SN4: $||z \cdot v|| = ||\ z \cdot v\ ||_1 \cdot ||\ z \cdot v\ ||_2 \leq (||z||_1 \cdot ||v||_1) \cdot (||z||_2 \cdot ||v||_2) = (||z||_1 \cdot ||z||_2) \cdot (||v||_1 \cdot ||v||_2) = ||z|| \cdot ||v||$ for any z and v from R because $||.||_1$ and $||.||_2$ are seminorms in the ring R and multiplication of real numbers is commutative.

N1: Let us assume that $||.||$ is a norm and $0 = ||v|| = ||v||_1 \cdot ||v||_2$, then either $||v||_1 = 0$ or $||v||_2 = 0$ because both values $||v||_1$ and $||v||_2$ are nonnegative. Consequently, $v = \mathbf{0}$ because both $||.||_1$ and $||.||_2$ are norms in R.

N4: $||z \cdot v|| = ||\ z \cdot v\ ||_1 \cdot ||\ z \cdot v\ ||_2 = (||z||_1 \cdot ||v||_1) \cdot (||z||_2 \cdot ||v||_2) = (||z||_1 \cdot ||z||_2) \cdot (||v||_1 \cdot ||v||_2) = ||z|| \cdot ||v||$ for any z and v from R because $||.||_1$ and $||.||_2$ are norms in the ring R and multiplication of real numbers is commutative.

Proposition is proved.

Corollary 3.4. If $||.||_1$ is a norm and $||.||_2$ is a seminorm but not a norm, then their product $||.|| = ||.||_1 \cdot ||.||_2$ is a seminorm but not a norm.

Corollary 3.5. Any polynomial with positive coefficients of seminorms (norms) in a ring R is a seminorm (norm).

Corollary 3.6. a) All seminorms in a ring R form a semialgebra over the semiring $\boldsymbol{R}^+$.

b) All norms in a ring R form an S-module over the S-ring $\boldsymbol{R}^{++}$.

Remark 3.1. It is possible to define the product of norms or seminorms in a vector space L but this product will not always be a norm or a seminorm in L

There are different relations in sets of norms and seminorms.

Let L be a vector space over a normed field K, while q and p are two seminorms in L.

Definition 3.15. a) The seminorm q is *stronger* than the seminorm p and p is *weaker* than q if there is a constant $k \geq 0$ such that for all $x \in L$,

$$p(x) \leq k \cdot q(x)$$

b) Seminorms q and p are *equivalent* if there are constants $k, h \geq 0$ such that for all $x \in L$,

$$h \cdot q(x) \leq p(x) \leq k \cdot q(x)$$

When the seminorm q is stronger than the seminorm p, it is denoted by

$$p \leq q$$

Lemma 3.9. Seminorms q and p in a vector space L are equivalentif and only if $p \leq q$ and $p \leq q$.

Proof is left as an exercise.

Lemma 3.10. For any seminorms q and p in a vector space $L, p \leq q$ if and only if one of the following conditions is satisfied:

(a) $p(x) \leq k \cdot q(x)$ for some real number $k \geq 0$ and all x from some p-sphere $S_{pc} = \{x; p(x) = c\}$ in L.

(b) $p(x) \leq k \cdot q(x)$ for some real number $k \geq 0$ and all x from some q-sphere $S_{qc} = \{x; q(x) = c\}$ in L and Ker q.

Proof. Necessity is clear in both cases and we need to prove only sufficiency.

(a) Let us assume that for two positive real numbers k and c, we have:

$$p(x) \leq k \cdot q(x) \text{ for all } x \text{ from the } p\text{-sphere } S_{pc} = \{x; p(x) = c\}$$

Taking an arbitrary element z from $L\backslash$ Ker p and the real number $a = (p(z))^{-1}c$, see that $az \in S_{pc}$. Then we have

$$p(az) \leq k \cdot q(az)$$

As p and q are seminorms and $a > 0$, we have

$$ap(z) \leq ka \cdot q(z)$$

and consequently,

$$p(z) \leq k \cdot q(z)$$

When $z \in$ Ker $p = \{x; p(z) = 0\}$, then $p(z) \leq k \cdot q(z)$ because $q(z) \geq 0$ for all $z \in L$.

Thus, for all $x \in L$, we have

$$p(x) \leq k \cdot q(x)$$

(b) Let us assume that for two positive real numbers k and c, we have:

$p(x) \le k \cdot q(x)$ for all x from the q-sphere $S_{qc} = \{x;\ q(x) = c\}$ and Ker q

Taking an arbitrary element z from $L \backslash$ Ker q and the real number $a = (q(z))^{-1}c$, see that $az \in S_{qc}$. Then we have

$$p(az) \le k \cdot q(az)$$

As p and q are seminorms and $a > 0$, we have

$$ap(z) \le ka \cdot q(z)$$

and consequently,

$$p(z) \le k \cdot q(z)$$

Thus, for all $x \in L$, we have

$$p(x) \le k \cdot q(x)$$

Lemma is proved.

Corollary 3.7. Any seminorms q and p in a vector space L are equivalent if and only if $h \cdot q(x) \le p(x) \le k \cdot q(x)$ for some real numbers $k, h \ge 0$ and all x from some p-sphere $S_{pc} = \{x;\ p(x) = c\}$ in L and Ker $q \cup$ Ker p.

Corollary 3.8. Any seminorms q and p in a vector space L are equivalent if and only if $h \cdot q(x) \le p(x) \le k \cdot q(x)$ for some real numbers $k, h \ge 0$ and all x from some q-sphere $S_{qc} = \{x;\ p(x) = c\}$ in L and Ker $q \cup$ Ker p.

As Ker $p = 0$ for any norm p, Lemma 3.10 implies the following results.

Corollary 3.9. For any norms q and p in a vector space L, $p \le q$ if and only if $p(x) \le k \cdot q(x)$ for some real number $k \ge 0$ and all x from some p-sphere $S_{pc} = \{x;\ p(x) = c\}$ in L.

Corollary 3.10. For any norms q and p in a vector space L, $p \le q$ if and only if $p(x) \le k \cdot q(x)$ for some real number $k \ge 0$ and all x from some q-sphere $S_{qc} = \{x;\ q(x) = c\}$ in L.

Corollary 3.11. Any norms q and p in a vector space L are equivalent if and only if $h \cdot q(x) \le p(x) \le k \cdot q(x)$ for some real numbers $k, h \ge 0$ and all x from some p-sphere $S_{pc} = \{x;\ p(x) = c\}$ in L.

Corollary 3.12. Any norms q and p in a vector space L are equivalent if and only if $h \cdot q(x) \le p(x) \le k \cdot q(x)$ for some real number $k \ge 0$ and all x from some q-sphere $S_{qc} = \{x;\ p(x) = c\}$ in L.

As any p-ball contains Ker p, Lemma 3.10 implies the following results.

Corollary 3.13. For any seminorms q and p in a vector space L, $p \le q$ if and only if $p(x) \le k \cdot q(x)$ for some real number $k \ge 0$ and all x from some p-ball $B = \{x; p(x) \le c\}$ in L.

Corollary 3.14. For any norms q and p in a vector space L, $p \le q$ if and only if $p(x) \le k \cdot q(x)$ for some real number $k \ge 0$ and all x from some p-ball $B = \{x; p(x) \le c\}$ in L.

Corollary 3.15. Any seminorms q and p in a vector space L are equivalent if and only if $h \cdot q(x) \le p(x) \le k \cdot q(x)$ for some real number $k \ge 0$ and all x from some p-ball $B = \{x; p(x) \le c\}$ in L.

Corollary 3.16. Any norms q and p in a vector space L are equivalent if and only if $h \cdot q(x) \le p(x) \le k \cdot q(x)$ for some real number $k \ge 0$ and all x from some p-ball $B = \{x; p(x) \le c\}$ in L.

As any q-ball contains Ker q, Lemma 3.10 implies the following results.

Corollary 3.17. For any seminorms q and p in a vector space L, $p \le q$ if and only if $p(x) \le k \cdot q(x)$ for some real number $k \ge 0$ and all x from some q-ball $B = \{x; q(x) \le c\}$ in L.

Corollary 3.18. For any norms q and p in a vector space L, $p \le q$ if and only if $p(x) \le k \cdot q(x)$ for some real number $k \ge 0$ and all x from some q-ball $B = \{x; q(x) \le c\}$ in L.

Corollary 3.19. Any seminorms q and p in a vector space L are equivalent if and only if $h \cdot q(x) \le p(x) \le k \cdot q(x)$ for some real number $k \ge 0$ and all x from some q-ball $B = \{x; q(x) \le c\}$ in L.

Corollary 3.20. Any norms q and p in a vector space L are equivalent if and only if $h \cdot q(x) \le p(x) \le k \cdot q(x)$ for some real number $k \ge 0$ and all x from some q-ball $B = \{x; q(x) \le c\}$ in L.

Some mathematicians introduced a relation between norms, which is stronger than the relation $\le$ [cf., e.g., (Maslov, 1976)]. Namely, the seminorm q is *essentially stronger* than the seminorm p if for all $x \in L$, we have

$$p(x) \le q(x)$$

When the seminorm q is essentially stronger than the seminorm p, it is denoted by

$$p \preccurlyeq q$$

We can easily see that relation $p \preccurlyeq q$ implies relation $p \le q$.

Lemma 3.11. For any seminorms q and p in a vector space $L, p \preccurlyeq q$ if and only if one of the following conditions is satisfied:

(a) $p(x) \leq q(x)$ for some real number $k \geq 0$ and all x from some p-sphere $S_{pc} = \{x; p(x) = c\}$ in L.

(b) $p(x) \leq q(x)$ for some real number $k \geq 0$ and all x from some q-sphere $S_{qc} = \{x; q(x) = c\}$ in L and Ker q.

Proof is similar to the proof of Lemma 3.10.

Corollary 3.21. For any norms q and p in a vector space $L, p \preccurlyeq q$ if and only if $p(x) \leq q(x)$ for all x from some p-sphere $S_{pc} = \{x; p(x) = c\}$ in L.

Corollary 3.22. For any seminorms q and p in a vector space $L, p \preccurlyeq q$ if and only if $p(x) \leq q(x)$ for all x from some p-ball $B = \{x; p(x) \leq c\}$ in L.

Corollary 3.23. For any norms q and p in a vector space $L, p \preccurlyeq q$ if and only if $p(x) \leq q(x)$ for all x from some p-ball $B = \{x; p(x) \leq c\}$ in L.

Corollary 3.24. For any norms q and p in a vector space $L, p \preccurlyeq q$ if and only if $p(x) \leq q(x)$ for all x from some q-sphere $S_{qc} = \{x; p(x) = c\}$ in L.

Corollary 3.25. For any seminorms q and p in a vector space $L, p \preccurlyeq q$ if and only if $p(x) \leq q(x)$ for all x from some q-ball $B = \{x; q(x) \leq c\}$ in L.

Corollary 3.26. For any norms q and p in a vector space $L, p \preccurlyeq q$ if and only if $p(x) \leq q(x)$ for all x from some q-ball $B = \{x; q(x) \leq c\}$ in L.

Any seminorm q in a vector space L induces convergence of sequences, namely, a sequence $\{x_i; x_i \in L, i = 1, 2, 3, \ldots\}$ q-converges to an element x from L, i.e., $\lim^q_{i\to\infty} x_i = x$, if $\lim_{i\to\infty} q(x_i - x) = 0$.

In turn, convergence of sequences induces order relation in sets of norms and seminorms.

Definition 3.16. a) A seminorm q is *sequentially stronger* than the seminorm p if q-convergence implies p-convergence, i.e., $\lim^q_{i\to\infty} x_i = x$ entails $\lim^p_{i\to\infty} x_i = x$ for any sequence $\{x_i; x_i \in L, i = 1, 2, 3, \ldots\}$.

b) Seminorms q and p are *sequentially equivalent* if each of them is sequentially stronger than the other.

When the seminorm q is sequentially stronger than the seminorm p, it is denoted by

$$p \sqsubseteq q$$

Note that for normed, seminormed, metric and quasimetric vector spaces, topology is defined by convergence of sequences (Kuratowski, 1966).

Theorem 3.1. For any seminorms q and p in a vector space L, the relation $p \leq q$ is valid is and only if $p \sqsubseteq q$ is valid.

Proof. Necessity. Let us consider seminorms q and p in a vector space L, such that $p \leq q$. It means that there is a constant $k \geq 0$ such that for all $x \in L$, we have

$$p(x) \leq k \cdot q(x)$$

Taking a sequence $\{x_i; x_i \in L, i = 1, 2, 3, \ldots\}$ such that $\lim^q_{i\to\infty} x_i = x$, we have

$$\lim_{i\to\infty} q(x_i - x) = 0$$

Then

$$\lim_{i\to\infty} p(x_i - x) \leq \lim_{i\to\infty} k \cdot q(x_i - x) = k \cdot \lim_{i\to\infty} q(x_i - x) = 0$$

As $p(x)$ is always larger than or equal to 0, this implies

$$\lim_{i\to\infty} p(x_i - x) = 0$$

Consequently, $p \sqsubseteq q$, i.e., the relation $p \leq q$ implies the relation $p \sqsubseteq q$.

Sufficiency. Let us consider seminorms q and p in a vector space L and the unit p-sphere $S_p = \{x; p(x) = 1\}$ in L supposing that it is not true that a seminorm q is stronger than a seminorm p. By Lemma 3.10, for any natural number n, there is an element x_n from S_p for which $n \cdot q(x_n) < p(x_n)$.

Let us consider the sequence $z = \{ z_n; n = 1, 2, 3, \ldots\}$ where $z_n = (1/n) \cdot x_n$ for all $n = 1, 2, 3, \ldots$ Then by properties of seminorms, we have

$$q(z_n) = q((1/n) \cdot x_n) = (1/n) \cdot q(x_n) < 1/n$$

Consequently,

$$\lim^q_{i\to\infty} z_i = 0$$

At the same time, as $x_n \in S_p$, we have

$$p(z_n) = p((1/n) \cdot x_n) = (1/n) \cdot p(x_n) \geq 1$$

This means that the sequence z does not p-converge to 0. By *reductio ad absurdum*, the relation $p \sqsubseteq q$ implies the relation $p \leq q$.

Theorem is proved.

Corollary 3.27. For any norms q and p in a vector space L, the relation $p \leq q$ is valid is and only if $p \sqsubseteq q$ is valid.

We remind that a sequence $\{a_i; i = 1, 2, 3, \dots\}$ in a vector space L with a norm (seminorm) q is a *Cauchy sequence with respect* to q if for any $\varepsilon \in \boldsymbol{R}^{++}$, there is $n \in \boldsymbol{N}$ such that for any $i, j \geq n$, we have

$$q(a_j - a_i) < \varepsilon$$

Lemma 3.12. Any q-convergent sequence is a Cauchy sequence with respect to q.

Proof is similar to the standard proof that any convergent sequence of real numbers is a Cauchy sequence [cf., e.g., (Burgin, 2008)].

Definition 3.17. Seminorms q and $p$are *compatible* if any Cauchy sequence with respect to one of them is also a Cauchy sequence with respect to the other.

Theorem 3.2. Any seminorms q and p in a vector space L are equivalent if and only if they are compatible.

Proof. Necessity. Let us consider equivalent seminorms q and p in a vector space L. By Lemma 3.9, we have $p \leq q$ and $p \leq q$. The inequality $p \leq q$ means that there is a constant $k \geq 0$ such that for all $x \in L$, we have

$$p(x) \leq k \cdot q(x)$$

Let $\boldsymbol{x} = \{x_i; x_i \in L, i = 1, 2, 3, \dots\}$ be a Cauchy sequence with respect to q. Then we know that given some $\varepsilon \in \boldsymbol{R}^{++}$, there is $n \in \boldsymbol{N}$ such that for any i, $j \geq n$, we have

$$q(a_j - a_i) < k^{-1} \cdot \varepsilon$$

Then

$$p(a_j - a_i) \leq k \cdot q(a_j - a_i) < k \cdot k^{-1} \cdot \varepsilon = \varepsilon$$

Thus, $\boldsymbol{x}$ is a Cauchy sequence with respect to p.

In a similar way, we show that any Cauchy sequence with respect to p is also a Cauchy sequence with respect to q.

Consequently, seminorms q and p are compatible.

Sufficiency. Let us consider compatible seminorms q and p in a vector space L and suppose that these seminorms are not equivalent. Then by

Lemma 3.9, either it is not true that a seminorm q is stronger than a seminorm p or it is not true that a seminorm p is stronger than a seminorm q. In the first case, by Lemma 3.10, for any natural number n, there is an element x_n from the unit p-sphere $S_p = \{x; p(x) = 1\}$ for which $n \cdot q(x_n) < p(x_n)$.

Let us consider the sequence $z = \{ z_n; n = 1, 2, 3, \ldots\}$ where $z_n = (1/n) \cdot x_n$ for all $n = 1, 2, 3, \ldots$ Then by properties of seminorms, we have

$$q(z_n) = q((1/n) \cdot x_n) = (1/n) \cdot q(x_n) < 1/n$$

Consequently,

$$\lim^q{}_{i\to\infty} z_i = 0$$

At the same time, as $x_n \in S_p$, we have

$$p(z_n) = p((1/n) \cdot x_n) = (1/n) \cdot p(x_n) \geq 1$$

Then the sequence $\boldsymbol{u} = \{ u_n; n = 1, 2, 3, \ldots\}$ where $u_{2n} = z_n$ and $z_{2n-1} = 0$ $(n = 1, 2, 3, \ldots)$ is by Lemma 3.12, a Cauchy sequence with respect to q. However, it is not a Cauchy sequence with respect to p because

$$p(u_{2n} - z_{2n-1}) \geq 1$$

This means that seminorms q and p are not compatible.

Similar arguments show that seminorms q and p are not compatible when it is not true that a seminorm p is stronger than a seminorm q. Thus, by *reductio ad absurdum*, compatibility of seminorms is sufficient for their equivalence.

Theorem is proved.

Corollary 3.28. Any norms q and p in a vector space L are equivalent if and only if they are compatible.

Normed and seminormed vector spaces are naturally related to another topological construction, which is called a metric space. Let X be a set.

Definition 3.18. a) A mapping $\mathbf{d}$: $X \times X \to \boldsymbol{R}^+$ is called a *metric* (or a *distance function*) if it satisfies the following axioms:

M1. $\mathbf{d}(x, y) = 0$ if and only if $x = y$.

M2. $\mathbf{d}(x, y) = \mathbf{d}(y, x)$ for all $x, y \in X$.

M3 (the triangle inequality)**.** $\mathbf{d}(x, y) \leq \mathbf{d}(x, z) + \mathbf{d}(z, y)$ for all $x, y, z \in X$.

b) A set X with a metric $\mathbf{d}$ is called a *metric space*.

c) The real number $\mathbf{d}(x, y)$ is called the *distance* between x and y in the metric space X.

For instance, the set of all complex numbers $\boldsymbol{C}$ is a metric space where $\mathbf{d}(x, y) = | (a + ib) - (d + ic) | = ((a - d)^2 + (b - c)^2)^{1/2}$ for all complex numbers $a + ib, d + ic \in \boldsymbol{C}$.

Definition 3.19. A mapping $\mathbf{d}$: $X \times X \to \boldsymbol{R}^+$ is called a ρ-*inframetric* (Fraigniaud, at al, 2008), also called *nearmetric* (Xia, 2008), if it satisfies Axioms M2, M3 and axiom:

IM3 (ρ-inframetric inequality). $\mathbf{d}(x, y) \leq \rho \cdot \max(\mathbf{d}(x, z), \mathbf{d}(z, y))$ for some positive real number ρ and all $x, y, z \in X$.

The ρ-inframetric inequality implies the ρ-relaxed triangle inequality (assuming the first axiom), and the ρ-relaxed triangle inequality implies the 2ρ-inframetric inequality.

Definition 3.20. A mapping $\mathbf{d}$: $X \times X \to \boldsymbol{R}^+$ is called an *ultrametric* (Lemin and Smirnov, 1986; Lemin, 2003) if it satisfies Axioms M2, M3 and axiom:

UM3 (ultrametric inequality). $\mathbf{d}(x, y) \leq \max(\mathbf{d}(x, z), \mathbf{d}(z, y))$ for some positive real number ρ and all $x, y, z \in X$.

Lemma 3.13. An ultranorm q in a vector space L induces an ultrametric $\mathbf{d}_q$ in this space.

Proof is left as an exercise.

Definition 3.21. a) A mapping $\mathbf{d}$: $X \times X \to \boldsymbol{R}^+$ is called a *pseudometric* also called a *semimetric* (Howes, 1995) if it satisfies Axioms M2, M3 and axiom:

$$\textbf{M1a.}\ \mathbf{d}(x, x) = 0.$$

b) A set X with a pseudometric $\mathbf{d}$ is called a *pseudometric space*.

c) The real number $\mathbf{d}(x, y)$ is called the *distance* between x and y in the pseudometric space X.

Any metric space is a pseudometric space.

Pseudometrics are used to define uniform spaces. Namely, a family $\{f_i; i \in I\}$ of pseudometrics on a set X defines uniform structure as the least upper bound of the uniform structures defined by the individual pseudometrics f_i.

Definition 3.22. If X and Y are pseudometric (metric) spaces, then a mapping f: $X \to Y$ is called an *isometry* if it preserves the distance between points, i.e., $\mathbf{d}(x, y) = \mathbf{d}(f(x), f(y))$ for any $x, y \in X$.

In other words, isometry preserves the distance between elements of a pseudometric or metric space.

Definition 3.23. a) A mapping **d**: $X \times X \to \boldsymbol{R}^+$ is called a *quasimetric* if it satisfies Axioms M1 and M3.

b) A set X with a quasimetric **d** is called a *quasimetric space*.

c) The real number $\mathbf{d}(x, y)$ is called the *distance* between x and y in the quasimetric space X.

Any metric space is a quasimetric space.

Quasimetrics are common in real life. For instance, given a set X of mountain villages, the typical walking times between elements of X form a quasimetric because travel uphill takes longer than travel downhill.

Another example is taxicab geometry when there are one-way streets. In this case, it is possible that a path from point A to point B comprises a different set of streets than a path from B to A. Nevertheless, quasimetric is rarely used in mathematics, and its name is not entirely standardized.

It is possible to define a quasimetric on the set $\boldsymbol{R}$ of real numbers by the following rule:

$$d(x, y) = x - y \text{ if } x \geq y, \text{ and } d(x, y) = 1 \text{ otherwise}$$

Definitely $d(x, x) = x - x = 0$ and $d(x, y) = x - y = 0$ implies $x = y$ for any x and y. To prove M3, we need to consider all possible relations between three arbitrary real numbers x, y and z. Here are some of them.

When $x \geq z \geq y$, we have

$$x - y = (x - z) + (z - y)$$

When $y \geq z \geq x$, we have

$$1 < 1 + 1$$

When $z \geq y \geq x$, we have

$$1 \leq 1 + (z - y)$$

When $z \geq x \geq y$, we have

$$x - y = 1 + (z - y)$$

Other cases also give the necessary inequality.

Lemma 3.14. For any quasimetric d on X, it is possible to define a metric **d** on X such that it preserves relations between distances, i.e., if $d(x, y) \leq d(z, v))$ and $d(y, x) \leq d(v, z))$, then $\mathbf{d}(x, y) \leq \mathbf{d}(x, y)$.

Indeed, we can put

$$\mathbf{d}(x, y) = \frac{1}{2}(d(x, y) + d(y, x)).$$

And obtain the necessary properties.

Definition 3.24. a) A mapping $\mathbf{d}: X \times X \to \boldsymbol{R}^+$ is called a *semimetric* if it satisfies Axioms M1 and M2.

b) A set X with a semimetric **d** is called a *semimetric space*.

c) The real number $\mathbf{d}(x, y)$ is called the *distance* between x and y in the semimetric space X.

Any metric space is a semimetric space.

Norms, seminorms, metrics and their generalizations use non-negative real numbers for valuation of elements in vector spaces, rings and modules. A natural development of these structures brings in real hypernumbers as a natural extension of real numbers setting off hypernorms, hyperseminorms and hypermetrics as a more general and thus, more comprising topological structures.

Let $\boldsymbol{R}_\omega$ be the set of all real hypernumbers, L be a vector (linear) space over a normed (valuated) field $\boldsymbol{F}$, and $\boldsymbol{R}_\omega^+$ be the set of all non-negative real hypernumbers (Burgin, 2012).

Definition 3.25. a) A mapping $q: L \to \boldsymbol{R}_\omega^+$ is called a *hypernorm* if it satisfies the following conditions:

N1. For any x from L, $q(x) = 0$ if and only if $x = \mathbf{0}$.

N2. $q(ax) = |a| \cdot q(x)$ for any x from L and any element a from $\boldsymbol{F}$.

N3 (the triangle inequality or subadditivity).

$$q(x + y) \leq q(x) + q(y) \text{ for any } x \text{ and } y \text{ from } L$$

b) A vector space L with a hypernorm is called a *hypernormed vector space* or simply, a *hypernormed space*.

c) The real hypernumber $q(x)$ is called the *hypernorm* of the element x from the hypernormed space L.

Note that *norms* in vector spaces coincide with those hypernorms that take values only in the set of real numbers.

There are different hypernorms and hyperseminorms (cf., Definition 3.29). Note that norms and seminorms in function spaces are non-negative subadditive functionals, while hypernorms and hyperseminorms in function spaces are non-negative subadditive hyperfunctionals (Burgin, 2004). Here are some examples of such hypernormed function spaces.

Example 3.8. As it is proved by Burgin (2012), the set of all real hypernumbers $\boldsymbol{R}_\omega$ is a hypernormed space where the hypernorm $||\cdot||$ is defined by the following formula:

If α is a real hypernumber, i.e., $\alpha = \mathrm{Hn}(a_i)_{i\in\omega}$ with $a_i \in \boldsymbol{R}$ for all $i\in\omega$, then $||\ \alpha\ || = \mathrm{Hn}(|a_i|)_{i\in\omega}$.

Note that this hypernorm coincides with the conventional norm on real numbers but it is impossible get the same topology by means of a conventional finite norm.

Example 3.9. As it is proved inBurgin (2002), the set of all complex hypernumbers $\boldsymbol{C}_\omega$ of all complex hypernumbers is a hypernormed space where the hypernorm $||\cdot||$ is defined by the following formula:

If α is a complex hypernumber, i.e., $\alpha = \mathrm{Hn}(a_i)_{i\in\omega}$ with $a_i \in \boldsymbol{C}$ for all $i\in\omega$, then $||\ \alpha\ || = \mathrm{Hn}(|a_i|)_{i\in\omega}$.

Note that this hypernorm coincides with the conventional norm on complex numbers but it is impossible get the same topology by means of a conventional finite norm.

There are hypernormed spaces that are not normed spaces.

Example 3.10. Let us take the space $C(\boldsymbol{R}, \boldsymbol{R})$ of all continuous functions on the real line $\boldsymbol{R}$. The hypernorm $||\cdot||$ in this space is defined by the following rule. If f: $\boldsymbol{R}\to\boldsymbol{R}$, then

$$||f||_\infty = \mathrm{Hn}(c_n)_{n\in\omega}$$

where $c_n = \max\{|f(x)|;\ x \in [-n, n]\}$. So, the set $\mathrm{C}(\boldsymbol{R}, \boldsymbol{R})$ of all continuous real functions is a hypernormed space. At the same time, it is known that $\mathrm{C}(\boldsymbol{R}, \boldsymbol{R})$ is not a normed space (Robertson and Robertson, 1964).

Example 3.11. Let us take the space $L^1(\boldsymbol{R}, \boldsymbol{R})$ of all locally absolutely Riemann integrable real functions on the real line $\boldsymbol{R}$. We remind that a function is locally Riemann integrable if it is Riemann integrable on an arbitrary interval $[a, b]$. The hypernorm in this space is defined by the following formula

$$||f||_{L^1} = \mathrm{Hn}(a_n)_{n\in\omega}$$

where $a_n = \int_{-n}^{n} |f(x)|\mathrm{d}x$.

Example 3.12. Let us take the space $L^2(\boldsymbol{R}, \boldsymbol{R})$ of all locally Riemann square-integrable real functions on the real line $\boldsymbol{R}$. The hypernorm in this space is defined by the following formula

$$||f||_{L^2} = \text{Hn}(b_n)_{n\in\omega}$$

where $b_n = [\int_{-n}^{n} |f(x)|^2 dx]^{1/2}$.

Example 3.13. Let us consider the space $L^p(\boldsymbol{R}, \boldsymbol{R})$ of all locally Riemann p-integrable ($p \geq 1$) functions on the real line $\boldsymbol{R}$. The hypernorm in this space is defined by the following formula

$$||f||_{L^p} = \text{Hn}(d_n)_{n\in\omega}$$

where $d_n = [\int_{-n}^{n} |f(x)|^p dx]^{1/p}$.

Example 3.14. Let us consider the space $B(\boldsymbol{R}, \boldsymbol{R})$ of all locally bounded functions on the real line $\boldsymbol{R}$. We remind that a function is locally bounded if it is bounded on an arbitrary interval $[a, b]$. The hypernorm in this space is defined by the following formula

$$||f||_{\infty} = \text{Hn}(a_n)_{n\in\omega}$$

where $a_n = \sup \{|f(x)|; x \in [-n, n]\}$.

Example 3.15. Let us take the generalized Sobolev space $H^1(\boldsymbol{R}, \boldsymbol{R})$ of all differentiable functions on the real line $\boldsymbol{R}$ such that they are locally square integrable and their derivatives are locally square integrable. The hypernorm in this space is defined by the following formula

$$||f||_{L^1} = ||f||_{L^2} + ||f'||_{L^2} = \text{Hn}(c_n)_{n\in\omega}$$

where $c_n = [\int_{-n}^{n} |f(x)|^2 dx]^{1/2} + [\int_{-n}^{n} f'(x)|^2 dx]^{1/2}$.

These examples show that the theory of hypernumbers and extrafunctions allows extending classical function spaces to more general function spaces and at the same time, preserving many good properties of classical function spaces.

Let us consider hypernormed vector spaces L and M with hypernorms $||x||_L$ and $||x||_M$ correspondingly.

Proposition 3.5. A mapping f: $L \rightarrow M$ is continuous if $|| f(x)||_M < a||x||_L$ for any element x from L.

Proof. A mapping $f: L \to M$ is continuous if the image of a converging sequence in L is a converging sequence in M. Let us take a sequence $(x_i)_{i \in \omega}$ in L and assume

$$\lim_{i \to \infty} x_i = x$$

It means that

$$\lim_{i \to \infty} ||x_i - x||_L = 0$$

By the initial conditions, we have

$$\lim_{i \to \infty} ||f(x_i) - f(x)||_M \leq \lim_{i \to \infty} a(||x_i - x||_L) = a(\lim_{i \to \infty} ||x_i - x||_L) = 0$$

It means that the sequence $(f(x_i))_{i \in \omega}$ converges in M.

Proposition is proved.

Corollary 3.29. Hypernormed vector spaces L and M with hypernorms $||x||_L$ and $||x||_M$ correspondingly, are isomorphic if they are isomorphic as vector spaces, there is a one-to-one linear mapping $f: L \to M$, and there are positive numbers a and b such that $||x||_L < a||f(x)||_M$ and $||y||_M < b||f^{-1}(y)||_L$ for any element x from L and any element y from M.

Proof is left as an exercise.

Theorem 3.3. A real finite dimensional hypernormed vector space L is isomorphic to the space $\boldsymbol{R}^n$ with the standard norm if and only if one of the following conditions is valid:

(a) L does not have one-dimensional subspaces with the discrete topology.

(b) There are no one-dimensional subspaces of L that have, at least, one isolated point.

(c) The hypernorm in L is finite.

Proof. Necessity. At first, we show that if the hypernorm $||x||_L$ of an element x from L is infinite, then x is an isolated point in the one-dimensional subspace $G = \{rx; r \in \boldsymbol{R}\}$. It means that absence of elements with infinite hypernorm implies absence of discrete points in L. Indeed,

$$||x - rx||_L = ||(1 - r)x||_L = |1 - r| \cdot ||x||_L$$

Then by Lemma 2.16, $||x - rx||_L$ is also an infinite hypernumber. Therefore, there is no element from G that belongs to the neighborhood $O_1(x) = \{y; ||x - y||_L < 1\}$. Consequently (Kuratowski, 1966), x is an isolated element in G.

Because x is an arbitrary element from G, the space G has the discrete topology.

Thus, we have also demonstrated that if the condition (c) is not true, then conditions (a) and (b) are also false. At the same time, the space $\boldsymbol{R}^n$ satisfies conditions (a), (b) and (c). Thus, if the space L is isomorphic to $\boldsymbol{R}^n$, it has also satisfy satisfies conditions (a), (b) and (c). This demonstrates necessity for each of these conditions and in particular, finiteness of the hypernorm in L.

Sufficiency. Let us assume that the dimension L of is n and the condition (c) is true for L. It means that the hypernorm in L is finite. Taking a basis $e_1, e_2, e_3, \dots, e_n$ of the vector space L, we can represent any element x from L in the form

$$x = a_1e_1 + a_2e_2 + a_3e_3 + \dots + a_ne_n$$

where all a_i are real numbers.

This allows us to assign the element $\underline{x} = (a_1, a_2, a_3, \dots, a_n)$ from $\boldsymbol{R}^n$ to the element x from L. This provides us with the one-to-one mapping f from the space L to the space $\boldsymbol{R}^n$. By definitions, this mapping f is an isomorphism of vector spaces and we need only to show that f is a homeomorphism, i.e., f and its inverse are continuous.

By properties of hypernorms and Cauchy inequality (Hardy et al., 1952), we have

$$\|x\|_L = \|a_1e_1 + a_2e_2 + a_3e_3 + \dots + a_ne_n\|_L \leq$$

$$|a_1| \cdot \|e_1\|_L + |a_2| \cdot \|e_2\|_L + |a_3| \cdot \|e_3\|_L + \dots + |a_n| \cdot \|e_n\|_L \leq$$

$$(\|e_1\|_L^2 + \|e_2\|_L^2 + \|e_3\|_L^2 + \dots + \|e_n\|_L^2)^{1/2} \cdot (a_1^2 + a_2^2 + a_3^2 + \dots + a_n^2)^{1/2}$$

Note that the operations of taking squares and square roots are well defined for finite hypernumbers (Burgin, 2012).

Thus, it is proved that the hypernorm in L is finite. Consequently (cf. Chapter 2), there is a positive number k such that

$$(\|e_1\|_L^2 + \|e_2\|_L^2 + \|e_3\|_L^2 + \dots + \|e_n\|_L^2)^{1/2} \leq k$$

As a result,

$$\|x\|_L \leq k \cdot (a_1^2 + a_2^2 + a_3^2 + \dots + a_n^2)^{1/2} = k \cdot \|\underline{x}\|$$

By Proposition 3.5, the mapping f^{-1} is continuous.

Let us consider the unit sphere S^n in $\boldsymbol{R}^n$:

$$S^n = \{\underline{x} = (a_1, a_2, a_3, \ldots, a_n);\ (a_1^2 + a_2^2 + a_3^2 + \ldots + a_n^2)^{1/2} \leq 1\}$$

We define the following function

$$f(\underline{x}) = f(a_1, a_2, a_3, \ldots, a_n) = \|x\|_L = \|a_1e_1 + a_2e_2 + a_3e_3 + \ldots + a_ne_n\|_L$$

As $\underline{x} \in S^n$, not all a_n are equal to 0. Consequently, x is not equal to $\mathbf{0}$. As a result,

$$f(\underline{x}) = \|x\|_L > 0$$

for all $\underline{x}$ from S^n. As S^n is a compact space, by the Weierstrass theorem, f reaches its minimum r on S^n and $r > 0$. This gives us

$$f(\underline{x}) = \|x\|_L = \|\underline{x}\| \cdot \|(a_1e_1)/(a_1^2 + a_2^2 + a_3^2 \ldots + a_n^2)^{1/2} + (a_2e_2)/(a_1^2 + a_2^2 + a_3^2 + \ldots + a_n^2)^{1/2} + \ldots + (a_ne_n)/(a_1^2 + a_2^2 + a_3^2 \ldots + a_n^2)^{1/2}\| \geq r \cdot \|\underline{x}\|$$

Thus, by Proposition 3.5, the mapping f is continuous.

Theorem is proved.

Corollary 3.30. A real finite dimensional normed vector space L is isomorphic to the space $\boldsymbol{R}^n$ with the standard norm.

Theorem 3.4. A complex finite dimensional hypernormed vector space L is isomorphic to the space $\boldsymbol{C}^n$ with the standard norm if and only if one of the following conditions is valid:

(a) L does not have one-dimensional subspaces with the discrete topology.
(b) There are no one-dimensional subspaces of L that have, at least, one isolated point.
(c) The hypernorm in L is finite.

Proof is similar to the proof of Theorem 3.3.

Corollary 3.31. A complex finite dimensional normed vector space L is isomorphic to the space $\boldsymbol{C}^n$ with the standard norm.

The proof of Theorem 3.3 gives us the following result.

Lemma 3.15. For a real (complex) hypernormed vector space L, the following conditions are equivalent:

(a) L does not have one-dimensional subspaces with the discrete topology.
(b) There are no one-dimensional subspaces of L that have, at least, one isolated point.
(c) The hypernorm in L is finite.

Definition 3.26. A mapping p: $L \rightarrow \boldsymbol{R}_{\omega}$ is called a *hyperpseudonorm* if it satisfies Axioms N1, N3 and the following condition:

PN2. $q(ax) \leq q(x)$ for any x from L and any number a from $\boldsymbol{R}$ with $\|a\| \leq 1$.

Any hypernorm, pseudonorm or norm is a hyperpseudonorm.

Lemma 3.16. If q: $L \rightarrow \boldsymbol{R}_{\omega}$ is a hyperpseudonorm, then $q(x) \geq 0$ and $q(-x) = q(x)$ for all $x \in L$.

Proof. By PN2, we have $q(-x) = q(x)$. At the same time, by N3, we have

$$q(x) + q(-x) \geq q(x + (-x)) = q(\mathbf{0})$$

and by N1, we have

$$q(\mathbf{0}) = 0$$

Thus, $q(x) + q(-x) = q(x) + q(x) = 2q(x) \geq 0$ and $q(x) \geq 0$.

Lemma is proved.

Hypernormed spaces are also hypermetric spaces.

Definition 3.27. a) A mapping $\mathbf{d}$: $X \times X \rightarrow \boldsymbol{R}_{\omega}^{+}$ is called a *hypermetric* (or a *hyperdistance function*) in a set X if it satisfies the following axioms:

M1. For any x and y from X, $\mathbf{d}(x, y) = 0$ if and only if $x = y$.

M2 (Symmetry)**.** $\mathbf{d}(x, y) = \mathbf{d}(y, x)$ for all $x, y \in X$.

M3 (the triangle inequality or subadditivity)**.**

$$\mathbf{d}(x, y) \leq \mathbf{d}(x, z) + \mathbf{d}(z, y) \text{ for all } x, y, z \in X.$$

b) A set X with a hypermetric $\mathbf{d}$ is called a *hypermetric space*.

c) The real hypernumber $\mathbf{d}(x, y)$ is called the *distance* between x and y in the hypermetric space X.

Note that the distance between two elements in a hypermetric space can be a real number, finite hypernumber or infinite hypernumber. When the distance between two elements of X is always a real number, $\mathbf{d}$ is a metric.

Proposition 3.6. a) A hyperpseudonorm q in a vector space L induces a hypermetric $\mathbf{d}_q$ in this space.

b) If q is a pseudonorm in L, then $\mathbf{d}_q$ is a metric.

Proof. If q: $X \rightarrow \boldsymbol{R}_{\omega}^{+}$ is a hyperpseudonorm in L and x and y are elements from L, then we can define $\mathbf{d}_q(x, y) = q(x - y)$. Properties of a hyperpseudonorm imply that $\mathbf{d}_q$ satisfies all axioms M1- M3.

Indeed, Axioms N1 for q implies Axioms M1 for $\mathbf{d}_q$. As by Lemma 3.16, $q(x - y) = q(-(x - y)) = q(y - x)$, Axiom M1 is true for $\mathbf{d}_q$. In addition, by definition and Axiom N3, we have

$$\mathbf{d}_q(x, y) = = q(x - y) \leq q(x - z) + q(z - y) = q(x - y) = \mathbf{d}_q(x, z) + \mathbf{d}_q(z, y)$$

The statement (b) directly follows from definitions.

Proposition is proved.

Corollary 3.32. a) A hypernorm q in a vector space L induces a hypermetric $\mathbf{d}_q$ in this space.

b) If q is a norm in L, then $\mathbf{d}_q$ is a metric.

Results from (Burgin, 2012) and Proposition 3.6 imply the following result (cf. Examples 3.1 and 3.2).

Corollary 3.33. $\boldsymbol{R}_\omega$ and $\boldsymbol{C}_\omega$ are hypermetric spaces.

Moreover, all spaces from Examples 3.1 – 3.8 are hypermetric spaces.

It is interesting to find what hypermetrics in vector spaces are induced by hypernorms and what metrics in vector spaces are induced by norms. To do this, let us consider additional properties of hypermetrics and metrics.

Definition 3.28. A hypermetric (metric) in a vector space L is called *linear* if it satisfies the following axioms:

LM1 (Translation invariance). $\mathbf{d}(x + z, y + z) = \mathbf{d}(x, y)$ for any $x, y, z \in L$.

LM2 (Multiplication invariance). $\mathbf{d}(ax, ay) = |a|\cdot\mathbf{d}(x, y)$ for all $x, y \in L$ and $a \in \boldsymbol{F}$.

Example 3.16. Let us take the space of all real numbers $\boldsymbol{R}$ as the space L. The natural metric in this space is defined as $\mathbf{d}(x, y) = |x - y|$. This metric is linear. Indeed,

$$\mathbf{d}(x + z, y + z) = |(x + z) - (y + z)| = |x - y| = \mathbf{d}(x, y)$$

and

$$\mathbf{d}(ax, ay) = |ax - ay| = |a(x - y)| = |a|\cdot|x - y| = |a|\cdot\mathbf{d}(x, y)$$

Example 3.17. Let us take the two-dimensional real vector space $\boldsymbol{R}^2$ as the space L. The natural metric in this space is defined by the conventional formula

If $x = (x_1, x_2)$ and $y = (y_1, y_2)$, then $\mathbf{d}(x, y) = \sqrt{(x_1 - y_1)^2 + (x_2 - y_2)^2}$

This metric is also linear. Indeed,

$\mathbf{d}(x + z, y + z) = \sqrt{((x_1 + z_1)^2 - (y_1 + z_1))^2 + (x_2 - y_2)^2} =$

$$\sqrt{(x_1 - y_1)^2 + (x_2 - y_2)^2} = \mathbf{d}(x, y)$$

and

$$\mathbf{d}(ax, ay) = \sqrt{(ax_1 - ay_1)^2 + (ax_2 - ay_2)^2} = \sqrt{a^2(x_1 - y_1)^2 + a^2(x_2 - y_2)^2} =$$

$$|a|\sqrt{(x_1 - y_1)^2 + (x_2 - y_2)^2} = |a|\cdot\mathbf{d}(x, y)$$

Example 3.18. Let us take the two-dimensional real vector space $\boldsymbol{R}^2$ as the space L. The natural metric in this space is defined by the following formula

If $x = (x_1, x_2)$ and $y = (y_1, y_2)$, then $\mathbf{d}(x, y) = |x_1 - y_1|^{1/2} + |x_2 - y_2|^{1/2}$

This is a metric. Indeed, $\mathbf{d}(x, x) = |x_1 - x_1|^{1/2} + |x_2 - x_2|^{1/2} = 0$ and $\mathbf{d}(x, y) = \mathbf{d}(y, x)$. Besides, for any real numbers a, b and c, we have

$$|a - b|^{1/2} = |a - c + c - b|^{1/2} \leq |a - c|^{1/2} + |c - b|^{1/2}$$

Consequently, we have

$$\mathbf{d}(a, b) \leq \mathbf{d}(a, c) + \mathbf{d}(c, b)$$

However, this metric is not linear. Indeed, let us take $x = (5, 5)$, $y = (1, 1)$, and $a = 2$. Then $\mathbf{d}(x, y) = 2 + 2 = 4$, while $\mathbf{d}(2x, 2y) = 2\sqrt{2} + 2\sqrt{2} = 4\sqrt{2}$.

These examples show that there are linear metrics (hypermetrics) in vector spaces and there are metrics (hypermetrics) in vector spaces that are not linear. The majority of popular metrics are induced by norms and thus, they are linear as the following result demonstrates.

Theorem 3.5. A hypermetric $\mathbf{d}$ is induced by a hypernorm if and only if $\mathbf{d}$ is linear.

Proof. Necessity. Let us consider a vector space L with a hypernorm q. By Pr4oposition 3.5, q induces the hypermetric $\mathbf{d}_q(x, y) = q(x - y)$. Then $\mathbf{d}_q(x + z, y + z) = q((x + z) - (y + z)) = q(x - y) = \mathbf{d}_q(x, y)$, i.e., Axiom LM1 is true. In addition, $\mathbf{d}_q(ax, ay) = q(ax - ay) = q(a(x - y)) = |a|\cdot q(x - y) = |a|\cdot\mathbf{d}_q(x, y)$, i.e., Axiom LM2 is also true.

Necessity. Let us consider a vector space L with a linear hypermetric $\mathbf{d}$. We define the hypernorm $q_{\mathbf{d}}$ by the following formula

$$q_{\mathbf{d}}(x) = \mathbf{d}(\mathbf{0}, x)$$

We show that $q_{\mathbf{d}}$ is a hypernorm. Indeed, $q_{\mathbf{d}}(\mathbf{0}) = \mathbf{d}(\mathbf{0},\mathbf{0}) = 0$. Besides, if $q_{\mathbf{d}}(x) = \mathbf{d}(\mathbf{0}, x) = 0$, then $x = \mathbf{0}$ by Axiom M1. This gives us Axiom N1 for $q_{\mathbf{d}}$.

In addition,

$$q_{\mathbf{d}}(ax) = \mathbf{d}(\mathbf{0}, ax) = \mathbf{d}(a\mathbf{0}, ax) = \mathbf{d}(a(\mathbf{0}, x)) = |a|\cdot\mathbf{d}(\mathbf{0}, x) = |a|\cdot q_{\mathbf{d}}(x)$$

by Axiom LM2. This gives us Axiom N2 for $q_{\mathbf{d}}$.

Likewise, by Axioms M3 and LM1, we have

$$q_{\mathbf{d}}(x+y) = \mathbf{d}(\mathbf{0}, x+y) \leq \mathbf{d}(\mathbf{0}, x) + \mathbf{d}(x, x+y) = \mathbf{d}(\mathbf{0}, x) + \mathbf{d}(\mathbf{0}, y) = q_{\mathbf{d}}(x) + q_{\mathbf{d}}(y)$$

This gives us the triangle inequality (Axiom N3) for $q_{\mathbf{d}}$.

Theorem is proved.

Corollary 3.34. A metric **d** is induced by a norm if and only if **d** is linear.

Theorem 3.6. a) Any linear hypermetric **d** in a vector space L induces a hypernorm q in L.

b) If **d** is a metric, then q is a norm.

Proof. Let us consider a vector space L with a linear hypermetric **d**. We define $q(x) = \mathbf{d}_q(x, 0)$ and check the axioms of hypernorm. Indeed, $q(\mathbf{0}) = \mathbf{d}(\mathbf{0},\mathbf{0}) = 0$. Besides, if $q(x) = \mathbf{d}(\mathbf{0}, x) = 0$, then $x = \mathbf{0}$ by Axiom M1. This gives us Axiom N1 for q.

In addition,

$$q(ax) = \mathbf{d}(\mathbf{0}, ax) = \mathbf{d}(a\mathbf{0}, ax) = \mathbf{d}(a(\mathbf{0}, x)) = |a| \cdot \mathbf{d}(\mathbf{0}, x) = |a| \cdot q(x)$$

by Axiom LM2. This gives us Axiom N2 for q.

Likewise, by Axioms M3 and LM1, we have

$$q(x+y) = \mathbf{d}(\mathbf{0}, x+y) \leq \mathbf{d}(\mathbf{0}, x) + \mathbf{d}(x, x+y) = \mathbf{d}(\mathbf{0}, x) + \mathbf{d}(\mathbf{0}, y) = q(x) + q(y)$$

This gives us the triangle inequality (Axiom N3) for q.

If **d** is a metric, it takes values only in $\boldsymbol{R}$. Thus, q also takes values only in $\boldsymbol{R}$ and is a norm.

Theorem is proved.

Taking only a part of the hypernorm properties, we come to the concept of a hyperseminorm.

Definition 3.29.

a) A mapping $q: L \to \boldsymbol{R}_\omega$ is called a *hyperseminorm* if it satisfies the following conditions:

N1. $q(\mathbf{0}) = 0$

N2. $q(ax) = |.a| \cdot q(x)$ for any x from L and any number a from $\boldsymbol{R}$.

N3 (the triangle inequality or subadditivity).

$$q(x+y) \leq q(x) + q(y) \text{ for any } x \text{ and } y \text{ from } L$$

b) A vector space L with a norm is called a *hyperseminormed vector space* or simply, a *hyperseminormed space*.
c) The real hypernumber $q(x)$ is called the *hyperseminorm* of an element x from the hyperseminormed space L.
d) A set $X \subseteq L$ is called *q-bounded* if there is a positive real number h such that for any element a from X, the inequality $q(a) < h$ is true.
e) A set $X \subseteq L$ is called *weakly q-bounded* if there is a positive real hypernumber α such that for any element a from X, the inequality $q(a) < \alpha$ is true.

Lemma 3.17. If $q: L \to \boldsymbol{R}_\omega$ is a hyperseminorm, then $q(x) \geq 0$ for all $x \in L$.

Proof. By N3, we have

$$q(x) + q(-x) \geq q(x + (-x)) = q(\mathbf{0})$$

At the same time, by N2, we have $q(\mathbf{0}) = 0 \cdot q(\mathbf{0}) = 0$ and $q(-x) = q(x)$. Thus, $q(x) + q(-x) = q(x) + q(x) = 2q(x) \geq 0$ and $q(x) \geq 0$.

Lemma is proved.

So, it is possible to define hyperseminorms as mappings $q: L \to \boldsymbol{R}_\omega^{\ +}$.

Corollary 3.35. For any element $x \in L$, the spectrum Spec $q(x)$ [cf., (Burgin, 2012)] does not contain negative real numbers.

Note that any seminorm is a hyperseminorm that takes values only in the set of real numbers.

Proposition 3.7. A mapping $q: L \to \boldsymbol{R}_\omega^{\ +}$ is a hyperseminorm if and only if one of the following conditions (a) or (b) is true:

(a) $q(ax + by) \leq |a|q(x) + |b|q(y)$ for any x and y from L and any numbers a and b from $\boldsymbol{R}$.
(b) $q(tx + (1 - t)y) \leq tq(x) + (1 - t)q(y)$ and $q(ax) = |a| \cdot q(x)$ for any x and y from L and any numbers t from the interval $[0, 1]$ and a from $\boldsymbol{R}$.

Proof is left as an exercise.

There are different hyperseminorms.

Example 3.19. Let us take the space $C(\boldsymbol{R}, \boldsymbol{R})$ of all continuous functions on the real line $\boldsymbol{R}$. Taking any natural number n, we obtain the following hyperseminorm

$$\|f\| = \text{Hn}(\|f\|_n)_{n \in \omega}$$

where

$$\|f\|_n = \max \{|f(x)|;\ x \in [-n, n]\}$$

Example 3.20. Let us take the space $L^1(\boldsymbol{R}, \boldsymbol{R})$ of all locally absolutely Riemann integrable real functions on the real line $\boldsymbol{R}$. We remind that a function is locally Riemann integrable if it is Riemann integrable on an arbitrary interval $[a, b]$. Taking any natural number n, we obtain the following hyperseminorm

$$||f||_{L^1} = \mathrm{Hn}(||f||_{n_{L^1}})_{n\in\omega}$$

where

$$||f||_{n_{L^1}} = \int_{-n}^{n} |f(x)|\mathrm{d}x$$

Example 3.21. Let us take the space $L^2(\boldsymbol{R}, \boldsymbol{R})$ of all locally Riemann square-integrable real functions on the real line $\boldsymbol{R}$. The hypernorm in this space is defined by the following formula

$$||f||_{L^2} = \mathrm{Hn}(||f||_{n_{L^2}})_{n\in\omega}$$

where

$$||f||_{n_{L^2}} = [\int_{-n}^{n} |f(x)|^2\mathrm{d}x]^{1/2}$$

Example 3.22. Let us consider the space $L^p(\boldsymbol{R}, \boldsymbol{R})$ of all locally Riemann p-integrable ($p \geq 1$) functions on the real line $\boldsymbol{R}$. The hypernorm in this space is defined by the following formula

$$||f||_{L^p} = \mathrm{Hn}(d_n)_{n\in\omega}$$

where

$$d_n = [\int_{-n}^{n} |f(x)|^p\mathrm{d}x]^{1/p}$$

Example 3.23. Let us take $B(\boldsymbol{R}, \boldsymbol{R})$ is the space of all locally bounded functions on the real line $\boldsymbol{R}$. We remind that a function is locally bounded if it is bounded on an arbitrary interval $[a, b]$. The hypernorm in this space is defined by the following formula

$$||f||_\infty = \mathrm{Hn}(a_n)_{n\in\omega}$$

where $a_n = \sup \{|f(x)|; x \in [-n, n]\}$

Example 3.24. Let us consider the generalized Sobolev space $H^1(\boldsymbol{R}, \boldsymbol{R})$ of all differentiable functions on the real line $\boldsymbol{R}$ such that they are locally square integrable and their derivatives are locally square integrable. The hypernorm in this space is defined by the following formula

$$\|f\|_{H^1} = \|f\|_{L^2} + \|f'\|_{L^2} = \mathrm{Hn}([\int_{-n}^{n} |f(x)|^2 dx]^{1/2})_{n\in\omega} + \mathrm{Hn}([\int_{-n}^{n} |f'(x)|^2 dx]^{1/2})_{n\in\omega}$$

These examples show that the theory of hypernumbers and extrafunctions allows extending classical function spaces to more general function spaces while preserving, at the same time, many useful properties of classical function spaces.

Proposition 3.8. If $q: L\to\boldsymbol{R}$ is a hyperseminorm, then it has the following properties:

(1) $q(x - y) = q(y - x)$ for any $x, y \in L$.

(2) $q(\mathbf{0}) = 0$.

(3) $|q(x) - q(y)| \leq q(x - y)$ for any $x, y \in L$.

(4) $|q(x) - q(y)| \leq q(x + y)$ for any $x, y \in L$.

Proof.

(1) By Axiom N2, we have

$$q(x - y) = q(-(y - x)) = |-1| \cdot q(y - x) = q(y - x)$$

(2) By Axiom N2, we have

$$q(\mathbf{0}) = q(0\cdot\mathbf{0}) = |0| \cdot q(\mathbf{0}) = 0$$

(3) By Axiom N3, we have

$$q(x) = q(x - y + y) \leq q(x - y) + q(y)$$

Thus,

$$q(x) - q(y) \leq q(x - y)$$

As q is symmetric (property (2)), we have

$$q(y) - q(x) \leq q(x - y)$$

Consequently,

$$|q(x) - q(y)| \le q(x - y)$$

Property (4) is a consequence of property (3).

Proposition is proved.

Corollary 3.36. If q: $L \to \boldsymbol{R}$ is a hypernorm (seminorm or norm), then it has the following properties:

(1) $q(x) \ge 0$ for any $x \in L$.

(2) $q(x - y) = q(y - x)$ for any $x, y \in L$.

(3) $q(\mathbf{0}) = 0$.

(4) $|q(x) - q(y)| \le q(x - y)$ for any $x, y \in L$.

(5) $|q(x) - q(y)| \le q(x + y)$ for any $x, y \in L$.

Lemma 3.18. Addition is continuous in vector spaces.

Proof is left as an exercise.

Lemma 3.19. Scalar multiplication is continuous with respect to the second argument from L in a hyperseminormed vector space L.

Proof is left as an exercise.

Let us assume that L is a linear algebra over $\boldsymbol{R}$.

Definition 3.30. a) The linear algebra L with a hyperseminorm q is called a (*submultiplicative*) *hyperseminormed algebra* if it satisfies Axioms SN1, N2, N3, and SN4.

b) The linear algebra L with a hyperseminorm q is called a *multiplicative algebra*, or *hypercomposition algebra* if it satisfies Axioms SN1, N2, N3, and N4.

For instance, the set of all continuous real functions form a hyperseminormed algebra over the field $\boldsymbol{R}$.

Note that any multiplicative hyperseminormed algebra is a submultiplicative hyperseminormed algebra, while any multiplicative (submultiplicative) seminormed algebra is a multiplicative (submultiplicative) hyperseminormed algebra.

It is important to understand that while multiplication of norms and seminorms coincides with multiplication of real numbers and thus, is defined in a unique way, multiplication of hypernorms and hyperseminorms coincides with multiplication of real hypernumbers, while this operation is not straightforward and has different definitions (Burgin, 2005a, 2015).

In the same way, it is possible to define hypernormed algebras.

Definition 3.31. a) The linear algebra L with a hypernorm q is called a (*submultiplicative*) *hypernormed algebra* if it satisfies Axioms N1, N2, N3, and SN4.

b) The linear algebra L with a hypernorm q is called a *multiplicative hypernormed algebra*, or *hypercomposition algebra*, if it satisfies Axioms N1, N2, N3, and N4.

For instance, algebras $\boldsymbol{R}_\omega$ and $\boldsymbol{C}_\omega$ are multiplicative hypernormed (hypercomposition) algebras over the field $\boldsymbol{R}$.

Note that any multiplicative hypernormed algebra is a submultiplicative hypernormed algebra, while any multiplicative (submultiplicative) normed algebra is a multiplicative (submultiplicative) hypernormed algebra.

Let us introduce and study operations with hypernorms and hyperseminorms.

Taking a vector space L over a topological (normed) field $\boldsymbol{F}$ with a system Q of hypernorms (hyperseminorms), we introduce the following operations:

Addition: if $||.||_1$ and $||.||_2$ belong to Q, then for any element v from L, we define their sum $||.|| = ||.||_1 + ||.||_2$ by the following rule

$$||v|| = ||v||_1 + ||v||_2$$

Scalar multiplication by nonnegative numbers from $\boldsymbol{R}^+$: if $||.||_1$ belongs to Q and a is a nonnegative number from $\boldsymbol{R}^+$, then for any element v from L, we define their scalar multiplication $a||.|| = ||.||_1$ by the following rule

$$||v||_1 = a||v||$$

Taking *maximum*: if $||.||_1$ and $||.||_2$ belong to Q, then for any element v from L, we define their maximum $||.|| = \max(||.||_1, ||.||_2)$ by the following rule

$$||v|| = \max(||v||_1, ||v||_2)$$

Taking *minimum*: if $||.||_1$ and $||.||_2$ belong to Q, then for any element v from L, we define their minimum $||.|| = \min(||.||_1, ||.||_2)$ by the following rule

$$||v|| = \min(||v||_1, ||v||_2)$$

We need the following property of maxima and minima in an arbitrary ordered additive semigroup K.

Lemma 3.20. For any elements x, y, u and w from K, we have:

(a) $\max(x + y, u + w) \leq \max(x, u) + \max(y, w)$

(b) $\min(x + y, u + w) \geq \min(x, u) + \min(y, w)$

Proof. (a) Let us assume that $u \leq x$ and $w \leq y$. Then $\max(x + y, u + w) = x + y$, $\max(x, u) = x$ and $\max(y, w) = y$. Consequently, $\max(x + y, u + w) \leq \max(x, u) + \max(y, w) = x + y$.

Let us assume that $u \leq x$ and $y \leq w$. Then $\max(x, u) = x$ and $\max(y, w) = w$. There are two possibilities: $\max(x + y, u + w) = x + y$ and $\max(x + y, u + w) = u + w$. In the first case, $\max(x, u) + \max(y, w) = x + w \geq x + y$ because $y \leq w$. In the second case, $\max(x, u) + \max(y, w) = x + w \geq u + w$ because $u \leq x$. Consequently, $\max(x + y, u + w) \leq \max(x, u) + \max(y, w)$.

All other possibilities ($x \leq u$ and $y \leq w$ or $x \leq u$ and $w \leq y$) are treated in a similar way.

(b) Let us assume that $u \leq x$ and $w \leq y$. Then $\min(x + y, u + w) = u + w$, $\min(x, u) = u$ and $\min(y, w) = w$. Consequently, $\min(x + y, u + w) \leq \min(x, u) + \min(y, w) = u + w$.

Let us assume that $u \leq x$ and $y \leq w$. Then $\min(x, u) = u$ and $\min(y, w) = y$. There are two possibilities: $\min(x + y, u + w) = x + y$ and $\min(x + y, u + w) = u + w$. In the first case, $\min(x, u) + \min(y, w) = u + y \leq x + y$ because $u \leq x$. In the second case, $\min(x, u) + \min(y, w) = u + y \leq u + w$ because $y \leq w$. Consequently, $\min(x + y, u + w) \geq \min(x, u) + \min(y, w)$.

All other possibilities ($x \leq u$ and $y \leq w$ or $x \leq u$ and $w \leq y$) are treated in a similar way.

Lemma is proved.

Proposition 3.9. a) If $\|.\|_1$ and $\|.\|_2$ are hypernorms (hyperseminorms), then their sum $\|.\| = \|.\|_1 + \|.\|_2$ is a hypernorm (hyperseminorm).

b) If $\|.\|_1$ is a hyperseminorm (hypernorm) and a is a nonnegative number from $\boldsymbol{R}^+$, then their scalar multiplication $a\|.\| = \|.\|_1$ is a hyperseminorm (a hypernorm when $a > 0$).

c) If $\|.\|_1$ and $\|.\|_2$ are hypernorms (hyperseminorms), then their maximum $\|.\| = \max(\|.\|_1, \|.\|_2)$ is a hypernorm (hyperseminorm).

Proof. We need to check axioms SN1 (N1), N2 and N3 for new functions.

(a) SN1: $\|\mathbf{0}\| = \|\mathbf{0}\|_1 + \|\mathbf{0}\|_2 = 0 + 0 = 0$ because $\|.\|_1$ and $\|.\|_2$ are hyperseminorms.

N2:

$$\|av\| = \|av\|_1 + \|av\|_2 = a\|v\|_1 + a\|v\|_2 = a(\|v\|_1 + \|v\|_2) = a\|v\|$$

for any v from L and any nonnegative element a from $\boldsymbol{R}$ because $\|.\|_1$ and $\|.\|_2$ are hyperseminorms.

N3:

$$\|z + v\| = \| z + v \|_1 + \| z + v \|_2 \leq (\|z\|_1 + \|v\|_1) + (\|z\|_2 + \|v\|_2) = (\|z\|_1 + \|z\|_2) + (\|v\|_1 + \|v\|_2) = \|z\| + \|v\|$$

for any z and v from L because $||.||_1$ and $||.||_2$ are hyperseminorms.

N1: Let us assume that $||.||$ is a norm and $0 = ||v|| = ||v||_1 + ||v||_2 = 0 + 0$, then $v = \mathbf{0}$ because $||v||_1$ and $||v||_2$ are nonnegative and $||.||_1$ and $||.||_2$ are hypernorms.

(b) SN1: $||\mathbf{0}||_1 = a||\mathbf{0}|| = a0 = 0$ because $||.||$ is a hyperseminorm.

N2:

$$||bv||_1 = a||bv|| = ab||v|| = b(a(||v||) = b||v||_1$$

for any v from L and any nonnegative element b from $\boldsymbol{F}$ because $||.||$ is a hyperseminorm.

N3:

$$||z + v||_1 = a||\ z + v\ || = ||\ az + av\ || \leq ||az|| + ||av|| = a||z|| + a||v|| = ||z||_1 + ||v||_1$$

for any z and v from L because $||.||$ is a hyperseminorm and $a \geq 0$.

N1: Let us assume that $||.||$ is a norm and $0 = ||v||_1 = a||v|| = a \cdot 0 = 0$, then $v = \mathbf{0}$ because $||.||$ is a hypernorm and $a > 0$.

(c) SN1: $||\mathbf{0}|| = \max(||\mathbf{0}||_1, ||\mathbf{0}||_2) = \max(0, 0) = 0$ because $||.||_1$ and $||.||_2$ are hyperseminorms.

N2: $||av|| = \max(||av||_1, ||av||_2) = \max(a||v||_1, a||v||_2) = a \cdot \max(||v||_1, ||v||_2) = a||v||$ for any v from L and any nonnegative element a from $\boldsymbol{R}$ because $||.||_1$ and $||.||_2$ are hyperseminorms.

N3: By Lemma 3.20, we have

$$||z + v|| = \max(||\ z + v\ ||_1, ||\ z + v\ ||_2) \leq \max[(||z||_1 + ||v||_1), (||z||_2 + ||v||_2)] \leq \max(||z||_1, ||z||_2) + \max(||v||_1, ||v||_2) = ||z|| + ||v||$$

for any z and v from L because $||.||_1$ and $||.||_2$ are hyperseminorms.

N1: Let us assume that $||.||$ is a norm and $0 = ||v|| = \max(||v||_1, ||v||_2)$, then $||v||_1 = ||v||_2 = 0$ and $v = \mathbf{0}$ because $||.||_1$ and $||.||_2$ are hypernorms.

Proposition is proved.

Note that if $a = 0$, then $a||.||$ is only a hyperseminorm but not a hypernorm.

Corollary 3.37. If $||.||_1$ is a hypernorm and $||.||_2$ is a hyperseminorms, then their sum $||.|| = ||.||_1 + ||.||_2$ is a hypernorm.

Corollary 3.38. Any linear combination with non-negative coefficients of hyperseminorms (hypernorms) in L is a hyperseminorms (hypernorm).

Corollary 3.39. a) All hyperseminorms in L form a semimodule over the semiring $\boldsymbol{R}^+$.

b) All hypernorms in L form an S-module over the S-ring $\boldsymbol{R}^{++}$.

Corollary 3.40. a) If $||.||_1$ and $||.||_2$ are norms (seminorms), then their sum $||.|| = ||.||_1 + ||.||_2$ is a norm (seminorm).

b) If $||.||_1$ is a norm (seminorm) and a is a nonnegative (positive) element from F, (a nonnegative number from $\boldsymbol{R}$) then their scalar multiplication $a||.||$ $= ||.||_1$ is a seminorm (norm).

Indeed, by Proposition 3.9, $||.||_1 + ||.||_2$ and $a||.|| = ||.||_1$ are hyperseminorms. They take values in $\boldsymbol{R}$ and so, they are seminorms. Besides, if $||.||_1$ or $||.||_2$ is a norm, then is also a norm and if $a > 0$ and $||.||_1$ is a norm, then $a||.||$ is also a norm.

Note that if $a = 0$, then $a||.||$ is only a seminorm but not a norm.

There are different relations in sets of hypernorms and hyperseminorms.

Let L be a vector space over a normed field K, while q and p are two hyperseminorms (hypernorms) in L.

Definition 3.32. a) The hyperseminorm (hypernorm)q is *stronger* than the hyperseminorm (hypernorm) p, while p is *weaker* than q if there is a real number $k > 0$ such that for all $x \in L$,

$$p(x) \leq k \cdot q(x)$$

b) Hyperseminorms (hypernorms) q and $p$are *equivalent* if there are constants k, $h > 0$ such that for all $x \in L$,

$$h \cdot q(x) \leq p(x) \leq k \cdot q(x)$$

When the hyperseminorm (hypernorm) q is stronger than the hyperseminorm (hypernorm) p, it is denoted by

$$p \leq q$$

Lemma 3.21. The relation $\leq$ defines a partial preorder in sets of hyperseminorms (hypernorms).

Proof is left as an exercise.

This partial preorder in sets of hyperseminorms (hypernorms) is compatible with operations introduced above as the following result demonstrates.

Let us consider three hypernorms (hyperseminorms) q_1, q_2 and q_3.

Proposition 3.10. a) If $q_1 \leq q_2$, then $q_1 + q_3 \leq q_2 + q_3$ for any hypernorm (hyperseminorm) q_3.

b) If $q_1 \leq q_2$, then $a \cdot q_1 \leq a \cdot q_2$ for any nonnegative element a from $\boldsymbol{F}$.

c) If $q_1 \leq q_2$, then their maximum $\max(q_1, q_3) \leq \max(q_2, q_3)$ for any hypernorm (hyperseminorm) q_3.

d) If $q_1 \leq q_2$, then their minimum $\min(q_1, q_3) \leq \min(q_2, q_3)$ for any hypernorm (hyperseminorm) q_3.

Proof. (a) The inequality $q_1 \leq q_2$ means that there is a real number $k \geq 0$ such that for all $x \in L$,

$$q_1(x) \leq k \cdot q_2(x)$$

Consequently, as $k \geq 0$, we have

$$q_1(x) + q_3(x) \leq k \cdot q_2(x) + q_3(x) \leq k \cdot q_2(x) + (1 + k) \cdot q_3(x) \leq (1 + k) \cdot (q_1(x) + q_3(x))$$

It means that $q_1 + q_3 \leq q_2 + q_3$.

(b) In a similar way, $q_1(x) \leq k \cdot q_2(x)$ implies $(a \cdot q_1)(x) \leq k(a \cdot q_2)(x)$ for all $x \in L$. Thus, $a \cdot q_1 \leq a \cdot q_2$.

Statements (c) and (d) are proved in a similar way.

Proposition is proved.

Corollary 3.41. a) If $q_1 \leq q_2$, then $q_1 + q_3 \leq q_2 + q_3$ for any norm (seminorm) q_3.

b) If $q_1 \leq q_2$, then $a \cdot q_1 \leq a \cdot q_2$ for any nonnegative element a from $\boldsymbol{F}$.

c) If $q_1 \leq q_2$, then their maximum $\max(q_1, q_3) \leq \max(q_2, q_3)$ for any norm (seminorm) q_3.

d) If $q_1 \leq q_2$, then their minimum $\min(q_1, q_3) \leq \min(q_2, q_3)$ for any norm (seminorm) q_3.

Proposition 3.10 and Corollary 3.41 mean that addition, scalar multiplication, taking maximum or minimum are monotonous operations in the systems of hyperseminorms, hypernorms, seminorms and norms in vector spaces.

Corollary 3.42. a) If hyperseminorms (hypernorms) q_1 and q_2 are equivalent, then hyperseminorms (hypernorms) $q_1 + q_3$ and $q_2 + q_3$ are equivalent for any hypernorm (hyperseminorm) q_3.

b) If hyperseminorms (hypernorms) q_1 and q_2 are equivalent, then $a \cdot q_1$ and $a \cdot q_2$ are equivalent for any nonnegative element a from $\boldsymbol{F}$.

c) If hyperseminorms (hypernorms) q_1 and q_2 are equivalent, then their maximum $\max(q_1, q_3)$ and $\max(q_2, q_3)$ are equivalent for any hypernorm (hyperseminorm) q_3.

d) If hyperseminorms (hypernorms) q_1 and q_2 are equivalent, then their minimum $\min(q_1, q_3)$ and $\min(q_2, q_3)$ are equivalent for any hypernorm (hyperseminorm) q_3.

Corollary 3.43. a) If seminorms (norms) q_1 and q_2 are equivalent, then seminorms (norms) $q_1 + q_3$ and $q_2 + q_3$ are equivalent for any norm (seminorm) q_3.

b) If seminorms (norms) q_1 and q_2 are equivalent, then $a \cdot q_1$ and $a \cdot q_2$ are equivalent for any nonnegative element a from $\boldsymbol{F}$.

c) If seminorms (norms) q_1 and q_2 are equivalent, then their maximum $\max(q_1, q_3)$ and $\max(q_2, q_3)$ are equivalent for any norm (seminorm) q_3.

d) If seminorms (norms) q_1 and q_2 are equivalent, then their minimum $\min(q_1, q_3)$ and $\min(q_2, q_3)$ are equivalent for any norm (seminorm) q_3.

Proposition 3.10 and Corollary 3.43 mean that addition, scalar multiplication, taking maximum or minimum are monotonous operations in the systems of hyperseminorms, hypernorms, seminorms and norms in vector spaces.

Corollary 3.44. a) If hyperseminorms (hypernorms) q_1 and q_2 are equivalent, then hyperseminorms (hypernorms) $q_1 + q_3$ and $q_2 + q_3$ are equivalent for any hypernorm (hyperseminorm) q_3.

b) If hyperseminorms (hypernorms) q_1 and q_2 are equivalent, then $a \cdot q_1$ and $a \cdot q_2$ are equivalent for any nonnegative element a from $\boldsymbol{F}$.

Let us consider two hypernorms (hyperseminorms) q_1 and q_2.

Proposition 3.11. $q_1 \leq q_1 + q_2$ for any hypernorms (hyperseminorms) q_1 and q_2.

Proof. By the definition of a hypernorm (hyperseminorm), $q_1(x) \geq 0$ and $q_2(x) \geq 0$ for all $x \in L$. Thus, we have

$$q_1(x) \leq 2q_1(x) + 2q_2(x) = 2(q_1(x) + q_2(x))$$

for all $x \in L$. It means that $q_1 \leq q_1 + q_2$.

Proposition is proved.

Corollary 3.45. $q_1 \leq q_1 + q_2$ for any norms (seminorms) q_1 and q_2.

Proposition 3.12. Hyperseminorms q_1 and $a \cdot q_1$ are equivalent for any hyperseminorm q_1 and any positive element a from $\boldsymbol{F}$.

Indeed, if $1 \leq a$, then $q_1(x) \leq (a \cdot q_1)(x)$ and $(a \cdot q_1)(x) \leq a \cdot q_1(x)$ for all $x \in L$. When $a \leq 1$, then$(a \cdot q_1)(x) \leq q_1(x)$ and $q_1(x) \leq a^{-1} \cdot (a \cdot q_1(x)) = q_1(x)$ for all $x \in L$. Thus, hyperseminorms $q_1 \leq a \cdot q_1$ are equivalent.

Corollary 3.46. Seminorms (norms) q_1 and $a \cdot q_1$ are equivalent for any seminorm (norm) q_1 and any positive element a from $\boldsymbol{F}$.

Let us consider three hypernorms (hyperseminorms) $\|.\|_1$, $\|.\|_2$ and $\|.\|_3$.

Proposition 3.13. $\|.\|_1 \leq \|.\|_3$ and $\|.\|_2 \leq \|.\|_3$ if and only if $\|.\|_1 + \|.\|_2 \leq \|.\|_3$.

Proof. Necessity. The inequality $\|.\|_1 \leq \|.\|_3$ means that there is a real number $k \geq 0$ such that for all $x \in L$,

$$||x||_1 \leq k \cdot ||x||_3$$

and the inequality $||.||_2 \leq ||.||_3$ means that there is a real number $h \geq 0$ such that for all $x \in L$,

$$||x||_2 \leq h \cdot ||x||_3$$

Consequently, as $k, h \geq 0$, we have

$$||.||_1 + ||.||_2 \leq k \cdot ||.||_3 + h \cdot ||.||_3 \leq (k + h) \cdot ||.||_3$$

It means that $||.||_1 + ||.||_2 \leq ||.||_3$.

Sufficiency. Let us assume $||.||_1 + ||.||_2 \leq ||.||_3$. It means that there is a real number $k \geq 0$ such that for all $x \in L$,

$$||x||_1 + ||x||_2 \leq k \cdot ||x||_3$$

Consequently, we have

$$||x||_1 \leq ||x||_1 + ||x||_2 \leq k \cdot ||x||_3$$

and the inequality $||.||_2 \leq ||.||_3$ means that there is a real number $h \geq 0$ such that for all $x \in L$,

$$||x||_2 \leq ||x||_1 + ||x||_2 \leq k \cdot ||x||_3$$

This implies inequalities $||.||_1 \leq ||.||_3$ and $||.||_2 \leq ||.||_3$.

Proposition is proved.

Let us consider three norms (seminorms) $||.||_1$, $||.||_2$ and $||.||_3$.

Corollary 3.47. $||.||_1 \leq ||.||_3$ and $||.||_2 \leq ||.||_3$ if and only if $||.||_1 + ||.||_2 \leq ||.||_3$.

Let us consider two hypernorms (hyperseminorms) $||.||_1$ and $||.||_3$.

Proposition 3.14. a) If $||.||_1 \leq ||.||_3$, then $a||.||_1 \leq ||.||_3$ for any non-negative number a.

b) If $a||.||_1 \leq ||.||_3$ for a positive number a, then $||.||_1 \leq ||.||_3$.

Proof. a) The inequality $||x||_1 \leq k \cdot ||x||_3$ implies the inequality $a \cdot ||x||_1 \leq ka \cdot ||x||_3$ for all $x \in L$. Thus, $a \cdot ||.||_1 \leq a \cdot ||.||_3$.

b) The inequality $a \cdot ||x||_1 \leq k \cdot ||x||_3$ implies the inequality $||x||_1 \leq (h)^{-1} k \cdot ||x||_3$, which means that $||.||_1 \leq ||.||_3$.

Proposition is proved.

Let us consider two norms (seminorms) $||.||_1$ and $||.||_3$.

Corollary 3.48. If $\|.\|_1 \leq \|.\|_3$, then $a\|.\|_1 \leq \|.\|_3$ for any non-negative number a.

Propositions 3.10 – 3.12 imply the following result.

Theorem 3.7. a) The set $\mathbf{HSN}_{\|\cdot\|}$ of all hyperseminorms that are weaker than a hyperseminorm $\|.\|$ is a semimodule over the semiring of all non-negative real numbers $\boldsymbol{R}^+$.

b) The set $\mathbf{HN}_{\|\cdot\|}$ of all hypernorms that are weaker than a hypernorm $\|.\|$ is an S-module over the S-ring of all positive real numbers $\boldsymbol{R}^{++}$.

Corollary 3.49. a) The set $\mathbf{SN}_{\|\cdot\|}$ of all seminorms in L that are weaker than a given seminorm $\|.\|$ is a semimodule over the semiring $\boldsymbol{R}^+$.

b) The set $\mathbf{N}_{\|\cdot\|}$ of all norms in L that are weaker than a given norm $\|.\|$ is an S-module over the S-ring $\boldsymbol{R}^{++}$.

Lemma 3.22. Two hyperseminorms (hypernorms) q and p in a vector space L are equivalent if and only if $p \leq q$ and $q \leq p$.

Proof. (a) By Definition 3.32. b, the hyperseminorms (hypernorms) q and $p$are equivalent if there are constants (positive real numbers) k, $h > 0$ such that for all $x \in L$,

$$h \cdot q(x) \leq p(x) \leq k \cdot q(x)$$

Thus, $p \leq q$ because $p(x) \leq k \cdot q(x)$ for all $x \in L$, and $p \leq q$ because $h \cdot q(x) \leq p(x)$ implies $p(x) \leq (h)^{-1} \cdot q(x)$ for all $x \in L$.

(b) if $p \leq q$ and $q \leq p$, then there are constants (positive real numbers) k, $h > 0$ such that for all $x \in L$,

$$p(x) \leq k \cdot q(x)$$

and

$$q(x) \leq h \cdot p(x)$$

This gives us two inequalities

$$(h)^{-1} \cdot q(x) \leq p(x) \leq k \cdot q(x)$$

which mean that hyperseminorms (hypernorms) q and p are equivalent.

Lemma is proved.

Corollary 3.50. Two seminorms (norms) q and p in a vector space L are equivalent if and only if $p \leq q$ and $q \leq p$.

Lemma 3.22 and Proposition 3.12 imply the following results

Corollary 3.51. Two hyperseminorms (hypernorms) q and p in a vector space L are equivalent to the third hyperseminorm (hypernorm) r if and only if $p + q$ is equivalent to r.

Corollary 3.52. Two seminorms (norms) q and p in a vector space L are equivalent to the third seminorm (norm) r if and only if $p + q$ is equivalent to r.

Lemma 3.22 and Proposition 3.12 imply the following results

Corollary 3.53. A hyperseminorm (hypernorm) q is equivalent to a hyperseminorm (hypernorm) r if and only if aq is equivalent to r for a positive number a.

Corollary 3.54. A seminorm (norm) q is equivalent to a seminorm (norm) r if and only if $p + q$ is equivalent to r.

Proposition 3.15. If a hyperseminorm q is stronger than a hypernorm p, then q is a hypernorm.

Indeed, a hyperseminorm is a hypernorm if it is not equal to 0 for all non-zero elements from L. Thus, if $x \in L$ and $x \neq \mathbf{0}$, then $p(x) \neq 0$ and $q(x) \neq 0$ because $p(x) \leq k \cdot q(x)$.

Corollary 3.55. If a hyperseminorm q is equivalent to a hypernorm p, then q is a hypernorm.

Corollary 3.56. If a seminorm q is stronger than a norm p, then q is a norm.

Corollary 3.57. If a seminorm q is equivalent to a hypernorm p, then q is a norm.

Definition 3.33. A hyperseminorm q is *essentially stronger* than a hyperseminorm p if for all $x \in L$, we have

$$p(x) \leq q(x)$$

When a hyperseminorm q is essentially stronger than a hyperseminorm p, it is denoted by

$$p \preccurlyeq q$$

We can easily see that relation $p \preccurlyeq q$ implies relation $p \leq q$.

Some properties of the relation $\preccurlyeq$ are similar to properties of the relation $\leq$, while other properties are different. Let us consider some examples.

Proposition 3.16. a) If $q_1 \preccurlyeq q_2$, then $q_1 + q_3 \preccurlyeq q_2 + q_3$ for any hypernorm (hyperseminorm) q_3.

b) If $q_1 \preccurlyeq q_2$, then $a \cdot q_1 \preccurlyeq a \cdot q_2$ for any nonnegative element a from $\boldsymbol{F}$.

Indeed, the inequality $q_1 \preccurlyeq q_2$ means that $q_1(x) \leq q_2(x)$ for all $x \in L$. Consequently, $q_1(x) + q_3(x) \leq q_2(x) + q_3(x)$ and $a \cdot q_1(x) \leq a \cdot q_2(x)$ for all $x \in L$ because $q_3(x) \geq 0$ and $a \geq 0$.

At the same time, a hyperseminorm q is not essentially stronger than the hyperseminorm $a \cdot q$ when $a > 1$, while by Proposition 3.12, the hyperseminorm q is always stronger than the hyperseminorm $a \cdot q$.

Similar to seminorms and norms, hyperseminorms and hypernorms determine such a topological property as convergence in vector spaces.

Definition 3.34. A sequence $\{x_i; x_i \in L, i = 1, 2, 3, \ldots\}$ *q-converges*, i.e., *converges with respect to* the hyperseminorm (hypernorm) q, to an element x from L, i.e., $\lim^q{}_{i\to\infty} x_i = x$, if $\lim_{i\to\infty} q(x_i - x) = 0$.

Convergence of real hypernumbers is an example of q-convergence where q is a natural hypernorm in the space of real hypernumbers (Burgin, 2012). Many properties of q-convergence are the same as properties of the classical convergence of number sequences. For instance we have the following result.

Lemma 3.23. If a sequence $\{x_i; x_i \in L, i = 1, 2, 3, \ldots\}$ q-converges to x, then any its infinite subsequence converges to the same element.

Proof is left as an exercise.

Convergence with respect to hyperseminorms induces relations between hyperseminorms.

Definition 3.35. a) A hyperseminorm q is *sequentially stronger* than a hyperseminorm p if q-convergence implies p-convergence, i.e., $\lim^q{}_{i\to\infty} x_i = x$ entails $\lim^p{}_{i\to\infty} x_i = x$ for any sequence $\{x_i; x_i \in L, i = 1, 2, 3, \ldots\}$.

b) Hyperseminorms q and p are *sequentially equivalent* if each of them is sequentially stronger than the other.

When a hyperseminorm q is sequentially stronger than a hyperseminorm p, it is denoted by

$$p \sqsubseteq q$$

Let us consider two hyperseminorms (hypernorms) q and p.

Proposition 3.17. If $p \leq q$ and $\lim^q{}_{i\to\infty} x_i = x$, then $\lim^p{}_{i\to\infty} x_i = x$, i.e., $p \leq q$ implies $p \sqsubseteq q$.

Proof is similar to the proof of sufficiency in Theorem 3.3.

It means that convergence with respect to a stronger (larger) hyperseminorm (hypernorm) implies convergence with respect to a weaker (smaller) hyperseminorm (hypernorm).

Remark 3.11. We have seen (Theorem 3.1) that the inverse of Proposition 3.17 is true for norms and seminorms. However, in contrast to norms and seminorms, the relation $p \sqsubseteq q$ for arbitrary hyperseminorms or hypernorms p and q does not always imply the relation $p \leq q$ as the following example demonstrates.

Example 3.25. Let us consider the set $\boldsymbol{R}_\omega$ of all real hypernumbers described in Chapter 2. It is a real vector space with a hypernorm $\|\cdot\|$ is defined by the formula:

If α is a real hypernumber, i.e., $\alpha = \text{Hn}(a_i)_{i\in\omega}$ with $a_i \in \boldsymbol{R}$ for all $i \in \omega$, then $\| \alpha \| = \text{Hn}(|a_i|)_{i\in\omega}$

As it is demonstrated in Burgin (2012), the space $\boldsymbol{R}_\omega$ is the disjoint union of components defined by the rule that the difference between two elements in any component is a finite hypernumber. For instance, hypernumbers $\text{Hn}(n)_{n\in\omega}$ and $\text{Hn}(n + 10)_{n\in\omega}$ belong to the same component, while hypernumbers $\text{Hn}(n)_{n\in\omega}$ and $\text{Hn}(2n)_{n\in\omega}$ belong to different components.

Let us take components C_k of the hypernumbers $\text{Hn}(n^k)_{n\in\omega}$ where $k = 1, 2, 3, \ldots$ and define a function q in each C_k by the formula:

$$\text{If } \alpha \in C_k, \text{ then } q(\alpha) = \text{Hn}(k|a_i|)_{i\in\omega}$$

In all other components of $\boldsymbol{R}_\omega$, the function q coincides with standard hypernorm in the space $\boldsymbol{R}_\omega$ (see Example 3.8).

It is easy to see that the function q satisfies axioms N1 and N2 of a hypernorm. Besides, as the sum of elements from different components C_k cannot belong to the component with a larger index k and the product of a hypernorm and a positive real number is a hypernorm (Proposition 3.10), the function q satisfies Axiom N3. Thus, the function q is a hypernorm in the space $\boldsymbol{R}_\omega$.

By Definition 3.35, the standard hypernorm $\|\cdot\|$ and the hypernorm q are sequentially equivalent in the space $\boldsymbol{R}_\omega$. In particular, $q \sqsubseteq \|\cdot\|$ but it is not true that $q \leq \|\cdot\|$ as for any natural number k, there is a hypernumber α such that $q(\alpha) > k\|\alpha\|$.

Let us consider two seminorms (norms) q and p.

Corollary 3.58. If $p \leq q$ and $\lim^q_{i\to\infty} x_i = x$, then $\lim^p_{i\to\infty} x_i = x$.

Definition 3.36. A sequence $\{a_i;\ i = 1, 2, 3, \ldots\}$ in a vector space L with a hypernorm (hyperseminorm) q is a *fundamental* or *Cauchy sequence* with respect to q if for any $\varepsilon \in \boldsymbol{R}^{++}$, there is $n \in \boldsymbol{N}$ such that for any $x \in X$ and any $i, j \geq n$, we have

$$q(a_j - a_i) < \varepsilon$$

As in the case of real numbers, we have the following result.

Lemma 3.24. Any converging with respect to q sequence is a Cauchy sequence with respect to q.

Proof is left as an exercise.

Lemma 3.25. Any subsequence of a Cauchy sequence with respect to q is Cauchy sequence with respect to q.

Proof is left as an exercise.

Cauchy sequences induce a relation between hyperseminorms.

Definition 3.37. Hyperseminorms (hypernorms) q and p are *compatible* if any Cauchy sequence with respect to one of them is also a Cauchy sequence with respect to the other.

Lemma 3.26. Equivalent hyperseminorms (hypernorms) are also compatible.

Proof is left as an exercise.

Remark 3.12. We have seen (Theorem 3.2) that the inverse of Lemma 3.26 is true for norms and seminorms. However, in contrast to norms and seminorms, compatibility of hyperseminorms or hypernorms does not imply their equivalence as Example 3.25 demonstrates.

Proposition 3.18. Two hyperseminorms (hypernorms) q and p in a vector space L are compatible with the third hyperseminorm (hypernorm) r if and only if $p + q$ is compatible with r.

Proof is left as an exercise.

Corollary 3.59. Two hyperseminorms (hypernorms) q and p in a vector space L are compatible with the third hyperseminorm (hypernorm) r if and only if $p + q$ is compatible with r.

Hyperseminormed spaces are also hyperpseudometric spaces.

Definition 3.38. A *hyperpseudometric* in a set X is a mapping $\mathbf{d}$: $X \times X \to \boldsymbol{R}_{\omega}^{+}$ that satisfies the following axioms:

P1. $\mathbf{d}(x, y) = 0$ if $x = y$,

i.e., the distance between an element and itself is equal to zero.

M2 (Symmetry). $\mathbf{d}(x, y) = \mathbf{d}(y, x)$ for all $x, y \in X$,

i.e., the distance between x and y is equal to the distance between x and y.

M3 (the triangle inequality or subadditivity).

$$\mathbf{d}(x, y) \leq \mathbf{d}(x, z) + \mathbf{d}(z, y) \text{ for all } x, y, z \in X.$$

When the distance between two elements of X is always a real number, $\mathbf{d}$ is a *pseudometric* (Kuratowski, 1966).

Note that although it would look natural, we do not use terms semimetric and hypersemimetric because according to the mathematical convention, semimetric is defined by a distance that satisfies only axioms M1 and M2.

Lemma 3.27. a) A hyperseminorm q in a vector space L induces a hyperpseudometric $\mathbf{d}_q$ in this space.

b) If q is a seminorm in L, then $\mathbf{d}_q$ is a pseudometric.

Indeed, if $q: X \to \boldsymbol{R}_\omega^{\ +}$ is a hyperseminorm in L and x and y are elements from L, then we can define $\mathbf{d}_q(x, y) = q(x - y)$. Properties of a hyperseminorm imply that $\mathbf{d}_q$ satisfies all axioms P1, M2 and M3. In addition, if q takes values only in $\boldsymbol{R}$, then the same is true for $\mathbf{d}_q$, i.e., $\mathbf{d}_q$ is a pseudometric.

It is interesting to find what hyperpseudometrics in vector spaces are induced by hyperseminorms and what pseudometrics in vector spaces are induced by seminorms. To do this, let us consider additional properties of hypermetrics and metrics.

Definition 3.39. A hyperpseudometric (hypermetric) $\mathbf{d}$ in a vector space L is called *linear* if it satisfies Axioms LM1 and LM2.

Examples 3.4–3.6 show that there are linear pseudometrics (hyperpseudometrics) in vector spaces and there are pseudometrics (hyperpseudometrics) in vector spaces that are not linear. The majority of popular pseudometrics are induced by seminorms and thus, they are linear as the following result demonstrates.

Theorem 3.8. A hyperpseudometric $\mathbf{d}$ is induced by a hyperseminorm if and only if $\mathbf{d}$ is linear.

Proof is similar to the proof of Theorem 3.1.

Corollary 3.60. A pseudometric $\mathbf{d}$ is induced by a seminorm if and only if $\mathbf{d}$ is linear.

Theorem 3.9. a) Any linear hyperpseudometric $\mathbf{d}$ in a vector space L induces a hyperseminorm q in L.

b) If $\mathbf{d}$ is a hypermetric, then q is a hypernorm.

Proof is similar to the proof of Theorem 3.4.

We define the kernel Ker q of a hyperseminorm q in L as

$$\text{Ker}\, q = \{x \in L;\ q(x) = 0\}$$

Theorem 3.10. The kernel Ker q of a hyperseminorm q in L is a closed vector subspace of L.

Proof. If $q(x) = 0$ and $a \in \boldsymbol{F}$, then by Axiom N2,

$$q(ax) = |a| \cdot q(x) = |a| \cdot 0 = 0$$

i.e., $ax \in \text{Ker } q$.

In addition, $q(x) = 0$ and $q(y) = 0$, then by Axiom N3,

$q(x + y) \leq q(x) + q(y) = 0 + 0 = 0$

and $q(x + y) = 0$ because by Proposition 3.7, $q(x + y) \geq 0$.

In addition, if a sequence $\{x_i; x_i \in \text{Ker } q, i = 1, 2, 3, \ldots\}$ converges to some element x from L, then by properties of hyperseminorms

$$q(x) \leq q(x - x_i) + q(x_i) = q(x - x_i)$$

As $q(x - x_i)$ converges to 0, $q(x)$ is equal to 0, i.e., x belongs to Ker q. Consequently, the space Ker q is closed in L.

Theorem is proved.

Theorem 3.10 allows factorization of the hyperseminormed space L by its subspace Ker q, obtaining the quotient space L_q. The hyperseminorm q induces the hypernorm p_q in the space L_q. This gives us the natural projection $\tau: L \to L_q$, which preserves the hyperseminorm q.

Example 3.26. Let us consider the set $C^\infty(\boldsymbol{R}, \boldsymbol{R})$ of all smooth real functions. The following seminorms are considered in the set $C^\infty(\boldsymbol{R}, \boldsymbol{R})$. For each point $a \in \boldsymbol{R}$ and $f \in C^\infty(\boldsymbol{R}, \boldsymbol{R})$, we define

$$q_k(f) = (f(a))^2 + (f''(a))^2 + \ldots + (f^{(k)}(a))^2$$

The factorization of the space $C^\infty(\boldsymbol{R}, \boldsymbol{R})$ by its subspace Ker q is called the k-th order jet space $J_a^{\,k}(\boldsymbol{R}, \boldsymbol{R})$ of $C^\infty(\boldsymbol{R}, \boldsymbol{R})$ at the point a. Jet spaces were introduced by Ehresmann (1952, 1953) and have various applications in the theory of differential equations and differential relations, as well as in the theory of manifolds (Gromov, 1986; Krasil'shchik et al., 1986).

It is possible to get the same quotient space using the following seminorm

$$m_k(f) = \max\{\ |f(a)|, |f''(a)|, \ldots, |f^{(k)}(a)|\ \}$$

Let us consider a Hausdorff space X that is a quotient space of L with the projection $\eta: L \to X$, preserves the hyperseminorm q. Then it is possible to define a projection $v: L_q \to X$ preserves the hyperseminorm q and for which $\eta = v\tau$, i.e., the following diagram is commutative:

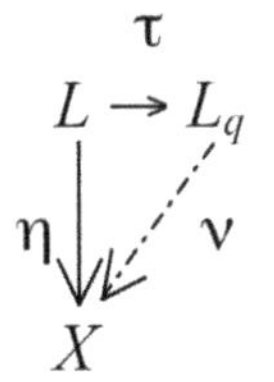

This gives us the following result.

Theorem 3.11. a) L_q is the largest Hausdorff quotient space of the topological space L that preserves the hyperseminorm q.

b) L_q is the largest quotient space of the topological space L in which the hyperseminorm q induces the hypernorm p_q.

It is possible to define two basic operators in a vector space L.

1. If a is an element from $\boldsymbol{F}$, then the translation operator T_a is defined by the formula:

$$T_z(x) = x + z \text{ where } x, z \in L$$

2. If $a \neq 0$ is an element from $\boldsymbol{F}$, then the multiplication operator M_a is defined by the formula:

$$M_a(x) = ax \text{ where } x \in L$$

Proposition 3.19. Operators T_a and M_a are homeomorphisms of the seminormed vector space L.

Proof. The axioms of a vector space imply that T_z and M_a are one-to-one mappings and their inverses are T_{-z} and M_{-a}, respectively. As addition is continuous in L, the operator T_z is also continuous. As scalar multiplication is continuous with respect to L, the operator M_a is also continuous.

Proposition is proved.

Definition 3.40. A system $\boldsymbol{U}$ of neighborhoods of a point x of L is called a *local base* of the topology in L if every neighborhood of x contains a neighborhood from $\boldsymbol{U}$.

Example 3.27. In the n-dimensional real space $\boldsymbol{R}^n$, the set $\boldsymbol{B}_0 = \{B_m = \{k; |k| < 1/m\}; = 1, 2, 3, \dots\}$ is a local base at $\boldsymbol{0}$.

Example 3.28. In the n-dimensional real space $\boldsymbol{R}^n$, the set $\boldsymbol{B}_x = \{B_m = \{x + k; |k| < 1/m\}; = 1, 2, 3, \dots\}$ is a local base at x for any real number x.

Example 3.29. As it is proved by Burgin (2012), the set of all real hypernumbers $\boldsymbol{R}_\omega$ is a hypernormed vector space where the hypernorm $\|\cdot\|$ is defined by the following formula.

If α is a real hypernumber, i.e., $\alpha = \mathrm{Hn}(a_i)_{i\in\omega}$ with $a_i \in \boldsymbol{R}$ for all $i\in\omega$, then the hypernorm has the form

$$\| \alpha \| = \mathrm{Hn}(|a_i|)_{i\in\omega}$$

Then the set $\boldsymbol{B}_0 = \{B_m = \{\alpha; |\alpha| < 1/m\}; = 1, 2, 3, \ldots\}$ is a local base at $\mathbf{0}$.

Corollary 3.61. The topology of a seminormed vector space L is translation-invariant, or simply invariant, i.e., a subset A from L is open if and only if any its translation $T_z(A) = A + z$ is open.

As a result, such a topology is completely determined by any local base and thus, by any local base at $\mathbf{0}$.

Definition 3.41. a) A mapping $\mathbf{d}: X \times X \to \boldsymbol{R}_\omega^{\ +}$ is called a *hyperquasimetric* if it satisfies Axioms M1 and M3.

b) A set X with a hyperquasimetric $\mathbf{d}$ is called a *hyperquasimetric space*.

c) The real number $\mathbf{d}(x, y)$ is called the *distance* between x and y in the hyperquasimetric space X.

Any hypermetric space is a hyperquasimetric space.

It is possible to define a non-trivial hyperquasimetric on the set $\boldsymbol{R}_\omega$ of real hypernumbers by the following rule:

$$d(x, y) = \begin{cases} x - y & \text{if } x \geq y \\ \mathrm{Max}(\|x\|, \|y\|) & \text{if } x \perp y \\ 1 & \text{otherwise} \end{cases}$$

Here *Max* is an operation with hypernumbers defined for hypernumbers $\alpha = \mathrm{Hn}(a_i)_{i\in\omega}$, $\beta = \mathrm{Hn}(b_i)_{i\in\omega} \in \boldsymbol{R}_\omega$, as

$$Max(\alpha, \beta) = \mathrm{Hn}(\max(a_i, b_i))_{i\in\omega}$$

Definitely $d(x, x) = x - x = 0$ and $d(x, y) = x - y = 0$ implies $x = y$ for any hypernumbers x and y. To prove M3, we need to consider all possible relations between three arbitrary real hypernumbers x, y and z. Here are some of them.

When $x \geq z \geq y$, we have

$$x - y = (x - z) + (z - y)$$

When $x \perp z$, $x \perp y$, and $z \perp y$, we have

$$Max\,(\|x\|, \|y\|) \leq Max(\|x\|, \|z\|) + Max(\|z\|, \|y\|)$$

When $y \geq z \geq x$, we have

$$1 < 1 + 1$$

When $y \geq z$, $x \geq y$, and $x \perp z$, we have

$$x - y \leq Max(\|x\|, \|z\|) + 1$$

When $x \geq z$, $x \geq y$, and $y \perp z$, we have

$$x - y \leq x - z + Max(\|y\|, \|z\|)$$

When $z \geq y \geq x$, we have

$$1 \leq 1 + (z - y)$$

When $z \geq x \geq y$, we have

$$x - y = 1 + (z - y)$$

Other cases also give the necessary inequality.

Definition 3.42. a) A mapping $\mathbf{d}: X \times X \rightarrow \boldsymbol{R}_{\omega}^{+}$ is called a *hypersemimetric* if it satisfies Axioms M1 and M3.

b) A set X with a hypersemimetric $\mathbf{d}$ is called a *hypersemimetric space*.

c) The real number $\mathbf{d}(x, y)$ is called the *distance* between x and y in the hypersemimetric space X.

Any hypermetric space is a hypersemimetric space.

Let us consider a set Q of seminorms in a vector space L. Such a space is called a *polynormed vector space* (see Helemski, 1989; Dosi, 2011). When there is only one seminorm in a vector space, it is a *seminormed vector space* considered above.

In a similar way, a vector space L with a set Q of hyperseminorms is called either a *polyhyperseminormed vector space* or *Q-hyperseminormed vector space* or simply, a *phs-normed space*.

Example 3.30. In the space $C(\boldsymbol{R}, \boldsymbol{R})$ of all continuous functions on the real line $\boldsymbol{R}$, the following system of seminorms is defined

$$\{\, \|f\|_n = \max\,\{|f(x)|\,;\, x \in [-n, n]\};\, n = 1, 2, 3, \ldots\}$$

Example 3.31. In the space $L^1(\boldsymbol{R}, \boldsymbol{R})$ of all locally absolutely Riemann integrable real functions on the real line $\boldsymbol{R}$, the following system of seminorms is defined

$$\{ \|f\|_{L^1} = \int_{-n}^{n} |f(x)|dx; n = 1, 2, 3, \ldots\}$$

Example 3.32. Let us take the space $L^2(\boldsymbol{R}, \boldsymbol{R})$ of all locally Riemann square-integrable real functions on the real line $\boldsymbol{R}$, the following system of seminorms is defined

$$\{ \|f\|_{L^2} = [\int_{-n}^{n} |f(x)|^2dx]^{1/2}; n = 1, 2, 3, \ldots\}$$

Example 3.33. In the space $L^p(\boldsymbol{R}, \boldsymbol{R})$ of all locally Riemann p-integrable ($p \geq 1$) functions on the real line $\boldsymbol{R}$, the following system of seminorms is defined

$$\{ \|f\|_{L^p} = [\int_{-n}^{n} |f(x)|^p dx]^{1/p}; n = 1, 2, 3, \ldots\}$$

Example 3.34. In the space $B(\boldsymbol{R}, \boldsymbol{R})$ of all locally bounded functions on the real line $\boldsymbol{R}$, the following system of seminorms is defined

$$\{ \|f\|_{\infty} = \sup \{|f(x)| ; x \in [-n, n]\}; n = 1, 2, 3, \ldots\}$$

Example 3.35. In the generalized Sobolev space $H^1(\boldsymbol{R}, \boldsymbol{R})$ of all differentiable functions on the real line $\boldsymbol{R}$, such that they are locally square integrable and their derivatives are locally square integrable, the following system of seminorms is defined

$$\{ \|f\|_{L^1} = \|f\|_{L^2} + \|f'\|_{L^2} = [\int_{-n}^{n} |f(x)|^2dx]^{1/2} + [\int_{-n}^{n} |f'(x)|^2dx]^{1/2}; n = 1, 2, 3, \ldots\}$$

Definition 3.43. a) A system Q of seminorms in a vector space L (*weakly*) *separates* an element f from L if for any element g from L such that $f \neq g$, there is a hyperseminorm q from Q such that $q(f) \neq q(g)$ ($q(f - g) \neq 0$).

b) A system Q of seminorms in a vector space L (*weakly*) *separates* L if for any different elements f and g from L, there is a hyperseminorm q from Q such that $q(f) \neq q(g)$ ($q(f - g) \neq 0$).

Separation property of seminorms was introduced by Stone (1948) and is similar to the property of the Hausdorff topology (cf. Appendix).

Let us consider a system Q of seminorms in a vector space L.

Lemma 3.28. The following properties are equivalent:

(a) Q weakly separates L.

(b) Q weakly separates $\mathbf{0}$ in L.

(c) Q weakly separates some element f from L.

(d) Q weakly separates all elements f from L.

Proof. (a)⇒(b) by Definition 3.43.

(b)⇒(a): Let us assume that Q weakly separates $\mathbf{0}$ in L. It means that a hyperseminorm q from Q such that $q(f)\neq 0$. Taking two different elements f and $g$from L, we see that $f - g \neq \mathbf{0}$. Thus a $q(f - g)\neq 0$, i.e., q weakly separates these two elements f and g.

The implication (b)⇒(a) is proved as elements f and g are arbitrary different elements f and g from L.

(b)⇒(c) by definition.

(a)⇒(d) by Definition 3.43.

(d)⇒(c) by definition.

(d)⇒(b) by definition.

Lemma is proved.

Lemma 3.29 (Burgin, 2012)**.** Any norm in L weakly separates L.

Indeed, if $f\neq g$, then $f - g \neq 0$ and consequently, $\|f - g\|\neq 0$ by Axiom N1.

Remark 3.2. A norm does not necessarily separate L as the following example demonstrates.

Example 3.36. Let us consider a finite dimensional real vector space L, i.e., $L = \boldsymbol{R}^n$ with the conventional Euclidean norm

$$\|x\| = \sqrt{x_1^{\,2} + x_2^{\,2} + \ldots + x_n^{\,2}}$$

if $x = x_1e_1 + x_2e_2 + \ldots + x_n e_n$ and $\{ e_1, e_2, \ldots, e_n\}$ is an orthonormal basis of $\boldsymbol{R}^n$.

Then $\| e_1\| = \|e_2\| = 1$ although e_1 and e_2 are different elements from L. Thus, the norm $\|.\|$ does not separate L.

Lemma 3.30. If a system Q of seminorms weakly separates (separates) L and $Q \subseteq T$, then the system T of seminorms weakly separates (separates) L.

Proof is left as an exercise.

Definition 3.44. A system Q of hyperseminorms in a vector space L *weakly separates* $\mathbf{0}$ *in* L if for any element $f\neq\mathbf{0}$ from L, there is a hyperseminorm q from Q such that $q(f)\neq 0$

Lemma 3.31. A system Q of hyperseminorms in a vector space L weakly separates L if and only if it weakly separates $\mathbf{0}$ in L.

Proof. Necessity is clear. Therefore, we need to prove only sufficiency.

Let us assume that Q weakly separates $\mathbf{0}$ in L. It means that a hyperseminorm q from Q such that $q(f) \neq 0$. Taking two different elements f and g from L, we see that $f - g \neq \mathbf{0}$. Thus a $q(f - g) \neq 0$, i.e., q separates these two elements f and g.

Lemma is proved as elements f and g are arbitrary different elements from L.

Lemma 3.32. Any hypernorm in L weakly separates L.

Proof is left as an exercise.

Lemma 3.5 implies the following result.

Lemma 3.33. If a system Q of hyperseminorms contains at least one hypernorm, then Q weakly separates L.

Proof is left as an exercise.

Lemma 3.34. If a system Q of hyperseminorms weakly separates (separates) L and $Q \subseteq T$, then the system T of hyperseminorms weakly separates (separates) L.

Proof is left as an exercise.

Theorem 3.12. For any finite system $Q = \{p_n; n = 1, 2, 3, \ldots, t\}$ of hyperseminorms in a vector space L, there is a hyperseminorm q in L, such that:

a) Q weakly separates L if and only if q weakly separates L.
b) For any elements f and g from L, we have $q(f - g) = 0$ if and only if $p(f - g) = 0$ for all p from Q.
c) Q is a hypernorm if, at least, one element from Q is a hypernorm.
d) Q weakly separates L if and only if q is a hypernorm.

Note that q does not necessarily belong to Q.

Proof. Let us consider a system $Q = \{p_n; n = 1, 2, 3, \ldots, t\}$ of hyperseminorms in a vector space L, and define $q = p_1 + p_2 + p_3 + \ldots + p_t$. By Proposition 3.1, q is a hyperseminorm in L.

Let us assume that Q weakly separates L. In particular, Q weakly separates $\mathbf{0}$. It means that for any element $f \neq \mathbf{0}$ from L, there is a hyperseminorm p_i from Q such that $p_i(f) \neq 0$. Then $q(f) \neq 0$ because all values $p_i(f)$ are positive. So, q weakly separates $\mathbf{0}$. By Lemma 3.31, this implies that q weakly separates L.

Now let us assume that Q does not weakly separates L. By Lemma 3.31, this implies that Q does not weakly separates $\mathbf{0}$, i.e., there is an element l from L such that $l \neq \mathbf{0}$ but $p_i(l) = 0$ for all hyperseminorms p_i from Q. As a result, $q(f) = 0$. Therefore, the hyperseminorm q also does not weakly separate $\mathbf{0}$.

In essence, Q weakly separates L if and only if q weakly separates L.

(a) In addition, for any elements f and g from L, $q(f-g) = 0$ if and only if $p_i(f-g) = 0$ for all p_i from Q.

(b) We need to check Axiom N1. Let us presuppose that, at least, one element from Q is a hypernorm. Because the used enumeration of elements from Q is arbitrary, it is possible to assume that p_1 a hypernorm. By construction of q, the equality $q(x) = 0$ implies $p_1(x) = 0$ and consequently, $x = \mathbf{0}$ because p_1 is a hypernorm. It means that q is a hypernorm.

Theorem is proved.

Corollary 3.62. For any finite system $Q = \{p_n; n = 1, 2, 3, \ldots, t\}$ of seminorms in a vector space L, there is a seminorm q in L, such that:

a) Q weakly separates L if and only if q weakly separates L.

b) For any elements f and g from L, $q(f-g) = 0$ if and only if $p(f-g) = 0$ for all p from Q.

c) Q is a norm if, at least, one element from Q is a norm.

d) Q weakly separates L if and only if $q$is a norm.

For infinite systems of seminorms we have a slightly weaker result.

Theorem 3.13. For any countable system Q of seminorms in a vector space L, there is a hyperseminorm q in L, such that:

a) Q weakly separates L if and only if q weakly separates L.

b) For any elements f and g from L, we have $q(f-g) = 0$ if and only if $p(f-g) = 0$ for all p from Q.

c) Q is a finite set if and only if q is a seminorm.

d) Q is a hypernorm if, at least, one element from Q is a norm.

Proof. Let us consider a system $Q = \{p_n; n = 1, 2, 3, \ldots\}$ of seminorms in a vector space L and define $t_m = p_1 + p_2 + p_3 + \ldots + p_m$ for all $m = 1, 2, 3, \ldots$ By Proposition 3.1, all t_m are seminorms. We define the following extrafunction on L

$$q(f) = \text{Hn}\ (t_i(f)))_{i\in\omega}$$

It is a hyperseminorm. Indeed, it satisfies all axioms for hyperseminorms.

SN1. $t_m(\mathbf{0}) = p_1(\mathbf{0}) + p_2(\mathbf{0}) + p_3(\mathbf{0}) + \ldots + p_m(\mathbf{0}) = 0 + 0 + 0 + \ldots + 0 = 0$ for all $m = 1, 2, 3, \ldots$ because all p_i are seminorms. Thus, $q(\mathbf{0}) = \text{Hn}\ (0)_{i\in\omega} = 0$.

N2. $q(af) = \text{Hn}\ (t_i(af))_{i\in\omega} = \text{Hn}\ (at_i(f))_{i\in\omega} = a\text{Hn}\ (t_i(f)))_{i\in\omega} = aq(f)$ because all t_i are seminorms.

N3. As all p_i are seminorms, we have

$$p_i(x + y) \leq p_i(x) + p_i(y)$$

for any x and y from L. Consequently,

$$t_m(x + y) \leq t_m(x) + t_m(y)$$

for any x and y from L and all $m = 1, 2, 3, \ldots$. Consequently,

$$q(x + y) = \text{Hn}\ (t_i(x + y)))_{i\in\omega} \leq \text{Hn}\ (t_i(x) + t_i(y)))_{i\in\omega} = q(x) + q(y)$$

We have proved that q is a hyperseminorm.

(a) Let us assume that the system Q weakly separates L. In particular, Q weakly separates $\mathbf{0}$. It means that for any element $f \neq \mathbf{0}$ from the space L, there is a hyperseminorm p_i from Q such that $p_i(f) \neq 0$. Then $t_m(f) \neq 0$ for all $m = i$, $i + 1$, $i + 2$, $i + 3$, …, $i + n$, … By definition, it means that $q(f) \neq 0$ because $t_m(f) \leq t_{m+1}(f)$ for all $m = 1, 2, 3, \ldots$ It means that q weakly separates $\mathbf{0}$. By Lemma 3.31, this implies that q weakly separates L.

Now let us assume that Q does not weakly separate L. By Lemma 3.31, this implies that Q does not weakly separate $\mathbf{0}$, i.e., there is an element l from L such that $l \neq \mathbf{0}$ but $p_i(l) = 0$ for all hyperseminorms p_i from Q. As a result, $t_m(f) = 0$ for all $m = 1, 2, 3, \ldots$ By properties of hypernumbers, q also does not weakly separate $\mathbf{0}$. By Lemma 3.31, this implies that q does not weakly separate L.

So, Q weakly separates L if and only if q weakly separates L.

(b) In addition, for any elements f and g from L, we have $q(f - g) = 0$ if and only if $p(f - g) = 0$ for all p from Q.

(c) follows from Proposition 3.1.

(d) Assuming that one element from Q is a norm, w e need to check Axiom N1 for q. Let us presuppose that, at least, one element from Q is a norm. Because the used enumeration of elements from Q is arbitrary, it is possible to assume that p_1 a norm.

Let us take an element x from L such that $q(x) = \text{Hn}\ (t_i(x)))_{i\in\omega} = 0$. If $t_1(x) = p_1(x) = k \neq 0$, then all $t_m(x) \geq k$ and $q(x) = \text{Hn}\ (t_i(x)))_{i\in\omega} \neq 0$ [cf. (Burgin, 2012)]. Thus, the equality $q(x) = 0$ implies $p_1(x) = 0$ and consequently, $x = \mathbf{0}$ because p_1 is a norm. It means that q is a hypernorm.

Theorem is proved.

An important case of polynormed spaces is countably-normed spaces (Kolmogorov and Fomin, 1999). Many important vector spaces, such as the vector space of all infinitely differentiable functions or the vector space of all infinite sequences of real numbers, cannot be defined by a single norm but have the structure of a countably-normed space in a natural way.

Definition 3.45. A vector space L with a countable system of pairwise compatible norms (seminorms) is called a countably-normed (countably-seminormed) space.

Theorem 3.13 implies the following result.

Corollary 3.63. Any countably-normed (countably-seminormed) vector space is a hypernormed (hyperseminormed) vector space.

In a similar way, we introduce countably-hypernormed and countably-hyperseminormed spaces.

Definition 3.46. A vector space L with a countable system of pairwise compatible hypernorms (hyperseminorms) is called a countably-hypernormed (countably-hyperseminormed) space.

It would be interesting to find if it is possible to define the same topology in a countably-hypernormed (countably-hyperseminormed) space by one hypernorm (hyperseminorm).

Definition 3.47. A system R of hypersemimetrics in a set X *weakly separates* X if for any different elements f and g from X, there is a hypersemimetric d from R such that d(f, g)≠0.

Lemma 3.35. Any hypermetric in X weakly separates X.

Proof is left as an exercise.

Proposition 3.20. If a system R of hypersemimetrics in a vector space L is induced by a system Q of hyperseminorms in L, then R weakly separates L if and only if Q weakly separates L.

Proof. Necessity. Let us consider a system Q of hyperseminorms in a vector space L. Then each hyperseminorm q from Q induces the hypersemimetric **d** in L by the following formula:

$$\mathbf{d}(x, y) = q(x - y)$$

In such a way, we obtain the system R of hypersemimetrics. The system Q weakly separates L when for any elements x and y from L, there is a hyperseminorm q from Q such that $x \neq y$ implies $q(x - y)\neq 0$. Consequently, $\mathbf{d}(x, y) = q(x - y) > 0$, i.e., the system R weakly separates L.

Sufficiency. Now let us suppose that R weakly separates L and consider different elements x and y from L. Then there is a hypersemimetric **d** from R such that $x \neq y$ implies $\mathbf{d}(x, y) > 0q(x - y) > 0$. Consequently, $q(x - y) = \mathbf{d}(x, y) > 0$, i.e., the system Q weakly separates L.

Proposition is proved.

Proposition 3.21. If a system Q of hyperseminorms in a vector space L is induced by a system R of hypersemimetrics in L, then R weakly separates L if and only if Q weakly separates L.

Proof. Necessity. Let us consider a system R of hypersemimetrics in a vector space L. Then each hypersemimetric $\mathbf{d}$ from R induces the hyperseminorm q in L by the following formula:

$$q(x) = \mathbf{d}(x, \mathbf{0})$$

In such a way, we obtain the system Q of hyperseminorms. The system R weakly separates L when for any elements x and y from L, there is a hypersemimetric $\mathbf{d}$ from R such that $x \neq y$ implies $\mathbf{d}(x, y) > 0$. Consequently, if $x \neq 0$, then $q(x) = \mathbf{d}(x, \mathbf{0}) \neq 0$, i.e., the system Q weakly separates L.

Sufficiency. Now let us suppose that Q weakly separates L and consider different elements x and y from L. Then there is a hyperseminorm q from Q such that $x \neq y$ implies $q(x - y) > 0$ because $x - y \neq 0$. Consequently, if a hypersemimetric $\mathbf{d}$ from R induces the hyperseminorm q, then (cf. Theorem 3.5) $\mathbf{d}(x, y) = q(x - y) > 0$, i.e., the system R weakly separates L.

Proposition is proved.

Theorem 3.14. In a hyperseminormed vector space L, there is a neighborhood V of $\mathbf{0}$ such that:

(a) V is a vector subspace of L;
(b) Operations of addition and scalar multiplication are continuous in V;
(c) V is closed;
(d) V is connected;
(e) V is convex.

Proof. Let us consider a hyperseminormed vector space L with a hyperseminorm q, defining V as

$$V = \{x;\, x \in L,\, q(x) \text{ is finite}\}$$

The set V is open in L because for any positive real number r with any element x, the set V contains its neighborhood

$$\mathrm{O}_r x = \{z;\, q(x - z) < r\}$$

(a) Taking two elements x and y from L and two real numbers a and b, let us consider the element $ax + by$. As by properties of hyperseminorms,

$$q(ax + by) \leq aq(x) + bq(y)$$

by properties of hypernumbers, belongs to V, and thus, V is a vector subspace of L.

(b) Now we show that addition is continuous and scalar multiplication is continuous in the second coordinate with respect to this topology.

Let us consider a sequence $\{x_i; x_i \in V, i = 1, 2, 3, \ldots\}$ that converges to x, a sequence $\{y_i; y_i \in V, i = 1, 2, 3, \ldots\}$ that converges to y, and the sequence $\{z_i = x_i + y_i; i = 1, 2, 3, \ldots\}$. By the part (a), all elements z_i belong to the space V. Convergence of these two sequences means that for any $k > 0$, there are a natural number n such that $q(x_i - x) < k$ for any $i > n$ and a natural number m such that $q(y_i - y) < k$ for any $i > m$. Then by properties of a hyperseminorm, we have

$$q(z_i - (x + y)) = q((x_i + y_i) - (x + y)) = q((x_i - x) + (y_i - y)) \leq q(x_i - x) + q(y_i - y) < k + k = 2k$$

when $i > \max\{n, m\}$. As k is an arbitrary positive real number, this means that the sequence $\{z_i = x_i + y_i; i = 1, 2, 3, \ldots\}$ converges to $x + y$. Consequently, addition is continuous in L.

Now for some number a from $\boldsymbol{F}$, we take a sequence $\{x_i; x_i \in V, i = 1, 2, 3, \ldots\}$ that converges to an element x, a sequence $\{a_i; i = 1, 2, 3, \ldots\}$ of real numbers that converges to a number a and the sequence $\{u_i = a_i x_i; i = 1, 2, 3, \ldots\}$. By (a), all elements u_i belong to the space V. Convergence of these two sequences means that for any $k > 0$, there are a natural number n such that $q(x_i - x) < k$ for any $i > n$ and a natural number m such that $q(a_i - a) < k$ for any $i > m$. Then we have

$$q(u_i - ax) = q(a_i x_i - ax) = q(a_i x_i - ax_i + ax_i - ax) \leq q(a_i x_i - ax_i) + q(ax_i - ax) =$$

$$q((a_i - a)x_i) + q(a(x_i - x)) = |a_i - a|\, q(x_i) + |a|\, q(x_i - x) < k + k = 2k$$

with $i > \max\{n, m\}$ because all values. $q(x_i)$ are finite when $x_i \in V$, $i = 1, 2, 3, \ldots$ As k is an arbitrary positive real number, this means that the sequence $\{u_i = a_i x_i; i = 1, 2, 3, \ldots\}$ converges to the element ax. Consequently, scalar multiplication is continuous in the space L.

(c) Let us show that V is closed. To do this, we take a sequence $\{x_i; x_i \in V, i = 1, 2, 3, \ldots\}$ that converges to an element x from L. It means that the hypernumbers $q(x_i - x)$ converge to 0, which means that for sufficiently big numbers i, these hypernumbers become less than some real number k. At the same time, by Proposition 3.6, we have

$$q(x) - q(x_i) \leq q(x_i - x)$$

If $q(x)$ is an infinite hypernumber with a positive infinite subhypernumber, then $q(x) - q(x_i)$ is also an infinite hypernumber with a positive infinite subhypernumber [cf. Chapter 2 or (Burgin, 2012)]. Properties of hypernumbers imply that $q(x_i - x)$ is also an infinite hypernumber with a positive infinite subhypernumber.

If $q(x)$ is an infinite hypernumber that does not have positive infinite subhypernumbers, then we consider implied by Proposition 3.6 inequality

$$q(x_i) - q(x) \leq q(x_i - x)$$

In this case, the same reasoning shows that $q(x_i - x)$ is also an infinite hypernumber with a positive infinite subhypernumber.

This brings us to a contradiction demonstrating that x has finite hyperseminorm and thus, belong to the space V. Because the sequence $\{x_i; x_i \in V, i = 1, 2, 3, \ldots\}$ is arbitrary, the space V is closed.

(d) Let us show that V is connected. Moreover, we demonstrate that V is linearly connected. Indeed, any vector space is linearly connected and V is a vector space by the part (a).

(e) Let us show that V is convex. Indeed, any vector space is convex and V is a vector space by the part (a).

Theorem is proved.

Theorem 3.15. A real finite dimensional polyhyperseminormed vector space L with a finite system Q of hyperseminorms is isomorphic to the space $\boldsymbol{R}^n$ with the standard norm if and only if condition (a) and one of the following conditions is valid (b) – (d):

(a) Q weakly separates points.

(b) L does not have one-dimensional subspaces with the discrete topology.

(c) There are no one-dimensional subspaces of L that have, at least, one isolated point.

(d) All hyperseminorms from Q are finite.

Proof. Necessity. At first, we show that if the hypernorm $||x||_L$ of an element x from L is infinite, then x is an isolated point in the one-dimensional subspace $G = \{rx; r \in \boldsymbol{R}\}$. Indeed,

$$||x - rx||_L = ||(1 - r)x||_L = |1 - r| \cdot ||x||_L$$

Then by Lemma 2.16, $||x - rx||_L$ is also an infinite hypernumber. So, there is no element from G that belongs to the neighborhood $O_1(x) = \{y;\ ||x - y||_L < 1\}$. Consequently (Kuratowski, 1966), x is an isolated element in G.

Because x is an arbitrary element from G, the space G has the discrete topology.

Thus, we have demonstrated that if the condition (c) is not true, then conditions (a) and (b) are also false. At the same time, the space $\boldsymbol{R}^n$ satisfies conditions (a), (b) and (c). Thus, if the space L is isomorphic to $\boldsymbol{R}^n$, it has also satisfy satisfies conditions (a), (b) and (c). This demonstrates necessity for each of these conditions.

Sufficiency. Let us assume that the dimension L of is n and the condition (c) is true for L. It means that the hypernorm in L is finite. Taking a basis e_1, e_2, e_3, …, e_n of the vector space L, we can represent any element x from L in the form

$$x = a_1e_1 + a_2e_2 + a_3e_3 + \dots + a_ne_n$$

where all a_i are real numbers.

This allows us to assign the element $\underline{x} = (a_1, a_2, a_3, \dots, a_n)$ from $\boldsymbol{R}^n$ to the element x from L. This provides us with the one-to-one mapping f from the space L to the space $\boldsymbol{R}^n$. By definitions, this mapping f is an isomorphism of vector spaces and we need only to show that f is a homeomorphism, i.e., f and its inverse are continuous.

By properties of hypernorms and Cauchy inequality, we have

$$||x||_L = ||a_1e_1 + a_2e_2 + a_3e_3 + \dots + a_ne_n||_L \le$$

$$|a_1| \cdot ||e_1||_L + |a_2| \cdot ||e_2||_L + |a_3| \cdot ||e_3||_L + \dots + |a_n| \cdot ||e_n||_L \le$$

$$(\,||e_1||_L^2 + ||e_2||_L^2 + ||e_3||_L^2 + \dots + ||e_n||_L^2)^{1/2} \cdot (a_1^2 + a_2^2 + a_3^2 + \dots + a_n^2)^{1/2}$$

It is proved that the norm in L is finite. Consequently (cf. Chapter 2), there is a positive number k such that

$$(\,||e_1||_L^2 + ||e_2||_L^2 + ||e_3||_L^2 + \dots + ||e_n||_L^2)^{1/2} \le k$$

As a result,

$$\|x\|_L \leq k \cdot (a_1^2 + a_2^2 + a_3^2 + \ldots + a_n^2)^{1/2} = k \cdot \|\underline{x}\|$$

By Proposition 3.4, the mapping f^{-1} is continuous.

Let us consider the unit sphere S^n in $\boldsymbol{R}^n$:

$$S^n = \{\underline{x} = (a_1, a_2, a_3, \ldots, a_n);\ (a_1^2 + a_2^2 + a_3^2 + \ldots + a_n^2)^{1/2} \leq 1\}$$

We define the following function

$$f(\underline{x}) = f(a_1, a_2, a_3, \ldots, a_n) = \|x\|_L = \|a_1e_1 + a_2e_2 + a_3e_3 + \ldots + a_ne_n\|_L$$

As $\underline{x} \in S^n$, not all a_n are equal to 0. Consequently, x is not equal to $\mathbf{0}$. As a result,

$f(\underline{x}) = \|x\|_L > 0$

for all $\underline{x}$ from S^n. As S^n is a compact space, by the Weierstrass theorem, f reaches its minimum r on S^n and $r > 0$. This gives us

$$f(\underline{x}) = \|x\|_L = \|\underline{x}\| \cdot \|(a_1e_1)/(a_1^2 + a_2^2 + a_3^2 + \ldots + a_n^2)^{1/2} + (a_2e_2)/(a_1^2 + a_2^2 + a_3^2 + \ldots + a_n^2)^{1/2} + \ldots + (a_ne_n)/(a_1^2 + a_2^2 + a_3^2 + \ldots + a_n^2)^{1/2}\| \geq r \cdot \|\underline{x}\|$$

Thus, by Proposition 3.4, the mapping f is continuous.

Theorem is proved.

Corollary 3.64. A real finite dimensional polyhyperseminormed vector space L with a finite system Q of hyperseminorms is isomorphic to the space $\boldsymbol{R}^n$ with the standard norm if, at least, one of the elements from Q is a hypernorm.

Corollary 3.65. A real finite dimensional polyseminormed vector space L with a finite system Q of hyperseminorms is isomorphic to the space $\boldsymbol{R}^n$ with the standard norm if, at least, one of the elements from Q is a norm.

Theorem 3.16. A complex finite dimensional polyhyperseminormed vector space L with a finite system Q of hyperseminorms is isomorphic to the space $\boldsymbol{C}^n$ with the standard norm if and only if condition (a) and one of the following conditions is valid (b) – (d):

(a) Q weakly separates points.

(b) L does not have one-dimensional subspaces with the discrete topology.

(c) There are no one-dimensional subspaces of L that have, at least, one isolated point.

(d) All hyperseminorms from Q are finite.

Proof is similar to the proof of Theorem 3.3.

Corollary 3.66. A complex finite dimensional polyhyperseminormed vector space L with a finite system Q of hyperseminorms is isomorphic to the space $\boldsymbol{C}^n$ with the standard norm if, at least, one of the elements from Q is a hypernorm.

Corollary 3.67. A complex finite dimensional polyseminormed vector space L with a finite system Q of hyperseminorms is isomorphic to the space $\boldsymbol{C}^n$ with the standard norm if, at least, one of the elements from Q is a norm.

Lemma 3.36. For a real (complex) hypernormed vector space L with a finite system Q of hyperseminorms, the following conditions are equivalent if Q weakly separates points:

(a) L does not have one-dimensional subspaces with the discrete topology.

(b) There are no one-dimensional subspaces of L that have, at least, one isolated point.

(c) All hyperseminorms from Q are finite.

Proof is left as an exercise.

Theorem 3.17. A real finite dimensional polyhyperseminormed vector space L with a countable system Q of hyperseminorms is isomorphic to the space $\boldsymbol{R}^n$ with the standard norm if and only if condition (a) and one of the following conditions (b) – (d) is valid:

(a) Q weakly separates points.

(b) L does not have one-dimensional subspaces with the discrete topology.

(c) There are no one-dimensional subspaces of L that have, at least, one isolated point.

(d) All hyperseminorms from Q are finite.

Proof. Necessity. At first, we show that if the hypernorm $||x||_L$ of an element x from L is infinite, then x is an isolated point in the one-dimensional subspace $G = \{rx; r \in \boldsymbol{R}\}$. Indeed,

$$||x - rx||_L = ||(1 - r)x||_L = |1 - r| \cdot ||x||_L$$

Then by Lemma 2.16, $||x - rx||_L$ is also an infinite hypernumber. So, there is no element from G that belongs to the neighborhood $\mathrm{O}_1(x) = \{y; ||x - y||_L < 1\}$. Consequently (Kuratowski, 1966), x is an isolated element in G.

Because x was taken as an arbitrary element from G, the space G has the discrete topology.

Thus, we have demonstrated that if the condition (c) is not true, then conditions (a) and (b) are also false. At the same time, the space $\boldsymbol{R}^n$ satisfies conditions (a), (b) and (c). Thus, if the space L is isomorphic to $\boldsymbol{R}^n$, it has also satisfy satisfies conditions (a), (b) and (c). This demonstrates necessity for each of these conditions.

Sufficiency. Let us assume that the dimension L of is n and the condition (c) is true for L. It means that the hypernorm in L is finite. Taking a basis e_1, e_2, e_3, …, e_n of the vector space L, we can represent any element x from L in the form

$$x = a_1e_1 + a_2e_2 + a_3e_3 + \dots + a_ne_n$$

where all a_i are real numbers.

This allows us to assign the element $\underline{x} = (a_1, a_2, a_3, \dots, a_n)$ from $\boldsymbol{R}^n$ to the element x from L. This provides us with the one-to-one mapping f from the space L to the space $\boldsymbol{R}^n$. By definitions, this mapping f is an isomorphism of vector spaces and we need only to show that f is a homeomorphism, i.e., f and its inverse are continuous.

By properties of hypernorms and Cauchy inequality, we have

$$||x||_L = ||a_1e_1 + a_2e_2 + a_3e_3 + \dots + a_ne_n||_L \leq$$

$$|a_1| \cdot ||e_1||_L + |a_2| \cdot ||e_2||_L + |a_3| \cdot ||e_3||_L + \dots + |a_n| \cdot ||e_n||_L \leq$$

$$(||e_1||_L^2 + ||e_2||_L^2 + ||e_3||_L^2 + \dots + ||e_n||_L^2)^{1/2} \cdot (a_1^2 + a_2^2 + a_3^2 + \dots + a_n^2)^{1/2}$$

It is proved that the norm in L is finite. Consequently (cf. Chapter 2), there is a positive number k such that

$$(||e_1||_L^2 + ||e_2||_L^2 + ||e_3||_L^2 + \dots + ||e_n||_L^2)^{1/2} \leq k$$

As a result,

$$||x||_L \leq k \cdot (a_1^2 + a_2^2 + a_3^2 + \dots + a_n^2)^{1/2} = k \cdot ||\underline{x}||$$

By Proposition 3.4, the mapping f^{-1} is continuous.

Let us consider the unit sphere S^n in $\boldsymbol{R}^n$:

$$S^n = \{\underline{x} = (a_1, a_2, a_3, \dots, a_n); (a_1^2 + a_2^2 + a_3^2 + \dots + a_n^2)^{1/2} \leq 1\}$$

We define the following function

$$f(\underline{x}) = f(a_1, a_2, a_3, \ldots, a_n) = \|x\|_L = \|a_1e_1 + a_2e_2 + a_3e_3 + \ldots + a_ne_n\|_L$$

As $\underline{x} \in S^n$, not all a_n are equal to 0. Consequently, x is not equal to $\mathbf{0}$. As a result,

$$f(\underline{x}) = \|x\|_L > 0$$

for all $\underline{x}$ from S^n. As S^n is a compact space, by the Weierstrass theorem, f reaches its minimum r on S^n and $r > 0$. This gives us

$$f(\underline{x}) = \|x\|_L = \|\underline{x}\| \cdot \|(a_1e_1)/(a_1^2 + a_2^2 + a_3^2 + \ldots + a_n^2)^{1/2} + (a_2e_2)/(a_1^2 + a_2^2 + a_3^2 + \ldots + a_n^2)^{1/2} + \ldots + (a_ne_n)/(a_1^2 + a_2^2 + a_3^2 + \ldots + a_n^2)^{1/2}\| \geq r \cdot \|\underline{x}\|$$

Thus, by Proposition 3.4, the mapping f is continuous.

Theorem is proved.

Corollary 3.68. A real finite dimensional normed vector space L is isomorphic to the space $\boldsymbol{R}^n$ with the standard norm.

Theorem 3.18. A complex finite dimensional polyhyperseminormed vector space L with a countable system Q of hyperseminorms is isomorphic to the space $\boldsymbol{C}^n$ with the standard norm if and only if condition (a) and one of the following conditions (b) – (d) is valid:

(a) Q weakly separates points.
(b) L does not have one-dimensional subspaces with the discrete topology.
(c) There are no one-dimensional subspaces of L that have, at least, one isolated point.
(d) All hyperseminorms from Q are finite.

Proof is similar to the proof of Theorem 3.17.

Corollary 3.69. For a real (complex) hypernormed vector space L hyperseminorms, the following conditions are equivalent:

(a) L does not have one-dimensional subspaces with the discrete topology.
(b) There are no one-dimensional subspaces of L that have, at least, one isolated point.
(c) All hyperseminorms from Q are finite.

It is well known that any infinite sequence of a compact subset K of a metric space has a Cauchy subsequence (Kuratowski, 1966). Let us extend this property of metric spaces to hyperpseudometric spaces.

Theorem 3.19. Any infinite sequence of a compact subset K of a hyperpseudometric space has a Cauchy subsequence.

Proof. Let us consider a compact subset K of a hyperpseudometric space M with a hyperpseudometric **d** and assume that for some real number $\varepsilon > 0$, there is a sequence $l = \{a_i; i = 1, 2, 3, \ldots\}$ such that for any $n \in \omega$, there are i, $j \geq n$ such that $\mathbf{d}(a_j, a_i) \geq \varepsilon$. This allows us to build a sequence $h = \{b_i; i = 1, 2, 3, \ldots\}$ such that for any natural numbers i and j, we have $\mathbf{d}(b_j, b_i) \geq \varepsilon$ and in particular, $\mathbf{d}(b_{i+1}, b_i) \geq \varepsilon$.

Let us take a positive real number $r = (⅓)\varepsilon$ the neighborhood $O_r(x) = \{z; \mathbf{d}(x, z) < r\}$ of a point x from M. Then if we take such neighborhoods $O_r(x)$ for all points x from K, we obtain a cover of K. By the definition of a compact set, there is a finite number of elements $x_1, x_2, x_3, \ldots, x_n$ from X such that the sets $O_r(x_1), O_r(x_2), O_r(x_3), \ldots, O_r(x_n)$ cover K. As the sequence $h = \{b_i; i = 1, 2, 3, \ldots\}$ is infinite, one of these sets, say $O_r(x_i)$, contains an infinite subsequence f of h. By the triangle inequality, the distance between any two elements from cannot be larger than $2r = (⅔)\varepsilon$. As a result, even two different elements from f cannot belong to the same neighborhoods $O_r(x_i)$ because the distance between them is larger than ε. This contradiction shows that any infinite sequence of a compact subset K of a hyperpseudometric space has a Cauchy subsequence.

Theorem is proved.

As any hyperseminormed vector space is also hyperpseudometric space (Lemma 3.26), we have the following result.

Corollary 3.70. Any infinite sequence of a compact subset K of a hyperseminormed vector space has a Cauchy subsequence.

Corollary 3.71. Any infinite sequence of a compact subset K of a complete hyperpseudometric space has a converging subsequence.

As any hyperseminorm induces hyperpseudometric (Lemma 3.26) and any compact set is closed (Kuratowski, 1966), Corollary 3.71 implies the following result.

Lemma 3.37. Any infinite sequence in a compact subset K of a complete hyperseminormed space L has a converging subsequence.

Definition 3.48. a) A subset X of a hyperseminormed (hyperpseudometric) vector space L with a hyperseminorm q is called *complete* if any Cauchy sequence with elements from X converges in X.

b) A hyperseminormed (hyperpseudometric) vector space L with a hyperseminorm q is called *complete* if any Cauchy sequence in L converges.

For instance, $\boldsymbol{R}$ is a complete normed space.

Lemma 3.38. Any closed subset X of a complete hyperseminormed vector space L is complete.

Indeed, if $l = \{a_i; i = 1, 2, 3, \ldots\}$ is a Cauchy sequence in X, then it has a limit a in L. As the set X is closed, a belongs to X. So, the space X is complete because l is an arbitrary sequence in X.

As any compact subset K of a topological space is closed (Kelly, 1955), we have the following result.

Lemma 3.39. Any compact subset K of a complete hyperseminormed space L is complete.

Proof is left as an exercise.

Let us consider seminormed vector spaces L_i ($i = 1, 2, 3, \ldots, n$). Then it is possible to build the Cartesian product $L_1 \times L_2 \times L_3 \times \ldots \times L_n$ of these spaces (cf. Appendix). We remind that elements of are vectors $(x_1, x_2, x_3, \ldots, x_n)$ where $x_i \in L_i$ ($i = 1, 2, 3, \ldots, n$).

There are different approaches to extending the seminorms in the vector spaces L_i to a seminorm in their Cartesian product. Let us study some of them.

Definition 3.49. The *Euclidean seminorm* (*norm*) on the Cartesian product $L_1 \times L_2 \times L_3 \times \ldots \times L_n$ is defined by the following formula

$$||x||_E = (||x_1||^2 + ||x_2||^2 + ||x_3||^2 + \ldots + ||x_n||^2)^{1/2} \tag{3.1}$$

Now we need to prove that Formula (3.1) really defines a seminorm or a norm on the Cartesian product,

Lemma 3.40. a) Formula (3.1) defines a seminorm on the Cartesian product $L_1 \times L_2 \times L_3 \times \ldots \times L_n$.

b) It is a norm if and only if all L_i are normed vector spaces ($i = 1, 2, 3, \ldots, n$).

Proof. (a) Let us check the axioms of seminorm.

N1: $q(\mathbf{0}) = q(\mathbf{0}_1, \mathbf{0}_2, \ldots, \mathbf{0}_n) = \max\{q_i(\mathbf{0}_i); i = 1, 2, 3, \ldots, n\} = 0$ as all q_i are seminorms.

N2: $q(ax) = q(ax_1, ax_2, \ldots, ax_n) = \max\{q_i(ax_i); i = 1, 2, 3, \ldots, n\} = \max\{|a|q_i(x_i); i = 1, 2, 3, \ldots, n\} = |a| \cdot q(x)$ for any x from L and any number a from $\boldsymbol{R}$ as all q_i are seminorms.

N3 (the triangle inequality or subadditivity):

$$q(x + y) = q_1(x_1 + y_1) + q_2(x_2 + y_2) + q_3(x_3 + y_3) + \dots + q_n(x_n + y_n) \leq$$

$$q_1(x_1) + q_1(y_1) + q_2(x_2) + q_2(y_2) + q_3(x_3) + q_3(y_3) + \dots + q_n(x_n) + q_n(y_n) = q(x) + q(y)$$

for any x and y from L

(b) If all q_i are norms, then $q(x) = 0$ implies all $q_i(x_i) = 0$. Thus, all $x_i = \mathbf{0}_i$ ($i = 1, 2, 3, \dots, n$). Consequently, $x = (\mathbf{0}_1, \mathbf{0}_2, \dots, \mathbf{0}_n) = \mathbf{0}$, i.e., q is a norm.

If at least, one of the seminorms, say q_i, is not a norm, then there are elements $x_i \neq y_i$ from L_i such that $q_i(x_i) = q_i(y_i)$. Then

$$q(\mathbf{0}_1, \mathbf{0}_2, \dots, x_i, \dots, \mathbf{0}_n) = q_i(x_i) = q_i(y_i) = q(\mathbf{0}_1, \mathbf{0}_2, \dots, x_i, \dots, \mathbf{0}_n)$$

although

$$(\mathbf{0}_1, \mathbf{0}_2, \dots, x_i, \dots, \mathbf{0}_n) \neq (\mathbf{0}_1, \mathbf{0}_2, \dots, x_i, \dots, \mathbf{0}_n)$$

Thus, q is not a norm.

(c) q is a seminorm if and only if it takes values only in the set $\boldsymbol{R}$ and this true if and only if all q_i take values only in the set $\boldsymbol{R}$ because the values of all q_i are non-negative, i.e., all L_i are seminormed vector spaces ($i = 1, 2, 3, \dots, n$).

Parts (b) and (c) imply (d).

Lemma is proved.

Example 3.37. A popular method in statistics is *ordinary least squares* or *linear least squares* used for estimation of the unknown parameters in linear regression models. This method minimizes the Euclidean norm of the difference between the observed responses in the dataset and the responses predicted by the linear approximation.

Remark 3.3. Defining squares and square roots for real hypernumbers, it is possible to build the Euclidean hyperseminorm on the Cartesian product of hyperseminormed vector spaces. There are different techniques for defining squares and square roots for real hypernumbers. Some are based on hypernumber extensions (Burgin, 2005a). Others utilize bundles with a hyperspace base (Burgin, 2010, 2012). However, this involves advanced fragments of the theory of hypernumbers, which go far beyond those presented in Chapter 2. That is why we consider here only Euclidean seminorms and norms.

Let us consider hyperseminormed vector spaces L_i ($i = 1, 2, 3, \dots, n$). There are different approaches to extending the hyperseminorms in the vector spaces L_i to a hyperseminorm in their Cartesian product. Let us study some of them.

Definition 3.50. The *Manhattan hyperseminorm* (*seminorm* or *norm*), also called *rectilinear hyperseminorm* (*seminorm* or *norm*), L_1 *hyperseminorm* (*seminorm* or *norm*), l_1 *hyperseminorm* (*seminorm* or *norm*), *city block hyperseminorm* (*seminorm* or *norm*), or *taxicab hyperseminorm* (*seminorm* or *norm*), on the Cartesian product $L_1 \times L_2 \times L_3 \times \ldots \times L_n$ is defined by the following formula

$$||x||_M = ||x_1|| + ||x_2|| + ||x_3|| + \ldots + ||x_n|| \quad (3.2)$$

The Manhattan, or taxicab, norm is often used for different applications (Krause, 1987).

Lemma 3.41. a) Formula (3.2) defines a hyperseminorm on the Cartesian product $L_1 \times L_2 \times L_3 \times \ldots \times L_n$.

b) It is a hypernorm if and only if all L_i are hypernormed vector spaces ($i = 1, 2, 3, \ldots, n$).

c) It is a seminorm if and only if all L_i are seminormed vector spaces ($i = 1, 2, 3, \ldots, n$).

d) It is a norm if and only if all L_i are normed vector spaces ($i = 1, 2, 3, \ldots, n$).

Proof. (a) Let us check the axioms of hyperseminorm.

N1: $q(\mathbf{0}) = q(\mathbf{0}_1, \mathbf{0}_2, \ldots, \mathbf{0}_n) = \max\{q_i(\mathbf{0}_i); i = 1, 2, 3, \ldots, n\} = 0$ as all q_i are hyperseminorms.

N2: $q(ax) = q(ax_1, ax_2, \ldots, ax_n) = \max\{q_i(ax_i); i = 1, 2, 3, \ldots, n\} = \max\{|a|q_i(x_i); i = 1, 2, 3, \ldots, n\} = |a|\cdot q(x)$ for any x from L and any number a from $\boldsymbol{R}$ as all q_i are hyperseminorms.

N3 (the triangle inequality or subadditivity):

$$q(x + y) = q_1(x_1 + y_1) + q_2(x_2 + y_2) + q_3(x_3 + y_3) + \ldots + q_n(x_n + y_n) \leq$$

$$q_1(x_1) + q_1(y_1) + q_2(x_2) + q_2(y_2) + q_3(x_3) + q_3(y_3) + \ldots + q_n(x_n) + q_n(y_n) = q(x) + q(y)$$

for any x and y from L.

(b) If all q_i are hypernorms, then $q(x) = 0$ implies all $q_i(x_i) = 0$. Thus, all $x_i = \mathbf{0}_i$ ($i = 1, 2, 3, \ldots, n$). Consequently, $x = (\mathbf{0}_1, \mathbf{0}_2, \ldots, \mathbf{0}_n) = q$, i.e., q is a hypernorm.

At the same time, if at least, one of the hyperseminorms, say q_i is not a hypernorm, then there are elements $x_i \neq y_i$ from L_i such that $q_i(x_i) = q_i(y_i)$. Then

$$q(\mathbf{0}_1, \mathbf{0}_2, \ldots, x_i, \ldots, \mathbf{0}_n) = q_i(x_i) = q_i(y_i) = q(\mathbf{0}_1, \mathbf{0}_2, \ldots, x_i, \ldots, \mathbf{0}_n)$$

although

$$(\mathbf{0}_1, \mathbf{0}_2, \ldots, x_i, \ldots, \mathbf{0}_n) \neq (\mathbf{0}_1, \mathbf{0}_2, \ldots, y_i, \ldots, \mathbf{0}_n)$$

Thus, q is not a hypernorm.

(c) q is a seminorm if and only if it takes values only in the set $\boldsymbol{R}$ and this true if and only if all q_i take values only in the set $\boldsymbol{R}$ because the values of all q_i are non-negative, i.e., all L_i are seminormed vector spaces ($i = 1, 2, 3, \ldots, n$).

Parts (b) and (c) imply (d).

Lemma is proved.

Let us consider hyperseminormed vector spaces L_i ($i = 1, 2, 3, \ldots, n$).

Definition 8.51. The *Chebyshev hyperseminorm* (*seminorm* or *norm*), also called *chessboard hyperseminorm* (*seminorm* or *norm*), *uniform hyperseminorm* (*seminorm* or *norm*), or *maximum hyperseminorm* (*seminorm* or *norm*), on the Cartesian product $L_1 \times L_2 \times L_3 \times \ldots \times L_n$ is defined by the following formula

$$||x||_{\mathrm{Ch}} = ||(x_1, x_2, x_3, \ldots, x_n)|| = \max\{||x_i||;\ i = 1, 2, 3, \ldots, n\} \quad (3.3)$$

Where $||x_i||$ is the hyperseminorm of the element x_i in the space L_i.

In particular, for $n = 2$, we have

$$||x||_{\mathrm{Ch}} = ||(x_1, x_2)|| = \max\{||x_1||, ||x_2||\}$$

The Chebyshev, or chessboard, norm is often used for different applications (Cantrell et al., 2000; Abello et al., 2002).

Proposition 3.22. a) Formula (3.3) defines a hyperseminorm q on the Cartesian product

$$L_1 \times L_2 \times L_3 \times \ldots \times L_n.$$

b) It is a hypernorm if and only if all L_i are hypernormed vector spaces ($i = 1, 2, 3, \ldots, n$).

c) It is a seminorm if and only if all L_i are seminormed vector spaces ($i = 1, 2, 3, \ldots, n$).

d) It is a norm if and only if all L_i are normed vector spaces ($i = 1, 2, 3, \ldots, n$).

Proof. (a) Let us check the axioms of hyperseminorm.

N1: $q(\mathbf{0}) = q(\mathbf{0}_1, \mathbf{0}_2, \ldots, \mathbf{0}_n) = \max\{q_i(\mathbf{0}_i); i = 1, 2, 3, \ldots, n\} = 0$ as all q_i are hyperseminorms.

N2: $q(ax) = q(ax_1, ax_2, \ldots, ax_n) = \max\{q_i(ax_i); i = 1, 2, 3, \ldots, n\} = \max\{|a|q_i(x_i); i = 1, 2, 3, \ldots, n\} = |a|\cdot q(x)$ for any x from L and any number a from $\boldsymbol{R}$ as all q_i are hyperseminorms.

N3 (the triangle inequality or subadditivity):

$$q(x+y) = \max\{ q_i(x_i + y_i); i = 1, 2, 3, \ldots, n\} \leq \max\{q_i(x_i) + q_i(y_i); = 1, 2, 3, \ldots, n\}$$

$$\leq \max\{q_i(x_i); i = 1, 2, 3, \ldots, n\} + \max\{| q_i(y_i); i = 1, 2, 3, \ldots, n\} = q(x) + q(y)$$

for any x and y from L

(b) If all q_i are hypernorms, then $q(x) = 0$ implies all $q_i(x_i) = 0$. Thus, all $x_i = \mathbf{0}_i (i = 1, 2, 3, \ldots, n)$. Consequently, $x = (\mathbf{0}_1, \mathbf{0}_2, \ldots, \mathbf{0}_n) = q$, i.e., q is a hypernorm.

At the same time, if at least, one of the hyperseminorms, say q_i is not a hypernorm, then there are elements $x_i \neq y_i$ from L_i such that $q_i(x_i) = q_i(y_i)$. Then

$$q(\mathbf{0}_1, \mathbf{0}_2, \ldots, x_i, \ldots, \mathbf{0}_n) = q_i(x_i) = q_i(y_i) = q(\mathbf{0}_1, \mathbf{0}_2, \ldots, x_i, \ldots, \mathbf{0}_n)$$

although

$$(\mathbf{0}_1, \mathbf{0}_2, \ldots, x_i, \ldots, \mathbf{0}_n) \neq (\mathbf{0}_1, \mathbf{0}_2, \ldots, x_i, \ldots, \mathbf{0}_n)$$

Thus, q is not a hypernorm.

(c) q is a seminorm if and only if it takes values only in the set $\boldsymbol{R}$ and this true if and only if all q_i take values only in the set $\boldsymbol{R}$, i.e., all L_i are seminormed vector spaces ($i = 1, 2, 3, \ldots, n$).

Parts (b) and (c) imply (d).

Proposition is proved.

One more useful topological structure in algebraic systems is a quasinorm. Although this structure was introduced in 1940s under the name *absolute value* (Aoki, 1942; Bourgin, 1941, 1943; Hyers, 1939, 1945), it became popular only recently [cf., e.g., (Kalton and Sik-Chung Tam, 1993; Gordon and Kalton, 1994; Kalton, 2003; Tabor, 2004; Rano and Bag, 2015)].

Let L be a vector (linear) space over a normed (valuated) field K, e.g., a real vector (linear) space over $\boldsymbol{R}$.

Definition 3.52. A mapping q: $L \to \boldsymbol{R}^+$ is called a *quasi-norm* if it satisfies the following conditions:

N1. $q(x) = 0$ if and only if $x = 0$ for any x from L

N2. $q(ax) = |a| \cdot q(x)$ for any x from L and any number a from $\boldsymbol{R}$.

QN3 (the generalized triangle inequality).

$q(x + y) \leq k(q(x) + q(y))$ for some number $k \geq 1$ and any x and y from L

Naturally, any norm a quasi-norm.

Let us consider some examples.

Example 3.38. Let us take $p \in (0, 1)$ and consider the space l^p of sequences of real numbers $l = \{x_i;\ i = 1, 2, 3, \ldots\}$ such that $(\Sigma_{i=1}^{\infty} x_i^{1/p})^p < \infty$. It is possible to define the mapping $\|.\|_l p$: $l \to \boldsymbol{R}$ by the following formula

$$\|x\|_l p = (\Sigma_{i=1}^{\infty} |x_i|^{1/p})^p$$

It is not a norm because as it is proved in Hardy et al (1952)

$\|x + y\|_l p = (\Sigma_{i=1}^{\infty} |x_i + y_i|^{1/p})^p > (\Sigma_{i=1}^{\infty} |x_i|^{1/p})^p + (\Sigma_{i=1}^{\infty} |x_i|^{1/p})^p) = (\|x\|_l p + \|y\|_l p)$

However, it is a quasi-norm in the space l^p since we have the following inequalities

$$\|x + y\|_l p = (\Sigma_{i=1}^{\infty} |x_i + y_i|^{1/p})^p \leq (\Sigma_{i=1}^{\infty} |x_i|^{1/p} + \Sigma_{i=1}^{\infty} |y_i|^{1/p})^p \leq$$

$$2^{p-1}(\Sigma_{i=1}^{\infty} |x_i|^{1/p})^p + (\Sigma_{i=1}^{\infty} |x_i|^{1/p})^p) = 2^{p-1}(\|x\|_l p + \|y\|_l p)$$

Remark 3.4. There are several other generalizations of the concept *norm*, which are not studied in this book. Examples of such generalizations are q-norms (Litvak, 1998), asymmetric norms (García-Raffi et al., 2002, 2003; Künzi, 2009; Cobzas, 2012), fuzzy norms (Felbin, 1992; Chang and Mordeson, 1994; Krishna and Sarma, 1994; Rhie et al., 1997; Bag and Samanta, 2003) and probabilistic norms (Schweizer and Sklar, 1983; Alsina et al., 1993, 1997; Guillen et al., 1997, 1998, 1999; Iqbal and Radhi, 2003; Lael and Nourouzi, 2009).

Remark 3.5. Some mathematicians used the name *quasi-norm* for other constructions. For instance, Brown and Pearcy (1997) define quasi-norm as a function q from a vector space L into the set $\boldsymbol{R}^+$ of non-negative real numbers that satisfies the following conditions:

1. $q(x) = 0$ if and only if $x = 0$ for any x from L
2. $q(ax) \leq q(x)$ for any x from L and any number $a \leq 1$ from $\boldsymbol{R}^+$.
3. $q(x + y) \leq q(x) + q(y)$ for any x and y from L
4. If $a_n \to 0$ implies $q(a_n \cdot x) \to 0$ for any a_n from $\boldsymbol{R}$ and x from L

Metzler and Nakano (1966) suggest a different concept. Namely, a real function q on a vector space L is called a quasi-norm if it satisfies the following conditions:

1. $0 \leq q(x) < +\infty$ for any x from L
2. $q(\mathbf{0}) = 0$
3. $|a| \leq |b|$ implies that $q(ax) \leq q(bx)$ for any x from L and any numbers a and b from $\boldsymbol{R}$
4. $q(x + y) \leq q(x) + q(y)$ for any x and y from L

One more definition of a quasi-norm is suggested by Yosida (1968).

It means that several distinct constructions of quasi-norms were introduced and studied by different authors. Thus, as Pietsch writes (2007), various concepts of pseudo-norms, quasi-norms, seminorms and quasi-seminorm appeared in functional analysis making the terminology exceedingly non-uniform. Here we study generalizations constructed by using the generalized triangle inequality.

Definition 3.53. A mapping q: $L \to \boldsymbol{R}$ is called a *quasi-seminorm* if it satisfies the following axioms:

SN1. $q(x) = 0$ if $x = \mathbf{0}$.

N2. $q(ax) = |a| \cdot q(x)$ for any x from L and any number a from $\boldsymbol{R}$.

QN3 (the generalized triangle inequality).

$q(x + y) \leq k(q(x) + q(y))$ for some number $k \geq 1$ and any x and y from L

Naturally, any quasi-norm and any seminorm are quasi-seminorms. However, there are quasi-seminorms that are not quasi-norms or seminorms.

Example 3.39. Let us take $p \in (0, 1)$ and consider the space $L^p(\boldsymbol{R}, \boldsymbol{R})$ of all absolutely p-integrable functions on the real line $\boldsymbol{R}$. It is possible to define the mapping $||.||_{L^p}$:$L^p(\boldsymbol{R}, \boldsymbol{R}) \to \boldsymbol{R}$ by the following formula

$$||f||_{L^p} = [\int_{-\infty}^{\infty} |f(x)|^p \, dx]^{1/p}$$

It is not a seminorm because as it is proved in Hardy et al. (1952)

$$||f + g||_{L^p} = [\int_{-\infty}^{\infty} |f(x) + g(x)|^p \, dx]^{1/p} > [\int_{-\infty}^{\infty} |f(x)|^p \, dx]^{1/p} + [\int_{-\infty}^{\infty} |g(x)|^p \, dx]^{1/p} = ||f||_{L^p} + ||g||_{L^p}$$

However, it is a quasi-seminorm in the space $L^p(\boldsymbol{R}, \boldsymbol{R})$ since we have the following inequality

$$||f + g||_{L^p} = [\int_{-\infty}^{\infty} |f(x) + g(x)|^p \, dx]^{1/p} \leq 2^{p-1}([\int_{-\infty}^{\infty} |f(x)|^p \, dx]^{1/p} + [\int_{-\infty}^{\infty} |g(x)|^p \, dx]^{1/p})$$
$$= 2^{p-1}(||f||_{L^p} + ||g||_{L^p})$$

Though, it is not a quasi-norm, and thus, not a norm, in $L^p(\boldsymbol{R}, \boldsymbol{R})$.because if a function h from $L^p(\boldsymbol{R}, \boldsymbol{R})$ is equal to zero almost everywhere, i.e., the measure of the set where h is not equal to zero is zero, then $||h||_{L^p} = 0$ although h is not zero in $L^p(\boldsymbol{R}, \boldsymbol{R})$.

At the same time, if we introduce equivalence of functions that different on a set with the zero measure, the introduced quasi-seminorm will be a quasi-norm in the quotient space of $L^p(\boldsymbol{R}, \boldsymbol{R})$.

Lemma 3.42. If q: $L \to \boldsymbol{R}$ is a quasi-seminorm, then $q(x) \geq 0$ and $q(-x) = q(x)$ for all $x \in L$.

Proof. By QN3, we have

$$k(q(x) + q(-x)) \geq q(x + (-x)) = q(\mathbf{0})$$

At the same time, by N2, we have $q(\mathbf{0}) = q(0 \cdot \mathbf{0}) = 0 \cdot q(\mathbf{0}) = 0$ and $q(-x) = |-1| \cdot q(x) = q(x)$. Thus,

$$k(q(x) + q(-x)) = k(q(x) + q(x)) = 2kq(x) \geq 0$$

and as $k \geq 1$, we have

$$q(x) \geq 0$$

Lemma is proved.

Corollary 3.72. If q: $L \to \boldsymbol{R}$ is a quasi-norm, then $q(x) \geq 0$ and $q(-x) = q(x)$ for all $x \in L$.

Taking a vector space L over a topological (normed) field $\boldsymbol{F}$ with a system Q of quasi-norms (quasi-seminorms), it is possible to define operations of addition and scalar multiplication of quasi-norms and quasi-seminorms in the same way as it is done above for of norms and seminorms.

Proposition 3.23. a) If q_1 and q_2 are quasi-norms (quasi-seminorms), then their sum $q = q_1 + q_2$ is a quasi-norm (quasi-seminorm).

b) If $q$is a quasi-norm (quasi-seminorm) and a is a nonnegative element from $\boldsymbol{R}$, then their scalar multiplication $aq = q_1$ is a quasi-norm (quasi-seminorm).

Proof is left as an exercise.

As in the case of norms and seminorms, we go to quasi-hypernorms and quasi-hyperseminorm from quasi-norms and quasi-seminorms.

Let L be a vector (linear) space over a normed (valuated) field K, e.g., a real vector (linear) space over $\boldsymbol{R}$.

Definition 3.54. A mapping q: $L \to \boldsymbol{R}_\omega$ is called a *quasi-hypernorm* if it satisfies the following conditions:

N1. $q(x) = 0$ if and only if $x = 0$ for any x from L

N2. $q(ax) = |a| \cdot q(x)$ for any x from L and any number a from $\boldsymbol{R}$.

QN3 (the generalized triangle inequality).

$q(x + y) \leq k(q(x) + q(y))$ for some number $k \geq 1$ and any x and y from L

Naturally, any hypernorm, quasi-norm or norm is a quasi-hypernorm.

Let us consider some examples.

Example 3.40. Let us take $p \in (0, 1)$ and consider the space l of all sequences of real numbers $x = \{x_i; i = 1, 2, 3, \ldots\}$. It is possible to define the mapping $||.||_{lp}$: $l \to \boldsymbol{R}$ by the following formula

$$||x||_{lp} = \text{Hn}(d_n)_{n\in\omega}$$

where $d_n = (\Sigma_{i=1}{}^{n}|x_i|^{1/p})^p$

It is not a norm because as it is proved in Hardy et al. (1952)

$$(\Sigma_{i=1}{}^{n}|x_i + y_i|^{1/p})^p > (\Sigma_{i=1}{}^{n}|x_i|^{1/p})^p + (\Sigma_{i=1}{}^{n}|x_i|^{1/p})^p)$$

and thus, by properties of hypernumbers (cf. Chapter 2), we have

$$||x + y||_{lp} > ||x||_{lp} + ||y||_{lp}$$

However, it is a quasi-norm in the space l^p since we have the following inequalities

$$||x + y||_{lp} = (\Sigma_{i=1}{}^{\infty}|x_i + y_i|^{1/p})^p \leq (\Sigma_{i=1}{}^{\infty}|x_i|^{1/p} + \Sigma_{i=1}{}^{\infty}|y_i|^{1/p})^p \leq$$

$$2^{p-1}(\Sigma_{i=1}{}^{\infty}|x_i|^{1/p})^p + (\Sigma_{i=1}{}^{\infty}|x_i|^{1/p})^p) = 2^{p-1}(||x||_{lp} + ||y||_{lp})$$

and thus, by properties of hypernumbers (cf. Chapter 2), we have

$$||x + y||_{lp} \leq 2^{p-1}(\,||x||_{lp} + ||y||_{lp})$$

Definition 3.55. A mapping $q: L \to \mathbf{R}_\omega$ is called a *quasi-hyperseminorm* if it satisfies the following axioms:

SN1. $q(x) = 0$ if $x = \mathbf{0}$.

N2. $q(ax) = |a| \cdot q(x)$ for any x from L and any number a from $\mathbf{R}$.

QN3 (the generalized triangle inequality).

$q(x + y) \leq k(q(x) + q(y))$ for some number $k \geq 1$ and any x and y from L

Naturally, any hypernorm, quasi-norm, hyperseminorm, seminorm or norm is a quasi-hyperseminorm.

Example 3.41. Let us take a number $p \in (0, 1)$ and consider the space $Loc^P(\mathbf{R}, \mathbf{R})$ of all locally absolutely p-integrable functions on the real line $\mathbf{R}$. It is possible to define the mapping $||.||_{L^P}$: $Loc^P(\mathbf{R}, \mathbf{R}) \to \mathbf{R}$ by the following formula

$$||f||_{Loc^p} = \text{Hn}(d_n)_{n \in \omega}$$

where $d_n = [\int_{-n}^{n} |f(x)|^p \, dx]^{1/p}$

It is not a norm because as it is proved in Hardy et al. (1952)

$$[\int_{-n}^{n} |f(x) + g(x)|^p \, dx]^{1/p} > [\int_{-n}^{n} |f(x)|^p \, dx]^{1/p} + [\int_{-n}^{n} |g(x)|^p \, dx]^{1/p}$$

and thus, by properties of hypernumbers (cf. Chapter 2), we have

$$||f + g||_{Loc^p} > ||f||_{Loc^p} + ||g||_{Loc^p}$$

However, it is a quasi-norm in the space $Loc^P(\mathbf{R}, \mathbf{R})$ since we have the following inequality

$$[\int_{-n}^{n} |f(x) + g(x)|^p \, dx]^{1/p} \leq 2^{p-1}([\int_{-n}^{n} |f(x)|^p \, dx]^{1/p} + [\int_{-n}^{n} |g(x)|^p \, dx]^{1/p})$$

and thus, by properties of hypernumbers (cf. Chapter 2), we have

$$||f + g||_{Loc^p} \leq 2^{p-1}(||f||_{L^p} + ||g||_{Loc^p})$$

Lemma 3.43. If $q: L \to \mathbf{R}_\omega$ is a quasi-hyperseminorm, then $q(x) \geq 0$ and $q(-x) = q(x)$ for all $x \in L$.

Proof. By QN3, we have

$$k(q(x) + q(-x)) \geq q(x + (-x)) = q(\mathbf{0})$$

At the same time, by N2, we have $q(\mathbf{0}) = q(0 \cdot \mathbf{0}) = 0 \cdot q(\mathbf{0}) = 0$ and $q(-x) = |-1| \cdot q(x) = q(x)$. Thus,

$$k(q(x) + q(-x)) = k(q(x) + q(x)) = 2kq(x) \geq 0$$

and as $k \geq 1$, we have

$$q(x) \geq 0$$

Lemma is proved.

Corollary 3.73. If q: $L \to \boldsymbol{R}$ is a quasi-hypernorm, then $q(x) \geq 0$ and $q(-x) = q(x)$ for all $x \in L$.

Definition 3.56. a) A vector/linear space L with a quasi-norm (quasi-seminorm) is called a *quasi-normed* (*quasi-seminormed*) *vector/linear space* or simply, a *quasi-normed* (*quasi-seminormed*) *space*.

b) A vector/linear space L with a quasi-hypernorm (quasi-hyperseminorm) is called a *quasi-hypernormed* (*quasi-hyperseminormed*) *vector/linear space* or simply, a *quasi-hypernormed* (*quasi-hyperseminormed*) *space*.

Taking a vector space L over a topological (normed) field $\boldsymbol{F}$ with a system Q of quasi-hypernorms (quasi-hyperseminorms), it is possible to define operations of addition and scalar multiplication of quasi-hypernorms and quasi-hyperseminorms in the same way as it is done above for of hypernorms and hyperseminorms.

Addition:

If q_1 and q_2 belong to Q, then for any element v from L, we define their sum $q = q_1 + q_2$ by the following rule

$$q(x) = q_1(x) + q_2(x),$$

Scalar multiplication by elements from $\boldsymbol{F}^{+}$:

If q belongs to Q and a is a nonnegative element from $\boldsymbol{F}$, then for any element v from L, we define their scalar multiplication $aq = q_1$ by the following rule

$$q_1(x) = a \cdot q(x)$$

Proposition 3.24. a) If q_1 and q_2 are quasi-hypernorms (quasi-hyperseminorms), then their sum $q = q_1 + q_2$ is a quasi-hypernorm (quasi-hyperseminorm).

b) If q is a quasi-hypernorm (quasi-hyperseminorm) and a is a nonnegative element from $\boldsymbol{F}$, then their scalar multiplication $aq = q_1$ is a quasi-hypernorm (quasi-hyperseminorm).

Proof. We need to check axioms SN1 (N1), N2 and QN3 for new functions. Proof for axioms SN1, N1, N2 and QN3 are the same as in Proposition 3.10 and we need only to verify axioms QN3 assuming that this axiom is true for q_1 and q_2.

(a) QN3 for the sum:

As $q_1(x + y) \leq k(q_1(x) + q_1(y))$ for some number $k \geq 1$ and all x and y from L and $q_2(x + y) \leq h(q_2(x) + q_2(y))$ for some number $h \geq 1$ and all x and y from L

When $r = \max(k, h)$, these inequalities imply

$$q(x + y) = q_1(x + y) + q_2(x + y) \leq k(q_1(x) + q_1(y)) + h(q_2(x) + q_2(y)) \leq$$

$$r(q_1(x) + q_1(y)) + r(q_2(x) + q_2(y)) = r(q_1(x) + q_1(y) + q_2(x) + q_2(y)) =$$

$$r(q_1(x) + q_2(x)) + r(q_1(y) + q_2(y)) = r(q(x) + q(y))$$

for any z and v from L.

(b) N3 for the scalar product:

As $a > 0$, we have

$$q_1(x + y) = aq(x + y) \leq ak(q(x) + q(y)) = k(aq(x) + aq(y)) = k(q_1(x) + q_1(y))$$

for any z and v from L because q is a quasi-hyperseminorm.

Proposition is proved.

Corollary 3.74. If q_1 is a quasi-hypernorm and q_2 is a quasi-hyperseminorms, then their sum $q = q_1 + q_2$ is a quasi-hypernorm.

Corollary 3.75. Any linear combination with positive coefficients of quasi-hyperseminorms (quasi-hypernorms) in L is a quasi-hyperseminorms (quasi-hypernorm).

Corollary 3.76. a) All quasi-hyperseminorms in L form a semimodule over the semiring $\boldsymbol{R}^+$.

b) All quasi-hypernorms in L form an S-module over the S-ring $\boldsymbol{R}^{++}$.

As it is done for norms, seminorms, hypernorms and hyperseminorms, it is possible to introduce a partial order in the set of al quasi-hypernorms (quasi-hyperseminorms) in a vector space. Namely, a quasi-hyperseminorm (quasi-hypernorm) q is *stronger* than the quasi-hyperseminorm (quasi-hypernorm)

p and p is *weaker* than q(it is denoted by $p \preceq q$) if there is a constant $k \geq 0$ such that for all $x \in L$,

$$p(x) \leq k \cdot q(x)$$

b) Quasi-hyperseminorm (quasi-hypernorm) q and p are *equivalent* if there are constants $k, h \geq 0$ such that for all $x \in L$,

$$h \cdot q(x) \leq p(x) \leq k \cdot q(x)$$

Lemma 3.44. Quasi-hyperseminorms (quasi-hypernorms) q and p in a vector space L are equivalent if and only if $p \preceq q$ and $q \preceq p$.

Proof is left as an exercise.

Lemma 3.45. The relation $\preceq$ defines a partial preorder in sets of quasi-hyperseminorms (quasi-hypernorms).

Proof is left as an exercise.

This partial preorder in sets of quasi-hyperseminorms (quasi-hypernorms) is compatible with operations introduced above as the following result demonstrates.

Let us consider three quasi-hypernorms (quasi-hyperseminorms) q_1, q_2 and q_3.

Proposition 3.25. a) If $q_1 \preceq q_2$, then $q_1 + q_3 \preceq q_2 + q_3$ for any quasi-hypernorm (quasi-hyperseminorm) q_3.

b) If $q_1 \preceq q_2$, then $a \cdot q_1 \preceq a \cdot q_2$ for any nonnegative element a from $\boldsymbol{F}$.

c) If $q_1 \preceq q_2$, then their maximum $\max(q_1, q_3) \preceq \max(q_2, q_3)$ for any quasi-hypernorm (quasi-hyperseminorm) q_3.

d) If $q_1 \preceq q_2$, then their minimum $\min(q_1, q_3) \preceq \min(q_2, q_3)$ for any quasi-hypernorm (quasi-hyperseminorm) q_3.

Proof. (a) The inequality $q_1 \preceq q_2$ means that there is a real number $k \geq 0$ such that for all $x \in L$,

$$q_1(x) \leq k \cdot q_2(x)$$

Consequently, as $k \geq 0$, we have

$$(q_1 + q_3)(x) = q_1(x) + q_3(x) \leq k \cdot q_2(x) + q_3(x) \leq k \cdot q_2(x) + (1 + k) \cdot q_3(x) \leq$$

$$(1 + k) \cdot (q_2(x) + q_3(x)) = (1 + k) \cdot (q_2 + q_3)(x)$$

It means that $q_1 + q_3 \leq q_2 + q_3$.

(b) In a similar way, $q_1(x) \leq k \cdot q_2(x)$ implies $a \cdot q_1(x) \leq ka \cdot q_2(x)$ for all $x \in L$. Thus, $a \cdot q_1 \leq a \cdot q_2$.

Statements (c) and (d) are proved in a similar way.

Proposition is proved.

Let us consider three quasi-norms (quasi-seminorms) q_1, q_2 and q_3.

Corollary 3.77. a) If $q_1 \leq q_2$, then $q_1 + q_3 \leq q_2 + q_3$ for any quasi-norm (quasi-seminorm) q_3.

b) If $q_1 \leq q_2$, then $a \cdot q_1 \leq a \cdot q_2$ for any nonnegative element a from $\boldsymbol{F}$.

c) If $q_1 \leq q_2$, then their maximum $\max(q_1, q_3) \leq \max(q_2, q_3)$ for any quasi-norm (quasi-seminorm) q_3.

d) If $q_1 \leq q_2$, then their minimum $\min(q_1, q_3) \leq \min(q_2, q_3)$ for any quasi-norm (quasi-seminorm) q_3.

Proposition 3.25 and Corollary 3.77 mean that addition, scalar multiplication, taking maximum or minimum are monotonous operations in the systems of quasi-hyperseminorms, quasi-hypernorms, quasi-seminorms and quasi-norms in vector spaces.

Let us consider three quasi-hypernorms (quasi-hyperseminorms) $||\cdot||_1$, $||\cdot||_2$ and $||\cdot||_3$.

Proposition 3.23. $||\cdot||_1 \leq ||\cdot||_3$ and $||\cdot||_2 \leq ||\cdot||_3$ if and only if $||\cdot||_1 + ||\cdot||_2 \leq ||\cdot||_3$.

Proof. Necessity. The inequality $||\cdot||_1 \leq ||\cdot||_3$ means that there is a real number $k \geq 0$ such that for all $x \in L$,

$$||x||_1 \leq k \cdot ||x||_3$$

and the inequality $||\cdot||_2 \leq ||\cdot||_3$ means that there is a real number $h \geq 0$ such that for all $x \in L$,

$$||x||_2 \leq h \cdot ||x||_3$$

Consequently, as $k, h \geq 0$, we have

$$||\cdot||_1 + ||\cdot||_2 \leq k \cdot ||\cdot||_3 + h \cdot ||\cdot||_3 \leq (k+h) \cdot ||\cdot||_3$$

It means that $||\cdot||_1 + ||\cdot||_2 \leq ||\cdot||_3$.

Sufficiency. Let us assume $||\cdot||_1 + ||\cdot||_2 \leq ||\cdot||_3$. It means that there is a real number $k \geq 0$ such that for all $x \in L$,

$$||x||_1 + ||x||_2 \leq k \cdot ||x||_3$$

Consequently, we have

$$||x||_1 \leq ||x||_1 + ||x||_2 \leq k \cdot ||x||_3$$

and the inequality $||.||_2 \leq ||.||_3$ means that there is a real number $h \geq 0$ such that for all $x \in L$,

$$||x||_2 \leq ||x||_1 + ||x||_2 \leq k \cdot ||x||_3$$

This implies inequalities $||.||_1 \leq ||.||_3$ and $||.||_2 \leq ||.||_3$.

Proposition is proved.

Let us consider three quasi-norms (quasi-seminorms) $||.||_1$, $||.||_2$ and $||.||_3$.

Corollary 3.78. $||.||_1 \leq ||.||_3$ and $||.||_2 \leq ||.||_3$ if and only if $||.||_1 + ||.||_2 \leq ||.||_3$.

Let us consider two quasi-hypernorms (quasi-hyperseminorms) $||.||_1$ and $||.||_3$.

Proposition 3.26. a) If $||.||_1 \leq ||.||_3$, then $a||.||_1 \leq ||.||_3$ for any non-negative number a.

b) If $a||.||_1 \leq ||.||_3$ for a positive number a, then $||.||_1 \leq ||.||_3$.

Proof. (a) The inequality $||x||_1 \leq k \cdot ||x||_3$ implies the inequality $a \cdot ||x||_1 \leq ka \cdot ||x||_3$ for all $x \in L$. Thus, $a \cdot ||.||_1 \leq a \cdot ||.||_3$.

(b) The inequality $a \cdot ||x||_1 \leq k \cdot ||x||_3$ implies the inequality $||x||_1 \leq (h)^{-1} k \cdot ||x||_3$, which means that $||.||_1 \leq ||.||_3$.

Proposition is proved.

Corollary 3.79. The set $\mathbf{QHSN}_{||\cdot||}$ of all quasi-hyperseminorms that are weaker than a quasi-hyperseminorm $||.||$ is a semimodule over the semiring of all positive real numbers $\boldsymbol{R}^{+}$.

Corollary 3.80. The set $\mathbf{QHN}_{||\cdot||}$ of all quasi-hypernorms that are weaker than a quasi-hypernorm $||.||$ is an S-module over the S-ring of all positive real numbers $\boldsymbol{R}^{++}$.

Let us consider two quasi-norms (quasi-seminorms) $||.||_1$ and $||.||_3$.

Corollary 3.81. If $||.||_1 \leq ||.||_3$, then $a||.||_1 \leq ||.||_3$ for any non-negative number a.

Corollary 3.82. The set $\mathbf{QSN}_{||\cdot||}$ of all quasi-seminorms that are weaker than a quasi-seminorm $||.||$ is a semimodule over the semiring of all positive real numbers $\boldsymbol{R}^{+}$.

Corollary 3.83. The set $\mathbf{QN}_{||\cdot||}$ of all quasi-norms in L that are weaker than a given quasi-norm $||.||$ is an S-module over the S-ring $\boldsymbol{R}^{++}$.

Let us consider a module L over a seminormed ring K.

Definition 3.57. The module L with a mapping q: $L \to \boldsymbol{R}_{\omega}^{+}$ is called a *hyperseminormed module*, while q is called a hyperseminorm in L, if q satisfies Axioms N3 and N5.

For instance, the set of all continuous real functions in $\boldsymbol{R}$ form a hyperseminormed module over the seminormed ring $\boldsymbol{Z}$.

Any seminormed module is a hyperseminormed module.

An interesting area of research is to study topological and algebraic properties of hyperseminormed modules.

It is also possible to introduce and study quasi-hyperseminormed and quasi-hypernormed modules.

KEYWORDS

- **field**
- **hypernorm**
- **hyperseminorm**
- **metric**
- **module**
- **norm**
- **quasi-norm**
- **quasi-seminorm**
- **quasimetric**
- **ring**
- **seminorm**
- **vector space**

CHAPTER 4

HYPEROPERATORS, HYPERFUNCTIONALS, AND EXTRAFUNCTIONS: CONSTRUCTIONS AND OPERATIONS

In this chapter, we further advance the theory of extrafunctions developed in Burgin (1993, 1995, 2002, 2012) extending it to the theory of hyperoperators and hyperfunctionals, which includes the classical theory of operators and functionals.

The usage of the word functional originated in the calculus of variations, where a functional was comprehended as a function with functions as its arguments and numbers as its values. The general concept of a functional was introduced by the Italian mathematician Volterra (1887, 1887a), who used the term *functions of curves* or *functions of other functions*. However, Hadamard was the first to use the name *functional* (Hadamard, 1903, 1910).

The theory of operators in normed spaces originated in the 20th century. Historians of mathematics assume that Riesz's work of 1910 marks the beginning of operator theory. However, the term *operator* and various operators were used in mathematics long before the 20th century. For instance, the creation of the Calculus brought forth differential and integral operators in the 17th century.

Note that extending the concept of a function, it would be natural to speak of hyperfunctions instead of extrafunctions because extrafunctions are generalized mappings of hypernumbers and not extranumbers. However, the term *hyperfunction* is already used in mathematics and has a different meaning (Sato, 1959, 1960). Therefore, we call mappings of hypernumbers by the name *extrafunction*. Here we define *general extrafunctions* and *norm-based extrafunctions*, which include conventional distributions, hyperdistributions, restricted pointwise extrafunctions, and compactwise extrafunctions, which have studied before in different publications.

Taking the definition of real functions and changing the set of real numbers $\boldsymbol{R}$ to the set of real hypernumbers $\boldsymbol{R}_\omega$, we come to the concept of a general extrafunction.

Definition 4.1. a) A partial mapping f: $\boldsymbol{R}_\omega \to \boldsymbol{R}_\omega$ is called a *real pointwise extrafunction* or a *real general extrafunction.*

b) A partial mapping h: $\boldsymbol{C}_\omega \to \boldsymbol{C}_\omega$ is called a *complex pointwise extrafunction* or a *complex general extrafunction.*

We denote by $F(\boldsymbol{R}_\omega, \boldsymbol{R}_\omega)$ the set of all real general (pointwise) extrafunctions and by $F(\boldsymbol{C}_\omega, \boldsymbol{C}_\omega)$ the set of all complex general (pointwise) extrafunctions.

Here we consider extrafunctions defined only for real numbers. The simplest class of such extrafunctions consists of restricted real pointwise extrafunctions, which form a subset of all general real extrafunctions.

Definition 4.2. a) A partial mapping f: $\boldsymbol{R} \to \boldsymbol{R}_\omega$ is called a *restricted real pointwise extrafunction.*

b) A partial mapping h: $\boldsymbol{C} \to \boldsymbol{C}_\omega$ is called a *restricted complex pointwise extrafunction.*

We denote by $F(\boldsymbol{R}, \boldsymbol{R}_\omega)$ the set of all restricted real pointwise extrafunctions and by $F(\boldsymbol{C}, \boldsymbol{C}_\omega)$ the set of all restricted complex pointwise extrafunctions.

Restricted real pointwise extrafunctions correspond to the topology of pointwise convergence in the space of all real functions. Other topologies on the space of all real functions allow us to define other classes of extrafunctions. To build these classes, we use norms and seminorms (cf. Appendix). Taking a set $\mathbf{T} \subseteq \mathrm{F}(\boldsymbol{R}, \boldsymbol{R})$ of real functions, it is possible to define different norms, seminorms, metrics, hypernorms, hyperseminorms, hypermetrics and topologies in $\mathbf{T}$. Note that norms and seminorms in function spaces are functionals, while hypernorms hyperseminorms in function spaces are hyperfunctionals.

The same is true for complex functions. However, here we also regard more general functions such as f: $\boldsymbol{R}^n \to \boldsymbol{R}^m$ or h: $\boldsymbol{C}^n \to \boldsymbol{C}^m$. We call such functions, as well as mappings of infinite dimensional vector spaces, by the name *operator* because are mappings between vector spaces. In the case when $m = 1$, operators are called functionals.

To define norm-based extrafunctions in a broad sense, we consider a set $\mathbf{F}$ of arbitrary functions with a set Q of seminorms in $\mathbf{F}$ and the set $\mathbf{F}^\omega$ of all sequences of functions from **F**. Note that Q may consist of a single norm or seminorm.

Definition 4.3. For arbitrary sequences $\boldsymbol{f} = (f_i)_{i\in\omega}$ and $\boldsymbol{g} = (g_i)_{i\in\omega}$ of functions from **F**,

$$\boldsymbol{F} \approx_Q \boldsymbol{g} \text{ means that } \lim_{i\to\infty} q(f_i - g_i) = 0 \text{ for any } q \in Q$$

Lemma 4.1. The relation $\approx_Q$ is an equivalence relation in $\mathbf{F}^\omega$.

Proof. By definition, this relation is reflexive. Besides, it is symmetric because $q(f-g) = q(g-f)$ for any seminorm q and we need only to show that the relation $\approx_Q$ is transitive. Taking three sequences $\boldsymbol{f} = (f_i)_{i\in\omega}$, $\boldsymbol{g} = (g_i)_{i\in\omega}$ and $\boldsymbol{h} = (h_i)_{i\in\omega}$ of real functions such that $\boldsymbol{f} \approx_Q \boldsymbol{g}$ and $\boldsymbol{g} \approx_Q \boldsymbol{h}$, for any seminorm q from Q, we have

$$\lim_{i\to\infty} q(f_i - g_i) = 0$$

and

$$\lim_{i\to\infty} q(g_i - h_i) = 0$$

By properties of seminorms and limits, we have

$$0 \le \lim_{i\to\infty} q(f_i - h_i) = \lim_{i\to\infty} q(f_i - g_i + g_i - h_i) \le$$

$$\lim_{i\to\infty} q(f_i - g_i) + \lim_{i\to\infty} q(g_i - h_i) = 0 + 0 = 0$$

Consequently,

$$\lim_{i\to\infty} q(f_i - h_i) = 0$$

i.e., $\boldsymbol{f} \approx_Q \boldsymbol{h}$.

Lemma is proved.

Definition 4.4. Classes of the equivalence relation $\approx_Q$ are called *Q-based real extrafunctions represented in* **F** and their set is denoted by $\mathbf{E}^{\mathbf{F}}_{\omega Q}$.

In such a way, any sequence $\boldsymbol{f} = (f_i)_{i\in\omega}$ of real (complex) functions from **F** determines (and represents) a Q-based real (complex) extrafunction $F = \mathrm{EF}_Q(f_i)_{i\in\omega}$ and this sequence is called a *defining sequence* or *representing sequence* or *representation* of the extrafunction $F = \mathrm{EF}_Q(f_i)_{i\in\omega}$. There is also a natural projection π_Q: $\mathbf{F}^\omega \to \mathbf{E}^{\mathbf{F}}_{\omega Q}$ where $\pi_Q((f_i)_{i\in\omega}) = \mathrm{EF}_Q(f_i)_{i\in\omega}$ for any sequence $\boldsymbol{f} = (f_i)_{i\in\omega} \in \mathbf{F}^\omega$.

Note that there are no special conditions from functions from **F** to represent an extrafunction. For instance, sequences $(i)_{i\in\omega}$, $(x^i)_{i\in\omega}$, $(2^{x_i})_{i\in\omega}$, and $(\sin ix)_{i\in\omega}$ represent real extrafunctions.

When **F** consists of multivariable functions, i.e., mappings of the form $f: \boldsymbol{R}^n \to \boldsymbol{R}$ (or $h: \boldsymbol{C}^n \to \boldsymbol{C}$), extrafunctions from $\mathbf{E}^{\mathbf{F}}_{\omega Q}$ are called multivariable extrafunctions.

There is a variety of extrafunction types. The most straightforward approach gives usreal (complex) pointwise extrafunctions also called general extrafunctions. Topological constructions lead to various classes of norm-based real (complex) extrafunctions, which include real (complex) distributions, extended distributions, hyperdistributions, restricted pointwise extrafunctions, and compactwise extrafunctions. Here we study only extrafunctions that are generated by real functions.

The simplest class of extrafunctions in finite and infinite dimensional vector spaces consists of hyperfunctionals.

Let us consider a real (complex) vector space L. It may be finite dimensional or infinite dimensional.

Definition 4.5. a) A (partial) mapping $f: L \to \boldsymbol{R}_\omega$ is called a *real general hyperfunctional*.

b) A (partial) mapping $f: L \to \boldsymbol{C}_\omega$ is called a *complex general hyperfunctional*.

We denote by $F(L, \boldsymbol{R}_\omega)$ the set of all real general hyperfunctionals and by $F(L, \boldsymbol{C}_\omega)$ the set of all complex general hyperfunctionals.

Note that in functional analysis, it is usually supposed that functionals are mappings of vector spaces of functions. However, here we develop a more general approach, in which functionals are mappings of arbitrary vector spaces, while hyperfunctionals are classes of functionals equivalent with respect to specific topological relations.

Restricted real pointwise extrafunctions studied in Burgin (1993, 1995, 2002, 2004a, 2012) form a particular case of finite dimensional real general hyperfunctionals, while restricted complex pointwise extrafunctions studied in Burgin and Ralston (2004) and Burgin (2010c, 2011c, 2013b) form a special case of finite dimensional complex general hyperfunctionals. General real extrafunctions studied in Burgin (2012) form an individual case of infinite dimensional real general hyperfunctionals and complex general extrafunctions studied in Burgin (2011c) form a distinctive case of infinite dimensional complex general hyperfunctionals.

In a similar way, it is possible to define $\boldsymbol{F}$-hyperfunctionals, e.g., p-adic hyperfunctionals. However, in this chapter, we do not want to achieve the highest possible generality and to simplify our exposition, we consider only extrafunctions, hyperfunctionals and hyperoperators in real infinite dimensional vector spaces, allowing the reader to derive similar results in more general cases or in similar but different situations, for example, for extrafunctions in complex infinite dimensional vector spaces.

Note that according to the terminology of the theory of extrafunctions a general hyperfunctional is a pointwise extrafunction in L (Burgin, 2012). Here we study only real general hyperfunctionals as an important type of hyperfunctionals.

Real hyperfunctionals correspond to the topology of pointwise convergence in the space of all real functionals. Other topologies on the space of all real functionals allow us to define other classes of hyperfunctionals. To build these classes, we use hypernorms and hyperseminorms (cf. Appendix).

To furnish a uniform exposition, we consider extrafunctions and hyperfunctionals in the context of hyperoperators.

Let us consider two (infinite-dimensional) vector spaces L and M and a set **K** of mappings (operators) from L into M. When L is a finite-dimensional vector space, then the set **K** consists of functions with values in M. When L is a space of functions and M is a set of numbers, e.g., real numbers, the set **K** consists of (real) functionals.

There are different norms, seminorms, metrics, hypernorms, hyperseminorms, hypermetrics and topologies in **K**. Note that norms and seminorms in function spaces are functionals, while hypernorms and hyperseminorms in function spaces are hyperfunctionals (Burgin, 2004). Here are some examples of such normed function spaces.

Example 4.1. Let us take **K** equal to the space $C(\boldsymbol{I}, \boldsymbol{R})$ of all continuous functions on an interval $\boldsymbol{I} = [a, b]$. The norm in this space is defined by the following formula

$$\|f\| = \max \{|f(x)|;\, x \in [a, b]\}.$$

Example 4.2. Let us take **K** equal to the space $C(\boldsymbol{R}, \boldsymbol{R})$ of all continuous functions on the real line $\boldsymbol{R}$. The hypernorm in this space is defined by the following formula

$$\|f\|_{\infty} = \mathrm{Hn}(c_n)_{n\in\omega}$$

where $c_n = \max \{|f(x)|;\, x \in [-n, n]\}$.

Example 4.3. Let us take **K** equal to the space $L^1(\boldsymbol{I}, \boldsymbol{R})$ of all absolutely integrable real functions on an interval $[a, b]$. The hypernorm in this space is defined by the following formula

$$\|f\|_{L^1} = \int_a^b |f(x)|\, \mathrm{d}x$$

Example 4.4. Let us take **K** equal to the space $Loc^1(\boldsymbol{R}, \boldsymbol{R})$ of all locally absolutely integrable real functions on the real line $\boldsymbol{R}$. We remind that a function is locally integrable if it is integrable on an arbitrary interval $[a, b]$. The hypernorm in this space is defined by the following formula

$$\|f\|_{Loc^1} = \mathrm{Hn}(a_n)_{n\in\omega}$$

where $a_n = \int_{-n}^{n} |f(x)|\, \mathrm{d}x$.

Example 4.5. Let us take **K** equal to the space $Loc^2(\boldsymbol{R}, \boldsymbol{R})$ of all locally square-integrable real functions on the real line $\boldsymbol{R}$. The hypernorm in this space is defined by the following formula

$$\|f\|_{Loc^2} = \mathrm{Hn}(b_n)_{n\in\omega}$$

where $b_n = [\int_{-n}^{n} |f(x)|^2\, \mathrm{d}x]^{1/2}$

Example 4.6. Let us take **K** equal to $Loc^p(\boldsymbol{R}, \boldsymbol{R})$ is the space of all locally absolutely p-integrable ($p \geq 1$) functions on the real line $\boldsymbol{R}$. The hypernorm in this space is defined by the following formula

$$\|f\|_{Loc^p} = \mathrm{Hn}(d_n)_{n\in\omega}$$

where $d_n = [\int_{-n}^{n} |f(x)|^p\, \mathrm{d}x]^{1/p}$

Example 4.7. Let us take **K** equal to $B(\boldsymbol{R}, \boldsymbol{R})$ is the space of all locally bounded functions on the real line $\boldsymbol{R}$. We remind that a function is locally bounded if it is bounded on an arbitrary interval $[a, b]$. The hypernorm in this space is defined by the following formula

$$\|f\|_{\infty} = \mathrm{Hn}(a_n)_{n\in\omega}$$

where $a_n = \sup \{|f(x)|;\ x \in [-n, n]\}$

Example 4.8. Let us take **K** equal to the generalized Sobolev space $H^1(\boldsymbol{R}, \boldsymbol{R})$ of all differentiable functions on the real line $\boldsymbol{R}$ such that they are locally square integrable and their derivatives are locally square integrable. The hypernorm in this space is defined by the following formula

$$\|f\|_{L^1} = \|f\|_{L^2} + \|f'\|_{L^2} = \mathrm{Hn}(c_n)_{n\in\omega}$$

where $c_n = [\int_{-n}^{n} |f(x)|^2 dx]^{1/2} + [\int_{-n}^{n} |f'(x)|^2 dx]^{1/2}$

These examples show that the theory of hypernumbers and extrafunctions allows extending classical function spaces to more general function spaces and at the same time, preserving many good properties of classical function spaces.

To define norm-based hyperfunctionals and hyperoperators, we treat functionals and functions as special cases of operators and consider a subset **K** of the polyhyperseminormed vector space (cf. Chapter 3) of all operators (mappings) from a vector space L into a vector space M and a set Q of hyperseminorms in **K**, denoting the set of all sequences of operators from **K** by $\mathbf{K}^\omega$. In the context of this chapter, **K** consists of operators (functions) in a (finite or infinite dimensional) real vector space L. In what follows, we assume that **K** is a vector space although it is possible to develop the theory of hyperfunctionals and hyperoperators in a more general case when **K** is only an abelian group. Note that considering systems of hyperseminorms, we automatically obtain many corresponding results from (Burgin, 2012) for seminormed and polyseminormed spaces because any seminorm also is a hyperseminorm.

Definition 4.6. For arbitrary sequences $\boldsymbol{f} = (f_i)_{i\in\omega}$ and $\boldsymbol{g} = (g_i)_{i\in\omega}$ of elements from **K**, we define the following relation

$$\boldsymbol{F} \approx_Q \boldsymbol{g} \text{ means that } \lim_{i\to\infty} q(f_i - g_i) = 0 \text{ for any } q \in Q$$

Lemma 4.2. The relation $\approx_Q$ is an equivalence relation in $\mathbf{K}^\omega$.

Proof. By definition, this relation is reflexive. Besides, it is symmetric because $q(f - g) = q(g - f)$ for any hyperseminorm q and we need only to show that the relation $\approx_Q$ is transitive. Taking three sequences $\boldsymbol{f} = (f_i)_{i\in\omega}$, $\boldsymbol{g} = (g_i)_{i\in\omega}$ and $\boldsymbol{h} = (h_i)_{i\in\omega}$ of elements from **K** such that $\boldsymbol{f} \approx_Q \boldsymbol{g}$ and $\boldsymbol{g} \approx_Q \boldsymbol{h}$, for any hyperseminorm q from Q, we have

$$\lim_{i\to\infty} q(f_i - g_i) = 0$$

and

$$\lim_{i\to\infty} q(g_i - h_i) = 0$$

By properties of hyperseminorms and limits, we have

$$0 \leq \lim_{i\to\infty} q(f_i - h_i) = \lim_{i\to\infty} q(f_i - g_i + g_i - h_i) \leq$$

$$\lim_{i\to\infty} q(f_i - g_i) + \lim_{i\to\infty} q(g_i - h_i) = 0 + 0 = 0$$

Consequently,

$$\lim_{i\to\infty} q(f_i - h_i) = 0$$

i.e., $\boldsymbol{f} \approx_Q \boldsymbol{h}$.

Lemma is proved.

Definition 4.7. Classes of the equivalence relation $\approx_Q$ are called *Q-based hyperoperators represented in* **K** and their space is denoted by $\mathbf{E}^{\mathbf{K}}_{\omega Q}$ and called a hyperspace (Burgin, 2010a; 2011, 2012a).

Note that in functional analysis, it is usually supposed that operators are mappings of vector spaces of functions. However, here we develop a more general approach, in which operators and hyperoperators are mappings of arbitrary vector spaces.

We will show that $\mathbf{E}^{\mathbf{K}}_{\omega Q}$ is a vector space if **K** is a vector space.

In such a way, any sequence $\boldsymbol{f} = (f_i)_{i\in\omega}$ of operators from **K** determines (and represents) a Q-based hyperoperator $F = \mathrm{EF}_Q(f_i)_{i\in\omega}$ and this sequence is called a *defining sequence* or *representing sequence* or *representation* of the hyperoperator $F = \mathrm{EF}_Q(f_i)_{i\in\omega}$. There is also a natural projection π_Q: $\mathbf{K}^\omega \to \mathbf{E}^{\mathbf{K}}_{\omega Q}$ where $\pi_Q((f_i)_{i\in\omega}) = \mathrm{EF}_Q(f_i)_{i\in\omega}$ for any sequence $\boldsymbol{f} = (f_i)_{i\in\omega} \in \mathbf{K}^\omega$.

When **K** is a set of (real) functionals, then $\mathbf{E}^{\mathbf{K}}_{\omega Q}$ is a set of (real) hyperfunctionals.

In particular, any sequence $\boldsymbol{f} = (f_i)_{i\in\omega}$ of (real) functions determines (and represents) a Q-based (real) extrafunction $F = \mathrm{EF}_Q(f_i)_{i\in\omega}$ and this sequence is called a *defining sequence* or *representing sequence* or *representation* of the extrafunction $F = \mathrm{EF}_Q(f_i)_{i\in\omega}$.

In a similar way, when **K** is a set of functionals, we obtain the set $\mathbf{E}^{\mathbf{K}}_{\omega Q}$ of hyperfunctionals.

Definite hyperintegrals studied in Burgin (1993, 2008a, 2012a) are examples of hyperfunctionals, while hyperderivatives studied in Burgin (1993, 2002, 2012) and indefinite hyperintegrals studied in Burgin (2010a) are examples of hyperoperators.

We see that the construction of norm-based hyperoperators is similar to the construction of hypernumbers.

Lemma 4.3. If **K** and **H** are spaces of operators and $\mathbf{K} \subseteq \mathbf{H}$, then for any system of hyperseminorms P, we have $\mathbf{E}^{\mathbf{K}}{}_{\omega P} \subseteq \mathbf{E}^{\mathbf{H}}{}_{\omega P}$.

Proof is left as an exercise.

An important question is when the space **K** is a natural subspace of the hyperspace $\mathbf{E}^{\mathbf{K}}{}_{\omega Q}$.

We remind that a system Q of hyperseminorms in **K** (weakly) *separates* **K** if for any different elements f and g from **K**, there is a hyperseminorm q from Q such that $q(f) \neq q(g)$ $(q(f-g) \neq 0)$ (Burgin, 2012).

Lemma 4.4. Any hypernorm h in **K** weakly separates **K**.

Indeed, if elements f and g are different elements from **K**, then $f - g \neq 0$. As h is a hypernorm (cf. Chapter 3), by Axiom N1, $h(f-g) \neq 0$.

Corollary 4.1 (Burgin, 2012). Any norm in **K** weakly separates **K**.

It is natural to define a mapping μ_Q: $\mathbf{K} \to \mathbf{E}^{\mathbf{K}}{}_{\omega Q}$ by the formula $\mu_Q(f) = \mathrm{EF}_Q(f_i)_{i\in\omega}$ where $f_i = f$ for all $i = 1, 2, 3, \ldots$.

Proposition 4.1. μ_Q is an injection (monomorphism of linear spaces when **K** is a linear space) if and only if the system Q of hyperseminorms weakly separates **K**.

Proof. Sufficiency. If f and g are different elements from **K** and the system Q of hyperseminorms in **K** weakly separates **K**, then there is a hyperseminorm q from Q such that $q(f-g) \neq 0$. Then by definition, $\mu_Q(f) \neq \mu_Q(g)$ and μ_Q is an injection.

If f and g are different elements from **K** and the system Q of hyperseminorms in **K** separates **K**, then there is a hyperseminorm q from Q such that $q(f) \neq q(g)$. Consequently, by properties of hyperseminorms, $q(f-g) \neq 0$, i.e., the system Q of hyperseminorms in **K** weakly separates **K**. Thus, as it is proved above, μ_Q is a monomorphism.

Necessity. If the system Q of hyperseminorms does not weakly separate **K**, then for some elements f and g from **K**, $q(f-g) = 0$ for all hyperseminorms q from Q. It means, by Definition 4.6, that $(f_i)_{i\in\omega} \approx_Q (f_i)_{i\in\omega}$ where $f_i = f$ and $g_i = g$ for all $i = 1, 2, 3, \ldots$.

Thus, $\mu_Q(f) = \mu_Q(g)$ and μ_Q is not an injection. Therefore, by the Law of Contraposition for propositions [cf., e.g., (Church, 1956)], the condition from Proposition 4.1 is necessary.

Proposition is proved.

Corollary 4.2. μ_Q is an injection (monomorphism of vector spaces when **K** is a vector space) if the system Q of hyperseminorms separates **K**.

Definition 4.8. a) A sequence $\{f_i; i = 1, 2, 3, \ldots\}$ of functions (operators) *Q-converges* to a function (operator) f, i.e., $\lim^Q_{i\to\infty} f_i = f$, if $\lim_{i\to\infty} q(f - f_i) = 0$ for all $q \in Q$. In this case, f is called a *Q-limit* of functions (operators) f_i.

b) A sequence $\{ f_i; i = 1, 2, 3, \ldots\}$ of functions (operators) *uniformly Q-converges* to a function (operator) f if for any $\varepsilon > 0$, there is a natural number n such that if $q \in Q$ and $i > n$, then $q(f - f_i) < \varepsilon$. In this case, f is called a *uniform Q-limit* of functions (operators) f_i.

The concept of Q-convergence generalizes the concept of conventional convergence and concept of Q-limit generalizes the concept of conventional limit.

Proposition 4.2. If **H** and **G** are sets of functions (operators) and any function (operator) f from **H** is a uniform Q-limit of functions (operators) from **G**, then any extrafunction (hyperoperator) represented in **H** is also represented in **G**.

Proof. Let us take an hyperoperator (extrafunction) $F = \mathrm{EF}_Q(f_i)_{i\in\omega}$ where all $f_i \in \mathbf{H}$. As any operator (function) f from **H** is a uniform Q-limit of operators (functions) from **G**, we see that for any f_i, there is an operator (function) h_i from **G** and natural number n such that if $q \in Q$ and $i > n$, then $q(h_i - f_i) < 1/i$. By Definitions 4.6 and 4.7, it means that the hyperoperator (extrafunction) $F = \mathrm{EF}_Q(h_i)_{i\in\omega}$, i.e., F is represented in **G**.

Proposition is proved.

Definition 4.9. A restricted real pointwise (general) extrafunction (real general hyperfunctional) F: $\boldsymbol{R} \to \boldsymbol{R}_\omega$ is *continuously represented* if there is a sequence $(f_i)_{i\in\omega}$ of continuous real functions (real functionals) f_i such that $F = \mathrm{EF}_{Q_{pt}}(f_i)_{i\in\omega}$.

For instance, the extrafunction defined by the formula $F(x) = \mathrm{Hn}(nx)_{n\in\omega}$ is continuously represented.

As we know, any continuous real function in an interval is uniform limit of a sequence of polynomials [cf., e.g., (Burgin, 2008)]. Thus, Proposition 4.2 gives us the following result.

Corollary 4.3. Any continuously represented compactwise extrafunction (Burgin, 2012) is also represented in the space of all real polynomials.

Relations between systems of hyperseminorms induce relations between corresponding classes of norm-based hyperoperators. Let us consider some of them, taking two systems of hyperseminorms $Q = \{q_i; i \in I\}$ and $P = \{p_j; j \in J\}$ in a space **K**. By the definition of $\mathbf{E}^{\mathbf{K}}{}_{\omega P}$ and $\mathbf{E}^{\mathbf{K}}{}_{\omega Q}$, there are natural projections $\pi_P: \mathbf{K}^\omega \to \mathbf{E}^{\mathbf{K}}{}_{\omega P}$ and $\pi_Q: \mathbf{K}^\omega \to \mathbf{E}^{\mathbf{K}}{}_{\omega Q}$.

Definition 4.10. a) The system P *dominates* the system Q in X (it is denoted by $P > Q$) if

$$\forall q \in Q \exists p \in P \exists k_q \in \boldsymbol{R}^{++} \forall x \in \mathbf{K}\, (q(x) \le k_q \cdot p(x))$$

b) The system P *regularly dominates* the system Q in X (it is denoted by $P \lhd Q$) if

$$\forall q \in Q \exists p \in P\, \forall x \in \mathbf{K}\, (q(x) \le p(x))$$

Lemma 4.5. If $P > Q$, then for any sequences $\boldsymbol{f}$ and $\boldsymbol{g}$ from $\mathbf{K}^{\omega}$, we have

$$\boldsymbol{f} \approx_P \boldsymbol{g} \text{ implies } \boldsymbol{f} \approx_Q \boldsymbol{g}$$

Corollary 4.4. If $P \lhd Q$, then for any sequences $\boldsymbol{f}$ and $\boldsymbol{g}$ from $\mathbf{K}^{\omega}$, we have

$$\boldsymbol{f} \approx_P \boldsymbol{g} \text{ implies } \boldsymbol{f} \approx_Q \boldsymbol{g}$$

These properties of operator (function) sequences imply corresponding properties of spaces of hyperoperators (extrafunctions).

Theorem 4.1. If $P > Q$, then there is a natural projection π: $\mathbf{E}^{\mathbf{K}}{}_{\omega P} \rightarrow \mathbf{E}^{\mathbf{K}}{}_{\omega Q}$ for which the following equality is valid $\pi_{\circ}\pi_P = \pi_Q$, i.e., the following diagram is commutative

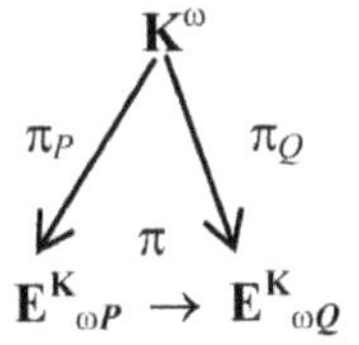

Proof is left as an exercise.

Corollary 4.5. If $P \lhd Q$, then there is a natural projection π: $\mathbf{E}^{\mathbf{K}}{}_{\omega P} \rightarrow \mathbf{E}^{\mathbf{K}}{}_{\omega Q}$ for which the following equality is valid $\pi_{\circ}\pi_P = \pi_Q$.

Different norms and seminorms allow us to build various classes of norm-based extrafunctions studied before by Burgin (2002, 2012) and Burgin and Ralston (2004). Here we consider several sets Q of seminorms, building important types of extrafunctions, which are related to types of extrafunctions, distributions and other generalized functions studied before.

Type 1 of extrafunctions. In the space $F(\boldsymbol{R})$ of all real functions, it is possible to define the seminorm q_{ptx} using an arbitrary real number x and the following formula

$$q_{ptx}(f) = |f(x)|$$

Indeed, properties of the absolute value imply that this functional satisfies both Conditions N2 and N3 from the definition of a seminorm (cf. Appendix).

We define $Q_{pt} = \{q_{ptx};\ x \in \boldsymbol{R}\}$. Definitions 4.6 and 4.7 show how this set of seminorms determines the equivalence relation $\approx_{pt}$ in the space of all sequences of real functions and defines Q_{pt}-based real extrafunctions represented in $F(\boldsymbol{R})$. A sequence $\boldsymbol{f} = (f_i)_{i\in\omega}$ of real functions determines (and represents) a Q_{pt}-based real extrafunction $F = \mathrm{EF}_{Qpt}(f_i)_{i\in\omega}$, which is also denoted by $F = \mathrm{Ep}(f_i)_{i\in\omega}$.

Proposition 4.3. The set $\mathbf{E}^{F(\boldsymbol{R})}{}_{\omega\, Qpt}$ of all Q_{pt}-based extrafunctions is isomorphic to the set $F(\boldsymbol{R}, \boldsymbol{R}_{\omega})$ of all restricted real pointwise extrafunctions.

Proof is left as an exercise.

Note that the set $\mathbf{E}^{F(\boldsymbol{R})}{}_{\omega\, Qpt}$ does not coincide with the set $F(\boldsymbol{R}, \boldsymbol{R}_{\omega})$ because elements of the first set are classes of functions, while elements of the second set are mappings.

Proposition 4.3 implies that it is possible to represent restricted real pointwise extrafunctions by sequences of ordinary real functions. Namely, a sequence $(f_i)_{i\in\omega}$ of real functions f_i represents a restricted pointwise extrafunction $F = \mathrm{EF}_{Qpt}(f_i)_{i\in\omega}$ such that $F(x) = \mathrm{Hn}(f_i(x))_{i\in\omega}$ for all x. However, in some cases, it is possible to find a representation with additional useful properties such as continuous representations (cf. Definition 4.9) or representations by polynomials.

Proposition 4.4. For any continuously represented restricted pointwise extrafunction F, there is a sequence $(g_i)_{i\in\omega}$ of bounded continuous real functions g_i such that $F = \mathrm{EF}_{Qpt}(g_i)_{i\in\omega}$.

Proof. Let $F = \mathrm{EF}_{Qpt}(f_i)_{i\in\omega}$ and all f_i are continuous real functions. Then we consider the sequence of intervals $[-k, k]$ with $k = 1, 2, \ldots, n, \ldots$. This makes it possible to define a function $g_k(x)$ equal to the function $f_k(x)$ when x belongs to the interval $[-k, k]$, equal to $f_k(-k)$ when $x < -k$, and equal to $f_k(k)$ when $x > k$. It means that the function $g_k(x)$ coincides with the function $f_k(x)$ inside the interval $[-k, k]$, is equal to the same value $f_k(-k)$ for all $x \leq -k$ and to the same value $f_k(k)$ for all $x \geq k$. As a function continuous in $\boldsymbol{R}$ is bounded in any finite interval, all $g_i(x)$ are bounded functions and it is possible to check that $F = \mathrm{EF}_{Qpt}(g_i)_{i\in\omega}$.

Proposition is proved.

Remark 4.1. Not all restricted pointwise real extrafunctions and even not all real functions can be continuously represented. To show this, let us consider the characteristic function h of the set of all rational numbers in the interval [0, 1]. It is equal to 1 for all rational numbers from the interval [0, 1] and is equal to 0 for all other real numbers.

In more detail, real pointwise extrafunctions are studied in Burgin (2002) and applied to the path integral in Burgin (2008, 2009). Complex pointwise extrafunctions are studied and applied to differential equations in Burgin and Ralston (2004) and Burgin (2010).

Type 2 of extrafunctions. Let us consider a class **K** of functions such that for any function $f(x)$ from the class **F** and any function $g(x)$ from **K**, the integral $\int f(x)g(x)\mathrm{d}x$ exists. If functions from both classes are defined in $\boldsymbol{R}$, then the integral is taken over the whole of $\boldsymbol{R}$, while if functions from both classes are defined in an interval $[a, b]$, then the integral is taken over the interval $[a, b]$. For instance, **K** consists of all continuous functions with the compact support and **F** consists of all continuous functions. For simplicity, we consider the Riemann integral although it is possible to use other kinds of integrals, e.g., Lebesgue integral, Stieltjes integral, Perron integral, Denjoy integral or gauge integral, for this purpose. For instance, taking integrals with respect to a system of measures on intervals of real numbers, we come to the concept of a real measure-wise extrafunction introduced and studied in Burgin (2002). In all cases, extended distributions are special kinds of extrafunctions.

Then it is possible to define the seminorm q_g in **F** using an arbitrary function $g(x)$ from **K** and the following formula

$$q_g(f) = |\int f(x)g(x)\mathrm{d}x\,|$$

Indeed, q_g satisfies Condition N3 from the definition of a seminorm because the integral is a linear functional and the absolute value satisfies the triangle inequality. Besides, it satisfies condition N2 from the definition of a seminorm because the absolute value of a real number is a norm.

We define $Q_{\mathbf{K}} = \{q_g;\ g \in \mathbf{K}\}$. Definitions 4.6 and 4.7 show how this set of seminorms determines the equivalence relation $\approx_{\mathbf{K}}$ in the space of all sequences of real functions from **F**, as well as $Q_{\mathbf{K}}$-based real extrafunctions represented in **F**. According to the tradition [cf., e.g., (Schwartz, 1950, 1951; Rudin, 1991)], we call **K** the *set of test functions*. A sequence $\boldsymbol{f} = (f_i)_{i\in\omega}$ of functions from **F** determines (and represents) a $Q_{\mathbf{K}}$-based real extrafunction $F = \mathrm{EF}_{Q_{\mathbf{K}}}(f_i)_{i\in\omega}$ represented in **F**.

The classes of sequences of continuous real functions equivalent with respect to **K** are called *real extended distributions* with respect to **K**. We denote the class of all real extended distributions with respect to **K** by $\mathbf{D}_{\mathbf{K}\omega}$, the expression $\mathrm{Ed}_{\mathbf{K}}(f_i)_{i\in\omega}$ denotes the real extended distribution with respect to **K** that is defined by the sequence of continuous functions $(f_i)_{i\in\omega}$, and the class of all such extended distributions is denoted by $\mathbf{E}^{C(\boldsymbol{R})}{}_{\omega Q\mathbf{K}}$.

In a natural way, extended distributions are connected to conventional distributions or generalized functions, which appeared in the following way. In classical analysis, continuous functions do not always have derivatives. To remedy this deficiency and to be able to solve various differential equations from physics, some distributions, such as Dirac delta-function $\delta(x)$, were used by physicists. Later the concept of a *distribution* was rigorously defined by mathematicians and theory of distributions was developed. There are different equivalent ways to define distributions. The definition of a distribution as a functional was historically the first (Schwartz, 1950, 1951). Another approach is called sequential because in it, distributions are defined as classes of equivalent sequences of ordinary functions (Mikusinski, 1948; Temple, 1953; Liverman, 1964; Antosik et al., 1973). Thus, it is natural that distributions are closely related to extrafunctions. Here we explain this relation.

Definition 4.11. A sequence $\{f_n; n\in\omega\}$ of real functions *converges almost uniformly* in $\boldsymbol{R}$ if it uniformly converges on every interval $[a, b]$.

For instance, the sequence $\{(1/n)x; n\in\omega\}$ of real functions converges almost uniformly in $\boldsymbol{R}$.

Definition 4.12. A sequence $\{f_n; n\in\omega\}$ of continuous real functions is called *fundamental* if for any interval $I = [a, b]$, there are a sequence $\{F_n; n\in\omega\}$ and a natural number k that satisfy the following conditions:

(1) $F_n^{(k)}(x) = f_n(x)$ for all $x \in I$ and $n\in\omega$ where $F_n^{(k)}(x) = (d^kF_n(x))/dx^k$ is the kth-order derivative of the function $F_n(x)$.

(2) The sequence $\{F_n; n\in\omega\}$ converges almost uniformly in $\boldsymbol{R}$.

For instance, the sequence $\{1/n; n\in\omega\}$ is fundamental.

Definition 4.13. Two fundamental sequences $\{f_n; n\in\omega\}$ and $\{g_n; n\in\omega\}$ are *equivalent* if the sequence $\{f_1, g_1, f_2, g_2, f_3, \ldots\}$ is fundamental, i.e., for any interval $I = [a, b]$, there are sequences $\{F_n; n\in\omega\}$ and $\{G_n; n\in\omega\}$ and a natural number k such that $F_n^{(k)}(x) = f_n(x)$, $G_n^{(k)}(x) = g_n(x)$ for all $x \in I$ and x

$\in \omega$ and sequences $\{F_n; n\in\omega\}$ and $\{G_n; n\in\omega\}$ converge almost uniformly in $\boldsymbol{R}$ to the same function.

Definition 4.14. A class of equivalent fundamental sequences is called a *distribution*.

If $\{f_n; n\in\omega\}$ is a fundamental sequence, then $\mathrm{D}\{f_n; n\in\omega\}$ denotes the distribution that contains $\{f_n; n\in\omega\}$, i.e., the class of all sequences equivalent to the sequence $\{f_n; n\in\omega\}$. The space of all distributions is denoted by D'.

Let us consider the space **CD** of all functions from $C^\infty(\boldsymbol{R})$ that have the compact support (Schwartz, 1950, 1951).

Theorem 4.2. There is a subspace **D** of the space $\mathbf{D}_{\mathbf{CD}\omega}$, which is isomorphic to the space D' of all distributions.

Proof. As the space $\mathbf{D}_{\mathbf{CD}\omega}$ is a set of all equivalent classes of function sequences, we can take the subspace of this space generated by fundamental sequences as **D**. Using definitions, it is possible to show that two fundamental sequences $\{f_n; n\in\omega\}$ and $\{g_n; n\in\omega\}$ are equivalent as extended distributions if and only if they are equivalent in the sense of Definition 4.12. This gives us the necessary isomorphism between spaces **D** and D'.

Theorem 4.2 shows that distributions are a special case of extrafunctions, namely, of extended distributions, in similar way as real numbers are a special case of real hypernumbers.

Distributions also form a special class of *ultradistributions* or *generalized distributions*. The basic idea of ultradistributions is taking smaller spaces of test functions than in the case of distributions, in such a way that important results of the theory of distributions remain true (in a generalized form), while the spaces of distributions are enlarged (Sebastiao e Silva, 1958; Pinto, 1989; Hoskins and Pinto, 2003).

For instance, Beurling (1961) and Björck (1966) built ultradistributions using spaces of test functions defined by the growth properties of their Fourier transforms. Roumieu (1960) utilized classes of ultradifferentiable functions taken from classical analysis to construct ultradistributions. Komatsu (1973) unified both approaches proving several structure theorems for studied ultradistributions.

Similar to distributions, ultradistributions find different applications in theoretical physics [cf., e.g., (Bollini et al., 1999; Bollini and Rocca, 2007)].

In any case, as the class of test functions **K** is sufficiently flexible in the construction of extended distributions, the following result shows that ultradistributions also are a special case of extrafunctions, namely, of extended distributions.

Let us consider a class *UD'* of ultradistributions, which are constructed using the space **U** of test functions.

Theorem 4.3. There is a subspace **E** of the space $\mathbf{D}_{\mathbf{U}\omega}$, which is isomorphic to the space *UD'*.

Proof is left as an exercise.

In comparison with ultradistributions, another direction in the theory of generalized functions was not to reduce the set of test functions but to increase it. In such a way, the space of tempered distributions was constructed (Schwartz, 1950, 1951). By Theorem 4.3, tempered distributions also are a special case of extended distributions.

Let us consider a class *TD'* of tempered distributions, which are constructed using the space **T** of test functions.

Theorem 4.4. There is a subspace **H** of the space $\mathbf{D}_{\mathbf{T}\omega}$, which is isomorphic to the space *TD'*.

Proof is left as an exercise.

In more detail, real extended distributions are studied in Burgin (2004) and applied to the Feynman path integral in Burgin (2008a). Complex extended distributions are studied and applied to differential equations in Burgin and Ralston (2004) and Burgin (2010).

Type 3 of extrafunctions. Let us consider the space $C(\boldsymbol{R})$ of all functions continuous in $\boldsymbol{R}$. Then it is possible to define the seminorm $q_{\max[a,\, b]}$ in $C(\boldsymbol{R})$ using an arbitrary interval $[a, b]$ and the following formula

$$q_{\max[a,\, b]}(f) = \max \{|f(x)|;\, x \in [a, b]\}$$

Indeed, we have

$$q_{\max[a,\, b]}(f + g) = \max \{|f(x) + g(x)|;\, x \in [a, b]\} \leq \max \{(|f(x)| + |g(x)|);\, x \in [a, b]\} =$$

$$\max \{|f(x)|;\, x \in [a, b]\} + \max \{|g(x)|;\, x \in [a, b]\} = q_{\max[a,\, b]}(f) + q_{\max[a,\, b]}(g)$$

and if a is a real number, then

$$q_{\max[a,\, b]}(af) = \max \{|af(x)|;\, x \in [a, b]\} = \max \{|a||f(x)|;\, x \in [a, b]\} =$$

$$|a|\cdot\max\{|f(x)|;\, x \in [a, b]\} = |a|\, q_{\max[a,\, b]}(f)$$

We define $Q_{\text{comp}} = \{q_{\max[a,\, b]}; a, b \in \boldsymbol{R}\}$. Definitions 4.6 and 4.7 show how this set of seminorms determines the equivalence relation $\approx_{\text{comp}}$ in the space of all sequences of continuous in $\boldsymbol{R}$ functions, as well as continuously represented Q_{comp}-based real extrafunctions. A sequence $\boldsymbol{f} = (f_i)_{i\in\omega}$ of continuous functions determines (and represents) a Q_{comp}-based real extrafunction $F = \text{EF}_{Q\text{comp}}(f_i)_{i\in\omega}$, which is also denoted by $F = \text{Ec}(f_i)_{i\in\omega}$ and called a *real continuously represented compactwise extrafunction*. Comp($\boldsymbol{R}$, $\boldsymbol{R}_\omega$) is the set/space of all real continuously represented compactwise extrafunctions.

Continuously represented compactwise extrafunctions are also related to distributions. For instance, as it proved in Burgin (2001), there is the linear subspace Comp D of Comp($\boldsymbol{R}$, $\boldsymbol{R}_\omega$) and a linear projection p: Comp D $\to D'$.

In more detail, real continuously represented compactwise extrafunctions are studied in Burgin (2001, 2002) and applied to the path integral in Burgin (2008, 2009). Complex continuously represented compactwise extrafunctions are studied and applied to differential equations in Burgin and Ralston (2004) and Burgin (2010).

Type 4 of extrafunctions. Let us consider the space $BI(\boldsymbol{R})$ of all real functions bounded in each interval $[a, b]$. Then it is possible to define the seminorm $q_{\sup[a,\, b]}$ in $BI(\boldsymbol{R})$ using an arbitrary interval $[a, b]$ and the following formula

$$q_{\sup[a,\, b]}(f) = \sup\ \{|f(x)|\ ;\ x \in [a, b]\}$$

Indeed, we have

$$q_{\sup[a,\, b]}(f + g) = \sup\ \{|f(x) + g(x)|;\ x \in [a, b]\} \leq \sup\ \{(|f(x)| + |g(x)|);\ x \in [a, b]\} =$$

$$\sup\ \{|f(x)|;\ x \in [a, b]\} + \sup\ \{|g(x)|;\ x \in [a, b]\} = q_{\sup[a,\, b]}(f) + q_{\sup[a,\, b]}(g)$$

and if a is a real number, then

$$q_{\sup[a,\, b]}(af) = \sup\ \{|af(x)\ |;\ x \in [a, b]\} = \sup\ \{|a||f(x)|;\ x \in [a, b]\} =$$

$$|a|\cdot\sup\{|f(x)|;\ x \in [a, b]\} = |a|\ q_{\sup[a,\, b]}(f)$$

We define $Q_{\text{cp}} = \{q_{\sup[a,\, b]};\ a, b \in \boldsymbol{R}\}$. Definitions 4.6 and 4.7 show how this set of seminorms determines the equivalence relation $\approx_{\text{cp}}$ in the space of all sequences of bounded in $\boldsymbol{R}$ functions, as well as continuously represented

Q_{cp}-based real extrafunctions. A sequence $f = (f_i)_{i\in\omega}$ of bounded functions determines (and represents) a Q_{cp}-based real extrafunction $F = \mathrm{EF}_{Q_{cp}}(f_i)_{i\in\omega}$, which is also denoted by $F = \mathrm{Ebc}(f_i)_{i\in\omega}$ and called a real boundedly represented compactwise extrafunction.

Theorem 4.1 allows us to find relations between the four introduced types of functions.

Proposition 4.5. $Q_{comp} \lhd Q_{pt}$ in the set $C(\boldsymbol{R})$ of all continuous real functions.

Proof is left as an exercise.

Corollary 4.6. There is a projection σ_{cont}: $\mathbf{E}^{C(\boldsymbol{R})}{}_{\omega Q_{comp}} \to \mathbf{E}^{C(\boldsymbol{R})}{}_{\omega Q_{pt}}$.

Proposition 4.6. $Q_{comp} = Q_{cp}$ in the set $C(\boldsymbol{R})$ of all continuous real functions.

Proof is left as an exercise.

Corollary 4.7. $\mathbf{E}^{C(\boldsymbol{R})}{}_{\omega Q_{comp}} = \mathbf{E}^{C(\boldsymbol{R})}{}_{\omega Q_{cp}}$.

Proposition 4.7. $Q_{cp} \lhd Q_{pt}$ in the set $BI(\boldsymbol{R})$ of all real functions that are bounded in each finite interval.

Proof is left as an exercise.

Corollary 4.8. There is a projection τ_{fbd}: $\mathbf{E}^{BI(\boldsymbol{R})}{}_{\omega Q_{cp}} \to \mathbf{E}^{BI(\boldsymbol{R})}{}_{\omega Q_{pt}}$.

Proposition 4.8. $Q_{comp} > Q_{\mathbf{K}}$ in the set $C(\boldsymbol{R})$ if all functions in $\mathbf{K}$ have a compact support.

Proof is left as an exercise.

Corollary 4.9. There is a projection τ_{cst}: $\mathbf{E}^{C(\boldsymbol{R})}{}_{\omega Q_{comp}} \to \mathbf{E}^{C(\boldsymbol{R})}{}_{\omega Q_{\mathbf{K}}}$.

Proposition 4.9. $Q_{cp} > Q_{\mathbf{K}}$ in the set $BI(\boldsymbol{R})$ if all functions in $\mathbf{K}$ have a compact support.

Proof is left as an exercise.

Corollary 4.10. There is a projection τ_{cst}: $\mathbf{E}^{C(\boldsymbol{R})}{}_{\omega Q_{cp}} \to \mathbf{E}^{C(\boldsymbol{R})}{}_{\omega Q_{\mathbf{K}}}$.

Now we study algebraic properties of hyperoperators.

Let us assume that the class $\mathbf{K}$ of operators with a system Q of hyperseminorms is closed with respect to addition and/or subtraction of operators. For instance, $\mathbf{K}$ is an abelian group if it is the set $C(\boldsymbol{R})$ of all continuous real functions. Then addition and subtraction of functions induce corresponding operations in sets of hyperoperators represented in $\mathbf{K}$.

Proposition 4.10. Operations of addition and/or subtraction are correctly defined for Q-based hyperoperators represented in $\mathbf{K}$.

Proof. Let us take two Q-based hyperoperators $F = \mathrm{EF}_Q(f_i)_{i\in\omega}$ and $G = \mathrm{EF}_Q(g_i)_{i\in\omega}$. We define $F + G = \mathrm{EF}_Q(f_i + g_i)_{i\in\omega}$. To show that this is an operation with hyperoperators, it is necessary to prove that $F + G$ does not depend on the choice of a representing sequences $(f_i)_{i\in\omega}$ and $(g_i)_{i\in\omega}$ for the hyperoperators

F and G. To do this, let us take another sequence $(h_i)_{i\in\omega}$ that represents F and show that $F + G = \mathrm{EF}_Q(h_i + g_i)_{i\in\omega}$. Note that all three sequences $(f_i)_{i\in\omega}$,$(h_i)_{i\in\omega}$ and $(g_i)_{i\in\omega}$ belong to $\mathbf{K}^\omega$.

In this case, for any hyperseminorm q from Q, we have

$$\lim_{i\to\infty} q((f_i + g_i) - (h_i + g_i)) = 0$$

because $\lim_{i\to\infty} q(f_i - h_i) = 0$ and $(f_i + g_i) - (h_i + g_i) = f_i - h_i$. Thus, addition is correctly defined for Q-based hyperoperators represented in **K**.

The proof for the difference of two Q-based hyperoperators is similar.

Proposition is proved.

The construction of addition and subtraction of Q-based hyperoperators represented in **K** implies the following result.

Proposition 4.11. Any identity that involves only operations of addition and/or subtraction and is valid for operators from **K** is also valid for Q-based hyperoperators represented in **K**.

Corollary 4.11. If **K** is a semigroup, then $\mathbf{E}^{\mathbf{K}}{}_{\omega Q}$ is also a semigroup.

Corollary 4.12. If **K** is an abelian group, then $\mathbf{E}^{\mathbf{K}}{}_{\omega Q}$ is also an abelian group.

Let us assume that the class **K** is closed with respect to multiplication by real numbers, i.e., if a is a real number and $f \in \mathbf{K}$, then the product af also belongs to **K**. In such a way, this transformation induces the corresponding operation in sets of represented in **K** hyperoperators.

Proposition 4.12. Multiplication by real numbers is correctly defined for Q-based hyperoperators represented in **K**.

Proof. Let us take a Q-based hyperoperator $F = \mathrm{EF}_Q(f_i)_{i\in\omega}$ and a real number a. We define $aF = \mathrm{EF}_Q(af_i)_{i\in\omega}$. Let us take another sequence $(h_i)_{i\in\omega}$ that represents F and show that $aF = \mathrm{EF}_Q(ah_i)_{i\in\omega}$. Indeed, for any hyperseminorm q from Q, we have

$$\lim_{i\to\infty} q((af_i - ah_i)) = \lim_{i\to\infty} |a|\cdot q((f_i - h_i)) = |a|\cdot \lim_{i\to\infty} q((f_i - h_i)) = 0$$

because $\lim_{i\to\infty} q(f_i - h_i) = 0$ and q satisfies Condition N2. Thus, multiplication by real numbers is correctly defined for Q-based hyperoperators represented in **K**.

Proposition is proved.

Note that this result is true not only for real hyperoperators but also for complex hyperoperators.

Proposition 4.13. Any identity that is valid for hyperoperators from **K** and involves only operations of addition and/or multiplication by real numbers is also valid for *Q*-based hyperoperators represented in **K**.

Proof is left as an exercise.

In particular, Proposition 4.7 gives the following identities for extrafunctions.

1. $F(x) + G(x) = G(x) + F(x)$
2. $F(x) + (G(x) + H(x)) = (F(x) + G(x)) + H(x)$
3. $a(F(x) + G(x)) = aF(x) + aG(x)$
4. $(a + b)F(x) = aF(x) + bF(x)$
5. $a(bF(x) = (ab)F(x)$
6. $1 \cdot F(x) = F(x)$

Corollary 4.13. Any identity that is valid for real functions and involves only operations of addition and/or multiplication by real numbers is also valid for *Q*-based real extrafunctions represented in **F**.

Propositions 4.4, 4.5 and 4.7 imply the following result.

Theorem 4.5. If **K** is a linear space over the field of real numbers $\boldsymbol{R}$, then $\mathbf{E}^{\mathbf{K}}{}_{\omega Q}$ is also a linear space over the field of real numbers $\boldsymbol{R}$.

Corollary 4.14. The set of all restricted extrafunctions is a linear space over the field of real numbers $\boldsymbol{R}$.

Corollary 4.15. The set of all continuously represented restricted real pointwise extrafunctions is a linear space over the field of real numbers $\boldsymbol{R}$.

Corollary 4.16. The set of all real compactwise extrafunctions is a linear space over the field of real numbers $\boldsymbol{R}$.

Corollary 4.17. The set of all real extended distributions is a linear space over the field of real numbers $\boldsymbol{R}$.

Theorems 4.1 and 4.5 imply the following result.

Theorem 4.6. If $P > Q$ and **K** is a linear space over $\boldsymbol{R}$, then there is a linear mapping π of the linear space $\mathbf{E}^{\mathbf{K}}{}_{\omega P}$ over $\boldsymbol{R}$ onto the linear space $\mathbf{E}^{\mathbf{K}}{}_{\omega Q}$ over $\boldsymbol{R}$.

Corollary 4.18. There is a linear mapping σ_{cont} of the linear space $\mathbf{E}^{C(\boldsymbol{R})}{}_{\omega Q_{\text{comp}}}$ onto the linear space $\mathbf{E}^{C(\boldsymbol{R})}{}_{\omega Q_{\text{pt}}}$.

Corollary 4.19. There is a linear mapping τ_{fbd} of the linear space $\mathbf{E}^{BI(\boldsymbol{R})}{}_{\omega Q_{\text{cp}}}$ onto the linear space $\mathbf{E}^{BI(\boldsymbol{R})}{}_{\omega Q_{\text{pt}}}$.

Corollary 4.20. There is a linear mapping τ_{cst} of the linear space $\mathbf{E}^{C(\boldsymbol{R})}{}_{\omega Q_{\text{comp}}}$ onto the linear space $\mathbf{E}^{C(\boldsymbol{R})}{}_{\omega Q_{\mathbf{K}}}$.

If A: $L \to M$ is an operator and D: $L \to \boldsymbol{R}$ is a functional, then it is possible to define multiplication of A by D by the following formula

$$(D \cdot A)(x) = D(x) \cdot A(x)$$

Note as $D(x)$ is a real number and $A(x)$ is an element from a real vector space, the product $D(x) \cdot A(x)$ is correctly defined.

Let us assume that the class of operators **K** is closed with respect to multiplication by functionals from a class of functionals **H**, i.e., if g is a functional from **H** and $f \in$ **K**, then $g \cdot f \in$ **K**, and for each q from Q, the space **H** is a seminormed ring and the space **K** is a hyperseminormed module over **H**. Then multiplication by functionals from **H** induces the corresponding operation in sets of hyperoperators represented in **K**.

Proposition 4.14. Multiplication by functionals from **H** is correctly defined for Q-based hyperoperators represented in **K**.

Proof. Let us take a Q-based hyperoperator $F = \mathrm{EF}_Q(f_i)_{i \in \omega}$ and a functional g from **H**. We define $g \cdot F = \mathrm{EF}_Q(g \cdot f_i)_{i \in \omega}$. Let us take another sequence $(h_i)_{i \in \omega}$ that represents F and show that $g \cdot F = \mathrm{EF}_Q(g \cdot h_i)_{i \in \omega}$. Note that both sequences $(f_i)_{i \in \omega}$ and $(h_i)_{i \in \omega}$ belong to $\mathbf{K}^{\omega}$. By conditions from the proposition, for each q from the set Q, the space **H** is a seminormed ring, the space **K** is a hyperseminormed module over **H** and thus, q satisfies Condition N5. Consequently, for any hyperseminorm q from Q, we have

$$\lim_{i \to \infty} q((g \cdot f_i - g \cdot h_i)) \leq \lim_{i \to \infty} q(g) \cdot q((f_i - h_i)) = 0$$

because $\lim_{i \to \infty} q(f_i - h_i) = 0$ and As q is a non-negative function, we have

$$\lim_{i \to \infty} q((g \cdot f_i - g \cdot h_i)) = 0$$

Thus, multiplication by functionals from **H** is correctly defined for Q-based hyperoperators represented in **K**.

Proposition is proved.

Let us assume that the class **F** is closed with respect to multiplication by functions from a class **H**, i.e., if g is a function from **H** and $f \in$ **F**, then $g \cdot f \in$ **F**, and for each q from Q, **H** is a seminormed ring and **F** is a hyperseminormed module over **H**. Then Proposition 4.14 implies that multiplication by functions from **H** induces the corresponding operation in sets of extrafunctions represented in **F**.

Corollary 4.21. Multiplication by functions from **H** is correctly defined for Q-based real extrafunctions represented in **F**.

Corollary 4.22. Multiplication by total continuous real functions is correctly defined for extended distributions.

Corollary 4.23 (Schwartz, 1950/1951)**.** Multiplication by total continuous real functions is correctly defined for distributions.

Propositions 4.13 and 4.14 imply the following result.

Theorem 4.7. If the class **K** operators is a Q-hyperseminormed module over a Q-seminormed ring **H** of functionals, then the class of hyperoperators $\mathbf{E}^{\mathbf{K}}{}_{\omega Q}$ is a module over the ring **H**.

Corollary 4.24. If the class **F** is a Q-hyperseminormed module over a Q-seminormed ring **H** of real functions, then $\mathbf{E}^{\mathbf{F}}{}_{\omega Q}$ is a module over the ring **H**.

As the class of all total real functions is a Q_{pt}-seminormed ring, we have the following result.

Corollary 4.25. The set of all restricted extrafunctions is a module over the ring of all total real functions.

As the class of all total real functions is a Q_{pt}-seminormed module over the Q_{pt}-seminormed ring of all total continuous real functions, we have the following results.

Corollary 4.26. The set of all continuously represented restricted real pointwise extrafunctions is a module over the ring of all total continuous real functions.

Corollary 4.27. The set of all real compactwise extrafunctions is a module over the ring of all total continuous real functions.

Corollary 4.28. The set of all real extended distributions is a module over the ring of all total continuous real functions.

Theorems 4.1 and 4.7 imply the following result.

Theorem 4.8. If $P > Q$ and **K** is $(P\cup Q)$-hyperseminormed module over the $(P\cup Q)$-seminormed ring **H**, then there is a homomorphism π of the P-hyperseminormed module $\mathbf{E}^{\mathbf{K}}{}_{\omega P}$ over **H** onto the Q-hyperseminormed module $\mathbf{E}^{\mathbf{K}}{}_{\omega Q}$ over **H**.

Corollary 4.29. There is a homomorphism σ_{cont} of the module $\mathbf{E}^{C(\boldsymbol{R})}{}_{\omega Q_{\mathrm{comp}}}$ onto the module $\mathbf{E}^{C(\boldsymbol{R})}{}_{\omega Q_{\mathrm{pt}}}$.

Corollary 4.30. There is a homomorphism τ_{fbd} of the module $\mathbf{E}^{BI(\boldsymbol{R})}{}_{\omega Q_{\mathrm{cp}}}$ onto the module $\mathbf{E}^{BI(\boldsymbol{R})}{}_{\omega Q_{\mathrm{pt}}}$.

Corollary 4.31. There is a homomorphism τ_{cst} of the module $\mathbf{E}^{C(\boldsymbol{R})}{}_{\omega Q_{\mathrm{comp}}}$ onto the module $\mathbf{E}^{C(\boldsymbol{R})}{}_{\omega Q_{\mathbf{K}}}$.

As before, we assume that q is a hyperseminorm and Q is a set of hyperseminorms.

Definition 4.15. A Q-based hyperoperator (extrafunction) $F = \text{EF}_Q(f_i)_{i\in\omega}$ is called *Q-bounded* if for any hyperseminorm q from Q, there is a positive real number c and a representation $(g_i)_{i\in\omega}$ of F, such that $q(g_i) < c$ for almost all $i = 1, 2, 3, \ldots$.

The set of all Q-bounded Q-based hyperoperators represented in **K** is denoted by $\mathbf{BE}^{\mathbf{K}}{}_{\omega Q}$ and the set of all Q-bounded Q-based extrafunctions represented in **F** is denoted by $\mathbf{BE}^{\mathbf{F}}{}_{\omega Q}$.

Lemma 4.6. For any representation $(f_i)_{i\in\omega}$ of a Q-bounded Q-based hyperoperator F and any hyperseminorm q from Q, there is a positive real number d, such that $q(f_i) < d$ for almost all $i = 1, 2, 3, \ldots$.

Proof. As F is a Q-bounded Q-based hyperoperator, there is a positive real number c and a representation $(g_i)_{i\in\omega}$ of F, such that $q(g_i) < c$ when $i > m$. As $(f_i)_{i\in\omega} \approx_Q (g_i)_{i\in\omega}$, we have $\lim_{i\to\infty} q(f_i - g_i) = 0$ for any $q \in Q$. Thus, there is a positive real number k, such that $q(f_i - g_i) < k$ when $i > n$.

As q is a hyperseminorm, we have

$$q(f_i) = q((f_i - g_i) + g_i) \leq q(f_i - g_i) + q(g_i) < c + k$$

when $i > \max\{m, n\}$.

Lemma is proved because we can take $d = c + k$.

Corollary 4.32. For any representation $(f_i)_{i\in\omega}$ of a Q-bounded Q-based hyperoperator F and any hyperseminorm q from Q, there is a positive real number d, such that $q(f_i) < d$ for all $i = 1, 2, 3, \ldots$.

Proof is left as an exercise.

Corollary 4.33. For any hyperseminorm q from Q, there is a positive real number d, such that for any representation $(f_i)_{i\in\omega}$ of a Q-bounded Q-based real extrafunction F, we have $q(f_i) < d$ for almost all $i = 1, 2, 3, \ldots$.

Proof is left as an exercise.

Let us assume that the class of real functionals **G** is closed with respect to multiplication, i.e., if real functionals $g: L \to \boldsymbol{R}$ and $f: L \to \boldsymbol{R}$ belong to **G**, then $g\cdot f \in \mathbf{G}$ where $(g\cdot f)(x) = g(x)\cdot f(x)$, and for each q from Q, **G** is a seminormed algebra over $\boldsymbol{R}$. Then multiplication in **G** induces the corresponding operation in sets of Q-bounded hyperfunctionals represented in **G**.

Proposition 4.15. Multiplication of Q-bounded Q-based real (complex) hyperfunctionals represented in **G** is correctly defined.

Proof. Let us take Q-bounded Q-based real (complex) hyperfunctionals $F = \mathrm{EF}_Q(f_i)_{i\in\omega}$ and $G = \mathrm{EF}_Q(g_i)_{i\in\omega}$ represented in **F**. We define $G{\cdot}F = \mathrm{EF}_Q(g_i{\cdot}f_i)_{i\in\omega}$. Take another sequence $(h_i)_{i\in\omega}$ that represents F, we show that $G{\cdot}F = \mathrm{EF}_Q(g_i{\cdot}h_i)_{i\in\omega}$. Note that all three sequences $(f_i)_{i\in\omega}$,$(h_i)_{i\in\omega}$ and $(g_i)_{i\in\omega}$ belong to $\mathbf{G}^\omega$. Then by Lemma 4.6, for any hyperseminorm q from Q and the representation $(g_i)_{i\in\omega}$ of G, there is a positive real number c, such that $q(g_i) < c$ for almost all $i = 1, 2, 3, \ldots$. Consequently, we have

$$\lim_{i\to\infty} q((g_i{\cdot}f_i - g_i{\cdot}h_i)) \leq \lim_{i\to\infty} q(g_i){\cdot}q((f_i - h_i)) \leq \lim_{i\to\infty} c{\cdot}\, q((f_i - h_i)) = 0$$

because $\lim_{i\to\infty} q(f_i - h_i) = 0$ and q satisfies Condition N4. As q is a non-negative function, we have

$$\lim_{i\to\infty} q((g_i{\cdot}f_i - g_i{\cdot}h_i)) = 0$$

As this multiplication is commutative, it is correctly defined for all Q-bounded Q-based real (complex) hyperfunctionals represented in **G**. Note that the product of Q-bounded real (complex) hyperfunctionals is also Q-bounded.

Proposition is proved.

Let us assume that the class of real functions **F** is closed with respect to multiplication, i.e., if $g, f \in \mathbf{F}$, then $g{\cdot} f \in \mathbf{F}$ where $(g{\cdot} f)(x) = g(x){\cdot} f(x)$, and for each q from Q, **F** is a seminormed algebra over $\boldsymbol{R}$. Then multiplication in **F** induces the corresponding operation in sets of Q-bounded extrafunctions represented in **F**. Then Proposition 4.6 implies the following result.

Corollary 4.34. Multiplication of Q-bounded Q-based real extrafunctions represented in **F** is correctly defined.

Proposition 4.16. Any identity that is valid for real (complex) functionals (functions) and involves only operations of addition and/or multiplication of functions and real numbers is also valid for Q-bounded Q-based real (complex) hyperfunctionals (extrafunctions) represented in **G** (in **F**).

Proof is left as an exercise.

In particular, Proposition 4.16 gives the following identities for Q-bounded hyperfunctionals (extrafunctions).

1. $F(x){\cdot}\, G(x) = G(x){\cdot}\, F(x)$
2. $F(x){\cdot}(G(x){\cdot}\, H(x)) = (F(x){\cdot}\, G(x)){\cdot}\, H(x)$
3. $F(x){\cdot}(G(x) + H(x)) = F(x){\cdot}G(x) + F(x){\cdot}H(x)$

Propositions 4.16 and 4.15 imply the following result.

Theorem 4.9. If the class **G** of bounded real functionals is closed with respect to multiplication and for each q from Q, the class **G** is a seminormed algebra over $\boldsymbol{R}$, then the class $\mathbf{BE}^{\mathbf{G}}{}_{\omega Q}$ is a linear algebra over $\boldsymbol{R}$.

Corollary 4.35. If the class **F** of bounded real functions is closed with respect to multiplication and for each q from Q, the class **F** is a seminormed algebra over $\boldsymbol{R}$, then the class $\mathbf{BE}^{\mathbf{F}}{}_{\omega Q}$ is a linear algebra over $\boldsymbol{R}$.

Corollary 4.36. The class $\mathbf{BE}^{\mathbf{G}}{}_{\omega Q}$ is a module over the linear algebra of all finite/bounded real hypernumbers $\mathbf{F}\boldsymbol{R}$

Corollary 4.37. The class $\mathbf{BE}^{\mathbf{F}}{}_{\omega Q}$ is a module over the linear algebra of all finite/bounded real hypernumbers $\mathbf{F}\boldsymbol{R}$

Lemma 4.7. A real function is bounded if and only if it is Q_{pt}-bounded.

Proof is left as an exercise.

Corollary 4.38. The set of all bounded restricted real extrafunctions is a linear algebra over $\boldsymbol{R}$.

Corollary 4.39. The set of all bounded continuously represented restricted real extrafunctions is a linear algebra over $\boldsymbol{R}$.

Corollary 4.40. The set of all bounded compactwise real extrafunctions is a linear algebra over $\boldsymbol{R}$.

Complex bounded extrafunctions and hyperfunctionals have similar properties.

KEYWORDS

- **addition**
- **distribution**
- **extrafunction**
- **function**
- **functional**
- **hyperfunctional**
- **hyperoperator**
- **linear algebra**
- **module**
- **multiplication**
- **operator**

CHAPTER 5

FROM TOPOLOGICAL VECTOR SPACES TO SEMITOPOLOGICAL VECTOR SPACES

We begin this chapter with the description of the classical concept of a topological vector space [cf., e.g., (Rudin, 1991)] and only then come to the new structure of a semitopological vector space.

Definition 5.1. A *topological vector space* L over a topological (normed or valuated) field $\boldsymbol{F}$ is a vector space over $\boldsymbol{F}$ with a topology in which addition and scalar multiplication by elements from $\boldsymbol{F}$ (dilation) are continuous operations.

Some authors additionally demand that the point $\mathbf{0}$ in a topological vector space is closed [cf., e.g., (Rudin, 1991)]. This condition results in the Hausdorff topology in topological vector spaces.

Many mathematicians studied properties of topological vector spaces. Thus, it is possible to read about them in many papers and books [cf., e.g., (Ali, 1965; Beckenstein et al., 1977; Bourbaki, 1953, 1955; Flood, 1984; Gierz, 1982; Grothendieck, 1992; Horváth, 1966; Husain, 1965; Khaleelulla, 1982; Köthe, 1969; Kurzweil, 2000; Narici and Beckenstein, 1985; Pantsulaia, 2007; Peressini, 1967; Pietsch, 1965; Robertson and Robertson, 1964; Rudin, 1991; Schaefer and Wolff, 1999; Trèves, 1995; Wilansky, 1978)].

The concept of a semitopological vector space is a natural extension of the concept of a topological vector space. This extension provides a possibility to expand the scope of the theory and its applications while preserving many useful results.

Definition 5.2. A *semitopological vector space* L over a topological (normed or valuated) field $\boldsymbol{F}$ is a vector space over $\boldsymbol{F}$ with a topology in which addition is continuous, while scalar multiplication by elements from

$\boldsymbol{F}$, which is called *dilation*, is continuous with respect to L, i.e., the scalar multiplication mapping m: $\boldsymbol{F}\times L \rightarrow L$ is continuous in the second coordinate.

When the multiplication mapping m: $\boldsymbol{F}\times L\rightarrow L$ is continuous, then L is a *topological vector space* over the field $\boldsymbol{F}$. Similar to topological vector spaces, when the point $\mathbf{0}$ in a semitopological vector space is closed, the space has Hausdorff topology (cf. Proposition 5.5).

Example 5.1. As it is proved by Burgin (2012), the set of all real hypernumbers $\boldsymbol{R}_\omega$ is a hypernormed vector space where the hypernorm $\|\cdot\|$ is defined by the following formula:

If α is a real hypernumber, i.e., $\alpha = \text{Hn}(a_i)_{i\in\omega}$ with $a_i \in \boldsymbol{R}$ for all $i\in\omega$, then

$$\| \alpha \| = \text{Hn}(|a_i|)_{i\in\omega}$$

Note that this hypernorm coincides with the conventional norm on real numbers but it is impossible get the same topology by means of a conventional finite norm.

Thus, according to Theorem 5.3, $\boldsymbol{R}_\omega$ is a semitopological vector space over the field $\boldsymbol{R}$. However, it is not a topological vector space. Indeed, taking the real hypernumber $\alpha = \text{Hn}(i)_{i\in\omega}$, we see that for any real number ε, however small it is, the hypernumber $\varepsilon\alpha = \text{Hn}(\varepsilon i)_{i\in\omega}$ does not belong to any neighborhood $O_k\alpha$ where k is a real number because $|\alpha - \varepsilon\alpha| = (1 - \varepsilon)|\alpha|$ is an infinite hypernumber and thus, it is larger than any real number, such as k. As a consequence, the multiplication mapping m: $\boldsymbol{R} \times \boldsymbol{R}_\omega \rightarrow \boldsymbol{R}_\omega$ is not continuous and thus, $\boldsymbol{R}_\omega$ is not a topological vector space.

Example 5.2. As it is proved by Burgin (2002), the set of all complex hypernumbers $\boldsymbol{C}_\omega$ of all complex hypernumbers is a hypernormed vector space where the hypernorm $\|\cdot\|$ is defined by the following formula:

If α is a complex hypernumber, i.e., $\alpha = \text{Hn}(a_i)_{i\in\omega}$ with $a_i \in \boldsymbol{C}$ for all $i\in\omega$, then

$$\| \alpha \| = \text{Hn}(|a_i|)_{i\in\omega}$$

Note that this hypernorm coincides with the conventional norm on complex numbers but it is impossible get the same topology by means of the conventional finite norm.

Thus, by Theorem 5.3, $\boldsymbol{C}_\omega$ is a semitopological vector space over the field $\boldsymbol{C}$. However, it is not a topological vector space. Indeed, taking the real hypernumber $\alpha = \text{Hn}(i)_{i\in\omega}$, we see that for any real number ε, however small it is, the hypernumber $\varepsilon\alpha = \text{Hn}(\varepsilon i)_{i\in\omega}$ does not belong to any neighborhood $O_k\alpha$ where k is a real number because $|\alpha - \varepsilon\alpha| = (1 - \varepsilon)|\alpha|$ is an infinite hypernumber and thus, it is larger than any real number, such as k. As a consequence, the multiplication mapping m: $\boldsymbol{R} \times \boldsymbol{C}_\omega \to \boldsymbol{C}_\omega$ is not continuous and thus, $\boldsymbol{C}_\omega$ is not a topological vector space.

In spite of scalar multiplication being not continuous in the first argument, note that for each number a, the multiplication mapping m_a: $a \times \boldsymbol{C}_\omega \to \boldsymbol{C}_\omega$ is continuous.

Let us consider a semitopological vector space L over $\boldsymbol{R}$ or $\boldsymbol{C}$ in which all points are closed.

Proposition 5.1. (a) The space L is convex.

(b) The space L is compact if and only if $L = \{\mathbf{0}\}$.

(c) Every finite dimensional subspace of L is closed in L.

Proof. Deduction of the part (a) is left as an exercise.

(b) Let us assume that the space L is compact and has a non-zero element u. Then if L is a vector space over $\boldsymbol{R}$, then it contains a subset R isomorphic to the real line $\boldsymbol{R}$. If L is a vector space over $\boldsymbol{C}$, then it contains a subset C isomorphic to $\boldsymbol{C}$. We consider only the first case because the second one is treated in a similar way.

Let us consider a net $l = \{u_i;\ u_i \in R,\ i\in I\}$. As R isomorphic to $\boldsymbol{R}$, there is a number net $f = \{r_i;\ r_i \in \boldsymbol{R},\ i\in I\}$ such that $u_i = r_i u$ for some non-zero element $u \in R$ and the net f converges to an element $x \in L$. The net $f = \{r_i;\ r_i \in \boldsymbol{R},\ i\in I\}$ is bounded or it contains a sequence $h = \{r_j;\ r_j \in \boldsymbol{R}, j = 1, 2, 3, \ldots\}$ such that numbers r_j grow without limit, i.e., $\lim_{j\to\infty} r_j = \infty$. As $\lim_{j\to\infty} r_j u = x$ because any subsequence of a converging net has the same limit, this implies the following equalities

$$u = \lim_{j\to\infty} (1/r_j)(r_j u) = \lim_{j\to\infty} (1/r_j) \cdot \lim_{j\to\infty} r_j u = 0 \cdot \lim_{j\to\infty} r_j u = 0 \cdot x = \mathbf{0}$$

This contradiction shows that the net $f = \{r_i;\ r_i \in \boldsymbol{R},\ i\in I\}$ is bounded, i.e., there is a number $u \in \boldsymbol{R}$ such that $|r_i| <$ for all $i\in I$. Then f has a converging subnet g, i.e., $\lim g = r$. Consequently,

$$x = \lim_{i\in I} r_i u = \lim_{j\to\infty} r_j u = ru$$

It means that $x \in R$, i.e., the set is closed. A closed subset of a compact set is a compact set (Alexandroff, 1961). However, R is not compact because $\boldsymbol{R}$ is not compact. This contradiction demonstrates that $L = \{\mathbf{0}\}$.

(c) is left as an exercise in applying induction on the dimension of L.

However, semitopological vector spaces do not have all properties of topological vector spaces. For instance, any semitopological vector space L is linearly connected, i.e., it is possible to connect any two points in L by a continuous line (Narici and Beckenstein, 1985).

Indeed, taking the semitopological vector space $\boldsymbol{R}_\omega$ (cf. Example 5.1), we see that it is impossible to connect any two points, one of which is finite, e.g., 7, and another is infinite, e.g., $\alpha = \text{Hn}(i)_{i\in\omega}$, in $\boldsymbol{R}_\omega$ by a continuous line because any infinite hypernumber cannot be a limit of finite hypernumbers, while any finite hypernumber cannot be a limit of infinite hypernumbers. Moreover, $\boldsymbol{R}_\omega$ is not even connected.

With respect of the algebraic structure, it is possible to define two basic operators in a vector space L over a field $\boldsymbol{F}$.

1. If a is an element from $\boldsymbol{F}$, then the *translation operator* T_z is defined by the formula:

$$T_z(x) = x + z \text{ where } x, z \in L$$

2. If $a \neq 0$ is an element from $\boldsymbol{F}$, then the *dilation operator* M_a is defined by the formula:

$$M_a(x) = ax \text{ where } x \in L$$

Proposition 5.2. Operators T_z and M_a are homeomorphisms of the semitopological vector space L.

Proof. The axioms of a vector space imply that T_z and M_a are one-to-one mappings. As addition is continuous in L, the operator T_z is also continuous. As scalar multiplication is continuous with respect to L, the operator M_a is also continuous.

Note that inverse operators T_z^{-1} and M_a^{-1} are also continuous because and they are equal to T_{-z} and $M_{1/a}$, respectively.

Proposition is proved.

Corollary 5.1. The topology of a semitopological vector space L is *translation-invariant*, or simply invariant, i.e., a subset A from L is open if and only if any of its translations $T_l(A) = A + l$ is open for any element l from L.

As a result, such a topology is completely determined by any local base and thus, by any local base at $\mathbf{0}$.

We remind that a *local base* at a point x from L is a set $\boldsymbol{LB}$ of neighborhoods of x such that any neighborhood of x contains an element from $\boldsymbol{LB}$.

Corollary 5.2. The topology of a semitopological vector space L is *dilation-invariant*, i.e., a subset A from L is open if and only if any set $M_a(A) = aA$ is open for any real number a.

It is possible to perform different operations with semitopological vector spaces. One of the most important is the *Cartesian product* $L = \prod_{i\in I} L_i$ of a family of semitopological vector spaces $\{L_i;\ i\in I\}$. It is built as the Cartesian product of the sets L_i (cf. Appendix) and becomes a semitopological vector space with the conventional *product topology*.

The open sets in the product topology are unions (finite or infinite) of sets of the form $\prod_{i\in I} U_i$, where each U_i is an open subset of the vector space L_i and $U_i \neq L_i$ for only finitely many indices i. Consequently, for a finite Cartesian product, in particular, for the Cartesian product of two semitopological vector spaces, the products of base elements of the vector spaces L_i give a basis for the product space $\prod_{i\in I} L_i$.

The product topology on X is the topology generated by sets of the form $\pi_i^{-1}(U_j)$, where π_i: $\prod_{i\in I} X_i \to X_i$ is the *canonical* i-th *projection mapping* (i-th *projection*), i is in I and U_j is an open subset of X_i. In other words, the sets $\{\pi_i^{-1}(U_j)\}$ form a subbase for the topology on X. A subset of X is open if and only if it is a (possibly infinite) union of intersections of finitely many sets of the form $\pi_i^{-1}(U_j)$. The $\pi_i^{-1}(U_j)$ are called *open cylinders*, and their intersections are cylinder sets.

The product topology is the coarsest topology (i.e., the topology with the fewest open sets) for which all the projections p_i are continuous.

There are other topologies in the Cartesian product $L = \prod_{i\in I} L_i$ of a family of semitopological vector spaces $\{L_i;\ i\in I\}$. For instance, the product of the topologies of each L_i forms a basis for what is called the *box topology* in L. In general, the box topology is finer than the product topology, but for finite products they coincide.

Example 5.3. Let us consider the set $F(\boldsymbol{R}, \boldsymbol{R})$ of all functions f: $\boldsymbol{R} \to \boldsymbol{R}$. The set $F(\boldsymbol{R}, \boldsymbol{R})$ has the natural product topology and is identified with the product space $\boldsymbol{R}^{\boldsymbol{R}}$. With this topology, $F(\boldsymbol{R}, \boldsymbol{R})$ becomes a topological vector space, called the *space of pointwise convergence*. The reason for this name is the following: if (f_n) is a sequence of elements in X, then f_n has limit f in X if and only if $f_n(x)$ has limit $f(x)$ for every real number x.

A semitopological vector space is said to be *normable* if its topology can be induced by a norm. The vector space $F(\boldsymbol{R}, \boldsymbol{R})$ is complete, but not normable: indeed, every neighborhood of $\mathbf{0}$ in the product topology contains lines, i.e., sets $\boldsymbol{R}f$ for an element $f \neq \mathbf{0}$.

Proposition 5.3. The Cartesian product of semitopological vector spaces is also a semitopological vector space.

Proof is left as an exercise.

Properties of topological structures and of vector (linear) spaces give us the following result.

Lemma 5.1. The intersection of any closed linear subspaces of a semitopological vector space $L$is a closed subspace of L.

Proof is left as an exercise.

Proposition 5.4. The closure CM of any linear subspace M of a semitopological vector space L is a linear subspace of L.

Proof. Let us consider a linear subspace M of a semitopological vector space L. Taking an element $y \in M$, we obtain the continuous mapping $A(x) = x + y$. As M is a linear subspace of L, we have $A(M) \subseteq M$ and $A(\mathrm{C}M) \subseteq \mathrm{C}M$ because A is a continuous mapping (Kuratowski, 1966). Thus, $x + y \in \mathrm{C}M$ when $y \in M$ and $x \in \mathrm{C}M$.

At the same time, it is possible to consider the continuous mapping $B(y) = x + y$ with $x \in \mathrm{C}M$. As M is a linear subspace of L, $B(M) \subseteq \mathrm{C}M$ and $B(\mathrm{C}M) \subseteq \mathrm{C}M$ because B is a continuous mapping (Kuratowski, 1966). Thus, $x + y \in \mathrm{C}M$ when $x \in \mathrm{C}M$ and $y \in \mathrm{C}M$. It means that CM is closed with respect to addition of vectors.

In a similar way, we can prove that $ax \in \mathrm{C}M$ for any $x \in \mathrm{C}M$ and any real number a. Thus, CM is closed with respect to multiplication of vectors by real numbers.

So, CM is a vector space.

Proposition is proved.

Definition 5.3. A subset C of a vector space L is called *circled* if $rC \subseteq C$ when $\|r\| \leq 1$.

Let L be a semitopological vector space.

Theorem 5.1. (a) If A, $B \subseteq L$ and the set B is open, then the set $A + B$ is also open.

(b) If $B \subseteq L$, r is a real number and the set B is open, then the set rB is also open.

(c) If $B \subseteq L$, r is a real number and the set B is closed, then the set rB is also closed.

(d) If $B \subseteq L$, $x \in L$ and the set B is open, then the set $x + B$ is also open.

(e) If $B \subseteq L$, $x \in L$ and the set B is closed, then the set $x + B$ is also closed.

(f) If A, $B \subseteq L$ and sets A and B are compact, then the set $A + B$ is also compact.

(g) If $A \subseteq L$ and $\boldsymbol{U}$ is the base neighborhood filter of $\mathbf{0}$, then the closure CA of A is given by the formula C$A = \bigcap_{i \in I}\{A + U;\ U \in \boldsymbol{U}\}$.

(h) If $A \subseteq L$ is a circled set, then the closure CA of A is also a circled set.

(i) If $A \subseteq L$ is a circled set and $\mathbf{0} \in A$, then the interior Int$A$ of A is also a circled set.

Proof. (a) Let us assume that the set B is open. Then $A + B = \bigcup_{x \in A} T_x(B)$. By Corollary 5.1, each set $T_x(B)$ is open. In addition, the union of open sets is open. Thus, the set $A + B$ is open.

(b) Let us assume that the set B is open. Then by Corollary 5.2, the set $rB = T_r(B)$ is open.

(c) Let us assume that the set B is closed. Then its complement CoB = $L\backslash B$ is open (Kelly, 1957). By Corollary 5.2, the set rCo$B = T_r$(CoB) is open. Distributivity of scalar multiplication implies the equality

$$r\mathrm{Co}B = \mathrm{Co}(rB)$$

As the set CoB is closed, the set B is open (Kelly, 1957).

Statements (d) and (e) follow from Proposition 5.1.

(f) Let us assume that sets A and B are compact. At the same time, $A + B = f(A \cdot B)$ where $f(x, y) = x + y$. The set $A \cdot B$ is compact and the mapping f is continuous in L. So, the set $A + B$ is also compact (Alexandroff, 1961).

(g) Let us take the set $A = \bigcap_{i \in I}\{A + U;\ U \in \boldsymbol{U}\}$. By the part (a), each set $\{x + U;\ U \in \boldsymbol{U}\}$ is a neighborhood of the element x for any $x \in L$. Consequently, if $x \in B$, then any neighborhood of x intersects A. It means that $B \subseteq$ CA (Kelly, 1955).

At the same time, if $x \in$ CA, then $x \in A + U$ for any neighborhood U of $\mathbf{0}$. Thus, C$A \subseteq B$. As a result, C$A = B$.

(h) Let us consider a circled set $A \subseteq L$ and take a positive real number r with $\|r\| \leq 1$. By Part (c), $rA \subseteq A$ implies rC$A \subseteq$ CA.Thus, the closureCA of A is also a circled set.

(i) As the scalar multiplication is continuous in L, if $r \neq 0$, then r(Int$A$) is the interior of rA. Consequently, r(Int$A$) $\subseteq$ Int$A$ as $rA \subseteq A$. The condition $\mathbf{0} \in A$ then shows that r(Int$A$) $\subseteq$ Int$A$ when $\|r\| \leq 1$. Thus, the interior Int$A$ of A is also a circled set.

Theorem is proved.

Note that a sum of two closed subsets in a semitopological vector space. L is not always closed (Khaleelulla, 1982).

Proposition 5.5. The following conditions are equivalent:

(1) The point $\mathbf{0}$ is closed.
(2) Some point x from L is closed.
(3) All points in L are closed.

Proof. (1)⇒(2): Let us assume that the point $\mathbf{0}$ is closed and take another point x from L. To prove that the point x from L is closed, we need to show that the complement of x is open. To achieve this, it is sufficient to take another point $y \neq x$ and to find a neighborhood of y that does not contain x.

Let us consider the element $z = y - x$. By construction, $z \neq \mathbf{0}$. As the point $\mathbf{0}$ is closed by the condition (1), there is a neighborhood Oz of z that does not contain $\mathbf{0}$. Then $y \in x + \text{O}z$ because $y = x + z$ and $z \in \text{O}z$, while $x + \text{O}z$ is a neighborhood of y by Theorem 5.1(d).

At the same time, $x \notin x + \text{O}z$ because otherwise $x = x + u$ for some element u from Oz. However, in this case, $u = \mathbf{0}$ what contradicts the choice of the neighborhood Oz. Thus, $x \notin x + \text{O}z$ and x is closed.

As x is an arbitrary point from L, this also proves (3), i.e., (1)⇒(3). The inverse implication (3)⇒(1) is evident.

(2)⇒(1): Let us assume that some point $x \neq \mathbf{0}$ from L is closed. Therefore for any point y from L, if $y \neq x$, then there is a neighborhood Oy of y that does not contain x.

Let us take an element $z \neq \mathbf{0}$. Then $y = x + z \neq z$. Taking a neighborhood Oy of y that does not contain x, we can consider the set $(-x) + \text{O}y$. Subsequently $z \in (-x) + \text{O}y$ because $z = y - x$ and $y \in \text{O}y$, while $(-x) + \text{O}y$ is a neighborhood of z by Theorem 5.1(d).

At the same time, $\mathbf{0} \notin (-x) + \text{O}y$ because otherwise $u - x = \mathbf{0}$ for some element u from Oy. However, in this case, it would be $u = x$ and this contradicts the choice of the neighborhood Oy. Thus, $\mathbf{0} \notin (-x) + \text{O}y$ and $\mathbf{0}$ is closed.

Proposition is proved.

We remind some useful concepts from linear algebra.

Definition 5.4. A subset A of the vector space L is called a *cone*, if $x \in A$ implies $ax \in A$ for any positive real number a.

Example 5.4. Let us consider the two-dimensional vector space $\boldsymbol{R}^2$ and two real numbers $0 < k < r$. Then the set $A = \{(x, y); y = tx, t \in \boldsymbol{R}, k < t < r\}$ is a cone.

By definition, any cone is a circled set.

Lemma 5.2. a) The union of cones is a cone.

b) The intersection of cones is a cone.

Proof. (a) Let us consider two cones A and B in a vector space L and take their union $D = A \cup B$. If $x \in D$, then either $x \in A$ or $x \in B$. In the first case,

$ax \in A$ for any positive real number a because A is a cone. In the second case, $ax \in B$ for any positive real number a because B is a cone. Thus, in both cases, $ax \in D$ for any positive real number a. Consequently, D is a cone.

(b) Now let us take the intersection $C = A \cap B$. If $x \in C$, then $x \in A$ and $x \in B$. Then $ax \in A$ for any positive real number a because A is a cone and $ax \in B$ for any positive real number a because B is a cone. Thus, $ax \in C$ for any positive real number a. Consequently, C is a cone.

Lemma is proved.

Definition 5.5. A subset A of the vector space L is called an *affine subspace* of L, if $A = x + M$ where M is a linear subspace of L.

Example 5.5. Let us consider the two-dimensional vector space $\boldsymbol{R}^2$. Then the set $A = \{(x, y); y = tx + 10, t \in \boldsymbol{R}\}$ is an affine subspace.

Note that any linear subspace of a vector space is also an affine subspace of the same vector space.

Let us find what properties are preserved by the closure operation.

Let L be a semitopological vector space and A, B be subsets of L.

Theorem 5.2. (a) $\mathrm{C}A + \mathrm{C}B \subseteq \mathrm{C}(A+B)$.

(b) If A is a convex set, then $\mathrm{C}A$ is also a convex set.

(c) If A is a cone, then $\mathrm{C}A$ is also a cone.

(d) If A is an affine subspace of L, then $\mathrm{C}A$ is also an affine subspace of L.

Proof. (a) Let us take elements $a \in \mathrm{C}A$ and $b \in \mathrm{C}B$. By the continuity of addition in L, for every neighborhood U of the point $a + b$, there are a neighborhood V of the point a and a neighborhood W of the point b, such that $V + W \subseteq U$.

By the definition of the closure $\mathrm{C}A$, every neighborhood of a intersects with A and by the definition of the closure $\mathrm{C}B$, every neighborhood of b intersects B. That is, there exist an element $x \in V \cap A$ and an element $y \in W \cap B$. Then, the memberships $x \in A$ and $y \in B$ imply the membership $x + y \in A + \mathrm{B}$, while the memberships $x \in V$ and $y \in W$ imply the membership $x + y \in V + W \subseteq U$.

In other words, every neighborhood of $a + b \in A + B$ intersects $A + B$, which implies that $a + b \in \mathrm{C}(A + B)$, and consequently, $\mathrm{C}A + \mathrm{C}B \subseteq \mathrm{C}(A + B)$.

(b) Let us take a convex set A and two points $x, y \in \mathrm{C}A$. We need to show that for any $0 < t < 1$, the point $tx + (1 - t)y$ also belongs to $\mathrm{C}A$.

Let us consider the mapping D_t of the Cartesian product $L \times L$ such that $D_t(x, y) = tx + (1 - t)y$.

As the set A is convex, we have

$$A \times A \subseteq D_t^{-1}(A) \subseteq D_t^{-1}(\mathrm{C}A)$$

As addition is continuous and scalar multiplication is continuous in the second coordinate, D_t is a continuous mapping and $D_t^{-1}(\mathrm{C}A)$ is a closed set. Consequently, we have

$$\mathrm{C}A \times \mathrm{C}A \subseteq D_t^{-1}(\mathrm{C}A)$$

and

$$D_t(\mathrm{C}A \times \mathrm{C}A) \subseteq \mathrm{C}A$$

Consequently, $tx + (1 - t)y \in \mathrm{C}A$ and as elements t, x and y are arbitrary, $\mathrm{C}A$ is a convex set.

(c) Proof is similar to the proof of Proposition 5.3.

(d) If A is an affine subspace of L, then $A = x + M$ where M is a linear subspace of L. Thus, $M = A - x = A + (-x)$. So, $\mathrm{C}M = \mathrm{C}A + (-x)$ is a linear subspace of L by Proposition 5.4. Consequently, $\mathrm{C}A = x + \mathrm{C}M$ is an affine subspace of L.

Theorem is proved.

In what follows, $\boldsymbol{F}$ stands either for the field $\boldsymbol{R}$ of all real numbers or for the field $\boldsymbol{C}$ of all complex numbers or for any subfield of $\boldsymbol{C}$ that contains $\boldsymbol{R}$, while $\mathbf{0}$ denotes the zero element of any vector space.

There are intrinsic relations between hyperseminorms and semitopological vector spaces.

Theorem 5.3. Any hypernormed vector space is a Hausdorff semitopological vector space in the topology induced by the hypernorm.

Proof. Let us consider a vector space L with a hypernorm q and describe the topology τ in L induced by the hypernorm q. Taking an element x from L and a positive real number k, we define the neighborhood $\mathrm{O}_k x$ of x by the following formula

$$\mathrm{O}_k x = \{y \in L;\ q(x-y) < k\}$$

At first, we show that the system of so defined neighborhoods determines a topology τ in L. To do this, it is necessary to check the following neighborhood axioms (Kuratowski, 1966):

NB1. Any neighborhood of a point $x \in X$ contains this point.

NB2. For any two neighborhoods O_1x and O_2x of a point $x \in X$, there is a neighborhood Ox of x that is a subset of the intersection $O_1x \cap O_2x$.

NB3. For any neighborhood Ox of a point $x \in X$ and a point $y \in Ox$, there is a neighborhood Oy of y that is a subset of Ox.

Let us consider a point x from X.

NB1: The point x belongs to O_kx because $q(x-x) = q(\mathbf{0}) = 0 < k$ for any positive real number k.

NB2: Taking two positive real numbers k and h, we see that the intersection $O_kx \cap O_hx = O_lx$ also is a neighborhood of x where $l = \min\{k, h\}$.

NB3: Let $y \in O_kx$. Then $q(x-y) < k$ and by properties of real numbers, there is a positive real number t such that $q(x-y) < k - t$. Then $O_ty \subseteq O_kx$. Indeed, if $z \in O_ty$, then $q(y-z) < t$. Consequently,

$$q(x-z) = q((x-y) + (y-z)) \leq q(x-y) + q(y-z) < (k-t) + t = k$$

It means that $z \in O_kx$.

Thus, we have a topology in L, and this topology τ is Hausdorff because any hypernorm weakly separates points, i.e., if $x \neq y$, then $x - y \neq \mathbf{0}$ and thus, $q(x-y) \neq 0$.

Now we show that addition is continuous and scalar multiplication is continuous in the second coordinate with respect to the topology τ.

Let us consider a sequence $\{x_i;\ i = 1, 2, 3, \ldots\}$ that converges to x, a sequence $\{y_i;\ i = 1, 2, 3, \ldots\}$ that converges to y, and the sequence $\{z_i = x_i + y_i;\ i = 1, 2, 3, \ldots\}$. Convergence of these two sequences means that for any $k > 0$, there are a natural number n such that $q(x_i-x) < k$ for any $i > n$ and a natural number m such that $q(y_i-y) < k$ for any $i > m$. Then by properties of a hypernorm, we have

$$q(z_i - (x + y)) = q((x_i + y_i) - (x + y)) = q((x_i-x) + (y_i-y)) \leq q(x_i-x) + q(y_i-y) < k + k = 2k$$

when $i > \max\{n, m\}$. As k is an arbitrary positive real number, this means that the sequence $\{z_i = x_i + y_i;\ i = 1, 2, 3, \ldots\}$ converges to the element $x + y$. Consequently, addition is continuous in L.

In addition, for any number a from $\boldsymbol{F}$, we have

$$q(u_i - ax) = q(ax_i - ax) = q(a(x_i-x)) \leq |a|\ q(x_i-x) < |a|k$$

where $u_i = ax_i$. As k is an arbitrary positive real number and $|a|$ is a constant, this means that the sequence $\{u_i = ax_i; i = 1, 2, 3, \ldots\}$ converges to ax. Consequently, scalar multiplication is continuous in the second coordinate.

Theorem is proved.

The topology τ in L induced by the hypernorm q is called the *standard topology* in L with respect to q.

Corollary 5.3. Any normed vector space is a Hausdorff topological vector space.

Proposition 5.2 gives us the following results.

Corollary 5.4. The topology of a hypernormed vector space L is translation-invariant.

As a result, such a topology is completely determined by any local base and thus, by any local base at **0**.

Corollary 5.5. The topology of a hypernormed vector space L is dilation-invariant.

However, in contrast to normed vector spaces, the topology constructed in the proof of Theorem 5.3 is not unique for hypernormed vector spaces.

Definition 5.6. (a) A hyperseminorm (hypernorm) q in L is called *full* if for each positive real hypernumber α, there is an element x from L such that $q(x) = \alpha$.

(b) A hyperseminorm (hypernorm) q in L is called *finitely full* if for each positive finite real hypernumber α, there is an element x from L such that $q(x) = \alpha$.

Naturally, any full hyperseminorm (hypernorm) is finitely full.

The standard hypernorm in the space $\boldsymbol{R}_\omega$ of all hypernumbers is an example of a full hypernorm. The standard hyperseminorm in the space $\boldsymbol{R}^\omega$ is an example of a full hyperseminorm.

Theorem 5.4. There is a topology υ defined by the hypernorm q in L, in which the hypernormed vector space L is also a Hausdorff semitopological vector space and which is strictly stronger than the topology τ when the hypernorm q is finitely full.

Proof. Let us consider a vector space L with a hypernorm q and describe the topology υ in L induced by the hypernorm q. Taking an element x from L and a positive finite real hypernumber α, we define the neighborhood $O_\alpha x$ of x by the following formula

$$O_\alpha x = \{y \in L; q(x-y) < \alpha\}$$

At first, we show that the system of so defined neighborhoods determines a topology υ in L. To do this, it is necessary to check the following neighborhood axioms (Kuratowski, 1966):

NB1. Any neighborhood of a point $x \in X$ contains this point.

NB2. For any two neighborhoods O_1x and O_2x of a point $x \in X$, there is a neighborhood Ox of x that is a subset of the intersection $O_1x \cap O_2x$.

NB3. For any neighborhood Ox of a point $x \in X$ and a point $y \in Ox$, there is a neighborhood Oy of y that is a subset of Ox.

Let us consider a point x from X.

NB1: The point x belongs to $O_\alpha x$ because $q(x–x) = q(\mathbf{0}) = 0 < \alpha$ for any positive finite real hypernumber α.

NB2: Taking two positive finite real hypernumbers α and β, we see that the intersection $O_\alpha x \cap O_\beta x = O_\delta x$ also is a neighborhood of x where $\delta = \min\{\alpha, \beta\}$.

NB3: Let $y \in O_\alpha x$. Then $q(x–y) < \alpha$ and by properties of real hypernumbers (cf. Chapter 2), there is a positive finite real hypernumber γ such that $q(x–y) < \alpha - \gamma$. Indeed, if $\alpha = \mathrm{Hn}(a_i)_{i\in\omega}$, then there is a representation $q(x–y) = \beta = \mathrm{Hn}(b_i)_{i\in\omega}$ such that $b_i < a_i$ for all $i\in\omega$. Consequently, there are positive real numbers such that $c_i + b_i < a_i$ for all $i\in\omega$ and taking $\gamma = \mathrm{Hn}(c_i)_{i\in\omega}$, we obtain the necessary hypernumber γ.

Then $O_\gamma y \subseteq O_k x$. Indeed, if $z \in O_\gamma y$, then $q(y–z) < \gamma$. Consequently,

$$q(x–z) = q((x–y) + (y–z)) \leq q(x–y) + q(y–z) < (\alpha–\gamma) + \gamma = \alpha$$

It means that $z \in O_\alpha x$.

Thus, we have a topology in L, and this topology υ is Hausdorff because any hypernorm weakly separates points, i.e., if $x \neq y$, then $x - y \neq \mathbf{0}$ and thus, $q(x–y) \neq 0$.

Now we show that addition is continuous and scalar multiplication is continuous in the second coordinate with respect to the topology υ.

Let us consider a sequence $\{x_i; i = 1, 2, 3, \ldots\}$ that converges to x, a sequence $\{y_i; i = 1, 2, 3, \ldots\}$ that converges to y, and the sequence $\{ z_i = x_i + y_i; i = 1, 2, 3, \ldots\}$. Convergence of these two sequences means that for any finite real hypernumber $\alpha > 0$, there are a natural number n such that $q(x_i–x) < \alpha$ for any $i > n$ and a natural number m such that $q(y_i–y) < \alpha$ for any $I > m$. Then by properties of a hypernorm, we have

$$q(z_i - (x + y)) = q((x_i + y_i) - (x + y)) = q((x_i–x) + (y_i–y)) \leq$$

$$q(x_i - x) + q(y_i - y) < \alpha + \alpha = 2\alpha$$

when $i > \max\{n, m\}$. As α is an arbitrary positive finite real hypernumber, this means that the sequence $\{ z_i = x_i + y_i;\ i = 1, 2, 3, \ldots\}$ converges to the element $x + y$. Consequently, addition is continuous in L.

In addition, for any number a from $\boldsymbol{F}$, we have

$$q(u_i - ax) = q(ax_i - ax) = q(a(x_i - x)) \leq |a|\ q(x_i - x) < |a|\alpha$$

where $u_i = ax_i$. As α is an arbitrary positive finite real hypernumber and $|a|$ is a constant, this means that the sequence $\{ u_i = ax_i;\ i = 1, 2, 3, \ldots\}$ converges to ax. Consequently, scalar multiplication is continuous in the second coordinate.

Let us show that the topology υ is strictly stronger than the topology τ. Indeed, any open set in the topology υ is also open in the topology τ because the system of neighborhoods that generate τ is a subset of the system of neighborhoods that generate υ as any real number is a finite real hypernumber (cf. Chapter 2).

At the same time, there are open sets in the topology υ that are not open in the topology τ. Here is one of them.

Let us consider the hypernumber $\beta = \mathrm{Hn}(b_i)_{i\in\omega}$ where $b_i = 1$ for $i = 2, 4, \ldots, 2n, \ldots$ and $b_i = 0$ for $i = 1, 3, \ldots, 2n - 1, \ldots$. Then open in the topology υ set $\mathrm{O}_\beta x$ is not open in the topology τ because it does not contain any set $\mathrm{O}_k x$ where k is a positive real number, while such sets form a neighborhood base of the topology τ.

Theorem is proved.

Let us find relations between hyperseminormed vector spaces and semitopological vector spaces.

Theorem 5.5. Any hyperseminormed vector space L is a semitopological vector space, which is Hausdorff if and only if L is a hypernormed vector space.

Proof. Let us consider a vector space L with a hyperseminorm q. Taking an element x from L and a positive real number k, we define the neighborhood $\mathrm{O}_k x$ of x by the following formula

$$\mathrm{O}_k x = \{y \in L;\ q(x - y) < k\}$$

To show that the system of so defined neighborhoods determines a topology π in L, we check the neighborhood axioms (Kuratowski, 1966).

NB1: The point x belongs to $O_k x$ because by definition, $q(x-x) = q(\mathbf{0}) = 0 < k$ for any positive real number k.

NB2: Taking two positive real numbers k and h, we see that the intersection $O_k x \cap O_h x = O_l x$ is also a neighborhood of x where $l = \min\{k, h\}$.

NB3: Let $y \in O_k x$. Then $q(x-y) < k$ and by properties of real numbers, there is a positive real number t such that $q(x-y) < k - t$. Then $O_t x \subseteq O_k x$. Indeed, if $z \in O_t x$, then $q(y-z) < t$. Consequently,

$$q(x-z) = q((x-y) + (y-z)) \leq q(x-y) + q(y-z) < (k-t) + t = k$$

It means that $z \in O_k x$.

Now we show that addition is continuous and scalar multiplication is continuous in the second coordinate with respect to the topology π.

Let us consider a sequence $\{x_i; i = 1, 2, 3, \ldots\}$ that converges to x, a sequence $\{y_i; i = 1, 2, 3, \ldots\}$ that converges to y, and the sequence $\{z_i = x_i + y_i; i = 1, 2, 3, \ldots\}$. Convergence of these two sequences means that for any $k > 0$, there are a natural number n such that $q(x_i - x) < k$ for any $i > n$ and a natural number m such that $q(y_i - y) < k$ for any $i > m$. Then by properties of a hyperseminorm, we have

$$q(z_i - (x + y)) = q((x_i + y_i) - (x + y)) = q((x_i - x) + (y_i - y)) \leq q(x_i - x) + q(y_i - y) < k + k = 2k$$

when $i > \max\{n, m\}$. As k is an arbitrary positive real number, this means that the sequence $\{z_i = x_i + y_i; i = 1, 2, 3, \ldots\}$ converges to $x + y$. Consequently, addition is continuous in L.

In addition, for any number a from $\boldsymbol{F}$, we have

$$q(u_i - ax) = q(ax_i - ax) = q(a(x_i - x)) \leq |a|\, q(x_i - x) < |a| k$$

where $u_i = ax_i$. As k is an arbitrary positive real number and $|a|$ is a constant, this means that the sequence $\{u_i = ax_i; i = 1, 2, 3, \ldots\}$ converges to ax. Consequently, scalar multiplication is continuous in the second coordinate. Thus, L is a semitopological vector space.

By Theorem 5.3, if q is a hypernorm, then the vector space L is Hausdorff. At the same time, if q is not a hypernorm, then there are x and y from L such that $x \neq y$ but $q(x-y) = 0$. According to definition, these points x and y cannot be separated in the topology π defined above. Thus, the space L is not Hausdorff in the topology π.

Theorem is proved.

The topology π in L induced by the hyperseminorm q is called the *standard topology* in L with respect to q.

As a seminormed vector space is a topological vector space (Rudin, 1991), Theorem 5.5 implies the following result.

Corollary 5.6. Any seminormed vector space is a Hausdorff topological vector space if and only if it is a normed space.

Proposition 5.2 gives us the following results.

Corollary 5.7. The topology of a hyperseminormed vector space L is translation-invariant, i.e., a subset A from L is open if and only if any of its translations $T_l(A) = A + l$ is open for any element l from L.

As a result, the topology in hyperseminormed vector spaces is completely determined by any local base and thus, by any local base at **0**.

Corollary 5.8. The topology of a hyperseminormed vector space L is dilation-invariant, i.e., a subset A from L is open if and only if any set $M_a(A) = aA$ is open for any real number a.

However, in contrast to seminormed vector spaces, the topology constructed in the proof of Theorem 5.5 is not unique for hyperseminormed vector spaces.

Theorem 5.6. There is a topology η defined by the hyperseminorm q in L, which is strictly stronger than the topology π when the hyperseminorm q is finitely full and is Hausdorff if and only if L is a hypernormed vector space.

Proof. Let us consider a vector space L with a hyperseminorm q and describe the topology η in L induced by the hypernorm q. Taking an element x from L and a positive finite real hypernumber α, we define the neighborhood $O_\alpha x$ of x by the following formula

$$O_\alpha x = \{y \in L;\ q(x-y) < \alpha\}$$

At first, we show that the system of so defined neighborhoods determines a topology ηin L. To do this, it is necessary to check the following neighborhood axioms (Kuratowski, 1966).

Let us consider a point x from X.

NB1:The point x belongs to $O_\alpha x$ because $q(x-x) = q(\mathbf{0}) = 0 < \alpha$ for any positive finite real hypernumber α.

NB2: Taking two positive finite real hypernumbers α and β, we see that the intersection $O_\alpha x \cap O_\beta x = O_\delta x$ also is a neighborhood of x where $\delta = \min\{\alpha, \beta\}$.

NB3: Let $y \in O_k x$. Then $q(x-y) < \alpha$ and by properties of real hypernumbers (cf. Chapter 2), there is a positive finite real hypernumber γ such that $q(x-y) < \alpha - \gamma$. Indeed, if $\alpha = \text{Hn}(a_i)_{i\in\omega}$, then there is a representation $q(x-y) = \beta = \text{Hn}(b_i)_{i\in\omega}$ such that $b_i < a_i$ for all $i\in\omega$. Consequently, there are positive real numbers $\{ c_i; i\in\omega\}$ such that $c_i + b_i < a_i$ for all $i\in\omega$, and taking $\gamma = \text{Hn}(c_i)_{i\in\omega}$, we obtain the necessary hypernumber γ.

Then $O_\gamma y \subseteq O_k x$. Indeed, if $z \in O_\gamma y$, then $q(y-z) < \gamma$. Consequently,

$$q(x-z) = q((x-y) + (y-z)) \leq q(x-y) + q(y-z) < (\alpha-\gamma) + \gamma = \alpha$$

It means that $z \in O_\alpha x$.

Now we show that addition is continuous and scalar multiplication is continuous in the second coordinate with respect to the topology η.

Let us consider a sequence $\{x_i; i = 1, 2, 3, \ldots\}$ that converges to x, a sequence $\{y_i; i = 1, 2, 3, \ldots\}$ that converges to y, and the sequence $\{ z_i = x_i + y_{i;}\ i = 1, 2, 3, \ldots\}$. Convergence of these two sequences means that for any finite real hypernumber $\alpha > 0$, there are a natural number n such that $q(x_i-x) < \alpha$ for any $i > n$ and a natural number m such that $q(y_i-y) < \alpha$ for any $i > m$. Then by properties of a hypernorm, we have

$$q(z_i - (x + y)) = q((x_i + y_i) - (x + y)) = q((x_i-x) + (y_i-y)) \leq q(x_i-x) + q(y_i-y) < \alpha + \alpha = 2\alpha$$

when $I > \max\{n, m\}$. As α is an arbitrary positive finite real hypernumber, this means that the sequence $\{ z_i = x_i + y_i; i = 1, 2, 3, \ldots\}$ converges to the element $x + y$. Consequently, addition is continuous in L.

In addition, for any number a from $\boldsymbol{F}$, we have

$$q(u_i - ax) = q(ax_i - ax) = q(a(x_i - x)) \leq |a|\ q(x_i-x) < |a|\alpha$$

where $u_i = ax_i$. As α is an arbitrary positive finite real hypernumber and $|a|$ is a constant, this means that the sequence $\{ u_i = ax_i; i = 1, 2, 3, \ldots\}$ converges to ax. Consequently, scalar multiplication is continuous in the second coordinate.

Let us show that the topology η is strictly stronger than the topology π. Indeed, any open set in the topology η is also open in the topology π because the system of neighborhoods that generate π is a subset of the system of neighborhoods that generate η as any real number is a finite real hypernumber (cf. Chapter 2).

At the same time, there are open sets in the topology η that are not open in the topology π. Here is one of them.

Let us consider the hypernumber $\beta = \text{Hn}(b_i)_{i\in\omega}$ where $b_i = 1$ for $i = 2, 4, \ldots, 2n, \ldots$ and $b_i = 0$ for $i = 1, 3, \ldots, 2n - 1, \ldots$. Then open in the topology η set $O_\beta x$ is not open in the topology π because it does not contain any set $O_k x$ where k is a positive real number, while such sets form a neighborhood base of the topology π.

By Theorem 5.5, if q is a hypernorm, then the vector space L is Hausdorff. At the same time, if q is not a hypernorm, then there are x and y from L such that $x \neq y$ but $q(x-y) = 0$. By definition, these points x and y cannot be separated in the topology η defined above. Thus, the space L is not Hausdorff in the topology η.

Theorem is proved.

Corollary 5.9. There are one-to-one continuous mappings of Hausdorff semitopological vector spaces that are not isomorphisms.

Indeed, if (L, η) is a semitopological vector space with the topology η and (L, π) is a semitopological vector space with the topology π, then the identical mapping E of (L, π) onto (L, η) is continuous and one-to-one but the inverse mapping is not continuous because the space (L, η) has more open sets than the space (L, π).

Let us find relations between hyperpseudonormed vector spaces and semitopological vector spaces.

Theorem 5.7. Any hyperpseudonormed vector space is a Hausdorff semitopological vector space.

Proof. Let us consider a vector space L with a hyperpseudonorm p. Taking an element x from L and a positive real number k, we define the neighborhood $O_k x$ of x by the following formula

$$O_k x = \{y \in L; p(x-y) < k\}$$

To show that the system of so defined neighborhoods determines a topology in L, we check the neighborhood axioms (Kuratowski, 1966).

NB1:The point x belongs to $O_k x$ because by definition, $p(x-x) = q(\mathbf{0}) = 0 < k$ for any positive real number k.

NB2: Taking two positive real numbers k and h, we see that the intersection $O_k x \cap O_h x = O_l x$ is also a neighborhood of x where $l = \min\{k, h\}$.

NB3: Let us assume that $y \in O_k x$. Then $p(x-y) < k$ and by properties of real numbers, there is a positive real number t such that $p(x-y) < k - t$. Then $O_t y \subseteq O_k x$. Indeed, if $z \in O_t y$, then $p(y-z) < t$. Consequently,

$$p(x-z) = p((x-y) + (y-z)) \leq p(x-y) + p(y-z) < (k-t) + t = k$$

It means that $z \in O_k x$.

Now we show that addition is continuous and scalar multiplication is continuous in the second coordinate with respect to this topology.

Let us consider a sequence $\{x_i;\ i = 1, 2, 3, \ldots\}$ that converges to x, a sequence $\{y_i; i = 1, 2, 3, \ldots\}$ that converges to y, and the sequence $\{ z_i = x_i + y_i;\ i = 1, 2, 3, \ldots\}$. Convergence of these two sequences means that for any $k > 0$, there are a natural number n such that $p(x_i - x) < k$ for any $i > n$ and a natural number m such that $p(y_i - y) < k$ for any $i > m$. Then by properties of a hyperpseudonorm, we have

$$p(z_i - (x + y)) = p((x_i + y_i) - (x + y)) = p((x_i - x) + (y_i - y)) \leq p(x_i - x) + p(y_i - y) < k + k = 2k$$

when $i > \max\{n, m\}$. As k is an arbitrary positive real number, this means that the sequence $\{ z_i = x_i + y_i;\ i = 1, 2, 3, \ldots\}$ converges to $x + y$. Consequently, addition is continuous in L.

In addition, for any element a from the field $\boldsymbol{F}$, we have

$$p(u_i - ax) = p(ax_i - ax) = p(a(x_i - x)) \leq |a| \cdot p(x_i - x) < |a| \cdot k$$

where $u_i = ax_i$. As k is an arbitrary positive real number and $|a|$ is a constant, this means that the sequence $\{ u_i = ax_i; i = 1, 2, 3, \ldots\}$ converges to ax. Consequently, scalar multiplication (dilation) is continuous in the second coordinate. Thus, L is a semitopological vector space.

Axiom N1 for p implies that the semitopological vector space L is Hausdorff.

Theorem is proved.

Corollary 5.10. Any pseudonormed vector space is a Hausdorff topological vector space.

Proposition 5.2 gives us the following results.

Corollary 5.11. The topology of a hyperpseudonormed vector space L is translation-invariant.

As a result, such a topology is completely determined by any local base and thus, by any local base at **0**.

Corollary 5.12. The topology of a hyperpseudormed vector space L is dilation-invariant.

Lemma 3.9 (cf. Chapter 3) implies the following results.

Corollary 5.13. The stronger hyperseminorm (hypernorm) defines a finer topology in the space L.

Corollary 5.14. Equivalent hyperseminorms (hypernorm) define the same topology in the space L.

Different properties of neighborhoods in topological vector spaces are studied in the literature. It is possible to consider similar properties for semi-topological vector spaces.

Definition 5.7. A neighborhood V of $\mathbf{0}$ is called *symmetric* if $V = -V$.

Lemma 5.3. In a semitopological vector space L, for any neighborhood V of $\mathbf{0}$, there is a symmetric neighborhood U of $\mathbf{0}$ such that $U + U \subseteq V$.

Proof. As $\mathbf{0} + \mathbf{0} = \mathbf{0}$ and addition is continuous in L, there are neighbor-hoods U_1 and U_2 of $\mathbf{0}$ such that $U_1 + U_2 \subseteq V$. Then we define

$$U = U_1 \cap U_2 \cap (-U_1) + (-U_2)$$

and have the necessary neighborhood because $U \subseteq U_1$ and $U \subseteq U_2$ while $U_1 + U_2 \subseteq V$.

Lemma is proved.

Corollary 5.15. In a semitopological vector space L, for any neighbor-hood V of $\mathbf{0}$ and any natural number n, there is a symmetric neighborhood U of $\mathbf{0}$ such that $U + U + \ldots + U + U \subseteq V$ where the number of neighborhoods U is equal to 2^n.

As $U + U \subseteq U + U + U$, we have the following result.

Corollary 5.16. In a semitopological vector space L, for any neighbor-hood V of $\mathbf{0}$ and any natural number n, there is a symmetric neighborhood U of $\mathbf{0}$ such that $U + U + \ldots + U + U \subseteq V$ where the number of neighborhoods U is equal to n.

Let us consider two subsets K and C of a semitopological vector space L.

Theorem 5.8. If C is compact, K is closed and $K \cap C = \varnothing$, then $\mathbf{0}$ has a neighborhood V such that

$$(K + V) \cap (C + V) = \varnothing$$

Proof. If the set K is empty, then for any subset V of the vector space L, we have

$$K + V = \varnothing$$

Consequently,

$$(K + V) \cap (C + V) = \varnothing$$

When the set K is not empty, we can take a point x that belongs to K. As the set C is closed, its complement is an open set that contains x because x does not belong to C. Thus, by Corollary 5.10, there is a symmetric neighborhood V_x such that $x + V_x + V_x + V_x$ does not intersect C, i.e.,

$$(x + V_x + V_x + V_x) \cap C = \varnothing$$

Then by the properties of vector spaces and symmetry of the set, we have

$$(x + V_x + V_x + V_x) + V_x = x + V_x + V_x + V_x + (-V_x) = x + V_x + V_x$$

and

$$(x + V_x + V_x) \cap (C + V_x) = \varnothing$$

Taking such neighborhoods V_x for all points x from K, we see that these neighborhoods cover K. As K is a compact set, we can take only a finite number of neighborhoods V_x such that the union of all sets $x + V_x$ cover K. Let us take points $x_1, x_2, x_3, \ldots, x_n$ that correspond to these neighborhoods V_{x_1}, V_{x_2}, V_{x_3}, …, V_{x_n}. Thus, we have

$$K \subseteq (x_1 + V_{x_1}) \cup (x_2 + V_{x_2}) \cup (x_3 + V_{x_3}) \cup \ldots \cup (x_n + V_{x_n})$$

Let us define

$$V = (x_1 + V_{x_1}) \cap (x_2 + V_{x_2}) \cap (x_3 + V_{x_3}) \cap \ldots \cap (x_n + V_{x_n})$$

Then

$$K + V \subseteq (x_1 + V_{x_1} + V) \cup (x_2 + V_{x_2} + V) \cup (x_3 + V_{x_3} + V) \cup \ldots \cup (x_n + V_{x_n} + V)$$
$$\subseteq$$
$$(x_1 + V_{x_1} + V_{x_1}) \cup (x_2 + V_{x_2} + V_{x_2}) \cup (x_3 + V_{x_3} + V_{x_3}) \cup \ldots \cup (x_n + V_{x_n} + V_{x_n})$$

By construction, for all = 1, 2, 3, …, n, we have

$$(x_i + V_{x_i} + V_{x_i}) \cap (C + V) = \varnothing$$

Thus,

$$(K+V)\bigcap(C+V)=\varnothing$$

Theorem is proved.

As topological vector spaces are special cases of semitopological vector spaces, Theorem 1.10 from (Rudin, 1991) is a corollary of Theorem 5.5. Namely, we have the following result.

Corollary 5.17. If K and C are subsets of a topological vector space L, C is compact, K is closed and $K\bigcap C=\varnothing$, then $\mathbf{0}$ has a neighborhood V such that

$$(K+V)\bigcap(C+V)=\varnothing$$

As by Theorem 5.5, any hypernormed vector space is a Hausdorff semitopological vector space in the induced topology, we have the following result.

Corollary 5.18. If K and C are subsets of a hypernormed vector space L, C is compact, K is closed and $K\bigcap C=\varnothing$, then $\mathbf{0}$ has a neighborhood V such that

$$(K+V)\bigcap(C+V)=\varnothing$$

As by Theorem 5.5, any hyperseminormed vector space is a Hausdorff semitopological vector space in the induced topology, we have the following result.

Corollary 5.19. If K and C are subsets of a hyperseminormed vector space L, C is compact, K is closed and $K\bigcap C=\varnothing$, then $\mathbf{0}$ has a neighborhood V such that

$$(K+V)\bigcap(C+V)=\varnothing$$

Let us consider some topological properties of semitopological vector spaces.

In a topological space X, the weakest separation axiom is $\mathbf{T}_0$ (Kelly, 1957) where:

$\mathbf{T}_0$ (the *Kolmogorov Axiom*). $\forall x, y \in X\,(\exists\, Ox\,(y \notin Ox) \vee \exists Oy\,(x \notin Oy))$.

Lemma 5.4. In a topological space X, all points are closed if and only if X satisfies the axiom $\mathbf{T}_0$.

Proof. Sufficiency. If X satisfies the axiom $\mathbf{T}_0$ and x is a point from X, then each point from the complement Cx of x has a neighborhood that does not contain x. Thus, all these neighborhoods are subsets of Cx. By definition, Cx is an open set (Kuratowski, 1966) and consequently, its complement x is a closed set.

Necessity. If $x, y \in X$ and the point x is closed, then y belongs to the complement Cx of x, which is open as the complement of a closed set (Kuratowski, 1966). Thus, y has a neighborhood Oy, which is a subset of the set Cx. Consequently, Oy does not contain x. As points x and y are arbitrary, X satisfies the axiom $\mathbf{T}_0$.

Lemma is proved.

We remind (Alexandroff, 1961) that T_3-*spaces*, or *regular spaces*, are topological spaces in which satisfy Axiom $\mathbf{T}_3$ (Hausdorff axiom):

$\mathbf{T}_3$. *For every point a and closed set B, there exist disjoint open sets which separately contain a and B*

It means that points and closed sets are separated.

Note that there are semitopological vector spaces in which not all points are closed. The space $\boldsymbol{R}^\omega$ of all sequences of real numbers is an example of such a semitopological vector space. Moreover, in $\boldsymbol{R}^\omega$, there are no closed points (Burgin, 2012).

As a point is a compact space, Theorem 5.5 implies the following result.

Corollary 5.20. Every semitopological vector space L in which all points are closed is a regular space.

Lemma 5.3 and Corollary 5.20 imply the following result.

Corollary 5.21. In semitopological vector spaces, L axiom $\mathbf{T}_0$ implies axiom $\mathbf{T}_3$.

As any regular space is a Hausdorff space (Alexandroff, 1961), we have the following result.

Corollary 5.22. Every semitopological vector space L in which all points are closed is a Hausdorff space.

As topological vector spaces are special cases of semitopological vector spaces, Theorem 1.12 from Rudin (1991) is a corollary of Corollary 5.22.

Lemma 5.3 and Corollary 5.22 imply the following result.

Corollary 5.23. In semitopological vector spaces, L axiom $\mathbf{T}_0$ implies axiom $\mathbf{T}_2$.

As both sets $K + V$ and $C + V$ in Theorem 5.5 are open, the closure of $K + V$ does not intersect $C + V$, while the closure of $C + V$ does not intersect $K + V$. As any point a from L is a compact space, we can take $K = \{a\}$. Applying Theorem 5.5 to this situation, we obtain the result, which has a considerable interest according to Rudin (1991).

Corollary 5.24. Any neighborhood O_a of any point a in a semitopological vector space L contains the closure of some neighborhood V_a of the same point a.

As topological vector spaces are special cases of semitopological vector spaces, Theorem 1.11 from Rudin (1991) is a corollary of Corollary 5.24.

Here we consider generalizations of Theorems 5.3 and 5.4 to the case when there are many hypernorms and or hyperseminorms.

Theorem 5.9. Any polyhypernormed vector space is a Hausdorff semitopological vector space in the induced topology.

Proof. Let us consider a vector space L with a system Q of hypernorms and define Q_+ as the set of all finite sums of elements from Q. By Corollary 3.5, all elements from Q_+ are hypernorms. Taking a hypernorm q from Q_+ an element x from L and a positive real number k, we define the neighborhood $O_k x$ of x by the following formula

$$O_{q,k}x = \{y \in L;\ q(x-y) < k\}$$

At first, we show that the system of so defined neighborhoods determines a topology in L. To do this, it is necessary to check the following neighborhood axioms NB1 – NB3 (Kuratowski, 1966).

Let us consider a point x from X.

NB1:The point x belongs to $O_{q,k}x$ because $q(x-x) = q(\mathbf{0}) = 0 < k$ for any positive real number k.

NB2: Let us consider two neighborhoods $O_{q,k}x$ and $O_{p,h}x$ of x, where k and h are positive real numbers, while q and p are hypernorms from Q_+. We see that $O_{(q+p),l}x$ where $l = \min\{k, h\}$ is a subset of the intersection $O_k x \cap O_h x = O_l x$ because

$$(q + p)(x-y) = q(x-y) + p(x-y) < l$$

implies $q(x-y) < k$ and $p(x-y) < h$ because $q(x-y)$ and $p(x-y)$ are nonnegative numbers.

In addition, $q + p$ is a hypernorm from Q_+ and thus, $O_{(q+p),l}x$ is a neighborhood of x

NB3: Let $y \in O_{q,k}x$. Then $q(x-y) < k$ and by properties of real numbers, there is a positive real number t such that $q(x-y) < k - t$. Then $O_{q,t}x \subseteq O_{q,k}x$. Indeed, if $z \in O_{q,t}x$, then $q(y-z) < t$. Consequently,

$$q(x-z) = q((x-y) + (y-z)) \leq q(x-y) + q(y-z) < (k-t) + t = k$$

It means that $z \in O_{q,k}x$.

Thus, we have a topology in L, and this topology is Hausdorff because any hypernorm weakly separates points, i.e., if $x \neq y$, then $q(x-y) \neq 0$.

Now we show that addition is continuous and scalar multiplication is continuous in the second coordinate with respect to this topology.

Let us consider a sequence $\{x_i;\ i = 1, 2, 3, \ldots\}$ that converges to x, a sequence $\{y_i;\ i = 1, 2, 3, \ldots\}$ that converges to y, and the sequence $\{z_i = x_i + y_i;\ i = 1, 2, 3, \ldots\}$. Convergence of these two sequences means that for any $k > 0$ and any hypernorm q from Q_+, there are a natural number n such that $q(x_i-x) < k$ for any $i > n$ and a natural number m such that $q(y_i-y) < k$ for any $i > m$. Then by properties of a hypernorm, we have

$$q(z_i - (x + y)) = q((x_i + y_i) - (x + y)) = q((x_i-x) + (y_i-y)) \leq q(x_i-x) + q(y_i-y) < k + k = 2k$$

when $i > \max\{n, m\}$. As k is an arbitrary positive real number, this means that the sequence $\{z_i = x_i + y_i;\ i = 1, 2, 3, \ldots\}$ converges to $x + y$. Consequently, addition is continuous in L.

In addition, for any number a from $\boldsymbol{F}$, where F is either $\boldsymbol{R}$ or $\boldsymbol{C}$, and any hypernorm q from Q_+, we have

$$q(u_i - ax) = q(ax_i - ax) = q(a(x_i-x)) \leq |a|\ q(x_i-x) < |a|k$$

where $u_i = ax_i$. As k is an arbitrary positive real number and $|a|$ is a constant, this means that the sequence $\{u_i = ax_i;\ i = 1, 2, 3, \ldots\}$ converges to ax. Consequently, scalar multiplication is continuous in the second coordinate.

Theorem is proved.

Corollary 5.25. Any normed vector space is a Hausdorff topological vector space.

Theorems 5.5 and 5.9 imply the following result.

Corollary 5.26. If K and C are subsets of a polyhypernormed vector space L, C is compact, K is closed and $K \cap C = \varnothing$, then $\mathbf{0}$ has a neighborhood V such that

$$(K + V) \cap (C + V) = \varnothing$$

Proposition 5.2 gives us the following results.

Corollary 5.27. The topology of a polyhypernormed vector space L is translation-invariant.

As a result, the topology in polyhypernormed vector spaces is completely determined by any local base and thus, by any local base at **0**.

Corollary 5.28. The topology of a polyhypernormed vector space L is dilation-invariant, i.e., a subset A from L is open if and only if any set $M_a(A) = aA$ is open for any real number a.

Theorem 5.10. Any polyhyperseminormed vector space L with a system Q of hyperseminorms is a semitopological vector space, which is Hausdorff in the induced topology if and only if Q weakly separates L.

Proof. Let us consider a vector space L with a system Q of hyperseminorms and define Q_+ as the set of all finite sums of elements from Q. By Corollary 3.5, all elements from Q_+ are hyperseminorms. Taking a hyperseminorm q from Q_+ an element x from L and a positive real number k, we define the neighborhood $O_{q,k}x$ of x by the following formula

$$O_{q,k}x = \{y \in L;\ q(x-y) < k\}$$

At first, we show that the system of so defined neighborhoods determines a topology in L. To do this, it is necessary to check the following neighborhood axioms NB1 – NB3 (Kuratowski, 1966).

Let us consider a point x from X.

NB1: The point x belongs to $O_{q,k}x$ because by properties of hyperseminorms, $q(x-x) = q(\mathbf{0}) = 0 < k$ for any positive real number k.

NB2: Let us consider two neighborhoods $O_{q,k}x$ and $O_{p,h}x$ of x, where k and h are positive real numbers, while q and p are hyperseminorms from Q_+. We see that $O_{(q+p),l}x$ where $l = \min\{k, h\}$ is a subset of the intersection $O_kx \cap O_hx = O_lx$ because

$$(q + p)(x-y) = q(x-y) + p(x-y) < l$$

implies $q(x-y) < k$ and $p(x-y) < h$ because $q(x-y)$ and $p(x-y)$ are nonnegative numbers.

In addition, $q + p$ is a hyperseminorm from Q_+ and thus, $O_{(q+p),l}x$ is a neighborhood of x

NB3: Let $y \in O_{q,k}x$. Then $q(x-y) < k$ and by properties of real numbers, there is a positive real number t such that $q(x-y) < k - t$. Then $O_{q,t}x \subseteq O_{q,k}x$. Indeed, if $z \in O_{q,t}x$, then $q(y-z) < t$. Consequently,

$$q(x-z) = q((x-y) + (y-z)) \leq q(x-y) + q(y-z) < (k-t) + t = k$$

It means that $z \in O_{q,k}x$.

Now we show that addition is continuous and scalar multiplication is continuous in the second coordinate with respect to this topology.

Let us consider a sequence $\{x_i;\ i = 1, 2, 3, \ldots\}$ that converges to x, a sequence $\{y_i;\ i = 1, 2, 3, \ldots\}$ that converges to y, and the sequence $\{ z_i = x_i + y_i;\ i = 1, 2, 3, \ldots\}$. Convergence of these two sequences means that for any $k > 0$ and any hyperseminorm q from Q_+, there are a natural number n such that $q(x_i-x) < k$ for any $i > n$ and a natural number m such that $q(y_i-y) < k$ for any $i > m$. Then by properties of hyperseminorm, we have

$$q(z_i - (x + y)) = q((x_i + y_i) - (x + y)) = q((x_i-x) + (y_i-y)) \leq q(x_i-x) + q(y_i-y) < k + k = 2k$$

when $i > \max\{n, m\}$. As k is an arbitrary positive real number, this means that the sequence $\{ z_i = x_i + y_i;\ i = 1, 2, 3, \ldots\}$ converges to $x + y$. Consequently, addition is continuous in L.

In addition, for any number a from $\boldsymbol{F}$ and any hypernorm q from Q_+, we have

$$q(u_i - ax) = q(ax_i-ax) = q(a(x_i-x)) \leq |a|\ q(x_i-x) < |a|k$$

where $u_i = ax_i$. As k is an arbitrary positive real number and $|a|$ is a constant, this means that the sequence $\{ u_i = ax_i;\ i = 1, 2, 3, \ldots\}$ converges to ax. Consequently, scalar multiplication is continuous in the second coordinate.

Sufficiency of separation. Thus, we have a topology in L. Let us assume that the system Q weakly separates points. Then by Lemma 3.14, the system Q_+ also weakly separates points. It means that if $x \neq y$, then $q(x-y) > 0$. By properties of hypernumbers, this implies that $q(x-y) > k$ for some positive real number k.

Let us take $h = ⅓k$ and consider a neighborhood $O_{q,h}x$ of x and a neighborhood $O_{q,h}y$ of y. Then these neighborhoods do not intersect. Indeed, let us suppose that $O_{q,h}x \cap O_{q,h}y \neq \varnothing$. It means that there is an element z from L

that belongs both to $O_{q,h}x$ and $O_{q,h}y$. By definition of $O_{q,h}x$ and $O_{q,h}y$, $q(x-z) < h$ and $q(y-z) < h$. Thus,

$$q(x-y) = q((x-z) + (z-y)) \leq q(x-z) + q(z-y) = q(x-z) + q(y-z) < 2h = \tfrac{2}{3}k$$

At the same time, by definition of k,

$$q(x-y) > k$$

Consequently, $O_{q,h}x$ and $O_{q,h}y$ do not intersect and as x and y were arbitrary elements from L, the topology defined by Q is Hausdorff.

Necessity of Separation. Let us assume that the system Q does not weakly separate points. It means that there are points x and y in L such that $x \neq y$ while $q(x-y) = 0$ for all hyperseminorms q from the system Q_+. Thus, the point y belongs to any neighborhood $O_{q,k}x$ of x and the point y belongs to any neighborhood $O_{q,h}y$ of y. According to definition, these points x and y cannot be separated in the topology defined above. Thus, the space L is not Hausdorff.

Theorem is proved.

This theorem solves Problem 11 from (Burgin, 2013).

As a seminormed vector space is a topological vector space (Rudin, 1991), Theorem 5.10 implies the following result.

Corollary 5.29. Any seminormed vector space is a Hausdorff topological vector space if and only if it is a normed space.

Theorems 5.8 and 5.10 imply the following result.

Corollary 5.30. If K and C are subsets of a polyhypesemirnormed vector space L, C is compact, K is closed and $K \cap C = \varnothing$, then $\mathbf{0}$ has a neighborhood V such that

$$(K + V) \cap (C + V) = \varnothing$$

Proposition 5.2 gives us the following results.

Corollary 5.31. The topology of a polyhyperseminormed vector space L is translation-invariant.

As a result, the topology in polyhyperseminormed vector spaces is completely determined by any local base and thus, by any local base at $\mathbf{0}$.

Corollary 5.32. The topology of a polyhyperseminormed vector space L is dilation-invariant.

Theorem 5.11. If $\boldsymbol{B}$ is a local base for Hausdorff semitopological vector space L, then every member of B contains the closure of some member of $\boldsymbol{B}$.

Proof. Let us consider an open neighborhood U of $\mathbf{0}$. Then the complement $\mathrm{Co}U$ of U is a closed set in L. As the space L is Hausdorff, $\mathbf{0}$ is a compact space. Thus, by Theorem 5.5, there is an open neighborhood V of $\mathbf{0}$, such that

$$V \cap (\mathrm{Co}U + V) = \varnothing$$

The closure CV of the neighborhood V does not intersect with the set $\mathrm{Co}U$ because $\mathrm{Co}U$ is a closed set. Thus, $CV \subseteq U$.

Theorem is proved.

As topological vector spaces are special cases of semitopological vector spaces, Theorem 1.11 from (Rudin, 1991) is a corollary of Theorem 5.11. Namely, we have the following result.

Corollary 5.33. If $\boldsymbol{B}$ is a local base for topological vector space L, then every member of B contains the closure of some member of $\boldsymbol{B}$.

By Theorem 5.6, any polyhypernormed vector space L is a Hausdorff semitopological vector space, Thus, Theorem 5.11 implies the following result.

Corollary 5.34. If B is a local base for polyhypernormed (hypernormed) vector space L, then every member of B contains the closure of some member of B.

By Theorem 5.11, any polyhyperseminormed vector space L is a Hausdorff semitopological vector space when its system of hyperseminorm Q weakly separates L. Thus, Theorem 5.11 implies the following result.

Corollary 5.35. If B is a local base for polyhyperseminormed vector space L, then every member of B contains the closure of some member of B when its system of hyperseminorm Q weakly separates L.

Let us consider special classes of semitopological vector spaces.

Definition 5.8. A semitopological vector space L over a topological (normed) field $\boldsymbol{F}$ is called:

(a) Locally convex if there is a local base whose members are convex.
(b) Locally bounded if $\mathbf{0}$ has a bounded neighborhood.
(c) Locally compact if $\mathbf{0}$ has a neighborhood with a compact closure.
(d) Metrizable if the topology in L is compatible with some metric, i.e., the topology in L is induced by some metric.
(e) Hypermetrizable if the topology in L is compatible with some hypermetric, i.e., the topology in L is induced by some hypermetric.

(f) Normable if the topology in L is compatible with some norm, i.e., the topology in L is induced by some norm.
(g) Hypernormable if the topology in L is compatible with some hypernorm, i.e., the topology in L is the standard topology induced by some hypernorm.
(h) Hyperseminormable if the topology in L is compatible with some hyperseminorm, i.e., the topology in L is the standard topology induced by some hyperseminorm.
(i) *SF*-space if the topology in L is compatible with some complete translation-invariant metric, i.e., the topology in L is induced by some complete translation-invariant metric.

Properties of semitopological vector spaces obtained in Chapter 3 show that:

- Any hypernormable semitopological vector space is hypermetrizable.
- Any normable semitopological vector space is metrizable.
- Any hypernormable semitopological vector space is locally convex.

Lemma 5.5. If L is a locally compact hyperseminormed vector space, then the closure $C(t \cdot O_1(x))$ of any space $t \cdot O_1(x)$ is a compact space.

Proof. The neighborhoods $k \cdot O_1(x)$ form a local base at the point x. Consequently, if a neighborhood V of the point x has the compact closure CV, then there is k such that $k \cdot O_1(x) \subseteq V$. As a result, the same is true for their closures, i.e., $C(k \cdot O_1(x)) \subseteq CV$. As a closed subspace of a compact space, $C(k \cdot O_1(x))$ is also a compact space (Kuratowski, 1966).

At the same time, $t \cdot O_1(x) = (t/k)k \cdot O_1(x)$ and $C(t \cdot O_1(x)) = ((t/k)k \cdot O_1(x)) = (t/k) \cdot C(k \cdot O_1(x))$. Thus, $C(t \cdot O_1(x))$ is also a compact space.

Lemma is proved.

Theorem 3.10 also gives the following result.

Corollary 5.36. In a hyperseminormed vector space L, there is a neighborhood V of $\mathbf{0}$ that is convex.

Theorem 5.12. Any polyhyperseminormed vector space L with a system Q of hyperseminorms is a locally convex semitopological vector space.

Proof. By Theorem 5.11, L is a semitopological vector space, and we need only to check existence of a convex local base in L. As it is demonstrated in the proof of Theorem 5.11, sets $O_{q,k}x = \{y \in L; q(x-y) < k\}$ form a local base for L. We show that all these sets are convex.

Indeed, let us take two points z and y from $O_{q,k}x$. Then we have

$$\| az + (1-a)y \| \leq \| az \| + \|(1-a)y \| = a\|z\| + (1-a)\|y \| < ak + (1-a)k = k$$

Because $||z|| < k$ and $||y|| < k$. Thus, the point $az + (1 - a)y$ also belongs to $O_{q,k}x$ and consequently, $O_{q,k}x$ is convex.

Theorem is proved.

Corollary 5.37. Any polyhypernormed vector space L with a system Q of hypernorms is a convex semitopological vector space.

Corollary 5.38. Any polynormed vector space L with a system Q of norms is a convex semitopological vector space.

Corollary 5.39. Any hyperseminormed vector space L with a system Q of hyperseminorms is a convex semitopological vector space.

Corollary 5.40. Any hypernormed vector space L with a system Q of hyperseminorms is a convex semitopological vector space.

Let us consider a semitopological vector space L and its linear subspace M. The quotient space L/M consists of equivalence classes $[x]$ of elements x from L where $[x] = \{y;\ x - y \in M\}$. It is proved that L/M is a vector space with operations $[x] + [y] = [x + y]$ and $r[x] = [rx]$ (Kurosh, 1963).

In addition, it is possible to define in L/M the quotient topology in which the sets $B + M$ with B from a local base of L form the local base of the vector space L/M. By Proposition 5.4(a), all sets $B + M$ are open in L.

Proposition 5.6. (a) The quotient space L/M with the quotient topology is a semitopological vector space.

(b) The canonical mapping $p: L \to L/M$ is continuous.

(c) The space L/M is Hausdorff if and only if M is closed.

Proof. Part (b) follows from the definition of the quotient topology and in addition, p is an open mapping, i.e., it maps open sets in L onto open sets in L/M.

(a) Taking two elements $[a]$ and $[b]$ from L/M, we consider a neighborhood U of $[a] + [b]$. Then by (b), $W = p^{-1}(U)$ is a neighborhood of $a + b$. As addition is continuous in L, there are a neighborhood X of a and a neighborhood Y of b, such that $X + Y \subseteq W$. As p is an open mapping, the sets $p(X)$ and $p(Y)$ are open and thus, they are neighborhoods of $[a]$ and $[b]$, respectively. By construction, $p(X) + p(Y) \subseteq p(W) = U$. It means that addition is continuous in L/M.

In the same way, we prove that scalar multiplication in L/M is continuous in the second coordinate. So, L/M with the quotient topology is a semitopological vector space.

(c) *Necessity.* If the space L/M is Hausdorff, then the point $[\mathbf{0}]$ is closed in it. As the mapping p is continuous, the set $p^{-1}([\mathbf{0}]) = M$ is closed in L.

Sufficiency. Let us assume that M is closed in L. Then the point $[\mathbf{0}] = p(M)$ is closed in L/M. Consequently, L/M is a Hausdorff topological space.

Proposition is proved.

We remind the following concepts of topology (Alexandroff, 1961). Let us take two topological spaces X and Y.

Definition 5.9. A mapping $f: X \to Y$ is called *continuous* if the inverse image of any open set in Y is an open set in X.

Continuity is a basic concept in topology, calculus and functional analysis. In turn, the basic property of continuous mappings is that the composition of continuous mappings is a continuous mapping (Alexandroff, 1961). Due to this fact, topological spaces and continuous mappings form a category.

A related concept is sequential continuity.

Definition 5.10. A mapping $f: X \to Y$ is called *sequentially continuous* if the image of any converging sequence in X is a converging sequence in Y.

The composition of sequentially continuous mappings is a sequentially continuous mapping. Due to this fact, topological spaces and sequentially continuous mappings form a category.

Proposition 5.7. In the category of hypernormable semitopological vector spaces, continuity coincides with sequential continuity.

Proof. Let us take two hypernormable semitopological vector spaces L and M and consider a mapping $f: L \to M$. Suppose that f is continuous. Then f is sequentially continuous because L and M are topological spaces and in topological spaces, continuity always implies sequential continuity (Kuratowski, 1966).

Now let us suppose that f is sequentially continuous mapping of the hypernormed vector spaces L and M. We check the following characteristic of continuity: a mapping f is continuous in a space L if and only if $f(\mathrm{C}X) \subseteq \mathrm{C}f(X)$ for the closure $\mathrm{C}X$ of an arbitrary subset X of L (cf. Proposition 6.1).

Let us take an element x from $\mathrm{C}X$. By properties of closed sets in hypernormed vector spaces, there is a converging sequence $\{x_i; i = 1, 2, 3, \ldots\}$ in L such that all x_i belong to $X (i = 1, 2, 3, \ldots)$ and $\lim_{i\to\infty} x_i = x$. As f is sequentially continuous, $\lim_{i\to\infty} f(x_i) = f(x)$. As all $f(x_i)$ belong to $f(X)$ $(i = 1, 2, 3, \ldots)$, $f(x)$ also belongs to $\mathrm{C}f(X)$. As x is an arbitrary element from $\mathrm{C}X$, we have the inclusion $f(\mathrm{C}X) \subseteq \mathrm{C}f(\mathrm{C}X)$, i.e., the mapping f is continuous in a space L.

Proposition is proved.

Thus, we see that many properties are the same both for topological vector spaces and for semitopological vector spaces. At the same time, it is necessary to understand that not all properties of topological vector spaces are valid for semitopological vector spaces. For instance, in a topological vector space, any compact set is bounded [cf., e.g., (Dunford and Schwartz, 1958)]. This is not true for semitopological vector spaces as Example 9.1 demonstrates.

Here is one more distinction. For topological vector spaces, it is proved that if L is a topological vector space and U is any neighborhood of **0**, then $\bigcup_{t \in \mathbf{R}^+} tU = L$. For semitopological vector spaces, this is not true as the example of the space $\mathbf{R}_\omega$ of real hypernumbers demonstrates. Indeed, taking the unit neighborhood $U = \{x; \|x\| < 1\}$ of **0**, we see that the union $\bigcup_{t \in \mathbf{R}^+} tU$ contains only finite hypernumbers and does not comprise the whole $\mathbf{R}_\omega$.

Corollary 5.41. In the category of normable semitopological vector spaces, continuity coincides with sequential continuity.

KEYWORDS

- continuity
- continuous mapping
- convex space
- neighborhood
- semitopological vector space
- sequential continuity
- topological vector space
- topology

CHAPTER 6

CONTINUITY AND BOUNDEDNESS

Here we study properties of operators (mappings) between polyhyperseminormed and in particular, polyhypernormed, polyseminormed and polynormed vector spaces. Obtained results for hyperseminorms and hypernorms have diverse corollaries for norms and seminorms. The majority of these corollaries are new containing some classical results from functional analysis as special cases.

The main emphasis is on linear operators and relations between their properties such as continuity and boundedness. We do this in the context of *relative continuity* and *relative boundedness*, which include classical continuity and boundedness as particular cases.

At first, let us recap one of the basic concepts in mathematics – continuity.

We remind that mapping $f: X \to Y$ of a topological space X into a topological space Y is *continuous* if the inverse image of any open set in Y is an open set in X.

Let us consider some useful properties of continuous mappings [cf., e.g., (Kelly, 1955; Kuratowski, 1966; Dunford and Schwartz, 1958)].

Proposition 6.1. The following properties are equivalent:

(1) f is continuous.

(2) If a set A is closed in Y, then set $f^{-1}(A)$ is closed in X.

(3) If $A \subseteq Y$, then $Cf^{-1}(A) \subseteq f^{-1}(CA)$.

(4) If $B \subseteq X$, then $f(CB) \subseteq Cf(B)$.

(5) If $B \subseteq X$, then $CB \subseteq f^{-1}(Cf(B))$.

(6) If $A \subseteq Y$, then $f(Cf^{-1}(A)) \subseteq CA$.

(7) If X and Y are metric (normed) spaces and $\lim_{i\to\infty} x_i = x$, then $\lim_{i\to\infty} f(x_i) = f(x)$.

Let us consider a Q-hyperseminormed vector space L, i.e., a vector space L with a system of hyperseminorms Q, a P-hyperseminormed vector space

M, i.e., a vector space M with a system of hyperseminorms P, a hyperseminorm q from Q, a hyperseminorm p from P, and a subset V of the space L.

Definition 6.1. (a) An operator (mapping) A: $L \to M$ is called (q, p)-*bounded at a point a* from L if for any positive real number k, there is a positive real number h such that for any element b from L, the inequality $q(a - b) < k$ implies the inequality $p(A(b) - A(a)) < h$.

(b) An operator (mapping) A: $L \to M$ is called (q, p)-*bounded* if it is (q, p)-bounded at all points of L.

(c) An operator (mapping) A: $L \to M$ is called *V-uniformly* (q, p)-*bounded* if for any positive real number k, there is a positive real number h such that for any element a from V and any element b from L, the inequality $q(a - b) < k$ implies the inequality $p(A(b) - A(a)) < h$.

(d) An operator (mapping) A: $L \to M$ is called *uniformly* (q, p)-*bounded in V* if for any positive real number k, there is a positive real number h such that for any elements a and b from V, the inequality $q(a - b) < k$ implies the inequality $p(A(b) - A(a)) < h$.

Note that when the set V contains only one point (say a), then V-uniform (q, p)-boundedness coincides with (q, p)-boundedness at the point a.

Definitions imply the following result.

Lemma 6.1. Any uniformly (q, p)-bounded in L operator is L-uniformly (q, p)-bounded and any L-uniformly (q, p)-bounded operator is (q, p)-bounded.

At the same time, as the following example demonstrates, there are (q, p)-bounded operators that are not L-uniformly (q, p)-bounded.

Example 6.1. Let us take $L = M = \mathbf{R}$ and assume that q and p are both equal to the absolute value, while $A(x) = x^2$. This mapping (operator) is (q, p)-bounded but not L-uniformly (q, p)-bounded.

However, for linear operators, the inverse of Lemma 6.1 is also true.

Proposition 6.2. The following conditions are equivalent for a linear operator (mapping) A:

(1) A is (q, p)-bounded.

(2) A is uniformly (q, p)-bounded in L.

(3) For some point a, the operator A is (q, p)-bounded at the point a.

(4) A is L-uniformly (q, p)-bounded.

Proof. Implications (2)⇒(1)⇒(3) directly follow from definitions. So, we need to prove only (3)⇒(2), namely, if A: $L \to M$ is (q, p)-bounded at a point a from L, then it is uniformly (q, p)-bounded.

Let us consider another point b from L and assume that $q(b - c) < k$ for some c from L. Then taking $d = c - (b - a)$, we have

$$q(a - d) = q(a - (c - (b - a)) = q(b - c) < k$$

As A is (q, p)-bounded at a, there is a positive real number h such that $p(A(a) - A(d)) < h$. As A is linear operator, we have

$$p(A(b) - A(c)) = p(A(b - c)) = p(A(a - (c - (b - a)) = p(A(a - d)) = p(A(a) - A(d)) < h$$

This shows that A is (q, p)-bounded at the point b because c is an arbitrary point for which $q(b - c) < k$. Thus, A is uniformly (q, p)-bounded in L because for a fixed number k, we have the same number h for all points in L.

In addition, we see that by definition, properties (2) and (4) always coincide.

Proposition is proved.

Corollary 6.1. A linear operator (mapping) A is (q, p)-bounded if and only if it is (q, p)-bounded at $\mathbf{0}$.

The above proof of Proposition 6.2 gives us the following result.

Corollary 6.2. Any (q, p)-bounded linear operator (mapping) $A: L \to M$ is L-uniformly (q, p)-bounded.

These results show that for linear operators, the concepts of a (q, p)-bounded at a point operator and of a (q, p)-bounded operator coincide.

For operators that are not linear, these results are true as the following examples demonstrate.

Example 6.2. Let us assume that $L = M = \boldsymbol{R}_\omega$ is the space of all real hypernumbers (cf. Example 2.1), while hyperseminorms q and p are both equal to the absolute value $\|\cdot\|$ of real hypernumbers. Actually the absolute value $\|\cdot\|$ is a norm in the space $\boldsymbol{R}_\omega$ (Burgin, 2012).

For the operator A, we define $A(x) = x$ for all real hypernumbers x but the hypernumber $\nu = \text{Hn}(i)_{i\in\omega}$ and put $A(\nu) = 1$. Then $\| \nu - (\nu + 1)\| = 1$ but $\| A(\nu) - A(\nu + 1)\| = \| 1 - (\nu + 1)\| = \| \nu \| = \nu$ and this hypernumber is larger than any positive real number (Burgin, 2012). Thus, operator A is (q, p)-bounded at any real number but it is not (q, p)-bounded at the hypernumbers ν.

This shows that an operator can be(q, p)-bounded at one point and not (q, p)-bounded at another point of L.

Example 6.3. Let us take $L = M = C(\boldsymbol{R}, \boldsymbol{R})$ where the space $C(\boldsymbol{R}, \boldsymbol{R})$ of all continuous real functions is a hypernormed vector space (cf. Example 2.10) with the hypernorm $\|\cdot\|$ is defined by the following formula:

If $f: \boldsymbol{R} \to \boldsymbol{R}$, then $\| f \| = \mathrm{Hn}(a_i)_{i\in\omega}$ where $a_i = \max\{|f(x)|; a_i \in [-i, i]\}$

We define $A(f) = f$ for all real functions f but the function $v(x) = x^2$ and put $A(x^2) = e(x)$ where $e(x) = 1$ for all $x \in \boldsymbol{R}$. This operator A is (q, p)-bounded at any constant function from L but it is not (q, p)-bounded at v. At the same time, taking $u(x) = x^2 + 1$, we have$\| v - u\| = 1$, while $\| A(v) - A(u)\| = \| e - u\| = \mathrm{Hn}(i)_{i\in\omega}$ and this hypernumber is larger than any positive real number (Burgin, 2012).

This also shows that an operator can be(q, p)-bounded at one point and not (q, p)-bounded at another point of L.

However, for norms and seminorms, we do not need additional conditions to establish the result of Proposition 6.1.

Proposition 6.3. If q is a seminorm, then an operator (mapping) $A: L \to M$ is (q, p)-bounded if and only if it is (q, p)-bounded, at least, at one point.

Proof. Let us consider two points a and c from L and assume that an operator $A: L \to M$ is (q, p)-bounded at the point a. Then taking a point b such that $q(c - b) < u$ where u is a positive real number.

As q is a seminorm, $q(a - c)$ is equal to some positive real number w. Thus, by properties of seminorms, we have

$$q(a - b) = q(a - c + c - b) \leq q(a - c) + q(c - b) < w + u$$

As the operator A is (q, p)-bounded at the point a and $q(a - c) < w + 1$, we have a positive real number h such that $p(A(a) - A(b)) < h$ and a positive real number k such that $p(A(a) - A(c)) < k$. Consequently,

$$p(A(c) - A(b)) \leq p(A(a) - A(c)) + p(A(a) - A(b)) < k + h$$

As b is an arbitrary point from L, A is (q, p)-bounded at the point c.

As c is an arbitrary point from L, the operator A is (q, p)-bounded.

Proposition is proved.

Proposition 6.3 implies the following results.

Corollary 6.3. The concepts of a (q, p)-bounded at a point operator and of a (q, p)-bounded operator coincide when q is a seminorm.

Note that Examples 6.2 and 6.3 show this is not true for the general case of hyperseminorms.

Corollary 6.4. When q is a seminorm, an operator (mapping) A is (q, p)-bounded if and only if it is (q, p)-bounded at $\mathbf{0}$.

The proof of Proposition 6.3 gives us the following result.

Corollary 6.5. If q is a seminorm, then any (q, p)-bounded operator (mapping) A: $L \rightarrow M$ is L-uniformly (q, p)-bounded.

Proposition 6.4. If q is a seminorm and there is a (q, p)-bounded operator (mapping) A of the vector space L onto the vector space M, then p is a finite hyperseminorm.

Proof. Let us take a point u from M. As A is a projection (surjection), there are points a and b such that $A(a) = \mathbf{0}$ and $A(b) = u$. As q is a seminorm, $q(b - a)$ is less than some positive real number w. As the operator A is (q, p)-bounded, there is a positive real number h such that $p(A(a) - A(b)) < h$

$$p(u) = p(u - \mathbf{0}) = p(A(b) - A(a)) < h$$

As u is an arbitrary point from M, the hyperseminorm p is finite.

Proposition is proved.

Note that a finite hyperseminorm is not always a seminorm and a finite hypernorm is not always a norm.

We remind (cf. Chapter 2) that a real hypernumber is *monotone* if it has a monotone, i.e., either increasing or decreasing, representative.

For instance, all real numbers are monotone hypernumbers (Burgin, 2012). At the same time, all finite monotone real hypernumbers are real numbers (Burgin, 2012). Thus, Proposition 6.4 implies the following result.

Corollary 6.6. If q is a seminorm, there is a (q, p)-bounded operator (mapping) A of the vector space L onto the vector space M and all values of p are monotone hypernumbers, then p is a seminorm.

Definitions imply the following results.

Lemma 6.2. If $W \subseteq V \subseteq L$, then any V-uniformly (q, p)-bounded operator is W-uniformly (q, p)-bounded and any uniformly (q, p)-bounded in V operator is uniformly (q, p)-bounded in W.

Lemma 6.3. Any V-uniformly (q, p)-bounded operator is (q, p)-bounded in V.

It is possible to define operations with operators in vector spaces.

If A: $L \rightarrow M$ and B: $L \rightarrow M$ are operators, then their *sum* $A + B$: $L \rightarrow M$ is defined as $(A + B)(x) = A(x) + B(x)$ for any element x from L.

If A: $L \rightarrow M$ is an operator and d is a real number, then the *scalar product* dA: $L \rightarrow M$ is defined as $(dA)(x) = dA(x)$ for any element x from L.

Proposition 6.5. (a) If A: $L \to M$ and B: $L \to M$ are (q, p)-bounded at a point a from L operators, then their sum $A + B$: $L \to M$ also is a (q, p)-bounded at the point a operator.

(b) If A: $L \to M$ is a (q, p)-bounded operator at a point a from L and d is a real number, then the scalar product dA is a (q, p)-bounded at the point a operator.

Proof. (a) Let us consider (q, p)-bounded at a point a from L operators A: $L \to M$ and B: $L \to M$. By definition, for any positive real number k, there are positive real numbers h and l such that for any element b from L, the inequality $q(a - b) < k$ implies the inequality $p(A(b) - A(a)) < h$ and the inequality $p(B(b) - B(a)) < l$. Then we have

$$p((A + B)(b) - (A + B)(a)) = p(A(b) - A(a) + B(b) - B(a)) \leq$$

$$p(A(b) - A(a)) + p(B(b) - B(a)) < h + l$$

As k is an arbitrary positive real number, the operator $A + B$: $L \to M$ is a (q, p)-bounded at the point a operator.

(b) Let us consider (q, p)-bounded at a point a from L operator A: $L \to M$. By definition, for any positive real number k, there are positive real numbers h such that for any element b from L, the inequality $q(a - b) < k$ implies the inequality $p(A(b) - A(a)) < h$. Then we have

$$p(dA(b) - dA(a)) = p(d(A(b) - A(a))) =$$

$$|d| \cdot p(A(b) - A(a)) < |d| \cdot h$$

As k is an arbitrary positive real number, the operator dA: $L \to M$ is a (q, p)-bounded at the point a operator.

Proposition is proved.

Corollary 6.7. (a) If A: $L \to M$ and B: $L \to M$ are (q, p)-bounded operators, then their sum $A + B$: $L \to M$ also is a (q, p)-bounded operator.

(b) If A: $L \to M$ is a (q, p)-bounded operator and d is a real number, then the scalar product dA is also a (q, p)-bounded operator.

Corollary 6.8. (a) If A: $L \to M$ and B: $L \to M$ are uniformly (q, p)-bounded operators, then their sum $A + B$: $L \to M$ also is a uniformly (q, p)-bounded operator.

(b) If A: $L \to M$ is a uniformly (q, p)-bounded operator and d is a real number, then the scalar product dA is also a uniformly (q, p)-bounded operator.

Let us consider a Q-hyperseminormed vector space L, i.e., a vector space L with a system of hyperseminorms Q, a P-hyperseminormed vector space M, i.e., a vector space M with a system of hyperseminorms P, two hyperseminorms r and q from Q, two hyperseminorms t and p from P, and a subset V of the space L.

Proposition 6.6. If $q \leq r$ and $t \leq p$, then any (q, p)-bounded at the point a operator is also (r, t)-bounded at the point a.

Proof. Let us consider a (q, p)-bounded at the point a operator $A: L \rightarrow M$ and a positive real number k. As $q \leq r$, we have $q(x) \leq d \cdot r(x)$ for some real number $d \geq 0$ and all $x \in L$. Thus, the inequality $r(a - b) < k$ implies the inequality $q(a - b) < dk$. By Definition 6.1, there is a positive real number l such that for any element b from L, the inequality $q(a - b) < dk$ implies the inequality $p(A(b) - A(a)) < l$. As $t \leq p$, we have $t(x) \leq c \cdot p(x)$ for some real number $c \geq 0$ and all $x \in L$. Consequently, if $p(A(b) - A(a)) < l$, then $t(A(b) - A(a)) < cl$. In other words, if $r(a - b) < k$, then $t(A(b) - A(a)) < h = cl$. This means that the operator A is (r, t)-bounded at the point a.

Proposition is proved.

Corollary 6.9. If $q \leq r$ and $t \leq p$, then any (q, p)-bounded operator is also (r, t)-bounded.

Corollary 6.10. If $q \leq r$ and $t \leq p$, then any V-uniformly (q, p)-bounded operator is also V-uniformly (r, t)-bounded.

Corollary 6.11. If $q \leq r$ and $t \leq p$, then any uniformly (q, p)-bounded operator is also uniformly (r, t)-bounded.

Corollary 6.12. If $q \leq r$, then any (q, p)-bounded at the point a operator is also (r, p)-bounded at the point a.

Corollary 6.13. If $q \leq r$, then any (q, p)-bounded operator is also (r, p)-bounded.

Corollary 6.14. If $q \leq r$, then any V-uniformly (q, p)-bounded operator is also V-uniformly (r, p)-bounded.

Corollary 6.15. If $q \leq r$, then any uniformly (q, p)-bounded operator is also uniformly (r, p)-bounded.

Corollary 6.16. For any positive real number c, any (q, p)-bounded at the point a operator is also (cq, p)-bounded at the point a.

Corollary 6.17. For any positive real number c, any (q, p)-bounded operator is also (cq, p)-bounded.

Corollary 6.18. For any positive real number c, any V-uniformly (q, p)-bounded operator is also V-uniformly (cq, p)-bounded.

Corollary 6.19. For any positive real number c, any uniformly (q, p)-bounded operator is also uniformly (cq, p)-bounded.

However, bounded with respect to one pair of hyperseminorms (hypernorms, seminorms or norms) operator can be unbounded with respect to another pair of hyperseminorms (hypernorms, seminorms or norms) in the same spaces. That is why we introduce a more general concept of boundedness in polyhyperseminormed vector spaces.

Let us consider a binary relation u between the system of hyperseminorms Q in a polyhyperseminormed vector space L and the system of hyperseminorms P in a polyhyperseminormed vector space M, taking an arbitrary subset V of the space L.

Definition 6.2. (a) An operator (mapping) A: $L \to M$ is called (Q, u, P)-*bounded at a point a* from L if for any hyperseminorms q and p such that $(q, p) \in u$, the operator (mapping) A is (q, p)-bounded at the point a. The triad (Q, u, P) is called the *extent of boundedness* of A at a.

(b) An operator (mapping) A: $L \to M$ is called *V-uniformly* (Q, u, P)-*bounded* if for any hyperseminorms q and p with $(q, p) \in u$ and any positive real number k, there is a positive real number h such that for any element a from V and any element b from L, the inequality $q(a - b) < k$ implies the inequality $p(A(b) - A(a)) < h$. The triad (Q, u, P) is called the *extent of V-uniform boundedness* of A.

(c) An operator (mapping) A: $L \to M$ is called *uniformly* (Q, u, P)-*bounded in V* if for any hyperseminorms q and p with $(q, p) \in u$ and any positive real number k, there is a positive real number h such that for any elements a and b from V, the inequality $q(a - b) < k$ implies the inequality $p(A(b) - A(a)) < h$. The triad (Q, u, P) is called the *extent of uniform boundedness* of A.

(d) An operator (mapping) A: $L \to M$ is called (Q, u, P)-*bounded* if it is (Q, u, P)-bounded at all points of L. The triad (Q, u, P) is called the *extent of boundedness* of A.

It means that an operator (mapping) A is (Q, u, P)-bounded if for any hyperseminorms q and p such that $(q, p) \in u$, the operator (mapping) A is (q, p)-bounded.

Note that when the set V contains only one point (say a), then V-uniform (Q, u, P)-boundedness coincides with (Q, u, P)-boundedness at the point a.

Lemma 6.4. If u and v are binary relations between the system of hyperseminorms Q in L and the system of hyperseminorms P in M and $u \subseteq v$, then:

(a) Any (Q, v, P)-bounded at a point a operator is (Q, u, P)-bounded at a.

(b) Any V-uniformly (Q, v, P)-bounded operator is V-uniformly (Q, u, P)-bounded.

(c) Any (Q, v, P)-bounded in V operator is (Q, u, P)-bounded in V.

(d) Any uniformly (Q, v, P)-bounded operator is uniformly (Q, u, P)-bounded.

Proof is left as an exercise.

Lemma 6.4 implies the following result.

Lemma 6.5. Any uniformly (Q, u, P)-bounded operator in L is L-uniformly (Q, u, P)-bounded, while any L-uniformly (Q, u, P)-bounded operator is (Q, u, P)-bounded.

At the same time, taking $L = M = \mathbf{R}$, $Q = \{q\}$, $P = \{p\}$, and assuming that q and p are both equal to the absolute value and $u = \{(q, p)\}$, we see that Example 6.1 demonstrates that there are (Q, u, P)-bounded operators that are not L-uniformly (Q, u, P)-bounded.

However, for linear operators, the inverse of Lemma 6.5 is also true because Proposition 6.2 implies the following result.

Proposition 6.7. The following conditions are equivalent for a linear operator (mapping) A:

(1) A is (Q, u, P)-bounded.

(2) A is uniformly (Q, u, P)-bounded in L.

(3) For some point a, the operator A is (Q, u, P)-bounded at the point a.

(4) A is L-uniformly (Q, u, P)-bounded.

Corollary 6.20. A linear operator (mapping) A is (Q, u, P)-bounded if and only if it is (Q, u, P)-bounded at $\mathbf{0}$.

Corollaries 6.2 and 6.20 imply the following result.

Corollary 6.21. Any (Q, u, P)-bounded linear operator (mapping) A: $L \to M$ is L-uniformly (Q, u, P)-bounded.

These results show that for linear operators, the concepts of a (Q, u, P)-bounded at a point operator and a (Q, u, P)-bounded operator coincide.

At the same time, taking $L = M = \mathbf{R}$, $Q = \{q\}$, $P = \{p\}$, and assuming that q and p are both equal to the absolute value and $u = \{(q, p)\}$, we see that Examples 6.2 and 6.3 demonstrate that there are operators that are (Q, u, P)-bounded at one point and not (Q, u, P)-bounded at another point.

However, for norms and seminorms, we do not need additional conditions to establish the result of Proposition 6.7. We remind that the *definability domain* of the relation u is defined as

$$\text{DDom } u = \{q; \text{ there is a pair } (q, p) \text{ that belongs to } u\}$$

Then Proposition 6.7 implies the following result.

Proposition 6.8. If all q from the definability domain DDom u of the relation u are seminorms, then an operator (mapping) A: $L \to M$ is (Q, u, P)-bounded if and only if it is (Q, u, P)-bounded, at least, at one point.

Proposition 6.8 implies the following result.

Corollary 6.22. The concepts of (Q, u, P)-bounded at a point operators and (Q, u, P)-bounded operator coincide when all q from the definability domain DDom u of the relation u are seminorms.

Note that Examples 6.2 and 6.3 show this is not true for the general case of hyperseminorms.

Corollary 6.23. When all q from the definability domain DDom u of the relation u are seminorms, an operator (mapping) A is (Q, u, P)-bounded if and only if it is (Q, u, P)-bounded at $\mathbf{0}$.

The above proof of Proposition 6.2 gives us the following result.

Corollary 6.24. If all q from the definability domain DDom u of the relation u are seminorms, then any (Q, u, P)-bounded operator (mapping) A: $L \to M$ is L-uniformly (Q, u, P)-bounded.

Proposition 6.8 implies the following result.

Proposition 6.9. If all q from the definability domain DDom u of the relation u are seminorms and there is a (Q, u, P)-bounded operator (mapping) A of the vector space L onto the vector space M, then all p from the range Rg u of u are finite hyperseminorms.

Corollary 6.25. If all q from the definability domain DDom u of u are seminorms and there is a (Q, u, P)-bounded operator (mapping) A of the vector space L onto the vector space M, and all values of all p from the range Rg u are monotone hypernumbers, then all such p are seminorms.

Definitions imply the following results.

Lemma 6.6. If $W \subseteq V \subseteq L$, then any V-uniformly (Q, u, P)-bounded operator is W-uniformly (q, p)-bounded and any uniformly (Q, u, P)-bounded in V operator is uniformly (q, p)-bounded in W.

Lemma 6.7. Any V-uniformly (Q, u, P)-bounded operator is (Q, u, P)-bounded in V.

Proposition 6.9 implies the following result.

Proposition 6.10. (a) If A: $L \to M$ and B: $L \to M$ are (Q, u, P)-bounded at a point a from L operators, then their sum $A + B$: $L \to M$ also is a (Q, u, P)-bounded at the point a operator.

(b) If A: $L \to M$ is a (Q, u, P)-bounded operator at a point a from L and d is a real number, then the scalar product dA is a (Q, u, P)-bounded at the point a operator.

Corollary 6.26. (a) If A: $L \to M$ and B: $L \to M$ are (Q, u, P)-bounded operators, then their sum $A + B$: $L \to M$ also is a (Q, u, P)-bounded operator.

(b) If A: $L \to M$ is a (Q, u, P)-bounded operator and d is a real number, then the scalar product dA is also a (Q, u, P)-bounded operator.

Corollary 6.27. (a) If A: $L \to M$ and B: $L \to M$ are uniformly (Q, u, P)-bounded operators, then their sum $A + B$: $L \to M$ also is a uniformly (Q, u, P)-bounded operator.

(b) If A: $L \to M$ is a uniformly (Q, u, P)-bounded operator and d is a real number, then the scalar product dA is also a uniformly (Q, u, P)-bounded operator.

Let us consider two systems of hyperseminorms Q and R in a vector space L, two systems of hyperseminorms P and T in a vector space M, a binary relation $u \subseteq Q \times P$ and a binary relation $W \subseteq R \times T$. Then Proposition 6.6 allows us to prove the following result.

Proposition 6.11. If for any pair (r, t) from w, there is a pair (q, p) from u such that $q \leq r$ and $t \leq p$, then any (Q, u, P)-bounded at the point a operator is also (R, w, T)-bounded at the point a.

Proof is left as an exercise.

Corollary 6.28. If for any pair (r, t) from w, there is a pair (q, p) from u such that $q \leq r$ and $t \leq p$, then any (Q, u, P)-bounded operator is also (R, w, T)-bounded.

Corollary 6.29. If for any pair (r, t) from w, there is a pair (q, p) from u such that $q \leq r$ and $t \leq p$, then any V-uniformly (Q, u, P)-bounded operator is also V-uniformly (R, w, T)-bounded.

Corollary 6.30. If for any pair (r, t) from w, there is a pair (q, p) from u such that $q \leq r$ and $t \leq p$, then any uniformly (Q, u, P)-bounded operator is also uniformly (R, w, T)-bounded.

Similar to boundedness, uniform boundedness of operators can in the whole space, in a subspace and at one point.

Let us take a subset V of the vector space L.

Definition 6.3. (a) An operator (mapping) A: $L \to M$ is called *uniformly* (*Q, u, P)-bounded at a point a* from L if for any positive real number k, there is a positive real number h such that for any hyperseminorms q and p with $(q, p) \in u$, and any element b from L, the inequality $q(a - b) < k$ implies the inequality $p(A(b) - A(a)) < h$.

(b) An operator (mapping) A: $L \to M$ is called *u-uniformly* (Q, u, P)*-bounded* if it is uniformly (Q, u, P)-bounded at all points of L.

(c) An operator (mapping) A: $L \to M$ is called *u-uniformly* (Q, u, P)*-bounded in* V if for any positive real number k, there is a positive real number h such that for any hyperseminorms q and p with $(q, p) \in u$, and any elements a and b from V, the inequality $q(a - b) < k$ implies the inequality $p(A(b) - A(a)) < h$.

(d) An operator (mapping) A: $L \to M$ is called *uV-uniformly* (Q, u, P)*-bounded* if for any positive real number k, there is a positive real number h such that for any hyperseminorms q and p with $(q, p) \in u$, and any elements a from V and b from L, the inequality $q(a - b) < k$ implies the inequality $p(A(b) - A(a)) < h$.

Asking whether any (Q, u, P)-bounded at a point operator (mapping) isuniformly (Q, u, P)-bounded at the same point, we find that the answer is negative.

Example 6.4. Let us take $L = M$ and equal to the vector space C($\boldsymbol{R}$, $\boldsymbol{R}$) of all continuous real functions. It is possible (Burgin, 2012) for all real numbers x, to define seminorms $q_{\text{pt}x} = p_{\text{pt}x}$ by the following formula

$$q_{\text{pt}x}(f) = p_{\text{pt}x}(f) = |f(x)|$$

We define $A(f) = xf(x)$ for all real functions f and $u = \{(q_{\text{pt}x}, p_{\text{pt}x}); x \in \boldsymbol{R}\}$. Taking the function $f(x) = x$ as the point a from L, we see that $A(f) = x^2$. Thus, taking some positive real number k, e.g., $k = 1$, the corresponding h from Definition 6.2 always exists but it grows with the growth of x. For instance, when $k = 1$, we have

$$q_{\text{pt}1}(f - g) < 1 \text{ implies } p_{\text{pt}1}(A(f)A(f)) = p_{\text{pt}1}(xf - xg) < 1$$

At the same time, $q_{\text{pt}10}(f - g) < 1$ does not imply $p_{\text{pt}10}(A(f)A(f)) < 1$. It only implies $p_{\text{pt}10}(A(f) - A(f)) = p_{\text{pt}10}(xf - xg) < 10$. This means that for any pair $(q_{\text{pt}x}, p_{\text{pt}x})$ of seminorms and a number k, we need to find a specific number h to satisfy Definition 6.3a. Consequently, the operator A is (Q, u, P)-bounded at f but it is not uniformly (Q, u, P)-bounded at f.

The same example shows that there are (Q, u, P)-bounded operators that are not uniformly (Q, u, P)-bounded.

It is also possible to ask whether Propositions 6.5 and 6.7 remain true for uniformly (Q, u, P)-bounded operators. In this case, the answer is positive.

Proposition 6.12. If all q from the definability domain DDom u of the relation u are seminorms, then an operator (mapping) A: $L \to M$ is uniformly (Q, u, P)-bounded if and only if it is uniformly (Q, u, P)-bounded, at least, at one point.

Indeed, Proposition 6.12 is a direct corollary of Proposition 6.7 because any uniformly (Q, u, P)-bounded at a point operator is (Q, u, P)-bounded at the same point and any uniformly (Q, u, P)-bounded operator is (Q, u, P)-bounded.

Proposition 6.12 implies the following result.

Corollary 6.31. The concepts of uniformly (Q, u, P)-bounded at a point operators and uniformly (Q, u, P)-bounded operators coincide when all q from the definability domain DDom u of u are seminorms.

Note that Examples 6.2 and 6.3 show this is not true for the general case of hyperseminorms.

Proposition 6.13. If all q from the definability domain DDom u of u are seminorms and there is a uniformly (Q, u, P)-bounded operator (mapping) A of the vector space L onto the vector space M, then all p from the range Rg u of u are finite hyperseminorms.

Indeed, Proposition 6.13 is a direct corollary of Proposition 6.8 because any uniformly (Q, u, P)-bounded operator is (Q, u, P)-bounded.

Corollary 6.32. If all q from the definability domain DDom u of u are seminorms and there is a uniformly (Q, u, P)-bounded operator (mapping) A of the vector space L onto the vector space M, and all values of all p from the range Rg u are monotone hypernumbers, then all such p are seminorms.

Definitions imply the following results.

Lemma 6.8. (a) Any uniformly (Q, u, P)-bounded at a point a operator A is (Q, u, P)-bounded at the point a.

(b) Any u-uniformly (Q, u, P)-bounded operator A is (Q, u, P)-bounded.

Lemma 6.9. Any u-uniformly (Q, u, P)-bounded in L operator is u-uniformly (Q, u, P)-bounded.

At the same time, taking $L = M = \boldsymbol{R}$, $Q = \{q\}$, $P = \{p\}$, and assuming that hyperseminorms q and p are both equal to the absolute value and $u = \{(q, p)\}$, we see that Example 6.1 demonstrates that there are u-uniformly (Q, u, P)-bounded operators that are not uniformly (Q, u, P)-bounded because if Q has only one hyperseminorm q, P also has only one hyperseminorm p and u is a complete relation, then any (Q, u, P)-bounded operator is u-uniformly (Q, u, P)-bounded.

However, for linear operators, this is impossible as Proposition 6.2 allows us to prove the following result.

Proposition 6.14. The following conditions are equivalent for a linear operator (mapping) A:

(1) A is u-uniformly (Q, u, P)-bounded.

(2) A is u-uniformly (Q, u, P)-bounded in L.

(3) For some point a, the operator A is uniformly (Q, u, P)-bounded at the point a.

Proof. Implications (2)⇒(1)⇒(3) directly follow from definitions. So, we need to prove only (3)⇒(2), namely, if A: $L \to M$ is uniformly (Q, u, P)-bounded at a point a from L, then it is uniformly (Q, u, P)-bounded in L.

Let us consider another point b from L, take two hyperseminorms q and p with $(q, p) \in u$, and assume that $q(b - c) < k$ for some c from L. Then taking $d = c - (b - a)$, we have

$$q(a - d) = q(a - (c - (b - a)) = q(b - c) < k$$

As A is uniformly (Q, u, P)-bounded at a, it is also (q, p)-bounded at a. Thus, there is a positive real number h such that $p(A(a) - A(d)) < h$. As A is linear operator, we have

$$p(A(b) - A(c)) = p(A(b - c)) = p(A(a - (c - (b - a)) = p(A(a - d)) = p(A(a) - A(d)) < h$$

This shows that A is (q, p)-bounded at the point b because c is an arbitrary point for which $q(b - c) < k$ and thus, A is u-uniformly (Q, u, P)-bounded because q and p are arbitrary hyperseminorms with $(q, p) \in u$. In addition, A is uniformly (q, p)-bounded in L because for a fixed number k, we have the same number h for all points in L.

Proposition is proved.

Corollary 6.33. A linear operator (mapping) A: $L \to M$ is u-uniformly (Q, u, P)-bounded if and only if it is uniformly (Q, u, P)-bounded at **0**.

Corollary 6.33 implies the following result.

Corollary 6.34. Any u-uniformly (Q, u, P)-bounded linear operator (mapping) A is u-uniformly (Q, u, P)-bounded in L.

These results show that for linear operators, different types of uniformly bounded operators coincide.

Proposition 6.15. If the relation u is finite, then an operator (mapping) A: $L \to M$ is uniformly (Q, u, P)-bounded (at a point a) if and only if it is (Q, u, P)-bounded (at the point a).

Proof. As any uniformly (Q, u, P)-bounded (at a point a) operator is (Q, u, P)-bounded (at the same point), we need only to show that when the relation u is finite, a (Q, u, P)-bounded (at a point a) operator A: $L \to M$ is uniformly (Q, u, P)-bounded (at the point a). At first, we consider local boundedness.

Indeed, by Definition 6.2, for any hyperseminorms q and p such that $(q, p) \in u$, the operator (mapping) A is (q, p)-bounded at the point a, that is, by Definition 6.1, the following condition is true:

Condition 1. For any positive real number k, there is a positive real number h such that for any element b from L, the inequality $q(a - b) < k$ implies the inequality $p(A(b) - A(a)) < h$.

This number h can be different for different pairs (q, p), but because u is finite, there is only a finite number of these pairs. So, we can take

$$l = \max \{h;\ h \text{ satisfies Condition 1 for a pair } (q, p) \in u\}$$

and this number l will satisfy the condition from Definition 6.3. Thus, the operator Ais uniformly (Q, u, P)-bounded at the point a.

The global case is proved in a similar way.

Proposition is proved.

Corollary 6.35. If systems of hyperseminorms Q and P are finite, then an operator (mapping) A is uniformly (Q, u, P)-bounded (at a point a) if and only if it is (Q, u, P)-bounded (at the point a).

Lemma 6.10. If u and v are binary relations between the system of hyperseminorms Q in L and the system of hyperseminorms P in M and $u \subseteq v$, then:

(a) Any uniformly (Q, v, P)-bounded at a point a operator is uniformly (Q, u, P)-bounded at a.
(b) Any u-uniformly (Q, v, P)-bounded operator is u-uniformly (Q, u, P)-bounded in L.
(c) Any u-uniformly (Q, v, P)-bounded operator in V is u-uniformly (Q, u, P)-bounded in V.
(d) Any uV-uniformly (Q, v, P)-bounded operator is uV-uniformly (Q, u, P)-bounded.

Proof is left as an exercise.

Now let us study different types of continuity in polyhyperseminormed vector spaces.

Definition 6.4. (a) An operator (mapping) A: $L \to M$ is called *(q, p)-continuous at a point a* from L if for any positive real number k, there is a positive real number h such that for any element b from L, the inequality $q(b - a) < h$ implies the inequality $p(A(b) - A(a)) < k$.

(b) An operator (mapping) A: $L \to M$ is called *(q, p)-continuous* if it is (q, p)-continuousat all points of L.

(c) An operator (mapping) A: $L \to M$ is called *uniformly (q, p)-continuous in $V \subseteq L$* if for any positive real number k, there is a positive real number h such that for any elements a and b from V, the inequality $q(b - a) < h$ implies the inequality $p(A(b) - A(a)) < k$.

(d) An operator (mapping) A: $L \to M$ is called *V-uniformly (q, p)-continuous* if for any positive real number k, there is a positive real number h such that for any element a from $V \subseteq L$ and any element b from L, the inequality $q(b - a) < h$ implies the inequality $p(A(b) - A(a)) < k$.

Note that when the set V contains only one point (say a), then V-uniform (q, p)-continuity coincides with (q, p)-continuity at the point a. Besides, to be *L-uniformly (q, p)-continuous* or to be *uniformly (Q, u, P)-continuous in L* means the same for all operators.

Definitions imply the following results.

Lemma 6.11. For any $V \subseteq L$, any V-uniformly (q, p)-continuous operator is (q, p)-continuous in V.

Lemma 6.12. Any L-uniformly (q, p)-continuous operator is (q, p)-continuous.

At the same time, as the following example demonstrates, there are (q, p)-continuous operators that are not L-uniformly (q, p)-continuous.

Example 6.5. Let us take $L = M = \mathbf{R}$ and assume that q and p are both equal to the absolute value, while $A(x) = x^2$. This mapping (operator) is (q, p)-continuous but not L-uniformly (q, p)-continuous.

However, for linear operators, the inverse of Lemma 6.11 is also true.

Proposition 6.16. The following conditions are equivalent for a linear operator (mapping) A:

(1) A is (q, p)-continuous.
(2) A is uniformly (q, p)-continuous in L.
(3) For some point a, the operator A is (q, p)-continuous at the point a.
(4) A is L-uniformly (q, p)-continuous.

Proof. Implications (2)⇒(1)⇒(3) directly follow from definitions. So, we need to prove only (3)⇒(2), namely, if A: $L \to M$ is (q, p)-continuous at a point a from L, then it is uniformly (q, p)-continuous in L.

Let us consider a positive real number k. Then because A is (q, p)-continuous at the point a, there is a positive real number h, such that the inequality $q(a - b) < h$ implies the inequality $p(A(b) - A(a)) < k$.

Let us take another point b from L and assume that $q(b - c) < h$ for some c from L. Then taking $d = c - (b - a)$, we have

$$q(a - d) = q(a - (c - (b - a)) = q(b - c) < h$$

As A is (q, p)-continuous at a, we have $p(A(a) - A(d)) < k$. As A is linear operator, we have

$$p(A(b) - A(c)) = p(A(b - c)) = p(A(a - (c - (b - a)) = p(A(a - d)) = p(A(a) - A(d)) < k$$

This shows that A is (q, p)-continuous at the point b because c is an arbitrary point for which $q(b - c) < h$. Thus, A is uniformly (q, p)-continuous in L because for a fixed number k, we have the same number h for all points in L.

In addition, we see that by definition, properties (2) and (4) always coincide.

Proposition is proved.

Corollary 6.36. A linear operator (mapping) A is (q, p)-continuous if and only if it is (q, p)-continuous at $\mathbf{0}$.

The proof of Proposition 6.16 gives us the following result.

Corollary 6.37. Any (q, p)-continuous linear operator (mapping) $A: L \to M$ is L-uniformly (q, p)-continuous.

These results show that for linear operators, the concepts of (q, p)-continuous at a point operators and (q, p)-continuous operators coincide.

For operators that are not linear, these results are not true as the following examples demonstrate.

Example 6.6. Let us take $L = M = \boldsymbol{R}_\omega$ (cf. Example 6.1) and assume that q and p are both equal to the absolute value $\|\cdot\|$ of real hypernumbers. We define $A(x) = x$ for all real hypernumbers x but the hypernumber $\nu = \text{Hn}(i)_{i\in\omega}$ and put $A(\nu) = 1$. Then $\| \nu - (\nu + 1)\| = 1$ but $\| A(\nu) - A(\nu + 1)\| = \| 1 - (\nu + 1)\| = \| \nu \| = \nu$ and this hypernumber is larger than any positive real number (Burgin, 2012). Thus, operator A is (q, p)-continuous at any real number but it is not (q, p)-continuous at ν.

This shows that an operator can be(q, p)-continuous at one point and not (q, p)-continuous at another point of L and thus not (q, p)-continuous in L, as well as not L-uniformly (q, p)-continuous.

Example 6.7. Let us take $L = M = C(\boldsymbol{R}, \boldsymbol{R})$, while the space $C(\boldsymbol{R}, \boldsymbol{R})$ of all continuous real functions is a hypernormed vector space (cf. Example 2.1) where the hypernorm $||\cdot||$ is defined by the following formula:

$$\text{If } f\text{: } \boldsymbol{R} \to \boldsymbol{R}, \text{ then } ||f|| = \text{Hn}(a_i)_{i\in\omega} \text{ where } a_i = \max\{|f(x)|;\ a_i \in [-i, i]\}$$

We define $A(f) = f$ for all real functions f but the function $v(x) = x^2$ and put $A(x^2) = e(x)$ where $e(x) = 1$ for all $x \in \boldsymbol{R}$. This operator A is (q, p)-continuous at any constant function from L. but it is not (q, p)-continuous at v. At the same time, taking $u(x) = x^2 + 1$, we have $||\ v - u|| = 1$, while $||\ A(v) - A(u)||$ $= ||\ e - u|| = \text{Hn}(i)_{i\in\omega}$ and this hypernumber is larger than any positive real number (Burgin, 2012).

This also shows that an operator can be (q, p)-continuous at one point and not (q, p)-continuous at another point of L and thus not (q, p)-continuous in L, as well as not L-uniformly (q, p)-continuous.

Definitions imply the following result.

Lemma 6.13. If $W \subseteq V \subseteq L$, then any V-uniformly (q, p)-continuous operator is W-uniformly (q, p)-continuous.

Proposition 6.17. (a) If A: $L \to M$ and B: $L \to M$ are (q, p)-continuous at a point a from L operators, then their sum $A + B$: $L \to M$ also is a (q, p)-continuous at the point a operator.

(b) If A: $L \to M$ is a (q, p)-continuous operator at a point a from L and d is a real number, then the scalar product dA is a (q, p)-continuous at the point a operator.

Proof. (a) Let us consider (q, p)-continuous at a point a from L operators A: $L \to M$ and B: $L \to M$. By definition, for any positive real number $\frac{1}{2}h$, there is a positive real number k such that for any element b from L, the inequality $q(a - b) < k$ implies the inequality $p(A(b) - A(a)) < \frac{1}{2}h$ and there is a positive real number l such that for any element b from L, the inequality $q(a - b) < l$ implies the inequality $p(B(b) - B(a)) < \frac{1}{2}h$. So, if $q(a - b) < t = \min \{k, l\}$, then we have

$$p((A + B)(b) - (A + B)(a)) = p(A(b) - A(a) + B(b) - B(a)) \leq$$

$$p(A(b) - A(a)) + p(B(b) - B(a)) < \tfrac{1}{2}h + \tfrac{1}{2}h = h$$

Thus, we have demonstrated that for any positive real number h, there is a positive real number k such that for any element b from L, the inequality

$q(a-b) < k$ implies the inequality $p((A+B)(b)-(A+B)(a)) < h$. It means the operator $A + B$: $L \to M$ is a (q, p)-continuous at the point a operator.

(b) Let us consider (q, p)-continuous at a point a from L operator A: $L \to M$ and a real number d. By definition, for any positive real number $(1/|d|) \cdot h$, there is a positive real number k such that for any element b from L, the inequality $q(a-b) < k$ implies the inequality $p(A(b)-A(a)) < (1/|d|) \cdot h$. Then we have

$$p(dA(b) - dA(a)) = p(d(A(b) - A(a))) =$$

$$|d| \cdot p(A(b) - A(a)) < |d| \cdot (1/|d|) \cdot h = h$$

Thus, we have demonstrated that for any positive real number h, there is a positive real number k such that for any element b from L, the inequality $q(a-b) < k$ implies the inequality $p(dA(b) - dA(a)) < h$. It means the operator dA: $L \to M$ is a (q, p)-continuous at the point a operator.

Proposition is proved.

Corollary 6.38. (a) If A: $L \to M$ and B: $L \to M$ are (q, p)-continuous operators, then their sum $A + B$: $L \to M$ also is a (q, p)-continuous operator.

(b) If A: $L \to M$ is a (q, p)-continuous operator and d is a real number, then the scalar product dA is also a (q, p)-continuous operator.

Corollary 6.39. (a) If A: $L \to M$ and B: $L \to M$ are uniformly (q, p)-continuous operators, then their sum $A + B$: $L \to M$ also is a uniformly (q, p)-continuous operator.

(b) If A: $L \to M$ is a uniformly (q, p)-continuous operator and d is a real number, then the scalar product dA is also a uniformly (q, p)-continuous operator.

Let us consider a Q-hyperseminormed vector space L, i.e., a vector space L with a system of hyperseminorms Q, a P-hyperseminormed vector space M, i.e., a vector space M with a system of hyperseminorms P, two hyperseminorms r and q from Q, two hyperseminorms t and p from P, and a subset V of the space L.

Proposition 6.18. If $q \le r$ and $t \le p$, then any (q, p)-continuous at the point a operator is also (r, t)-continuous at the point a.

Proof. Let us consider a (q, p)-continuous at the point a operator A: $L \to M$ and a positive real number k. As $t \le p$, we have $t(x) \le c \cdot p(x)$ for some real number $c \ge 0$ and all $x \in L$. Consequently, if $p(A(b) - A(a)) < c^{-1}k$, then $t(A(b) - A(a)) < k$.

Besides, by Definition 6.1, there is a positive real number l such that for any element b from L, the inequality $q(a-b) < l$ implies the inequality $p(A(b) - A(a)) < c^{-1}k$. As $q \le r$, we have $q(x) \le d \cdot r(x)$ for some real number $d \ge 0$ and all $x \in L$. Thus, the inequality $r(a-b) < d^{-1}l$ implies the inequality $q(a-b) < l$. Consequently, if $r(a-b) < d^{-1}l$, then $t(A(b) - A(a)) < k$. This means that the operator A is (r, t)-continuous at the point a because k is an arbitrary positive real number.

Proposition is proved.

Corollary 6.40. If $q \le r$ and $t \le p$, then any (q, p)-continuous operator is also (r, t)-continuous.

Corollary 6.41. If $q \le r$ and $t \le p$, then any V-uniformly (q, p)-continuous operator is also V-uniformly (r, t)-continuous.

Corollary 6.42. If $q \le r$ and $t \le p$, then any uniformly (q, p)-continuous operator is also uniformly (r, t)-continuous.

Corollary 6.43. If $q \le r$, then any (q, p)-continuous at the point a operator is also (r, p)-continuous at the point a.

Corollary 6.44. If $q \le r$, then any (q, p)-continuous operator is also (r, p)-continuous.

Corollary 6.45. If $q \le r$, then any V-uniformly (q, p)-continuous operator is also V-uniformly (r, p)-continuous.

Corollary 6.46. If $q \le r$, then any uniformly (q, p)-continuous operator is also uniformly (r, p)-continuous.

Corollary 6.47. For any positive real number c, any (q, p)-continuous at the point a operator is also (cq, p)-continuous at the point a.

Corollary 6.48. For any positive real number c, any (q, p)-continuous operator is also (cq, p)-continuous.

Corollary 6.49. For any positive real number c, any V-uniformly (q, p)-continuous operator is also V-uniformly (cq, p)-continuous.

Corollary 6.50. For any positive real number c, any uniformly (q, p)-continuous operator is also uniformly (cq, p)-continuous.

However, continuous with respect to one pair of hyperseminorms (hypernorms, seminorms or norms) operator can be discontinuous with respect to another pair of hyperseminorms (hypernorms, seminorms or norms) in the same spaces. That is why we introduce a more general concept of continuity in polyhyperseminormed vector spaces, namely, continuity of operators with respect to a binary relation u between systems of hyperseminorms.

Definition 6.5. (a) An operator (mapping) A: $L \to M$ is called *(Q, u, P)-continuous at a point a* from L if for any hyperseminorms q and p such that $(q, p) \in u$, the operator (mapping) A is (q, p)-continuous at the point a. The triad (Q, u, P) is called the *extent of continuity* of A at a.

(b) An operator (mapping) A: $L \rightarrow M$ is called *(Q, u, P)-continuous* if it is (Q, u, P)-continuous at all points of L. The triad (Q, u, P) is called the *extent of continuity* of A.

(c) An operator (mapping) A: $L \rightarrow M$ is called *uniformly (Q, u, P)-continuous in $V \subseteq L$* if for any hyperseminorms q and p such that $(q, p) \in u$ and any positive real number k, there is a positive real number h such that for any elements a and b from V, the inequality $q(b - a) < h$ implies the inequality $p(A(b) - A(a)) < k$. The triad (Q, u, P) is called the *extent of uniform continuity* of A.

(d) An operator (mapping) A: $L \rightarrow M$ is called *V-uniformly (Q, u, P)-continuous* if for any hyperseminorms q and p such that $(q, p) \in u$ and for any positive real number k, there is a positive real number h such that for any element a from $V \subseteq L$ and any element b from L, the inequality $q(b - a) < h$ implies the inequality $p(A(b) - A(a)) < k$. The triad (Q, u, P) is called the *extent of V-uniform continuity* of A.

Note that to be *L-uniformly (Q, u, P)-continuous* or to be *uniformly (Q, u, P)-continuous in L* means the same for all operators.

Let us consider two systems of hyperseminorms Q and R in a vector space L, two systems of hyperseminorms P and T in a vector space M, a binary relation $u \subseteq Q \times P$ and a binary relation $W \subseteq R \times T$. Then Proposition 6.18 allows us to prove the following result.

Proposition 6.19. If for any pair (r, t) from w, there is a pair (q, p) from u such that $q \leq r$ and $t \leq p$, then any (Q, u, P)-continuous at the point a operator is also (R, w, T)-continuous at the point a.

Proof is left as an exercise.

Corollary 6.51. If for any pair (r, t) from w, there is a pair (q, p) from u such that $q \leq r$ and $t \leq p$, then any (Q, u, P)-continuous operator is also (R, w, T)-continuous.

Corollary 6.52. If for any pair (r, t) from w, there is a pair (q, p) from u such that $q \leq r$ and $t \leq p$, then any V-uniformly (Q, u, P)-continuous operator is also V-uniformly (R, w, T)-continuous.

Corollary 6.53. If for any pair (r, t) from w, there is a pair (q, p) from u such that $q \leq r$ and $t \leq p$, then any uniformly (Q, u, P)-continuous operator is also uniformly (R, w, T)-continuous.

Lemma 6.14. If u and v are binary relations between the system of hyperseminorms Q in L and the system of hyperseminorms P in M and $u \subseteq v$, then:

(a) Any (Q, v, P)-continuousat a point a operator is (Q, u, P)-continuous at a.
(b) Any V-uniformly (Q, v, P)-continuous operator is V-uniformly (Q, u, P)-continuous.
(c) Any (Q, v, P)-continuous in V operator is (Q, u, P)-continuous in V.
(d) Any uniformly (Q, v, P)-continuous operator is uniformly (Q, u, P)-continuous.

Proof is left as an exercise.

Lemma 6.14 implies the following result.

Lemma 6.15. Any uniformly (Q, u, P)-continuous in L operator is (Q, u, P)-continuous.

Proof is left as an exercise.

At the same time, taking $L = M = \mathbf{R}$, $Q = \{q\}$, $P = \{p\}$, and assuming that q and p are both equal to the absolute value and $u = \{(q, p)\}$, we see that Example 6.5 demonstrates that there are (Q, u, P)- continuous operators that are not L-uniformly (Q, u, P)-continuous.

However, for linear operators, the inverse of Lemma 6.15 is also true as Proposition 6.12 implies the following result.

Proposition 6.20. The following conditions are equivalent for a linear operator (mapping) A:

(1) A is (Q, u, P)-continuous.
(2) A is uniformly (Q, u, P)-continuous in L.
(3) For some point a, the operator A is (Q, u, P)-continuous at the point a.
(4) A is L-uniformly (Q, u, P)-continuous.

Corollary 6.54. A linear operator (mapping) A is (Q, u, P)-continuous if and only if it is (Q, u, P)-continuous at $\mathbf{0}$.

Corollary 6.54 implies the following result.

Corollary 6.55. Any (Q, u, P)-continuous linear operator (mapping) A: $L \to M$ is L-uniformly (Q, u, P)-continuous.

These results show that for linear operators, the concepts of (Q, u, P)-continuous at a point operators and (Q, u, P)-continuous operators coincide.

At the same time, taking $L = M = \mathbf{R}_\omega$, $Q = \{q\}$, $P = \{p\}$, and assuming that q and p are both equal to the absolute value of real hypernumbers and $u = \{(q, p)\}$, we see that Example 6.6 demonstrates that there are operators that are (Q, u, P)-continuous at one point and not (Q, u, P)-continuous at another point. A similar situation is also presented in Example 6.7.

Definitions and Lemma 6.12 imply the following result.

Lemma 6.16. If $W \subseteq V \subseteq L$, then any V-uniformly (Q, u, P)-continuous operator is W-uniformly (Q, u, P)-continuous.

Lemma 6.11 implies the following result.

Lemma 6.17. For any $V \subseteq L$, any V-uniformly (Q, u, P)-continuous operator is (Q, u, P)-continuous in V.

Proposition 6.11 implies the following result.

Proposition 6.21. (a) If $A: L \to M$ and $B: L \to M$ are (Q, u, P)-continuous at a point a from L operators, then their sum $A + B: L \to M$ also is a (Q, u, P)-continuous at the point a operator.

(b) If $A: L \to M$ is a (Q, u, P)-continuous operator at a point a from L and d is a real number, then the scalar product dA is a (Q, u, P)-continuous at the point a operator.

Proof. (a) Let us consider (Q, u, P)-continuous at a point a from L operators $A: L \to M$ and $B: L \to M$ and take a pair $(q, p) \in u$. By definition, for any positive real number $\frac{1}{2}h$, there is a positive real number k such that for any element b from L, the inequality $q(a - b) < k$ implies the inequality $p(A(b) - A(a)) < \frac{1}{2}h$ and there is a positive real number l such that for any element b from L, the inequality $q(a - b) < l$ implies the inequality $p(B(b) - B(a)) < \frac{1}{2}h$. So, if $q(a - b) < t = \min \{k, l\}$, then we have

$$p((A + B)(b) - (A + B)(a)) = p(A(b) - A(a) + B(b) - B(a)) \leq$$

$$p(A(b) - A(a)) + p(B(b) - B(a)) < \tfrac{1}{2}h + \tfrac{1}{2}h = h$$

Thus, we have demonstrated that for any positive real number h, there is a positive real number k such that for any element b from L, the inequality $q(a - b) < k$ implies the inequality $p((A + B)(b) - (A + B)(a)) < h$. As (q, p) is an arbitrary element from u, it means the operator $A + B: L \to M$ is a (Q, u, P)-continuous at the point a operator.

(b) Let us consider (Q, u, P)-continuous at a point a from L operator $A: L \to M$, a real number d and take a pair $(q, p) \in u$. By definition, for any positive real number $(1/|d|) \cdot h$, there is a positive real number k such that for any element b from L, the inequality $q(a - b) < k$ implies the inequality $p(A(b) - A(a)) < (1/|d|) \cdot h$. Then we have

$$p(dA(b) - dA(a)) = p(d(A(b) - A(a))) =$$

$$|d| \cdot p(A(b) - A(a)) < |d| \cdot (1/|d|) \cdot h = h$$

Thus, we have demonstrated that for any positive real number h, there is a positive real number k such that for any element b from L, the inequality $q(a - b) < k$ implies the inequality $p(dA(b) - dA(a)) < h$. As (q, p) is an arbitrary element from u, it means the operator dA: $L \rightarrow M$ is a (Q, u, P)-continuous at the point a operator.

Proposition is proved.

Corollary 6.56. (a) If A: $L \rightarrow M$ and B: $L \rightarrow M$ are(Q, u, P)-continuous operators, then their sum $A + B$: $L \rightarrow M$also is a (Q, u, P)-continuous operator.

(b) If A: $L \rightarrow M$is a (Q, u, P)-continuous operator and d is a real number, then the scalar product dA is also a (Q, u, P)-continuous operator.

Corollary 6.57. (a) If A: $L \rightarrow M$ and B: $L \rightarrow M$ areuniformly (Q, u, P)-continuous operators, then their sum $A + B$: $L \rightarrow M$also is a uniformly (Q, u, P)-continuous operator.

(b) If A: $L \rightarrow M$is a uniformly (Q, u, P)-continuous operator and d is a real number, then the scalar product dA is also a uniformly (Q, u, P)-continuous operator.

One of the basic theorems of the theory of topological vector spaces describes connections between continuity and boundedness of linear operators. Continuous linear operators are homomorphisms of topological and semitopological vector spaces, forming the class of morphisms in the categories of such spaces. Here we study relations between relative continuity and relative boundedness.

Let us consider a Q-hyperseminormed vector space L, i.e., a vector space L with a system of hyperseminorms Q, a P-hyperseminormed vector space M, i.e., a vector space M with a system of hyperseminorms P, a hyperseminorm q from Q, a hyperseminorm p from P, and a subset V of the space L.

Theorem 6.1. A linear operator (mapping) A: $L \rightarrow M$is (Q, u, P)-continuous if and only if it is (Q, u, P)-bounded.

Proof. Sufficiency. Let us consider a (Q, u, P)-bounded linear operator (mapping) A: $L \rightarrow M$ and suppose that A is not (Q, u, P)-continuous. It means that for some pair $(q, p) \in u$ of hyperseminorms q and p, the operator A is not (q, p)-continuous. By Corollary 6.54, A is not (q, p)-continuous at $\mathbf{0}$. Consequently, there is a positive real number k such that for any natural number n, there is an element x_n from L for which $q(x_n) < 1/n$ while $p(A(x_n)) > k$.

Let us consider the set $Z = \{ z_n; n = 1, 2, 3, \ldots\}$ where $z_n = n \cdot x_n$ for all $n = 1, 2, 3, \ldots$ Then

$$q(z_n) = q(n \cdot x_n) = n \cdot q(x_n) < 1$$

i.e., Z is a q-bounded set. At the same time, as A is a linear operator, we have

$$p(A(z_n)) = p(A(n \cdot x_n)) = n \cdot p(A(x_n)) > kn$$

Thus, the image of Z is not a p-bounded set and A is not a (Q, u, P)-bounded operator. This contradicts our assumption and by *reductio ad absurdum*, A is (Q, u, P)-continuous.

Necessity. Let us consider a (Q, u, P)-continuous linear operator (mapping) $A: L \rightarrow M$ and suppose that A is not (Q, u, P)-bounded. It means that for some pair $(q, p) \in u$ of hyperseminorms q and p, the operator A is not (q, p)-bounded. By Corollary 6.3, A is not (q, p)-bounded at **0**. Consequently, there is a positive real number k such that for any natural number n, there is an element x_n from L for which $q(x_n) < k$ while $p(A(x_n)) > n$.

Let us consider the set $Z = \{ z_n; n = 1, 2, 3, \ldots\}$ where $z_n = (1/n) \cdot x_n$ for all $n = 1, 2, 3, \ldots$ Then

$$q(z_n) = q((1/n) \cdot x_n) = (1/n) \cdot q(x_n) < k/n$$

At the same time, as A is a linear operator, we have

$$p(A(z_n)) = p(A((1/n) \cdot x_n)) = (1/n) \cdot p(A(x_n)) > k$$

This violates conditions from Definition 6.4 and shows A is not a (Q, u, P)-continuous operator. Thus, we have a contradiction with our assumption that A is a (Q, u, P)-continuous operator. By *reductio ad absurdum*, A is (Q, u, P)-bounded.

Theorem is proved.

Corollary 6.58. A linear operator (mapping) $A: L \rightarrow M$ is (q, p)-continuous if and only if it is (q, p)-bounded.

Corollary 6.58 implies the following result.

Corollary 6.59. A linear operator (mapping) $A: L \rightarrow M$ is L-uniformly (Q, u, P)-continuous if and only if it is L-uniformly (Q, u, P)-bounded.

As topology of topological vector spaces is determined by system of seminorms (Rudin, 1991), Theorem 6.1 gives us the following classical result (Dunford and Schwartz, 1958; Rudin, 1991).

Corollary 6.60. A linear mapping A of a topological vector space L into a topological vector space M is continuous if and only if it is bounded.

As for linear operators (mappings) continuity at a point coincides with continuity and boundedness at a point coincides with boundedness, we have the following results.

Corollary 6.61. A linear operator (mapping) A: $L \to M$ is (q, p)-continuous at a point a if and only if it is (q, p)-bounded at a.

Corollary 6.62. A linear operator (mapping) A: $L \to M$ is (Q, u, P)-continuous at a point a if and only if it is (Q, u, P)-bounded at a.

Let us take a vector subspace V of L and consider uniform (Q, u, P)-continuity in V.

Theorem 6.2. A linear operator (mapping) A: $L \to M$ is uniformly (Q, u, P)-continuous in V if and only if it is uniformly (Q, u, P)-bounded in V.

Proof. Sufficiency. Let us consider a vector subspace V of L and a uniformly (Q, u, P)-bounded in V linear operator (mapping) A: $L \to M$ and suppose that A is not uniformly (Q, u, P)-continuous in V. It means that for some pair $(q, p) \in u$ of hyperseminorms q and p, the operator A is not uniformly (q, p)-continuous. Consequently, there is a positive real number k such that for any natural number n, there are elements x_n and y_n from V for which $q(x_n - y_n) < 1/n$ while $p(A(x_n) - A(y_n)) > k$.

Let us consider two sets $Z = \{ z_n; n = 1, 2, 3, \ldots\}$ and $U = \{ u_n; n = 1, 2, 3, \ldots\}$ where $z_n = n \cdot x_n$ and $u_n = n \cdot y_n$ for all $n = 1, 2, 3, \ldots$ As V is a vector subspace of L, then Z and U are subsets of V. Besides,

$$q(z_n - u_n) = q(n \cdot x_n - n \cdot y_n) = q(n \cdot (x_n - y_n)) = n \cdot q(x_n - y_n) < 1$$

At the same time, as A is a linear operator, we have

$$p(A(z_n - u_n)) = p(A(n \cdot x_n - n \cdot y_n)) = n \cdot p(A(x_n) - A(y_n)) > kn$$

Thus, A is not a uniformly (Q, u, P)-bounded in V operator. This contradicts our assumption and by *reductio ad absurdum*, A is uniformly (Q, u, P)-continuous in V.

Necessity. Let us consider a uniformly (Q, u, P)-continuous in V linear operator (mapping) A: $L \to M$ and suppose that A is not uniformly (Q, u, P)-bounded in V. It means that for some pair $(q, p) \in u$ of hyperseminorms q and p, the operator A is not uniformly (q, p)-bounded in V. By Corollary 6.3, A is not (q, p)-bounded in V at $\mathbf{0}$ as V is a vector subspace of L. Consequently, there is a positive real number k such that for any natural number n, there is an element x_n from V for which $q(x_n) < k$, while $p(A(x_n)) > n$.

Let us consider the set $Z = \{ z_n; n = 1, 2, 3, \dots\}$ where $z_n = (1/n) \cdot x_n$ for all $n = 1, 2, 3, \dots$ Then

$$q(z_n) = q((1/n) \cdot x_n) = (1/n) \cdot q(x_n) < k/n$$

At the same time, as A is a linear operator, we have

$$p(A(z_n)) = p(A((1/n) \cdot x_n)) = (1/n) \cdot p(A(x_n)) > k$$

This violates conditions from Definition 6.5 and shows A is not a uniformly (Q, u, P)-continuous in V operator. Thus, we have a contradiction with our assumption that A is a uniformly (Q, u, P)-continuous in V operator. By *reductio ad absurdum*, A is uniformly (Q, u, P)-bounded in V.

Theorem is proved.

Corollary 6.63. For any vector subspace V of L, a linear operator (mapping) A: $L \to M$ is uniformly (q, p)-continuous in V if and only if it is uniformly (q, p)-bounded in V.

As before, V is a vector subspace of L and we study V-uniform (Q, u, P)-continuity.

Theorem 6.3. A linear operator (mapping) A: $L \to M$ is V-uniformly (Q, u, P)-continuous if and only if it is V-uniformly (Q, u, P)-bounded.

Proof. Sufficiency. Let us consider a vector subspace V of L and a V-uniformly (Q, u, P)-bounded linear operator (mapping) A: $L \to M$ and suppose that A is not V-uniformly (Q, u, P)-continuous. It means that for some pair $(q, p) \in u$ of hyperseminorms q and p, the operator A is not V-uniformly (q, p)-continuous. Consequently, there is a positive real number k such that for any natural number n, there are elements x_n from V and y_n from L for which $q(x_n - y_n) < 1/n$ while $p(A(x_n) - A(y_n)) > k$.

Let us consider two sets $Z = \{ z_n; n = 1, 2, 3, \dots\}$ and $U = \{ u_n; n = 1, 2, 3, \dots\}$ where $z_n = n \cdot x_n$ and $u_n = n \cdot y_n$ for all $n = 1, 2, 3, \dots$ As V is a vector subspace of L, then Z is a subset of V. Besides,

$$q(z_n - u_n) = q(n \cdot x_n - n \cdot y_n) = q(n \cdot (x_n - y_n)) = n \cdot q(x_n - y_n) < 1$$

At the same time, as A is a linear operator, we have

$$p(A(z_n - u_n)) = p(A(n \cdot x_n - n \cdot y_n)) = n \cdot p(A(x_n) - A(y_n)) > kn$$

Thus, A is not a V-uniformly (Q, u, P)-bounded operator. This contradicts our assumption and by *reductio ad absurdum*, A is V-uniformly (Q, u, P)-continuous.

Necessity. Let us consider a V-uniformly (Q, u, P)-continuous linear operator (mapping) A: $L \to M$ and suppose that A is not V-uniformly (Q, u, P)-bounded. It means that for some pair $(q, p) \in u$ of hyperseminorms q and p, the operator A is not V-uniformly (q, p)-bounded. By Corollary 6.3, A is not (q, p)-bounded at **0**. Consequently, there is a positive real number k such that for any natural number n, there is an element x_n from L for which $q(x_n) < k$ while $p(A(x_n)) > n$.

Let us consider the set $Z = \{ z_n; n = 1, 2, 3, \ldots\}$ where $z_n = (1/n) \cdot x_n$ for all $n = 1, 2, 3, \ldots$ Then

$$q(z_n) = q((1/n) \cdot x_n) = (1/n) \cdot q(x_n) < k/n$$

At the same time, as A is a linear operator, we have

$$p(A(z_n)) = p(A((1/n) \cdot x_n)) = (1/n) \cdot p(A(x_n)) > k$$

This violates conditions from Definition 6.5 and shows A is not a V-uniformly (Q, u, P)-continuous operator. Thus, we have a contradiction with our assumption that A is a V-uniformly (Q, u, P)-continuous operator. By *reductio ad absurdum*, A is V-uniformly (Q, u, P)-bounded.

Theorem is proved.

Corollary 6.64. For any subset V of L, a linear operator (mapping) A: $L \to M$ is V-uniformly (q, p)-continuous if and only if it is V-uniformly (q, p)-bounded.

Let us take a subset V of the vector space L.

Definition 6.6. (a) An operator (mapping) A: $L \to M$ is called *uniformly* (Q, u, P)*-continuous at a point a* from L if for any positive real number k, there is a positive real number h such that for any hyperseminorms q and p with $(q, p) \in u$, and any element b from L, the inequality $q(a - b) < h$ implies the inequality $p(A(b) - A(a)) < k$.

(b) An operator (mapping) A: $L \to M$ is called *u-uniformly* (Q, u, P)*-continuous* if it is uniformly (Q, u, P)-continuous at all points of L.

(c) An operator (mapping) A: $L \to M$ is called *u-uniformly* (Q, u, P)*-continuous in* V if for any positive real number k, there is a positive real number h such that for any elements a and b from V and any hyperseminorms

q and p with $(q, p) \in u$, the inequality $q(a - b) < h$ implies the inequality $p(A(b) - A(a)) < k$.

(d) An operator (mapping) A: $L \to M$ is called *uV-uniformly* (Q, u, P)*-continuous* if for any positive real number k, there is a positive real number h such that for any elements a from V and b from L, and any hyperseminorms q and p with $(q, p) \in u$, the inequality $q(a - b) < h$ implies the inequality $p(A(b) - A(a)) < k$.

Note that to be *uL-uniformly* (Q, u, P)*-continuous* or to be *u-uniformly* (Q, u, P)*-continuous in L* means the same for all operators.

Lemma 6.18. If u and v are binary relations between the system of hyperseminorms Q in L and the system of hyperseminorms P in M and $u \subseteq v$, then:

(a) Any uniformly (Q, v, P)-continuous at a point a operator A is uniformly (Q, u, P)-continuous at a.

(b) Any u-uniformly (Q, v, P)-continuous operator A is u-uniformly (Q, u, P)-continuous in L.

(c) Any u-uniformly (Q, v, P)-continuous in V operator A is u-uniformly (Q, u, P)-continuous in V.

(d) Any uV-uniformly (Q, v, P)-continuous operator is uV-uniformly (Q, u, P)-continuous.

Proof is left as an exercise.

It is possible to ask a question how u-uniform (Q, u, P)-continuity is connected to (Q, u, P)-continuity. The following example and Lemma 6.7 clarify this situation.

Example 6.8. Let us take $L = M$ and equal to the vector space C(***R***, ***R***) of all continuous real functions. It is possible (Burgin, 2012) for all real numbers x, to define seminorms $q_{\text{pt}x} = p_{\text{pt}x}$ by the following formula

$$q_{\text{pt}x}(f) = p_{\text{pt}x}(f) = |f(x)|$$

We define $A(f) = xf(x)$ for all real functions f and $u = \{(q_{\text{pt}x}, p_{\text{pt}x}); x \in \boldsymbol{R}\}$. Taking the function $f(x) = x$ as the point a from L, we see that $A(f) = x^2$. Thus, taking some positive real number k, e.g., $k = 1$, the corresponding h from Definition 6.2 always exists but it decreases with the growth of x. For instance, when $k = 1$, we have

$$q_{\text{pt}1}(f - g) < 1 \text{ implies } p_{\text{pt}1}(A(f) - A(g)) = p_{\text{pt}1}(xf - xg) < 1$$

At the same time, $q_{pt10}(f - g) < 1$ does not imply $p_{pt10}(A(f) - A(g)) < 1$. It only implies $p_{pt10}(A(f) - A(g)) = p_{pt10}(xf - xg) < 10$. To have $p_{pt10}(A(f) - A(g)) < 1$, we need $q_{pt10}(f - g) < 0.1$.

It means that for any pair (q_{ptr}, p_{ptr}) of seminorms and a number k, we need to find a specific number h to satisfy Definition 6.6a. Consequently, the operator A is (Q, u, P)-continuous at $f = x$ but it is not uniformly (Q, u, P)-continuous at f.

The same example shows that there are (Q, u, P)-continuous operators that are not u-uniformly (Q, u, P)-continuous.

Definitions imply the following results.

Lemma 6.19. (a) Any uniformly (Q, u, P)-continuous at a point a operator A is (Q, u, P)-continuous at the point a.

(b) Any u-uniformly (Q, u, P)-continuous operator A is (Q, u, P)-continuous.

Lemma 6.20. Any u-uniformly (Q, u, P)-continuous in L operator is u-uniformly (Q, u, P)-continuous.

For linear operators, the inverse of Lemma 6.20 is also true.

Proposition 6.22. The following conditions are equivalent for a linear operator (mapping) A:

(1) A is u-uniformly (Q, u, P)-continuous.
(2) A is u-uniformly (Q, u, P)-continuous in L.
(3) For some point a, the operator A is uniformly (Q, u, P)-continuous at the point a.
(4) A is uL-uniformly (Q, u, P)-continuous.

Proof is left as an exercise.

Corollary 6.65. A linear operator (mapping) A is u-uniformly (Q, u, P)-continuous in L if and only if it is (Q, u, P)- continuous at $\mathbf{0}$.

Corollary 6.65 implies the following result.

Corollary 6.66. Any u-uniformly (Q, u, P)-continuous linear operator (mapping) A: $L \to M$ is u-uniformly (Q, u, P)-continuous in L.

These results show that for linear operators, the concepts of uniformly (Q, u, P)-continuous at a point operators and u-uniformly (Q, u, P)-continuous operators coincide.

At the same time, taking $L = M = \boldsymbol{R}_\omega$, $Q = \{q\}$, $P = \{p\}$, and assuming that q and p are both equal to the absolute value of real hypernumbers and $u = \{(q, p)\}$, we see that Example 6.6 demonstrates that there are operators that are (Q, u, P)-continuous at one point and not (Q, u, P)-continuous at another point. A similar situation is also presented in Example 6.7.

Definitions and Lemma 6.11 imply the following result.

Lemma 6.21. If $W \subseteq V \subseteq L$, then any u-uniformly (Q, u, P)-continuous in V operator is u-uniformly (Q, u, P)-continuous in W.

For finite relations u, different concepts of uniform continuity coincide.

Proposition 6.23. If the relation u is finite, then, an operator (mapping) A: $L \rightarrow M$ is u-uniformly (Q, u, P)-continuous (u-uniformly (Q, u, P)-continuous at a point a if and only if it is (Q, u, P)-continuous ((Q, u, P)-continuous at a point a).

Proof. As any u-uniformly (Q, u, P)-continuous (u-uniformly (Q, u, P)-continuous at a point a) operator is (Q, u, P)-continuous ((Q, u, P)-continuous at the same point), we need only to show that when the relation u is finite, a (Q, u, P)-continuous (at a point a) operator A: $L \rightarrow M$ is uniformly (Q, u, P)-continuous (at the point a). At first, we consider local boundedness.

Indeed, by Definition 6.5, for any hyperseminorms q and p such that $(q, p) \in u$, the operator (mapping) A is (q, p)-continuous at the point a, that is, by Definition 6.3, the following condition is true:

Condition 2. For any positive real number k, there is a positive real number h such that for any element b from L, the inequality $q(a - b) < h$ implies the inequality $p(A(b) - A(a)) < k$.

This number h can be different for different pairs (q, p), but because u is finite, there is only a finite number of these pairs. So, we can take

$$l = \min \{h;\ h \text{ satisfies Condition 2 for a pair } (q, p) \in u\}$$

and this number l will satisfy the condition from Definition 6.6. Thus, the operator A is u-uniformly (Q, u, P)-continuous at the point a.

The global case is proved in a similar way.

Proposition is proved.

Corollary 6.67. If systems of hyperseminorms Q and P are finite, then an operator (mapping) A is uniformly (Q, u, P)-continuous if and only if it is (Q, u, P)-continuous.

There are connections between uniform with respect to systems of hyperseminorms continuity and uniform boundedness that are similar to the connections between nonuniform with respect to systems of hyperseminorms continuity and nonuniform boundedness described in Theorems 6.1–6.3. Namely, we have the following results.

Theorem 6.4. A linear operator (mapping) $A: L \to M$ is uniformly (Q, u, P)-continuous at a point a if and only if it is uniformly (Q, u, P)-bounded at a.

Proof is similar to the proof of Theorem 6.1.

Let us take a vector subspace V of the vector space L.

Theorem 6.5. A linear operator (mapping) $A: L \to M$ is u-uniformly (Q, u, P)-continuous in V if and only if it is u-uniformly (Q, u, P)-bounded in V.

Proof is similar to the proof of Theorem 6.2.

Theorem 6.6. A linear operator (mapping) $A: L \to M$ is uV-uniformly (Q, u, P)-continuous if and only if it is uV-uniformly (Q, u, P)-bounded.

Proof is similar to the proof of Theorem 6.3.

Now let us apply the obtained results to functionals and hyperfunctionals on polyhyperseminormed and polyseminormed vector spaces. To do this, we specify Definitions 6.1 – 6.6 for functionals and hyperfunctionals.

Let us consider a Q-hyperseminormed vector space L and a subset G of the set Q. Then a real functional on L is a mapping $T: L \to \boldsymbol{R}$ and a real hyperfunctional on L is a mapping $K: L \to \boldsymbol{R}_\omega$, while a complex functional on L is a mapping $T: L \to \boldsymbol{C}$ and a complex hyperfunctional on L is a mapping $K: L \to \boldsymbol{C}_\omega$. In what follows, $\boldsymbol{F}$ denotes either $\boldsymbol{R}$ or $\boldsymbol{C}$.

Definition 6.7. a) A functional $T: L \to \boldsymbol{F}$ (a hyperfunctional $T: L \to \boldsymbol{F}_\omega$) is called *$G$-bounded at a point a* from L if for any hyperseminorm q from G and any positive real number k, there is a positive real number h such that for any element b from L, the inequality $q(a - b) < k$ implies the inequality $|T(b) - T(a)| < h$.

b) A functional $T: L \to \boldsymbol{F}$ (a hyperfunctional $T: L \to \boldsymbol{F}_\omega$) is called *$G$-bounded* if it is G-bounded at all points of L.

c) A functional $T: L \to \boldsymbol{F}$ (a hyperfunctional $T: L \to \boldsymbol{F}_\omega$) is called *$V$-uniformly G-bounded* if for any hyperseminorm q from G and any positive real number k, there is a positive real number h such that for any element a from V and any element b from L, the inequality $q(a - b) < k$ implies the inequality $|T(b) - T(a)| < h$.

d) A functional $T: L \to \boldsymbol{F}$ (a hyperfunctional $T: L \to \boldsymbol{F}_\omega$) is called *uniformly G-bounded in V* if for any hyperseminorm q from G and any positive real number k, there is a positive real number h such that for any elements a and b from V, the inequality $q(a - b) < k$ implies the inequality $|T(b) - T(a)| < h$.

Proposition 6.6 (or Proposition 6.22) implies the following result.

Proposition 6.24. The following conditions are equivalent for a linear functional $T: L \to \boldsymbol{F}$ (linear hyperfunctional $T: L \to \boldsymbol{F}_\omega$):

(1) T is G-bounded.
(2) T is uniformly G-bounded in L.
(3) For some point a, the hyperfunctional (functional)T is G-bounded at the point a.
(4) T is L-uniformly G-bounded.

In a similar way, we define continuity of functionals and hyperfunctionals.

Definition 6.8. (a) A functional T: $L \to \boldsymbol{F}$ (a hyperfunctional T: $L \to \boldsymbol{F}_\omega$) is called *G-continuous at a point a* from L if for any hyperseminorm q from G and any positive real number k, there is a positive real number h such that for any element b from L, the inequality $q(b - a) < h$ implies the inequality $|T(b) - T(a)| < k$.

(b) A functional T: $L \to \boldsymbol{F}$ (a hyperfunctional T: $L \to \boldsymbol{F}_\omega$) is called *G-continuous* if it is G-continuous at all points of L.

(c) A functional T: $L \to \boldsymbol{F}$ (a hyperfunctional T: $L \to \boldsymbol{F}_\omega$) is called *uniformly G-continuous in* $V \subseteq L$ if for any hyperseminorm q from G and any positive real number k, there is a positive real number h such that for any elements a and b from V, the inequality $q(b - a) < h$ implies the inequality $|T(b) - T(a)| < k$.

(d) A functional T: $L \to \boldsymbol{F}$ (a hyperfunctional T: $L \to \boldsymbol{F}_\omega$) is called *V-uniformly G-continuous* if for any hyperseminorm q from G and any positive real number k, there is a positive real number h such that for any element a from $V \subseteq L$ and any element b from L, the inequality $q(b - a) < h$ implies the inequality $|T(b) - T(a)| < k$.

Proposition 6.22 (or Proposition 6.24) implies the following result.

Proposition 6.25. The following conditions are equivalent for a linear functional T: $L \to \boldsymbol{F}$ (hyperfunctional T: $L \to \boldsymbol{F}_\omega$):

(1) T is G-continuous.
(2) T is uniformly G-continuous in L.
(3) For some point a, the hyperfunctional (functional) T is G-continuous at the point a.
(4) T is L-uniformly G-continuous.

As $\boldsymbol{R}$, $\boldsymbol{C}$, $\boldsymbol{R}_\omega$ and $\boldsymbol{C}_\omega$ are hypernormed vector spaces (cf. Chapter 3), while functionals and hyperfunctionals are special cases of operators, Theorem 6.1 allows finding relations between continuity and boundedness for linear functionals and hyperfunctionals.

Theorem 6.7. A linear functional T: $L \to \boldsymbol{F}$ (linear hyperfunctional T: $L \to \boldsymbol{F}_\omega$) is G-continuous if and only if it is G-bounded.

In the same way, we obtain the following result from Theorem 6.2.

Theorem 6.8. A linear functional T: $L\to\mathbf{F}$ (linear hyperfunctional T: $L\to\mathbf{F}_\omega$) is uniformly G-continuous in V if and only if it is uniformly G-bounded in V.

Theorem 6.3 allows us to prove the following result because functionals and hyperfunctionals are special cases of operators.

Theorem 6.9. A linear functional T: $L\to\mathbf{F}$ (linear hyperfunctional T: $L\to\mathbf{F}_\omega$) is V-uniformly G-continuous if and only if it is V-uniformly G-bounded.

To conclude, let us study some properties of hyperseminorms, hypernorms, seminorms and norms as functionals. It is easy to see that hyperseminorms and hypernorms are non-negative subadditive hyperfunctionals, while seminorms and norms are non-negative subadditive functionals. In addition, they have the following properties.

Proposition 6.26. If a hyperseminorm (hypernorm) q is stronger than a hyperseminorm (hypernorm) p, then the hyperfunctional p is (q, st)-continuous where st is the standard norm in the space of real numbers.

Proof. Let us assume that a hyperseminorm q is stronger than a hyperseminorm p. It means there is a constant $k \geq 0$ such that for all $x \in L$,

$$p(x) \leq k{\cdot}q(x)$$

Let us take a sequence $\{x_i;\ x_i \in L,\ i = 1, 2, 3, \ldots\}$ that q-converges to an element x from L. It means that $q(x_i - x)\to 0$. Taking the sequence $\{p(x_i);\ x_i \in L,\ i = 1, 2, 3, \ldots\}$. We see that by Proposition 3.9,

$$0 \leq |p(x_i) - p(x)| \leq p(x - x_i) \leq k{\cdot}q(x_i - x)$$

As $q(x_i - x) \to 0$, we have $|p(x_i) - p(x)| \to 0$ (Burgin, 2005). It means that $p(x_i)$ converges to $p(x)$ with respect to the standard norm in the space of real numbers. Consequently, the hyperfunctional p is (q, st)-continuous.

The statement for hypernorms is the direct corollary of the statement for hyperseminorms.

Proposition is proved.

Corollary 6.68. If a seminorm (norm) q is stronger than a seminorm (norm) p, then the hyperfunctional p is (q, st)-continuous.

Proposition 6.26 and Corollary 6.69 mean that any norm, seminorm, hypernorm or hyperseminorm is continuous in the topology determined by a stronger norm, seminorm, hypernorm or hyperseminorm.

Corollary 6.69. Any hyperseminorm (hypernorm) q is a non-negative continuous hyperfunctional.

Corollary 6.70. Any seminorm (norm) q is a non-negative continuous functional.

Corollaries 6.69 and 6.70 mean that any norm, seminorm, hypernorm or hyperseminorm is a continuous functional or hyperfunctional.

As continuity implies boundedness, we have the following results.

Corollary 6.71. If a hyperseminorm (hypernorm) q is stronger than a hyperseminorm (hypernorm) p, then the hyperfunctional p is (q, st)-bounded.

Corollary 6.72. If a seminorm (norm) q is stronger than a seminorm (norm) p, then the hyperfunctional p is (q, st)-bounded.

KEYWORDS

- **bounded**
- **hyperfunctional**
- **hyperoperator**
- **operator**
- **polyhypernormed vector space**
- **polyhyperseminormed vector space**
- **polynormed vector space**
- **polyseminormed vector space**

CHAPTER 7

FROM CONTINUITY TO FUZZY CONTINUITY

In this and next chapters, we further develop neoclassical analysis in the framework of hyperseminormed and hypernormed vector spaces. Neoclassical analysis (Burgin, 2008) extends methods of classical calculus reflecting imprecision and uncertainties that arise in computations and measurements. In it, ordinary structures of analysis, that is, functions, sequences, series, and operators, are studied by means of fuzzy concepts such as fuzzy limits, fuzzy continuity, and fuzzy derivatives. For instance, continuous operators, which are studied in the classical functional analysis, become a part of the set of the fuzzy continuous operators studied in neoclassical analysis. Aiming at representation of uncertainty, vagueness and imprecision, neoclassical analysis makes, at the same time, methods of the classical calculus more precise with respect to real life applications and extends the scope of the classical calculus and functional analysis.

There are three approaches to the definition of fuzzy continuous functionals and operators in hyperseminormed and hypernormed vector spaces:

- using fuzzy limits, which can be either fuzzy limits of sequences (Burgin, 1995a; 2000) or fuzzy limits of functions (Burgin, 2006a);
- by the most popular in calculus (ε, δ)-construction (Ross, 1996);
- deriving and applying discontinuity measures (Burgin, 1999, 2008).

The first two approaches are represented in Definitions 7.2 and 7.3.

Let us consider hyperseminormed vector spaces L and M over a field $\boldsymbol{F}$ where $\boldsymbol{F}$ is either the field $\boldsymbol{R}$ of all real numbers or the field $\boldsymbol{C}$ of all complex numbers. The hyperseminorms in L and M are denoted by the same symbol $||.||$. Assume that $r \in \boldsymbol{R}^+$, i.e., r is a non-negative real number, and $l = \{a_i; i = 1, 2, 3, \ldots\}$ is a sequence of elements from L.

Definition 7.1. a) An element a is called an *r-limit of a sequence l* (it is denoted by $a = r\text{-}\lim_{i\to\infty} a_i$ or $a = r\text{-}\lim l$) if for any $\varepsilon \in \mathbf{R}^{++}$, the inequality $\| a - a_i \| < r + \varepsilon$ is valid for almost all a_i, i.e., there is such n that for any $i > n$, we have $\| a - a_i \| < r + \varepsilon$.

b) A sequence l that has an r-limit is called *r-convergent* and it is said that l *r-converges* to its r-limit a. It is denoted by $l \to_r a$.

c) An element a is called a *fuzzy limit of a sequence l* if it is an r-limit of l for some non-negative real number r.

d) The number $r_{l,a} = \inf \{r;\ l$ *r-converges* to $a\}$ (if it exists) is called the *measure of convergence* of l to a.

e) A sequence l that has a fuzzy limit is called *fuzzy convergent* and it is said that l *fuzzy converges* to its fuzzy limit.

f) The number $r_l = \inf \{ r_{l,a};\ a \in L\}$(if it exists) is called the *measure of convergence* of l.

Informally, a is an r-limit of a sequence l if for an arbitrarily small ε, the distance between a and all but a finite number of elements from l is smaller than $r + \varepsilon$.

Remark 7.1. Measures of convergence r_l and $r_{l,a}$ are not defined for all sequences but only for fuzzy convergent sequences. The sequence $l = \{1, 2, 3, \ldots\}$ is an example of the situation when r_l and $r_{l,a}$ are not defined.

Example 7.1. Let us take $L = \mathbf{R}$ and define $l = \{1/i;\ i = 1, 2, 3, \ldots\}$.Then 1 and –1 are 1-limits of l; $0 = 0\text{-}\lim l$; and ½ is a (½)-limit of l, but 1 is not a (½)-limit of l. Thus, the sequence l 1-converges and 0-converges.

Example 7.2. Let us take $L = \mathbf{R}$ and define $l = \{(-1)^i;\ i = 1, 2, 3, \ldots\}$.Then this sequence does not have the classical limit. At the same time, points 1 and –1 are 2-limits of l because all elements from this sequence starting with the second one are less than $-1 - \varepsilon$ and smaller than $1 + \varepsilon$ for any positive numbersε. For the same reason, 0 is a 1-limit of l, but 1 is not a 1-limit of l as this sequence has infinitely many elements that are smaller than 1 – (1/2) –¼. Thus, the sequence l 2-converges and 1-converges, but does not 0-converge.

The following result shows that fuzzy convergence is a natural extension of the conventional convergence.

Lemma 7.1. A sequence l converges to a point a if and only if it 0-converges to a, i.e., $a = \lim l$ if and only if $a = 0\text{-}\lim l$.

Proof is left as an exercise.

Corollary 7.1. A sequence l converges to a point a if and only if $r_{l,a} = 0$.

Corollary 7.2. If $r_l = 0$, then l is a Cauchy sequence.

Corollary 7.3. In a complete hyperseminormed vector space, a sequence l converges if and only if $r_l = 0$.

As rule, one sequence has several r-limits when $r > 0$. For 0-limits, the situation is different.

Proposition 7.1. A 0-limit (limit) of any sequence in L is unique if and only if $\|.\|$ is a hypernorm.

Proof. Necessity. Let us suppose that the hyperseminorm $\|.\|$ is not a hypernorm. Then there is an element x from L such that $x \neq 0$ and $\|x\| = 0$. Taking the sequence $l = \{x, x, x, \ldots, x, \ldots\}$, we see that $x = 0\text{-lim } l$ and $0 = 0\text{-lim } l$, i.e., the limit is not unique. Thus, the condition that the 0-limit (limit) of any sequence in L is unique implies that $\|.\|$ is a hypernorm, i.e., it satisfies Axiom N1.

Sufficiency. Let us assume that $\|.\|$ is a hypernorm, while $x = 0\text{-lim } l$ and $y = 0\text{-lim } l$ for some sequence $l = \{a_i \in L; i = 1, 2, 3, \ldots\}$ in L. By the definition of a 0-limit, for any number $\varepsilon > 0$, there is such a number n that for any $i > n$, we have $\| x - a_i \| < \varepsilon$ and there is such m that for any $i > m$, we have $\| y - a_i \| < \varepsilon$. Taking $p = \max \{m, n\}$ and any natural number $i > p$, we have $\| x - a_i \| < \varepsilon$ and $\| y - a_i \| < \varepsilon$. Consequently, for any $i > p$, we have

$$\| x - y \| = \| x - a_i + a_i - y \| \leq \| x - a_i \| + \| y - a_i \| < 2\varepsilon$$

Because ε can be as small as we want,

$$\| x - y \| = 0$$

As $\|.\|$ is a hypernorm, $x = y$. It means that the 0-limit is unique.

Proposition is proved.

Lemma 7.2. If a sequence l r-converges to a point a, then it q-converges to a for any $q > r$, i.e., if $a = r\text{-lim } l$, then $a = q\text{-lim } l$ for any $q > r$.

Proof is left as an exercise.

Lemma 7.3. A sequence $l = \{a_i; i = 1, 2, 3, \ldots\}$ r-converges to a point a and d is an element from L if and only if the sequence $l + d = \{a_i + d; i = 1, 2, 3, \ldots\}$, which called a *shift* of l, r-converges to the point $a + d$.

Proof. Necessity. Let us take a sequence $l = \{a_i; i = 1, 2, 3, \ldots\}$ that r-converges to a point a and an element d from L. It means that for any $\varepsilon \in \boldsymbol{R}^{++}$, the inequality $\| a - a_i \| < r + \varepsilon$ is valid for almost all a_i, i.e., there is such n that for any $i > n$, we have $\| a - a_i \| < r + \varepsilon$. Then for the same index i, we have

$$\| (a + d) - (a_i + d) \| = \| a - a_i \| < r + \varepsilon$$

Consequently, the sequence $l + d = \{a_i + d;\ i = 1, 2, 3, \dots\}$ r-converges to the point $a + d$.

Necessity is proved.

Sufficiency. As we have proved that r-convergence of a sequence implies r-convergence of any its shift, r-convergence of $l + d$ implies r-convergence of l because l is a shift of $l + d$.

Lemma is proved.

Let us consider two sequences $l = \{a_i;\ i = 1, 2, 3, \dots\}$ and $h = \{b_i;\ i = 1, 2, 3, \dots\}$ with elements from L and a real number d.

Proposition 7.2. (a) If $a = r\text{-lim } l$ and $b = q\text{-lim } h$, then $a + b = (r + q)\text{-lim } (l + h)$.

(b) If $a = r\text{-lim } l$ and $b = q\text{-lim } h$, then $a - b = (r + q)\text{-lim } (l - h)$

(c) If $a = r\text{-lim } l$, then $da = (|d| \cdot r)\text{-lim } dl$.

Proof is left as an exercise.

Proposition 7.3. If h is a subsequence of a sequence l and $a = r\text{-lim } l$, then $a = r\text{-lim } h$.

Proof is left as an exercise.

Proposition 7.3 implies the following result.

Proposition 7.4. If h is a subsequence of a sequence l, then $r_{h,a} \leq r_{l,a}$.

Corollary 7.4. If h is a subsequence of a sequence l, then $r_h \leq r_l$.

In some cases, Proposition 7.4 and Corollary 7.4 give us exact equalities, while there are situations when $r_{h,a} < r_{l,a}$ and $r_h < r_l$ as the following example demonstrates.

Example 7.3. Let us take $L = \mathbf{R}$, a sequence $l = \{(-1)^i;\ i = 1, 2, 3, \dots\}$ and its subsequence $h = (1, 1, 1, \dots)$.Then 1 and -1 are 2-limits of l and of h; $0 = 0\text{-lim } l = 0\text{-lim } l$; $r_{h,1} = 0 < r_{l,1} = 2$ and $r_h = 0 < r_l = 2$.

We apply the concept of a fuzzy limit to define fuzzy continuity of operators, functionals and functions.

Definition 7.2. a) A partial operator (mapping) A: $L \rightarrow M$ is called *r-continuous at* a point $a \in L$ if A is defined at a and for any sequence $l = \{a_i \in L;\ i = 1, 2, 3, \dots\}$ that converges to a, the point $A(a)$ is an r-limit of the sequence $\{A(a_i) \in M;\ i = 1, 2, 3, \dots\}$.

b) The number $r_{A,a} = \inf \{r;\ A \text{ is } r\text{-continuous at } a\}$ (if it exists) is called the *measure of discontinuity* (or *continuity defect* or *parameter of discontinuity*) of the operator A at the point a.

c) A partial operator (mapping) A: $L \rightarrow M$ is called *fuzzy continuous at* a point $a \in L$ if A is r-continuous at a for some number $r \in \mathbf{R}^+$.

Informally, r-continuity at a means that the gap of the operator (mapping) A at the point a cannot be larger than r.

Example 7.4. Let us consider the one-dimensional operator A defined by the following formula

$$A(x) = \begin{cases} x^2 & \text{if } x \in [0, 1/2]; \\ x & \text{otherwise.} \end{cases}$$

This operator is (1/4)-continuous at the point ½ and 0-continuous at all other points.

There is also an (ε, δ)-definition of r-continuity similar to the traditional (ε, δ)-definition of continuity.

Definition 7.3. An operator (mapping) $A: L \to M$ is called *r-continuous at* a point $a \in L$ if A is defined at a and for any $\varepsilon > 0$, there is $\delta > 0$ such that the inequality $\| a - x \| < \delta$ implies the inequality $\| A(x) - A(a) \| < r + \varepsilon$, or in other words, for any x with $\| a - x \| < \delta$, we have $\| A(x) - A(a) \| < r + \varepsilon$.

It means that the gap of A at the point a cannot be larger than r.

Taking the operator A from Example 7.4, we see that it is (1/4)-continuous at the point ½ and 0-continuous at all other points with respect to both definitions. This is a general case as the following result shows.

Let us consider two hyperseminormed vector spaces L and M over a field $\boldsymbol{F}$.

Proposition 7.5. Definitions 7.2 and 7.3 are equivalent, i.e., an operator $A: L \to M$ is r-continuous at a point a according to Definition 7.2 if and only if A is r-continuous at the point a according to Definition 7.3.

Proof. a) Definition 7.2 Definition 7.3. Let us assume that an operator A is not r-continuous at the point a according to Definition 7.3. It means that there is $\varepsilon > 0$ such that for any $\delta > 0$, there is a point x from the space L with $\| x - a \| < \delta$ but $\| A(x) - A(a) \| \geq r + \varepsilon$. Let us take a sequence of numbers δ_i equal to ½, $^1/_3$, ¼, …, $^1/_n$, …

This gives us the sequence $l = \{x_i;\ i = 1, 2, 3, \ldots\}$ in which $\| x_i - a \| < 1/i$, but $\| A(x_i) - A(a) \| \geq r + \varepsilon$. Consequently (cf., Sections 2.1 and 2.2), $a = \lim l$, but the sequence $h = \{ A(x_i);\ i = 1, 2, 3, \ldots\}$ does not r-converge to $A(a)$. This violates the condition from Definition 7.2. So, Definition 7.2 implies Definition 7.3 because we have demonstrated that it is impossible that for an arbitrary operator $A: L \to M$, Definition 7.2 is true for the operator A at the point a and at the same time, Definition 7.3 is not true for A at the point a.

b) Definition 7.3 Definition 7.2. Let us assume that A is not r-continuous at the point a operator according to Definition 7.2. It means that there is a sequence $l = \{x_i; i = 1, 2, 3, \ldots\}$ such that $a = \lim l$, but the sequence $h = \{ A(x_i); i = 1, 2, 3, \ldots\}$ does not r-converge to $A(a)$. By Definition 7.1, this implies that for some real number $k > 0$, there are infinitely many elements from the sequence h outside the neighborhood $Oa = (a - r - k, a + r + k)$ of the point a.

It means that the condition from Definition 7.3 is not satisfied and A is not an r-continuous at the point a operator according to Definition 7.3. So, Definition 7.3 implies Definition 7.2 because we have demonstrated that it is impossible that for an arbitrary operator A, Definition 7.3 is true for A at the point a and at the same time, Definition 7.2 is not true for A at the point a.

Proposition is proved.

Proposition 7.5 shows that it is possible to take only the conditions either from Definition 7.2 or from Definition 7.3 to define r-continuity and to consider other conditions as derived properties of r-continuous operators or as criteria of r-continuity.

Remark 7.2. It is possible to define fuzzy continuity utilizing three different approaches: fuzzy limits, neighborhoods and by the (ε, δ)-definition. For operators in hyperseminormed vector spaces, all these definitions result in the same concept of fuzzy continuity although here we prove this only for the first and third approaches. However, in a more general situation of topological spaces, these three definitions bring us to different structures. Neighborhood fuzzy continuity is directly extended to fuzzy continuous mappings of scalable topological spaces in the most general case (Burgin, 2005, 2006). Fuzzy limits define fuzzy continuity only for sequential scalable topology (Burgin, 2004, 2006), while the (ε, δ)-definition can be extended only to metric spaces (Burgin, 2008) but not to general topological spaces.

Remark 7.3. It is possible to define fuzzy continuity at a point a even when an operator A is not defined at this point.

Remark 7.4. It is possible to argue that fuzzy continuity is interesting (and meaningful) only when the parameter of discontinuity r is small. However, understanding what quantities are small and what are big is relative. For instance, for a human being, 1000 years is a very big period of time. At the same time, like any good scientific measurement, every dated boundary in the geological time scale has an uncertainty associated with it. This uncertainty is expressed as "± X millions of years" [cf. (Harland et al.,

1990)]. It means that in the case of the geological time scale, the "small" parameter r is measured in millions of years.

One more example of the situation when the parameter of discontinuity r can be very big is given by Henri Poincaré. In his book (Poincaré, 1913), he describes the result of Perrin, who found that there are 683×10^{21} atoms in one gram of hydrogen. Another experimental method gives 650×10^{21} atoms in one gram of hydrogen. Poincaré writes that it is an excellent correlation because the difference is not larger than several thousand of billions of billions. This number is the measure of discontinuity r of physical continuity analyzed by Poincaré (1902, 1908). We can see that by an average estimation this number is extremely big, while in the scale of numbers of atoms, this number is sufficiently small to define indiscernability, which determines physical continuity.

If $A: L \to M$ is operator (mapping) of hyperseminormed vector spaces L and M, it is possible to build a *shift* (also called *translation*) of the operator A taking an element u from M and defining the operator $A + u$ by the following formula:

$$(A + u)(x) = A(x) + u$$

Lemma 7.4. An operator (mapping) $A: L \to M$ is r-continuous at a point a if and only if any its shift $A + u$ is r-continuous at the point a.

Proof. Necessity. Let us take an r-continuous at a point a operator (mapping) $A: L \to M$. It means that A is defined at a and for any sequence $l = \{a_i \in L; i = 1, 2, 3, \ldots\}$ that converges to a, the point $A(a)$ is an r-limit of the sequence $A(l) = \{ A(a_i) \in M; i = 1, 2, 3, \ldots\}$. By Lemma 7.3, the point $A(a) + u$ is an r-limit of the sequence $\{(A + u)(a_i) = A(a_i) + u \in M; i = 1, 2, 3, \ldots\}$. Thus, the operator $A + u$ is r-continuous at the point a.

Sufficiency. As we have proved that r-continuity of an operator implies r-continuity of any its shift, r-continuity of the operator $A + u$ implies r-continuity of the operator A because A is a shift of $A + u$.

Lemma is proved.

Corollary 7.5. A functional $A: L \to \boldsymbol{R}$ is r-continuous at a point a if and only if any its shift $A + u$ is r-continuous at the point a.

Corollary 7.6. If $B = A + u$ and $b = a + u$, then $r_{A, a} = r_{B, b}$.

The following result shows that local fuzzy continuity of operators is a natural extension of the conventional local continuity.

Proposition 7.6. An operator (mapping) $A: L \to M$ is continuous at a point $a \in L$ if and only if one of the following conditions is valid:

(a) A is 0-continuous at the point a.

(b) $r_{A,a} = 0$.

Proof. (a) If A is 0-continuous at the point a, then for any $\varepsilon > 0$, there is $\delta > 0$ such that the inequality $\| a - x \| < \delta$ implies the inequality $\| A(x) - A(a)\| < 0 + \varepsilon = \varepsilon$. This means that the operator A is continuous at the point a.

The inverse is also true as continuity gives us exact conditions of 0-continuity. Thus, the condition of 0-continuity coincides with the condition of continuity for operators in vector spaces.

(b) If A is continuous at the point a, then by (a), A is 0-continuous at the point a and thus, $r_{A,a} = 0$.

Now let us assume that $r_{A,a} = 0$ and some $\varepsilon > 0$ is given. Then by Definition 7.2, there is $\alpha > 0$ such that $\alpha < \varepsilon/2$ and A is α-continuous at the point a. By Definition 7.3, it means that there is $\delta > 0$ such that the inequality $\| a - x \| < \delta$ implies the inequality $\| A(x) - A(a) \| < \alpha+\varepsilon/2$, or in other words, for any x with $\| a - x \| < \delta$, we have $\| A(x) - A(a)\| < \alpha+\varepsilon/2 < \varepsilon$. Consequently, the operator A is continuous at the point a.

Proposition is proved.

Definitions imply the following result.

Proposition 7.7. If an operator A is r-continuous at a point a, then it is q-continuous at a for any $q > r$.

Indeed, if that the inequality $\| a - x \| < \delta$ implies the inequality $\| A(x) - A(a) \| < r + \varepsilon$, then the inequality $\| a - x \| < \delta$ implies the inequality $\| A(x) - A(a) \| < q + \varepsilon$ because $q > r$. Thus, by Definition 7.3, if A is an r-continuous at a point a operator, then it is also q-continuous at a.

Proposition 7.8. An operator (mapping) A: $L \to M$ is r-continuous at a point a for some $r > 0$ if and only if $r_{A,a}$ is defined and $r_{A,a} \leq r$.

Proof. Necessity. Let us consider an r-continuous at a point a from the space L operator (mapping) A: $L \to M$. Then $r_{A,a}$ is defined because any bounded set of real numbers that is not empty has its infimum.

Sufficiency. Suppose that $r_{A,a}$ is defined. Then the operator (mapping) A is r-continuous at a point a for some $r > 0$ because the set of numbers r that define is not empty.

Proposition is proved.

We remind that if A: $L \to M$ and B: $L \to M$ are operators, then their *sum* $A + B$: $L \to M$ is defined as $(A + B)(x) = A(x) + B(x)$ for any element x from

L and their *difference* $A - B$: $L \to M$ is defined as $(A - B)(x) = A(x) - B(x)$ for any element x from L.

Besides, if A: $L \to M$ is an operator and d is a real number, then the *scalar product* dA: $L \to M$ is defined as $(dA)(x) = dA(x)$ for any element x from L.

Proposition 7.9. a) If A: $L \to M$ is an r-continuous at a point a from L operator and B: $L \to M$ is a q-continuous at the same point operator, then their sum $A + B$: $L \to M$ and difference $A - B$: $L \to M$ are $(r + q)$-continuous at the point a operators.

b) If A: $L \to M$ is an r-continuous at a point a operator from L and d is a real number, then the scalar product dA is a $(|d| \cdot r)$-continuous at the point a operator.

Proof. a) Let us consider an r-continuous at a point a from the space L operator (mapping) A: $L \to M$ and a q-continuous at the point a operator (mapping) B: $L \to M$. Then taking a number $\varepsilon > 0$, by Definition 7.3, we have a number $\rho > 0$ such that for any element x from L, the inequality $\| x - a \| < \rho$ implies the inequality $\|A(x) - A(a)\| < r + \varepsilon/2$ and a number $\beta > 0$ such that for any element x from L, the inequality $\| x - a \| < \beta$ implies the inequality $\|B(x) - B(a)\| < q + \varepsilon/2$. Taking $\delta = \min \{\beta, \rho\}$, we see that for any element x from L, the inequality $\| x - a \| < \delta$ implies inequalities $\|A(x) - A(a)\| < r + \varepsilon/2$ and a $\|B(x) - B(a)\| < q + \varepsilon/2$. Consequently, we have

$$\|(A + B)(x) - (A + B)(a)\| = \|A(x) - A(a) + B(x) - B(a)\| \leq$$

$$\|A(x) - A(a)\| + \| B(x) - B(a)\| < r + \varepsilon/2 + q + \varepsilon/2 = r + q + \varepsilon$$

It means that $A + B$: $L \to M$ is a $(r + q)$-continuous at the point a operator.

In the case, of the difference $A - B$, we have

$$\|(A - B)(x) - (A - B)(a)\| = \|A(x) - A(a) - (B(x) - B(a))\| = \|A(x) - A(a) + B(a) - B(x)\| \leq$$

$$\|A(x) - A(a)\| + \| B(a) - B(x)\| < r + \varepsilon/2 + q + \varepsilon/2 = r + q + \varepsilon$$

It means that $A - B$: $L \to M$ is a $(r + q)$-continuous at the point a operator.

b) Let us consider an r-continuous at a point a from the space L operator (mapping) A: $L \to M$. Then taking a number $\varepsilon > 0$, by Definition 7.3, we have a number $\delta > 0$ such that for any element x from L, the inequality $\| x - a \| < \delta$ implies the inequality $\|A(x) - A(a)\| < r + \varepsilon/|d|$. Consequently, we have

$$||(dA)(x) - (dA)(a)|| = ||dA(x) - dA(a)|| \leq$$

$$||d(A(x) - A(a))|| = |d|\cdot||(A(x) - A(a))|| < |d|\cdot(r + \varepsilon/|d|) = |d|\cdot r + \varepsilon$$

It means that dA is a $(|d|\cdot r)$-continuous at the point a operator.

Proposition is proved.

Remark 7.5. Even if the conditions of Proposition 7.9 are satisfied, the operator $A - B$ is not necessarily $(r - q)$-continuousat a. However, in some cases, it is so. Besides, operators $A + B$ and $A - B$ can be continuous at some point where neither of them is continuous.

Example 7.5. Let us consider the following one-dimensional operators:

$$A(x) = \begin{cases} 0 \text{ if } x \text{ is a rational number;} \\ 1 \text{ if } x \text{ is an irrational number} \end{cases}$$

and

$$B(x) = \begin{cases} 0 \text{ if } x \text{ is a rational number;} \\ -1 \text{ if } x \text{ is an irrational number.} \end{cases}$$

Both operators A and B are 1-continuous but not continuous at the point 0. At the same time, the operator $A + B$ is continuous at 0 and the operator $A - A$ is only $(1 + 1)$-continuous, that is, 2-continuous at 0.

Theorem 7.3 imply the following result because any (0, 0)-continuous function (operator) is continuous.

Corollary 7.7.(any course of the calculus, cf., e.g., (Ribenboim, 1964)). If the functions $f(x)$ and $g(x)$ are continuous at a point a, then:

a) the function $(f + g)(x)$ is continuous at the point a;

b) the function $(f - g)(x)$ is continuous at the point a;

c) the function kA is continuous at the point a.

Corollary 7.8. a) If A: $L \to \boldsymbol{R}$ is an r-continuous at a point a from L functional and B: $L \to \boldsymbol{R}$ is a q-continuous at the same point functional, then their sum $A + B$: $L \to \boldsymbol{R}$ and difference $A - B$: $L \to M$ are $(r + q)$-continuous at the point a functionals.

b) If A: $L \to \boldsymbol{R}$ is an r-continuous at a point a from L functional and d is a real number, then the scalar product dA is a $|d|\cdot r$-continuous at the point a functional.

Corollary 7.9. If A: $L \to M$ is an r-continuous at a point a operator and B: $L \to M$ is a q-continuous at a operator, then $r_{A+B,\, a} \le r_{A,\, a} + r_{B,\, a}$.

It is possible that the inequality in Corollary 7.9 is strict.

Example 7.6. Let us consider one-dimensional operators A and B defined by the following formulas

$$A(x) = \begin{cases} 0 & \text{if } x \le 0 \\ 1 & \text{if } x > 0 \end{cases}$$

and

$$B(x) = \begin{cases} 0 & \text{if } x \le 0 \\ -1 & \text{if } x > 0 \end{cases}$$

Then we have

$$r_{A+B,0} = 0 < r_{A,0} = r_{B,0} = 1 < r_{A,0} + r_{B,0} = 2$$

Corollary 7.10. If A: $L \to M$ is an r-continuous at a point a operator and d is a real number, then $r_{dA,\, a} \le |d| \cdot r_{A,\, a}$.

Now let us consider the composition of operators.

Proposition 7.10. If A: $L \to M$ is a 0-continuous, i.e., continuous, at a point a from L operator and B: $M \to N$ is a q-continuous at the point $A(a)$ from M operator, then their composition BA: $L \to N$ is a q-continuous at the point a operator.

Proof. Let us consider an r-continuous at a point a from the space L operator (mapping) A: $L \to M$ and a q-continuous at a point $A(a)$ from the space M operator (mapping) B: $M \to N$. Then taking a number $\varepsilon > 0$, by Definition 7.3, there is a number $\delta > 0$ such that for any element x from L, the inequality $\|A(x) - A(a)\| < \delta$ implies the inequality $\|BA(x) - BA(a)\| < q + \varepsilon$.

In addition, by Definition 7.3, there is a number $\tau > 0$ such that for any element x from L, the inequality $\| a - x \| < \tau$ implies the inequality $\| A(x) - A(a)\| < \delta$. Thus, for any element x from L, the inequality $\| a - x \| < \tau$ implies the inequality $\|BA(x) - BA(a)\| < q + \varepsilon$. Thus, BA is a q-continuous at the point a operator.

Proposition is proved.

Taking $q = 0$, we obtain the following result.

Corollary 7.11 (Any course of functional analysis). The composition BA: $L \to N$ of continuous at a point a operators A: $L \to M$ and B: $M \to N$ is continuous at the same point.

Remark 7.6. In a general case, the composition BA: $L \to N$ of an r-continuous at a point a operator A: $L \to M$ and a q-continuous at a point $A(a)$ operator B: $M \to N$ is not always t-continuous at the point a for some number t as the following example demonstrates. It also means that the composition of fuzzy continuous at a point operators is not always fuzzy continuous at the initial point.

Example 7.7. Let us consider the one-dimensional operators A: $\boldsymbol{R} \to \boldsymbol{R}$ and B: $\boldsymbol{R} \to \boldsymbol{R}$ defined by the following formulas

$$A(x) = \begin{cases} 0 & \text{if } x \leq 0 \\ x + 1 & \text{if } 0 < x \end{cases}$$

and

$$B(x) = \begin{cases} 0 & \text{if } x < 0 \\ [1/(1 - x)] - 1 & \text{if } 0 \leq x < 1 \\ 1 & \text{if } x = 1 \\ 1/(x - 1) & \text{if } x > 1 \end{cases}$$

The operator A is 1-continuous at the point 0 and the operator B is even 0-continuous at the point $A(0) = 0$. However, for any number r, the composition BA: $\boldsymbol{R} \to \boldsymbol{R}$ is not an r-continuous at a point 0 operator.

This example also shows that the condition that the operator A is even 0-continuous at the point a is essential for Proposition 7.10.

Remark 7.8. Topological spaces with continuous mappings as morphisms form a category (Herrlich and Strecker, 1973; Lowen et al., 2009). In particular, taking the space $\boldsymbol{R}$ with continuous mappings as its morphisms, we obtain a category. However, Example 7.7 shows that we cannot build a category with fuzzy continuous mappings as morphisms.

Let us consider operators that take values in hyperseminormed linear algebra. Then there is the standard way to define the product of such operators.

Let A: $L \to M$ and B: $L \to M$ be two operators, L is a hyperseminormed vector space, M is a hyperseminormed linear algebra (cf. Chapter 3) and $\|.\|$ denotes the hyperseminorms in L and M.

Definition 7.4. The product $A \cdot B$ of operators A and B is defined by the formula

$$(A \cdot B)(x) = A(x) \cdot B(x)$$

Proposition 7.11. If A: $L \to M$ is an r-continuous at a point a from L operator, B: $L \to M$ is a q-continuous at the point a operator and the hyperseminorms of both operators at a are finite, i.e., $\|A(a)\| < w$ and $\|B(a)\| < s$ for some positive real numbers w and s, then their product $A \cdot B$: $L \to M$ is a u-continuous at the point a operator where $u = qr + r + q + w + s$.

Proof. Let us consider an r-continuous at a point a from the space L operator (mapping) A: $L \to M$ and a q-continuous at a operator (mapping) B: $L \to M$. Then taking a number $\eta > 0$, by Definition 7.3, there is a number $\delta > 0$ such that for any element x from L, the inequality $\| x - a \| < \delta$ implies the inequality $\|B(x) - B(a)\| < q + \eta$ and there is a number $\tau > 0$ such that for any element x from L, the inequality $\| a - x \| < \tau$ implies the inequality $\| A(x) - A(a)\| < r + \eta$. Thus, for any element x from L, the inequality $\| a - x \| < \gamma = \min \{\tau, \delta\}$ implies the inequality $\| A(x) - A(a)\| \cdot \| B(x) - B(a)\| < (q + \eta) \cdot (r + \eta) = qr + \eta r + \eta q + \eta^2$.

At the same time, we have

$$(A(x) - A(a)) \cdot (B(x) - B(a)) = A(x) \cdot B(x) - A(a) \cdot B(x) - A(x) \cdot B(a) + A(a) \cdot B(a) =$$

$$(A(x) \cdot B(x) - A(a) \cdot B(a)) + (A(a) \cdot B(a) - A(a) \cdot B(x)) + (A(a) \cdot B(a) - A(x) \cdot B(a)) =$$

$$(A(x) \cdot B(x) - A(a) \cdot B(a)) + A(a) \cdot (B(a) - B(x)) + B(a) \cdot (A(a) - A(x))$$

and therefore,

$$(A(x) \cdot B(x) - A(a) \cdot B(a)) =$$

$$(A(x) - A(a)) \cdot (B(x) - B(a)) - A(a) \cdot (B(a) - B(x)) - B(a) \cdot (A(a) - A(x)) =$$

$$(A(x) - A(a)) \cdot (B(x) - B(a)) + A(a) \cdot (B(x) - B(a)) + B(a) \cdot (A(x) - A(a))$$

Consequently, denoting $\| A(a) \|$ by w and $\| B(a) \|$ by s and assuming that $\eta < 1$ we have

$$\| ((A \cdot B)(x) - (A \cdot B)(a)) \| = \| (A(x) \cdot B(x) - A(a) \cdot B(a)) \| =$$

$$\| (A(x) - A(a)) \cdot (B(x) - B(a)) + A(a) \cdot (B(x) - B(a)) + B(a) \cdot (A(x) - A(a)) \| \leq$$

$$\| (A(x) - A(a)) \cdot (B(x) - B(a)) \| + \|A(a) \cdot (B(x) - B(a)) \| + \|B(a) \cdot (A(x) - A(a)) \| \leq$$

$$\| (A(x) - A(a)) \| \cdot \| (B(x) - B(a)) \| + \|A(a) \| \cdot \|(B(x) - B(a) \| + \|B(a) \| \cdot \|A(x) - A(a) \| <$$

$$qr + \eta r + \eta q + \eta^2 + w(q + \eta) + s(r + \eta) = qr + \eta(r + q + w + s) + \eta^2 <$$

$$qr + r + q + w + s + \eta^2 = qr + r + q + w + s + \varepsilon$$

It means that he inequality $\| a - x \| < \gamma$ implies the inequality

$$\| ((A \cdot B)(x) - (A \cdot B)(a)) \| < u + \varepsilon$$

where $u = qr + r + q + w + s$.

Thus, the product $A \cdot B$: $L \rightarrow M$ is a u-continuous at the point a operator.

Proposition is proved.

Note that the hyperseminorm or the hypernorm of an element in a vector space can be finite even when it is not a real number. For instance, it is possible to define a hyperseminorm in the set of real functions such that the hyperseminorm of the function sin x will be equal to the proper hypernumber defined by the sequence (0, 1, 0, 1, 0, 1, …). Indeed, let us define the hyperseminorm of a real function $f(x)$ as

$$\|f(x)\| = \mathrm{Hn}(|f(\pi n)|)_{n \in \omega}$$

Then $\|\sin x\| = (0, 1, 0, 1, 0, 1, \ldots)$.

Corollary 7.12. The product $A \cdot B$: $L \rightarrow M$ of r-continuous at a point a operators A: $L \rightarrow M$ and B: $L \rightarrow M$ with finite hyperseminorms at a, i.e., $\|A(a) \| < w$ and $\|B(a) \| < s$ for some positive real numbers w and s, is u-continuous at the same point where $u = r^2 + 2r + w + s$.

Remark 7.7. Condition that the hyperseminorms of both operators at a are finite is essential as the following example demonstrates.

Example 7.8. Let us take the one-dimensional vector space $L = \boldsymbol{R}$ and the vector space $M = V\{\boldsymbol{R}, \alpha, \alpha^2, \alpha^3, \ldots\}$ generated by elements 1 and $\alpha^n = \mathrm{Hn}(i^n)_{i \in \omega}$ ($n = 1, 2, 3, \ldots$) over the field $\boldsymbol{R}$. Elements from M are polynomials from α and are multiplied as polynomials. Besides, if $\beta = \Sigma_{i=0}^{k} a_i\alpha^i$ is an element from *M*, then its norm is defined as

$$\| \beta \| = \Sigma_{i=0}^{k} | a_i | \alpha^i$$

This makes *M* a hypernormed algebra and it is possible to define the product of operators with the range *M*.

We define operators A: $L \to M$ and B: $L \to M$ by the following rules:

$$A(x) = \begin{cases} 0 & \text{if } x \leq 0 \\ 1 & \text{if } x > 0 \\ B(x) = \alpha & \text{for all } x \text{ from } L \end{cases}$$

The operator *A* is 1-continuous at the point 0 and the operator *B* is even 0-continuous at the point $A(0) = 0$. However, for any positive real number *r*, the product $A \cdot B$: $L \to M$ is not an *r*-continuous at the point 0 operator because the hypernumberα is larger than any natural number *r*, while

$$(A \cdot B)(0) = A(0) \cdot B(0) = 0 \cdot \alpha = 0$$

and for any $k > 0$,

$$(A \cdot B)(1 + k) = A(1 + k) \cdot B(1 + k) = 1 \cdot \alpha = \alpha$$

If both operators *A* and *B* are continuous and have finite norms at a point *a*, then $p = q = 0$ and by Lemma 7.4, it is possible to shift both operators so that $\|A(a)\| = \|B(a)\| = 0$. Note that the second condition is always true when *L* and *M* are seminormed (normed) vector spaces. This gives us the following result.

Corollary 7.13 (Any course of functional analysis). The product $A \cdot B$: $L \to M$ of continuous at a point *a* operators A: $L \to M$ and B: $L \to M$ is continuous at the same point.

For functions, we also have the classical result.

Corollary 7.14 (Any course of calculus). The product $A \cdot B$: $\boldsymbol{R} \to \boldsymbol{R}$ of continuous at a point *a* functions A: $\boldsymbol{R} \to \boldsymbol{R}$ and B: $\boldsymbol{R} \to \boldsymbol{R}$ is continuous at the same point.

In contrast to fuzzy continuity, continuity is a local property, i.e., an operator is continuous in a space if it is continuous at each point of this space. Thus, Corollary 7.4 implies the following result.

Corollary 7.15 (Any course of functional analysis). The product $A \cdot B$: $L \to M$ of continuous operators A: $L \to M$ and B: $L \to M$ is also continuous.

For functions, we have a similar result.

Corollary 7.16 (Any course of calculus). The product $A \cdot B$: $\boldsymbol{R} \to \boldsymbol{R}$ of continuous functions A: $\boldsymbol{R} \to \boldsymbol{R}$ and B: $\boldsymbol{R} \to \boldsymbol{R}$ is also continuous.

From fuzzy continuity at a point, we naturally come to the concept of *fuzzy continuity in a set.*

Definition 7.5. (a) An operator A: $L \to M$ is called *r-continuous in* a set $X \subseteq L$ if A is r-continuous at each point a from X.

(b) The number $r_{A, X} = \sup\{ r_{A, a;}\ a \in X\}$ (if it exists) is called the *measure of discontinuity*(or *continuity defect* or *parameter of discontinuity*) of A in the set X.

(c) The number $r_{A, \mathrm{u}X} = \inf\{ r;\ A$ is an r-continuous in X operator $\}$(if it exists) is called the *uniform measure of discontinuity* (or *uniform continuity defect* or *uniform parameter of discontinuity*) of A in the set X.

For an operator A, r-continuity means that the gap of A at all points from X cannot be larger than r.

The measure of discontinuity and the uniform measure of discontinuity are not defined for all operators.

Proposition 7.7 implies the following result.

Lemma 7.5. If an operator A: $L \to M$ is r-continuous in X, then it is q-continuous in X for any $q > r$.

Indeed, if an operator A: $L \to M$ is r-continuous in X, then A is r-continuous at each point of X. By Proposition 7.7, A is q-continuous at each point of X if $q > r$. Thus, by definition, the operator A is q-continuous in X.

Lemma 7.6. If an operator A: $L \to M$ is r-continuous in X and $Y \subseteq X$, then it is r-continuous in Y.

Indeed, ifan operator A is r-continuous at each point of X, then A is r-continuous at each point of Y.

Lemma 7.6 directly implies the following result.

Lemma 7.7. If $Y \subseteq X$, then $r_{A, Y} \leq r_{A, X}$.

Measures of discontinuity characterize operators in hypernormed spaces.

Lemma 7.8. An operator (mapping) A: $L \to M$ is r-continuous in a set X for some $r > 0$ if and only if the $r_{A, X}$ is defined and $r_{A, X} \leq r$.

Proof. Necessity. Let us consider an r-continuous in a set X from the space L operator (mapping) A: $L \to M$. Then by definition, $r_{A,\, a} \leq r$ for all points a from X. Consequently, $r_{A,\, X}$ is defined because any bounded set of real numbers has its supremum.

Sufficiency. Suppose that $r_{A,X}$ is defined. Then by Lemma 7.5, the operator (mapping) A is $r_{A,\, a}$-continuous at all points a from X.

Lemma is proved.

Lemma 7.9. For an operator (mapping) A: $L \to M$, the measure $r_{A,\, uX}$ is defined if and only if the measure $r_{A,X}$ is defined and $r_{A,X} = r_{A,\, uX}$.

Indeed, by Lemma 7.5 if $r_{A,X}$ is defined, then by Lemma 7.5, the operator (mapping) A is $r_{A,\, a}$-continuous at all points a from X. Consequently, the measure $r_{A,\, uX}$ is defined and $r_{A,X} = r_{A,\, uX}$.

Proposition 7.9 implies the following result.

Proposition 7.12. a) If A: $L \to M$ is an r-continuous in X operator and B: $L \to M$ is a q-continuous in X operator, then their sum $A + B$: $L \to M$ and difference $A - B$: $L \to M$ are $(r + q)$-continuous in X operators.

b) If A: $L \to M$ is an r-continuous in X operator and d is a real number, then the scalar product dA is a $|d| \cdot r$-continuous in X operator.

Proof is left as an exercise.

Corollary 7.17. If A: $L \to M$ is an r-continuous in X operator and B: $L \to M$ is a q-continuous operator in X, then $r_{A+B,\, X} \leq r_{A,\, X} + r_{B,\, X}$.

It is possible that the inequality is strict in Corollary 7.17.

Example 7.9. Let us take operators A and B from Example 7.6 and assume $X = [0, 1]$. Then we have

$$r_{A+B,\, X} = 0 < r_{A,X} = r_{B,X} = 1 < r_{A,X} + r_{B,X} = 2$$

Corollary 7.18. If A: $L \to M$ is an r-continuous in X operator and d is a real number, then $r_{dA,\, X} \leq |d| \cdot r_{A,\, X}$.

Corollary 7.19. a) If A: $L \to \mathbf{R}$ is an r-continuous in $X \subseteq L$ functional and B: $L \to \mathbf{R}$ is a q-continuous in X functional, then their sum $A + B$: $L \to \mathbf{R}$ and difference $A - B$: $L \to \mathbf{R}$ are $(r + q)$-continuous in X functionals.

b) If A: $L \to \mathbf{R}$ is an r-continuous in X functional and d is a real number, then the scalar product dA is a $|d| \cdot r$-continuous in X functional.

Corollary 7.20. a) If f: $\mathbf{R} \to \mathbf{R}$ is an r-continuous in $X \subseteq L$ function and g: $\mathbf{R} \to \mathbf{R}$ is a q-continuous in X function, then their sum $f + g$: $\mathbf{R} \to \mathbf{R}$ and difference $A - B$: $\mathbf{R} \to \mathbf{R}$ are $(r + q)$-continuous in X functions.

b) If f: $\boldsymbol{R} \to \boldsymbol{R}$ is an r-continuous in $X \subseteq L$ function and d is a real number, then the scalar product df is a $|d| \cdot r$-continuous in X function.

Lemma 7.10. An operator (mapping) A: $L \to M$ is r-continuous in a set X if and only if any its shift $A + u$ is r-continuous in the set X.

Proof is left as an exercise.

Corollary 7.21. A functional A: $L \to \boldsymbol{R}$ is r-continuous in a set X if and only if any its shift $A + u$ is r-continuous in the set X.

Proposition 7.13. If $X \subseteq Z$, then $r_{A, X} \leq r_{A, Z}$.

Proof is left as an exercise.

Corollary 7.22. If $a \in X$, then $r_{A, a} \leq r_{A, X}$.

Proposition 7.10 implies the following result.

Proposition 7.14. If A: $L \to M$ is a 0-continuous, i.e., continuous, in $X \subseteq L$ operator and B: $M \to N$ is a q-continuous in $A(X) \subseteq M$ operator, then their composition BA: $L \to N$ is a q-continuous in X operator.

Proof is left as an exercise.

Corollary 7.23. If A: $L \to \boldsymbol{R}$ is a 0-continuous in $X \subseteq L$ functional and B: $\boldsymbol{R} \to \boldsymbol{R}$ is a q-continuous in $A(X)$ function, then their composition BA: $L \to \boldsymbol{R}$ is a q-continuous in X functional.

Corollary 7.24. If f: $\boldsymbol{R} \to \boldsymbol{R}$ is a 0-continuous in $X \subseteq L$ function and g: $\boldsymbol{R} \to \boldsymbol{R}$ is a q-continuous in $A(X)$ function, then their composition gf: $\boldsymbol{R} \to \boldsymbol{R}$ is a q-continuous in X function.

However, if A: $L \to M$ is a q-continuous and B: $M \to N$ is a 0-continuous in $A(X) \subseteq M$ operator, then their composition BA: $L \to N$ is not necessarily a q-continuous in X operator as the following example demonstrates.

Example 7.10. Let us consider the one-dimensional operators A: $\boldsymbol{R} \to \boldsymbol{R}$ and B: $\boldsymbol{R} \to \boldsymbol{R}$ defined by the following formulas

$$A(x) = \rfloor x \lfloor, \text{ i.e., } A(x) = n \text{ when } x \in [n, n+1), n \in \boldsymbol{Z}$$

$$B(x) = e^x$$

By definition, the operator A is 1-continuous and the operator B is continuous. However, the operator BA is not q-continuous in $\boldsymbol{R}$ for any q because the quantity $e^{n+1} - e^n = e(e-1)$ can become larger than any real number.

Proposition 7.11 implies the following results.

Proposition 7.15. If A: $L \to M$ is an r-continuous in X operator, B: $L \to M$ is a q-continuous in X operator, and the hyperseminorms of both operators at a are finite for all points a from X, i.e., $\|A(a)\| < w$ and $\|B(a)\| < s$ for some positive real numbers w and s, then their product $A \cdot B$: $L \rightarrow M$ is a u-continuous in X operator where $u = qr + r + q + w + s$.

Proof is left as an exercise.

Corollary 7.25 (Any course of calculus). The product $A \cdot B$: $L \rightarrow M$ of continuous in X operators A: $L \rightarrow M$ and B: $L \rightarrow M$ is continuous in X.

Corollary 7.26. The product $A \cdot B$: $L \rightarrow M$ of r-continuous in X operators A: $L \rightarrow M$ and B: $L \rightarrow M$ is u-continuous in X where $\|A(a)\| < w$, $\|B(a)\| < s$ for all points a from X and $u = r^2 + 2r + w + s$.

Remark 7.9. Example 7.8 shows that finiteness of the hyperseminorms of operators A and B is an essential condition in Proposition 7.15 and Corollary 7.26.

From fuzzy continuity in a set, we naturally come to fuzzy continuity inside a set.

Definition 7.6. (a) An operator A: $L \rightarrow M$ is called *r-continuous inside* a set $X \subseteq L$ if A is defined at all points from X and for any sequence $l = \{ a_i \in X; i = 1, 2, 3, \dots\}$ that converges to a point a from X, the point $A(a)$ is an r-limit of the sequence $\{ A(a_i) \in M; i = 1, 2, 3, \dots\}$.

(b) The number $r_{A, iX} = \inf\{ r; A$ is an r-continuous inside X operator $\}$ is called the *measure of discontinuity* (or *continuity defect* or *parameter of discontinuity*) inside the set X of the operator A.

It means that gaps of the restriction of A on the set X cannot be larger than r.

Remark 7.10. In a general case, r-continuity in a set and r-continuity inside the same set do not coincide as the following example demonstrates.

Example 7.10. Let us consider the one-dimensional operator A: $\boldsymbol{R} \rightarrow \boldsymbol{R}$ defined by the following formula

$$A(x) = \begin{cases} 0 & \text{if } x < 0 \\ 1 & \text{if } 0 \leq x \leq 1 \\ 2 & \text{if } x > 1 \end{cases}$$

The operator A is 0-continuous (continuous) inside the segment [0, 1] but only 1-continuous in [0, 1] and not 0-continuous in [0, 1].

Moreover, it is possible that an r-continuous (continuous) inside some X operator is not q-continuous in X for any number q.

Example 7.11. Let us consider the one-dimensional operator A: $\boldsymbol{R} \rightarrow \boldsymbol{R}$ defined by the following formula

$$A(x) = \begin{cases} 1/x & \text{if } x < 0 \\ 1 & \text{if } 0 \leq x \leq 1 \\ 1/(x-1) & \text{if } x > 1 \end{cases}$$

The operator A is 0-continuous (continuous) inside the segment [0, 1] but only 1-continuous in [0, 1] and not 0-continuous in [0, 1].

At the same time, "r-continuity in" is stronger than "r-continuity inside."

Lemma 7.11. If an operator A: $L \to M$ is r-continuous in a set X, then A is r-continuous inside X.

Proof is left as an exercise.

However, in some cases, the concepts of "r-continuity in" and "r-continuity inside" coincide.

Proposition 7.16. If X is an open subset of the space L, then an operator A: $L \to M$ is r-continuous in a set X if and only if A is r-continuous inside X.

Proof. Necessity follows from Lemma 7.11.

Sufficiency. Let us consider an r-continuous inside X operator A: $L \to M$ and a point a from X. Taking an arbitrary sequence $l = \{a_i \in L; i = 1, 2, 3, \ldots\}$ that converges to a, we see that starting from some number n, all elements from this sequence belong to because X is an open set. Thus, the sequence $h = \{a_i \in L; i = n, n + 1, n + 2, n + 3, \ldots\}$ belongs to X and also converges to a as a subsequence of l.

Because the operator A is r-continuous inside, the point $A(a)$ is an r-limit of the sequence $A(h) = \{ A(a_i) \in M; i = n, n + 1, n + 2, n + 3, \ldots\}$. Thus, the point $A(a)$ is also an r-limit of the sequence $A(l) = \{ A(a_i) \in M; i = 1, 2,3, \ldots\}$ because the limit of a converging sequence coincides with the limit of any of its subsequence. As a is an arbitrary point from X and l is an arbitrary sequence which converges to a, A is r-continuous in X.

Proposition is proved.

We remind that the interior *intX* of a set X is the largest open subset of X (Alexandroff, 1961).

Lemma 7.6 and Proposition 7.16 imply the following result.

Corollary 7.27. If an operator A: $L \to M$ is r-continuous in a set X, then A is r-continuous inside *intX*.

Lemma 7.12. An operator (mapping) A: $L \to M$ is r-continuous inside a set X if and only if any its shift $A + u$ is r-continuous inside the set X.

Proof is left as an exercise.

Corollary 7.28. A functional A: $L \to \boldsymbol{R}$ is r-continuous inside a set X if and only if any its shift $A + u$ is r-continuous inside the set X.

Proposition 7.7 implies the following result.

Lemma 7.13. If an operator A: $L \to M$ is r-continuous inside X, then it is q-continuous inside X for any $q > r$.

Proof is similar to the proof of Proposition 7.7.

Lemma 7.14. If an operator $A: L \to M$ is r-continuous inside X and $Y \subseteq X$, then it is r-continuous inside Y.

Proof is similar to the proof of Lemma 7.6.

Lemma 7.15. If $Y \subseteq X$, then $r_{A,\, iY} \leq r_{A,\, iX}$.

Proof is left as an exercise.

Lemma 7.16. An operator (mapping) $A: L \to M$ is r-continuous inside a set X for some $r > 0$ if and only if $r_{A,\, iX}$ is defined and $r_{A,\, iX} \leq r$.

Proof is similar to the proof of Lemma 7.6.

Proposition 7.9 implies the following result.

Proposition 7.17. a) If $A: L \to M$ is an r-continuous inside X operator and $B: L \to M$ is a q-continuous inside X operator, then their sum $A + B: L \to M$ and difference $A - B: L \to M$ are $(r + q)$-continuous inside X operators.

b) If $A: L \to M$ is an r-continuous inside X operator and d is a real number, then the scalar product dA is a $|d| \cdot r$-continuous inside X operator.

Proof is similar to the proof of Proposition 7.9.

Corollary 7.29. a) If $A: L \to \boldsymbol{R}$ is an r-continuous functional inside X and $B: L \to \boldsymbol{R}$ is a q-continuous functional inside X, then their sum $A + B: L \to \boldsymbol{R}$ and difference $A - B: L \to \boldsymbol{R}$ are $(r + q)$-continuous inside X functionals.

b) If $A: L \to \boldsymbol{R}$ is an r-continuous functional inside X and d is a real number, then the scalar product dA is a $|d| \cdot r$-continuous inside X functional.

Corollary 7.30. a) If $f: \boldsymbol{R} \to \boldsymbol{R}$ is an r-continuous inside $X \subseteq L$ function and $g: \boldsymbol{R} \to \boldsymbol{R}$ is a q-continuous in X function, then their sum $f + g: \boldsymbol{R} \to \boldsymbol{R}$ and difference $A - B: \boldsymbol{R} \to \boldsymbol{R}$ are $(r + q)$-continuous inside X functions.

b) If $f: \boldsymbol{R} \to \boldsymbol{R}$ is an r-continuous in $X \subseteq L$ function and d is a real number, then the scalar product df is a $|d| \cdot r$-continuous inside X function.

Corollary 7.31. If $A: L \to M$ is an r-continuous inside X operator and $B: L \to M$ is a q-continuous operator inside X, then $r_{A+B,\, iX} \leq r_{A,\, iX} + r_{B,\, iX}$.

It is possible that the inequality is strict in Corollary 7.31.

Example 7.12. Let us take operators A and B from Example 7.5 and define $X = [-1, 1]$. Then we have

$$r_{A+B,\, iX} = 0 < r_{A,\, iX} = r_{B,\, iX} = 1 < r_{A,\, iX} + r_{B,\, iX} = 2$$

Corollary 7.32. If $A: L \to M$ is an r-continuous inside X operator and d is a real number, then $r_{dA,\, iX} \leq |d| \cdot r_{A,\, iX}$.

Now let us study fuzzy continuity of compositions of operators.

Proposition 7.18. If A: $L \to M$ is a 0-continuous, i.e., continuous, inside X operator and B: $M \to N$ is a q-continuous inside the subset $A(X)$ of M operator, then their composition BA: $L \to N$ is a q-continuous inside X operator.

Proof is similar to the proof of Proposition 7.10.

Corollary 7.33. If A: $L \to \boldsymbol{R}$ is a 0-continuous, i.e., continuous, inside X functional and B: $\boldsymbol{R} \to \boldsymbol{R}$ is a q-continuous inside $A(X)$ function, then their composition BA: $L \to \boldsymbol{R}$ is a q-continuous inside X functional.

Corollary 7.34. If f: $\boldsymbol{R} \to \boldsymbol{R}$ is a 0-continuous inside $X \subseteq L$ function and g: $\boldsymbol{R} \to \boldsymbol{R}$ is a q-continuous in $A(X)$ function, then their composition gf: $\boldsymbol{R} \to \boldsymbol{R}$ is a q-continuous inside X function.

Proposition 7.11 implies the following results.

Proposition 7.19. If A: $L \to M$ is an r-continuous inside X operator, B: $L \to M$ is a q-continuous inside X operator and the hyperseminorms of both operators at a are finite for all points a from X, i.e., $\|A(a)\| < w$ and $\|B(a)\| < s$ for some positive real numbers w and s, then their product $A \cdot B$: $L \to M$ is a u-continuous inside X operator where $u = qr + r + q + w + s$.

Proof is left as an exercise.

Corollary 7.35 (Any course of calculus). The product $A \cdot B$: $L \to M$ of continuous inside X operators A: $L \to M$ and B: $L \to M$ is continuous inside X.

Corollary 7.36. The product $A \cdot B$: $L \to M$ of r-continuous inside X operators A: $L \to M$ and B: $L \to M$ with finite hyperseminorms at all points a from X is u-continuous inside X where $\|A(a)\| < w$, $\|B(a)\| < s$ at all points a from X and $u = r^2 + 2r + w + s$.

Remark 7.11. Example 7.7 shows that finiteness of the hyperseminorms of operators A and B is an essential condition for validity of Proposition 7.19 and Corollary 7.36.

Local and regional continuity of operators are related to their global continuity.

Definition 7.7.(a) An operator A: $L \to M$ is called *globally r-continuous* if A is r-continuous at each point a from L.

(b) The number $gr_A = \inf \{r;\ A$ is a globally r-continuous operator$\}$ is called the *global measure of discontinuity*(or the *global continuity defect* or the *global parameter of discontinuity*) of A.

(c) An operator (mapping) A: $L \to M$ is *globally fuzzy continuous* if A is globally r-continuous for some number r.

Proposition 7.6 implies the following result.

Proposition 7.20. An operator (mapping) A: $L \to M$ is continuous if and only if one of the following conditions is valid:

(a) A is globally 0-continuous.

(b) $gr_A = 0$.

Proof is left as an exercise.

This result shows that global fuzzy continuity of operators is a natural extension of the conventional global continuity.

Under the assumption that an operator $A: L \to M$ is globally r-continuous for some $r > 0$, Corollary 7.6 implies the following result.

Corollary 7.37. If $B = A + u$, then $gr_A = gr_B$.

Lemma 7.17. An operator (mapping) $A: L \to M$ is globally r-continuous if and only if any its shift $A + u$ is globally r-continuous.

Proof is similar to the proof of Lemma 7.4.

Corollary 7.38. A functional $A: L \to \boldsymbol{R}$ is globally r-continuous if and only if any its shift $A + u$ is globally r-continuous.

Corollary 7.39. An operator (mapping) $A: L \to M$ is globally fuzzy continuous if and only if any its shift $A + u$ is globally fuzzy continuous.

Corollary 7.40. A functional $A: L \to \boldsymbol{R}$ is globally fuzzy continuous if and only if any its shift $A + u$ is globally fuzzy continuous.

Lemma 7.14 implies the following result.

Lemma 7.18. For any subset X of the space L, $r_{A, X} \leq gr_A$.

Proposition 7.17 implies the following result.

Proposition 7.21. a) If $A: L \to M$ is a globally r-continuous operator and $B: L \to M$ is a globally q-continuous operator, then their sum $A + B: L \to M$ and difference $A - B: L \to M$ are globally $(r + q)$-continuous operators.

b) If $A: L \to M$ is a globally r-continuous operator and d is a real number, then the scalar product dA is a globally $|d| \cdot r$-continuous operator.

Proof is similar to the proof of Proposition 7.17.

Corollary 7.41. a) If $A: L \to \boldsymbol{R}$ is a globally r-continuous functional and $B: L \to \boldsymbol{R}$ is a globally q-continuous functional, then their sum $A + B: L \to \boldsymbol{R}$ and difference $A - B: L \to \boldsymbol{R}$ are globally $(r + q)$-continuous functionals.

b) If $A: L \to \boldsymbol{R}$ is a globally r-continuous functional and d is a real number, then the scalar product dA is a globally $|d| \cdot r$-continuous functional.

Corollary 7.42. If $A: L \to M$ is a globally r-continuous operator and $B: L \to M$ is a globally q-continuous operator, then $gr_{A+B} \leq gr_A + gr_B$ and $gr_{A-B} \leq gr_A + gr_B$.

Corollary 7.43. If $A: L \to M$ is a globally r-continuous operator and d is a real number, then $gr_{dA} \leq |d| \cdot gr_A$.

Corollary 7.44. If A: $L \to M$ and B: $L \to M$ are globally fuzzy continuous operators, then their sum $A + B$: $L \to \boldsymbol{R}$ and difference $A - B$: $L \to \boldsymbol{R}$ are globally fuzzy continuous operators.

Corollary 7.45. If A: $L \to M$ is a globally fuzzy continuous operator and d is a real number, then the scalar product dA is a globally fuzzy continuous operator.

Proposition 7.18 implies the following result.

Proposition 7.22. If A: $L \to M$ is a globally 0-continuous, i.e., continuous, operator and B: $M \to N$ is a globally q-continuous operator, then their composition BA: $L \to N$ is a globally q-continuous operator.

Proof is similar to the proof of Proposition 7.18.

Corollary 7.46. The composition BA: $L \to N$ of a continuous operator A: $L \to M$ and globally fuzzy continuous operator B: $M \to N$ is also globally fuzzy continuous.

Corollary 7.47 (Any course of functional analysis). The composition BA: $L \to N$ of continuous operators A: $L \to M$ and B: $M \to N$ is also continuous.

Corollary 7.48 (Any course of calculus). The composition $f{\cdot}g$ of continuous functions f and g is continuous.

Remark 7.15. One of the principal results of the classical analysis (and topology) states that the product of continuous real or complex functions is continuous [cf., e.g., (Ross, 1996; Burgin, 2008)]. For globally fuzzy continuous operators and functions, this result is not true in a general case.

Example 7.17. Let us consider the one-dimensional operator (function) $A(x) = n$ when $x \in [n, n + 1)$, $n \in \boldsymbol{Z}$ and the operator $A(x) = x$. The operator $A(x)$ is 1-continuous and the operator $A(x)$ is continuous and consequently, 1-continuous and 0-continuous on $\boldsymbol{R}$. However, their product $(B{\cdot}A)(x) = n^2$ when $x \in [n, n + 1)$, $n \in \boldsymbol{Z}$ is not globally fuzzy continuous on $\boldsymbol{R}$. It means that the result of multiplication of a fuzzy continuous operator and a continuous operator can be not globally fuzzy continuous.

However, globally fuzzy continuous have a weaker property. Namely, Proposition 7.19 implies the following results.

Proposition 7.23. If A: $L \to M$ is a globally r-continuous operator, B: $L \to M$ is a globally q-continuous bounded operator, and the hyperseminorms of both operators are finitely bounded, i.e., $\|A(a)\| < w$ and $\|B(a)\| < s$ for some positive real numbers w and s at all points from L, then their product $A{\cdot}B$: $L \to M$ is a globally u-continuous operator where $u = qr + r + q + w + s$.

Proof is similar to the proof of Proposition 7.19.

Corollary 7.49. The product $A \cdot B$: $L \to M$ of globally fuzzy continuous operators A: $L \to M$ and B: $L \to M$ with finitely bounded hyperseminorms is globally fuzzy continuous.

Corollary 7.50. The product $A \cdot B$: $L \to M$ of globally r-continuous operators A: $L \to M$ and B: $L \to M$ with finite hyperseminorms is globally u-continuous where $\|A(a)\| < w$, $\|B(a)\| < s$ for all points a in L and $u = r^2 + 2r + w + s$.

Let us assume that L and M are seminormed (normed) vector spaces.

Corollary 7.51 (Any course of functional analysis)**.** The product $A \cdot B$: $L \to M$ of continuous operators A: $L \to M$ and B: $L \to M$ is continuous.

Corollary 7.52 (Any course of calculus)**.** The product $f \cdot g$: $\boldsymbol{R} \to \boldsymbol{R}$ of continuous functions f: $\boldsymbol{R} \to \boldsymbol{R}$ and g: $\boldsymbol{R} \to \boldsymbol{R}$ is continuous.

Remark 7.10. Example 7.8 shows that finite boundedness of the hyperseminorms of operators A and B is an essential condition in Proposition 7.23 and Corollaries 7.49 and 7.50.

Proposition 7.7 implies the following result.

Proposition 7.24. If an operator A: $L \to M$ is globally r-continuous, then it is globally q-continuous for any $q > r$.

Proof is left as an exercise.

Example 7.13. The operator A from Example 7.4 is (1/4)-continuous in $\boldsymbol{R}$.

Exploring operators from one hyperseminormed (normed) space into another, we find that in contrast to continuity, for which local continuity coincides with global continuity, global fuzzy continuity of operators is different from their local fuzzy continuity. This observation brings us to the new concepts of neoclassical analysis.

Definition 7.8. a) An operator A: $L \to M$ is called *locally fuzzy continuous in X* if for each point a from X, the operator A is r-continuous at a for some $r \in \boldsymbol{R}^+$.

b) An operator A: $L \to M$ is called *locally fuzzy continuous inside X* if for each point a from X, there is a number $r \in \boldsymbol{R}^+$ such that the restriction of A on X is r-continuous at a.

c) An operator A: $L \to M$ is called *locally fuzzy continuous* if for each point a from L, A is r-continuous at a for some number $r \in \boldsymbol{R}^+$.

d) The number $\mathrm{l}r_A = \sup\{ r_{A,a};\ a \in L \}$ (if it exists) is called the *measure of local discontinuity* (or the *local continuity defect* or the *parameter of local discontinuity*) of A.

e) The number $\mathrm{l}r_{A,X} = \sup\{ r_{A,a};\ a \in X \}$ (if it exists) is called the *measure of local discontinuity* (or the *local continuity defect* or the *parameter of local discontinuity*) of A in X.

Naturally any continuous at a operator is fuzzy continuous at a and any continuous operator is locally fuzzy continuous. At the same time, an operator can be locally fuzzy continuous but not globally fuzzy continuous.

Example 7.14. Let us consider the one-dimensional operator $A(x) = x^n$ when $x \in [n, n + 1)$ for all $n \in \boldsymbol{Z}$ and the one-dimensional operator $B(x) = (\lfloor x \rfloor)^n$. These operators are locally fuzzy continuous, i.e., fuzzy continuous at each point of $\boldsymbol{R}$, but they are not globally fuzzy continuous (in $\boldsymbol{R}$).

The measure of local discontinuity is not defined for all operators.

Lemma 7.19. For an operator (mapping) A: $L \rightarrow M$, the measure $\mathrm{l}r_{A, X}$ is defined if and only if $r_{A, \mathrm{u}X}$ is defined and $\mathrm{l}r_{A, X} = r_{A, \mathrm{u}X}$.

Proof. Necessity. Let us consider an operator (mapping) A: $L \rightarrow M$ and assume the measure $\mathrm{l}r_{A, X}$ is defined. Then by Lemma 7.5, the operator (mapping) A is $\mathrm{l}r_{A, X}$-continuous at all points a from X, and by definition, $r_{A, \mathrm{u}X} = \mathrm{l}r_{A, X}$.

Sufficiency. Let us consider an operator (mapping) A: $L \rightarrow M$ and assume the measure $r_{A, \mathrm{u}X}$ is defined. Then by Lemma 7.6, the operator (mapping) A is $r_{A, \mathrm{u}X}$-continuous at all points a from X. Thus, the measure $\mathrm{l}r_{A, X}$ exists and $\mathrm{l}r_{A, X} \leq r_{A, \mathrm{u}X}$. Then A is $\mathrm{l}r_{A, X}$-continuous at all points a from X, and by definition, it is impossible to have $\mathrm{l}r_{A, X} < r_{A, \mathrm{u}X}$. Consequently, $\mathrm{l}r_{A, X} = r_{A, \mathrm{u}X}$.

Lemma is proved.

Corollary 7.53. An operator (mapping) A: $L \rightarrow M$ is globally r-continuous if and only if $\mathrm{l}r_{A, X}$ is defined and $\mathrm{l}r_{A, X} \leq r$.

Proposition 7.12 implies the following results.

Proposition 7.25. a) If A: $L \rightarrow M$ and B: $L \rightarrow M$ are locally fuzzy continuous operators, then their sum $A + B$: $L \rightarrow M$ and difference $A - B$: $L \rightarrow M$ are locally fuzzy continuous operators.

b) If A: $L \rightarrow M$ and B: $L \rightarrow M$ are locally fuzzy continuous in X operators, then their sum $A + B$: $L \rightarrow M$ and difference $A - B$: $L \rightarrow M$ are locally fuzzy continuous in X operators.

c) If A: $L \rightarrow M$ and B: $L \rightarrow M$ are locally fuzzy continuous inside X operators, then their sum $A + B$: $L \rightarrow M$ and difference $A - B$: $L \rightarrow M$ are locally fuzzy continuous inside X operators.

Proof is left as an exercise.

Proposition 7.26. a) If A: $L \rightarrow M$ is a locally fuzzy continuous operator and d is a real number, then the scalar product dA is a locally fuzzy continuous operator.

b) If A: $L \to M$ is a locally fuzzy continuous in X operator and d is a real number, then the scalar product dA is a locally fuzzy continuous in X operator.

c) If A: $L \to M$ is a locally fuzzy continuous inside X operator and d is a real number, then the scalar product dA is a locally fuzzy continuous inside X operator.

Proof is left as an exercise.

Proposition 7.14 gives us the following results.

Proposition 7.27. a) If A: $L \to M$ is a continuous operator and B: $M \to N$ is a locally fuzzy continuous operator, then their composition BA: $L \to N$ is a locally fuzzy continuous operator.

b) If A: $L \to M$ is a continuous in X operator and B: $M \to N$ is a locally fuzzy continuous in the subset $A(X)$ of M operator, then their composition BA: $L \to N$ is a locally fuzzy continuous in X operator.

c) If A: $L \to M$ is a continuous inside X operator and B: $M \to N$ is a locally fuzzy continuous inside the subset $A(X)$ of M operator, then their composition BA: $L \to N$ is a locally fuzzy continuous inside X operator.

Proof is left as an exercise.

Proposition 7.15 gives us the following results.

Proposition 7.28. a) The product $A \cdot B$: $L \to M$ of locally fuzzy continuous operators A: $L \to M$ and B: $L \to M$ with finite hyperseminorms is a locally fuzzy continuous operator.

b) The product $A \cdot B$: $L \to M$ of locally fuzzy continuous in X operators A: $L \to M$ and B: $L \to M$ with finite hyperseminorms in X is a locally fuzzy continuous in X operator.

c) The product $A \cdot B$: $L \to M$ of locally fuzzy continuous inside X operators A: $L \to M$ and B: $L \to M$ with finite hyperseminorms in X is a locally fuzzy continuous inside X operator.

Proof is left as an exercise.

Let us assume that L and M are seminormed vector spaces.

Corollary 7.21. a) The product $A \cdot B$: $L \to M$ of locally fuzzy continuous operators A: $L \to M$ and B: $L \to M$ is a locally fuzzy continuous operator.

b) The product $A \cdot B$: $L \to M$ of locally fuzzy continuous in X operators A: $L \to M$ and B: $L \to M$ is a locally fuzzy continuous in X operator.

c) The product $A \cdot B$: $L \to M$ of locally fuzzy continuous in X operators A: $L \to M$ and B: $L \to M$ is a locally fuzzy continuous in X operator.

From local regional fuzzy continuity, we naturally come to global regional fuzzy continuity.

Definition 7.9. a) An operator A: $L \to M$ is called *globally fuzzy continuous in* $X \subseteq L$ if A is r-continuous in X for some $r \in \mathbf{R}^+$.

b) An operator A: $L \to M$ is called *globally fuzzy continuous inside* $X \subseteq L$ if A is r-continuous inside X for some $r \in \mathbf{R}^+$.

Remark 7.11. If X contains more than one point, then for each $r \in \mathbf{R}^+$ every r-continuous in X operator A is globally and locally fuzzy continuous in X, but not every locally fuzzy continuous in X operator C is r-continuous in X for some $r \in \mathbf{R}^+$ and thus, not, globally fuzzy continuous in X. It means that in contrast to continuous operators where operators continuous at each point are also globally continuous, it is possible that a fuzzy continuous at each point operator D is not globally fuzzy continuous, i.e., local fuzzy continuity does not always coincide with global fuzzy continuity.

Example 7.15. Let us consider the one-dimensional operator $D(x) = (\lfloor \text{tg } x \rfloor)^2$. This operator is locally fuzzy continuous in the interval $(-\pi, \pi)$, i.e., fuzzy continuous at each point of $(-\pi, \pi)$, but it is not globally fuzzy continuous in $(-\pi, \pi)$, i.e., there is no number r such that D is r-continuous in $(-\pi, \pi)$.

Proposition 7.29. a) If A: $L \to M$ and B: $L \to M$ are globally fuzzy continuous in X (inside X) operators, then their sum $A + B$: $L \to M$ and difference $A - B$: $L \to M$ are globally fuzzy continuous in X (inside X) operators.

b) If A: $L \to M$ is a locally fuzzy continuous in X (inside X) operator and d is a real number, then the scalar product dA is a locally fuzzy continuous in X (inside X) operator.

Proof is left as an exercise.

Proposition 7.14 gives us the following results.

Proposition 7.30. a) If A: $L \to M$ is a continuous in X (inside X) operator and B: $M \to N$ is a globally fuzzy continuous in the subset $A(X)$ of M (inside $A(X)$) operator, then their composition BA: $L \to N$ is a globally fuzzy continuous in X (inside X) operator.

Proof is left as an exercise.

Proposition 7.15 implies the following results.

Proposition 7.31. The product $A \cdot B$: $L \to M$ of globally fuzzy continuous in X (inside X) operators A: $L \to M$ and B: $L \to M$ with finitely bounded in X (inside X) hyperseminorms is globally fuzzy continuous in X (inside X).

Proof is left as an exercise.

There are concepts closely related to fuzzy continuity in the sense of this chapter. For instance, Klee (1961) introduced r-continuous (or ε – continuous or nearly continuous) real functions and studied fixed points of such functions. We consider constructions from Klee, (1961) and Klee and Yandl (1974) with the necessary modifications for operators and functionals in hyperseminormed vector spaces.

Let us consider hyperseminormed vector spaces L and M over a field $\boldsymbol{F}$ where $\boldsymbol{F}$ is either the field $\boldsymbol{R}$ of all real numbers or the field $\boldsymbol{C}$ of all complex numbers, taking a fixed positive real number ε and a set X in the hyperseminormed vector space L.

Definition 7.10. a) An operator (mapping) $A: L \to M$ is called *ε-continuous at a point a* from L if there is a real number $\delta > 0$ such that the inequality $\| a - x \| \leq \delta$ implies the inequality $\| A(x) - A(a) \| \leq \varepsilon$, or in other words, for any x with $\| a - x \| \leq \delta$, we have $\| A(x) - A(a) \| \leq \varepsilon$.

b) An operator (mapping) $A: L \to M$ is called *ε-continuous* in a set X if there is $\delta > 0$ such that for any x and y from X, the inequality $\| x - y \| \leq \delta$ implies the inequality $\| A(x) - A(y) \| \leq \varepsilon$, or in other words, for any x and y from X with $\| x - y \| \leq \delta$, we have $\| A(x) - A(y) \| \leq \varepsilon$.

c) An operator (mapping) $A: L \to M$ is called *globally ε-continuous* in a set X if it is ε-continuous at each point a from X.

d) An operator (mapping) $A: L \to M$ is called *ε-continuous at a point a* $\in L$ if A is ε-continuous at a for some $\varepsilon \in \boldsymbol{R}^+$.

e) An operator (mapping) $A: L \to M$ is called *nearly continuous* in a set X if A isε-continuous in X for some $\varepsilon \in \boldsymbol{R}^+$.

f) An operator (mapping) $A: L \to M$ is called *globally nearly continuous* in a set X if A is globally ε-continuous in X for some $\varepsilon \in \boldsymbol{R}^+$.

In this work, Klee also defined a stability condition for ε-continuous and nearly continuous mappings (Klee, 1961). Namely, a compact space X satisfies the stability condition if for any nearly continuous mapping A of X into itself, there is a point in X nearly invariant (or fuzzy invariant in the context of neoclassical analysis) under A. Such a point is called an approximate (fuzzy) fixed point. Approximate (fuzzy) fixed points were studied in Klee (1961), as well as in the work of Yandl (1965).

In addition, ε-continuous mappings found interesting applications in topology based on the concept of an ε – homotopy of a topological space into a metric space (Klee and Yandl, 1974). Namely, an ε – *homotopy* of a topological space T into a metric space L is an ε-continuous mapping $h: T \times [0, 1] \to L$. Klee and Yandl developed some of the basic parts of the theory of absolute retracts in a proximate form where continuous functions are replaced by nearly continuous functions, fixed points by approximate fixed points, etc.

Interesting results in this direction were obtained by Felt and Sanjurjo, who used ε-continuous mappings and ε – homotopy to study the shape category of compact spaces (Felt, 1974; Sanjurjo, 1988, 1992). In Giraldo, et al. (2001), the concept of ε-continuity is used to prove that the Čech homology groups are in fact inverse limits of certain homology groups. It provides an intrinsic characterization of Čech homology, eliminating all external elements, like embeddings, inverse sequences of simplicial complexes, etc.

There is a transparent relation between ε-continuous operators (mappings) and fuzzy continuous operators (mappings).

Proposition 7.32. An operator $A: L \to M$ is r-continuous at a point $a \in L$ if and only if A is ε-continuous at a point a for any $\varepsilon > r$.

Proof. Sufficiency. Let us assume that an operator $A: L \to M$ is an ε-continuous at a point $a \in L$ for any $\varepsilon > r$. We will show that A satisfies Definition 7.3. Indeed, given a number $k > 0$, we can take a number $\varepsilon > r$ such that $\varepsilon - r < k$. Then by Definition 7.10, there is $\delta > 0$ such that the inequality $\| a - x \| \leq \delta$ implies the inequality $\| A(x) - A(a)\| \leq \varepsilon$. Consequently, $\| A(x) - A(a)\| < r + k$. As k is an arbitrary positive number, by Definition 7.3, the operator A: $L \to M$ is r-continuous at the point a.

Necessity. Let us assume that an operator $A: L \to M$ is r-continuous at a point a and take some $\varepsilon > r$. Then given a number k with $\varepsilon - r > k > 0$, by Definition 7.11, there is $\delta > 0$ such that the inequality $\| a - x \| < \delta$ implies the inequality $\| A(x) - A(a)\| < r + k$. Consequently, $\| A(x) - A(a)\| < \varepsilon$. Thus, the operator A is ε-continuous at the point a.

Proposition is proved.

For real functions, a similar result was obtained in Burgin (2008).

Fuzzy continuity extends the concept of continuity for functions, functionals and operators. However, if we want to understand continuity of functions and operators on discrete sets, we need a more general construction than fuzzy continuity. It is called *extended* or *twofold fuzzy continuity* and was studied for real functions in Burgin (2008, 2012c). Here we introduce and explore twofold fuzzy continuity for functionals and operators in the new realm of hyperseminormed vector spaces and algebras.

There are three ways to define twofold fuzzy continuous functions, functionals and operators in hyperseminormed vector spaces: the sequential approach, the most popular in calculus (ε, δ)-construction, and the topological definition based on the concept of a scalable topological space (Burgin, 2004, 2005, 2006). Here we consider only definitions based on two first approaches because scalable topological spaces are highly abstract mathematical objects, exposition of which will demand essential space unnecessarily increasing the length and complexity of this book.

At first, we give the sequential definition of twofold fuzzy continuity.

Definition 7.11. a) An operator A: $L \rightarrow M$ is called (q, r)-*continuous at a point* $a \in L$ if for any sequence $l = \{a_i \in L; i = 1, 2, 3, \dots\}$, for which a is an q-limit, the point $A(a)$ is an r-limit of the sequence $\{ f(a_i) \in M; i = 1, 2, 3, \dots\}$.

b) The pair (q, r) is called the *dyadic measure of discontinuity* (or *dyadic continuity defect* or *dyadic parameter of discontinuity*) of the operator A at the point a.

c) An operator A: $L \rightarrow M$ is called *twofold fuzzy continuous at a point* $a \in L$ if it is (q, r)-continuous at a for some q and r.

Example 7.16. The one-dimensional operator $A(x) = x^2$ is $(1, 5)$-continuous at the point 2 but it is not $(1, 3)$-continuous at the same point.

Example 7.17. The one-dimensional operator $A(x)$ defined as

$$A(x) = \begin{cases} 0 & \text{if } x \leq 0 \\ 1 + x^2 & \text{if } x > 0 \end{cases}$$

is $(1, 1)$-continuous at the point 0, $(0, 0)$-continuous at the point 2 but it is not $(0, 0)$-continuous at the point 0 and not $(1, 1)$-continuous at the point 1.

Before exploring properties of 2-fuzzy continuous functionals and operators, we give the second definition, in which (q, r)-continuous operators are defined not by explicit utilization of limits but by means of the (ε, δ)-construction.

Definition 7.12. An operator A: $L \rightarrow M$ is called (q, r)-*continuous at a point* $a \in L$ if for any $\varepsilon > 0$ there is $\delta > 0$ such that the inequality $\| a - x \| < q + \delta$ implies the inequality $\| A(x) - A(a)| < r + \varepsilon$, or in other words, for any x with $\| a - x \| < q + \delta$, we have $\| A(x) - A(a)\| < r + \varepsilon$.

b) The pair (q, r) is called the *dyadic measure of discontinuity*(or *dyadic continuity defect* or *dyadic parameter of discontinuity*) of the operator A at the point a.

c) An operator A: $L \rightarrow M$ is called *twofold fuzzy continuous at a point* $a \in L$ if it is (q, r)-continuous at a for any some non-negative real numbers q and r.

In a similar way as we did in Proposition 7.5, we can prove the following statement.

Proposition 7.33. Definitions 7.12 and 7.11 are equivalent, i.e., they define the same concepts.

This result shows that investigating properties of (q, r)-continuity and twofold fuzzy continuity, we can use any of these definitions.

Twofold fuzzy continuity is a natural generalization of fuzzy continuity as the following results demonstrate.

Lemma 7.20. a) An operator $A: L \to M$ is r-continuous at a point $a \in L$ if and only if it is $(0, r)$-continuous at the point a.

b) An operator $A: L \to M$ is continuous at a point $a \in L$ if and only if it is $(0, 0)$-continuous at the point a.

Proof is left as an exercise.

These results show that the concept of (q, r)-continuity is a natural extension of the concept of conventional continuity, as well as of r-continuity.

Definition 7.12 implies the following result.

Lemma 7.21. If $t > r$, and $p < q$, then any (q, r)-continuous at a point a operator $A: L \to M$ is also (p, t)-continuous at a.

Indeed, if that the inequality $| a - x | < q + \delta$ implies the inequality $\| A(x) - A(a)\| < r + \varepsilon$, then the inequality $| a - x | < p + \delta$ implies the inequality $\| a - x \| < q + \delta$ because $p < q$. The inequality $\| a - x \| < q + \delta$implies the inequality $| A(x) - A(a)| < r + \varepsilon$ as A is (q, r)-continuous at a, which, in turn, implies the inequality $\| A(x) - A(a)\| < t + \varepsilon$ because $t > r$. Thus, by Definition 7.12, if A is (q, r)-continuous at a point a operator, then it is also (p, t)-continuous at a.

Corollary 7.4. If $q > l(r < p)$, then any (q, r)-continuous at a point $a \in L$operator is also(l, r)-continuous and (q, p)-continuous at the point a.

Corollary 7.5. If $q > l$ and $r < p$, then any (q, r)-continuous operator is also (l, p)-continuous.

Corollary 7.6. If an operator $A: L \to M$ is (q, r)-continuous at a point $a \in L$, then A is r-continuous at the point a.

Lemmas 7.20 and 7.21 show that (q, r)-continuity is stronger that r-continuity. Namely, we have the following result.

Lemma 7.22. Any (q, r)-continuous at a point a operator $A: L \to M$ is also r-continuous at a for any number $r > 0$ and $q \geq 0$.

Indeed, by Lemma 7.21, (q, r)-continuity implies $(0, r)$-continuity and by Lemma 7.20, $(0, r)$-continuity implies r-continuity.

Note that if $q < p$, then it is possible that a (q, r)-continuous at a point a operator is not(p, r)-continuous at a. For instance, the one-dimensional operator $A(x) = x$ is $(0, 0)$-continuous at the point 0, but for any $p > 0$, it is not $(p, 0)$-continuous at 0.

Lemma 7.23. An operator (mapping) A: $L \to M$ is (q, r)-continuous at a point a if and only if any its shift $A + u$ is (q, r)-continuous at a.

Proof is left as an exercise.

Corollary 7.3. A functional A: $L \to \boldsymbol{R}$ is (q, r)-continuous at a point a if and only if any its shift $A + u$ is (q, r)-continuous at a point a.

Let A: $L \to M$ and B: $L \to M$ be two operators and k is a real number.

Proposition 7.34. If the operator A is (q, r)-continuous at a point a and the operator B is (p, h)-continuous at the point a, then:

a) the operator $A + B$ is (u, $r + h$)-continuous at the point a where $u = \min \{q, p\}$;

b) the operator $A - B$ is (u, $r + h$)-continuous at the point a where $u = \min \{q, p\}$;

c) the operator kA is (q, $|k| \cdot r$)-continuous at the point a.

Proof. a) Let us take a (q, r)-continuous at a point a operator A, a (p, h)-continuous at a point a operator B and a sequence $l = \{a_i; i = 1, 2, 3, \ldots\}$ such that $a = q$-lim l and $a = p$-lim l. Then by Lemma 7.2, $a = u$-lim l where $u = \min \{q, p\}$. By Definition 7.11, the point $A(a)$ is an r-limit of the sequence $\{ A(a_i) \in M; i = 1, 2, 3, \ldots\}$ and the point $g(a)$ is an h-limit of the sequence $\{ B(a_i) \in M; i = 1, 2, 3, \ldots\}$. By Proposition 7.2, the point $A(a) + B(a)$ is an ($r + h$)-limit of the sequence $\{ A(a_i) + B(a_i); i = 1, 2, 3, \ldots\}$. Thus, by Definition 7.11, the operator $A + B$ is (u, $r + h$)-continuous at the point a.

Proofs of parts b) and c) are similar and also use properties of fuzzy limits obtained in Proposition 7.2.

Theorem 7.3 and Lemma 7.10 imply the following result.

Corollary 7.5. If an operator (function) A is r-continuous at a point a and an operator (function) B is h-continuous at a point a, then:

a) the operator $A + B$ is ($r + h$)-continuous at a point a;

b) the operator $A - B$ is ($r + h$)-continuous at a point a;

c) the operator kA is $|k| \cdot r$-continuous at a point a.

Corollary 7.8. If operators A and B are fuzzy (twofold fuzzy) continuous at a point a, then operators $A + B$, $A - B$, and kA are fuzzy (twofold fuzzy) continuous at the point a.

Corollary 7.9. If functions $f(x)$ and $g(x)$ are fuzzy (twofold fuzzy) continuous in (inside) X, then functions $(f + g)(x)$, $(f - g)(x)$, and $(kf)(x)$ are fuzzy (twofold fuzzy) continuous in (inside) X.

Corollary 7.10. The set of all fuzzy (twofold fuzzy) continuous at a point a operators is a linear space.

Corollary 7.11. The set of all fuzzy (twofold fuzzy) continuous (inside) X operators is a linear space.

From twofold fuzzy continuity at a point, we naturally come to twofold fuzzy continuity in a set. We introduce here several concepts of twofold fuzzy continuity for further study. Different properties determine corresponding classes of functions and operators that more or less preserve some properties of continuous functions and operators. At first, we consider twofold local fuzzy continuity in a set.

Definition 7.13. a) An operator A: $L \to M$ is called *(q, r)-continuous in* a set $X \subseteq L$ if A is (q, r)-continuous at each point a from X.

b) An operator A: $L \to M$ is called *twofold fuzzy continuous in* a set X if it is (q, r)-continuous in X for any some non-negative real numbers q and r.

Let us find some properties of twofold fuzzy continuity.

Lemma 7.24. a) An operator A: $L \to M$ is r-continuous in a set X if and only if it is $(0, r)$-continuous in X.

b) An operator A: $L \to M$ is continuous in a set X if and only if it is $(0, 0)$-continuous in X.

Proof is left as an exercise.

Lemma 7.25. An operator (mapping) A: $L \to M$ is (q, r)-continuous in a set X if and only if any its shift $A + u$ is (q, r)-continuous in the set X.

Proof is left as an exercise.

Corollary 7.3. A functional A: $L \to \mathbf{R}$ is (q, r)-continuous in a set X if and only if any its shift $A + u$ is (q, r)-continuous in a set X.

Lemma 7.26. If $t \leq q$ and $r \leq p$, then any (q, r)-continuous in X operator is also locally fuzzy (t, p)-continuous in X.

Proof is similar to the proof of Lemma 7.9.

Lemma 7.27. Any (q, r)-continuous in X (inside X) operator A is r-continuous in X.

Proof is similar to the proof of Lemma 7.10.

Let A: $L \to M$ and B: $L \to M$ be two operators and k is a real number.

Proposition 7.35. If the operator A is (q, r)-continuous in a set X and the operator B is (p, h)-continuous in X, then:

a) the operator $A + B$ is $(u, r + h)$-continuous in a set X where $u = \min \{q, p\}$;

b) the operator $A - B$ is $(u, r + h)$-continuous in a set X where $u = \min \{q, p\}$;

c) the operator kA is $(q, |k| \cdot r)$-continuous in a set X.

Proof is similar to the proof of Proposition 7.34.

As in the case of fuzzy continuity, we come to twofold fuzzy continuity inside a set.

Definition 7.6. a) An operator A: $L \to M$ is called *(q, r)-continuous inside* a set $X \subseteq L$ if A is defined at all points from X and for any sequence $l = \{a_i \in X; i = 1, 2, 3, \dots\}$ that q-converges to a point a from X, the point $A(a)$ is an r-limit of the sequence$\{ A(a_i) \in M; i = 1, 2, 3, \dots\}$.

b) An operator A: $L \to M$ is called *twofold fuzzy continuous inside* a set X if it is (q, r)-continuous inside X for any some non-negative real numbers q and r.

It is possible to define twofold fuzzy continuity inside a set in a different way.

Definition 7.7. a) An operator A: $L \to M$ is called *(q, r)-continuous inside* a set $X \subseteq L$ if the restriction of A on X is (q, r)-continuous at each point a from X.

b) An operator A: $L \to M$ is called *twofold fuzzy continuous inside* a set X if it is (q, r)-continuousinside X for any some non-negative real numbers q and r.

Proposition 7.36. Definitions 7.6 and 7.7 are equivalent.

Proof is left as an exercise.

The following results give us examples of (q, r)-continuous inside X one-dimensional operators.

Lemma 7.28. If $X = \{ ku; k = 0, \pm 1, \pm 2, \pm 3, \dots\}$ or $X = \{ ku; k \in \boldsymbol{Z}$ and $m < k < n\}$ and $q < u$, then any function f: $\boldsymbol{R} \to \boldsymbol{R}$ is (q, r)-continuous inside X.

Corollary 7.8. Any operator A: $L \to M$ defined only on a discrete set $X = \{ ku; u \in L, k = 0, \pm 1, \pm 2, \pm 3, \dots\}$ or $X = \{ ku; u \in L, k \in \boldsymbol{Z}$ and $m < k < n\}$ is (q, r)-continuous if $q < \|u\|$.

At the same time, "(q, r)-continuity in" is stronger than "(q, r)-continuity inside."

Lemma 7.29. If an operator A: $L \to M$ is (q, r)-continuous in a set X, then A is (q, r)-continuous inside X.

Proof is left as an exercise.

However, in some cases, the concepts of "r-continuity in" and "r-continuity inside" coincide.

Proposition 7.37. If X is an q-open subset of the space L, then an operator A: $L \to M$ is (q, r)-continuous in a set X if and only if A is (q, r)-continuous inside X.

Sufficiency. Let us consider an r-continuous inside X operator A: $L \to M$ and a point a from X. Taking an arbitrary sequence $l = \{a_i \in L; i = 1, 2, 3, \dots\}$

that converges to a, we see that starting from some number n, all elements from this sequence belong to because X is an open set. Thus, the sequence $h = \{a_i \in L; i = n, n + 1, n + 2, n + 3, \ldots\}$ belongs to X and also converges to a as a subsequence of l.

Because the operator A is r-continuous inside X, the point $A(a)$ is an r-limit of the sequence $A(h) = \{ A(a_i) \in M; i = n, n + 1, n + 2, n + 3, \ldots\}$. Thus, the point $A(a)$ is also an r-limit of the sequence $A(l) = \{ A(a_i) \in M; i = 1, 2, 3, \ldots\}$ because the limit of a converging sequence coincides with the limit of any of its subsequence. As a is an arbitrary point from X and l is an arbitrary sequence which converges to a, A is r-continuous in X.

Proposition is proved.

Lemma 7.30. An operator (mapping) $A: L \to M$ is (q, r)-continuous inside a set X if and only if any its shift $A + u$ is (q, r)-continuous inside the set X.

Lemma 7.31. a) An operator $A: L \to M$ is r-continuous inside a set X if and only if it is $(0, r)$-continuous inside X.

b) An operator $A: L \to M$ is continuous inside a set X if and only if it is $(0, 0)$-continuous inside X.

Proof is left as an exercise.

Lemma 7.32. An operator (mapping) $A: L \to M$ is (q, r)-continuous inside a set X if and only if any its shift $A + u$ is (q, r)-continuous inside the set X.

Proof is left as an exercise.

Corollary 7.3. A functional $A: L \to \mathbf{R}$ is (q, r)-continuous inside a set X if and only if any its shift $A + u$ is (q, r)-continuous inside a set X.

Lemma 7.33. If $t \leq q$ and $r \leq p$, then any (q, r)-continuous inside X operator is also locally (t, p)-continuous inside X.

Proof is similar to the proof of Lemma 7.9.

Lemma 7.34. Any (q, r)-continuous inside X operator A is r-continuous inside X.

Proof is similar to the proof of Lemma 7.10.

Let $A: L \to M$ and $B: L \to M$ be two operators and k is a real number.

Proposition 7.35. If the operator A is (q, r)-continuous inside a set X and the operator B is (p, h)-continuous inside X, then:

a) the operator $A + B$ is $(u, r + h)$-continuous inside a set X where $u = \min \{q, p\}$;

b) the operator $A - B$ is $(u, r + h)$-continuous inside a set X where $u = \min \{q, p\}$;

c) the operator kA is $(q, |k| \cdot r)$-continuous inside a set X.

Proof is similar to the proof of Proposition 7.34.

These results show that parameters q and r play complimentary roles for (q, r)-continuous mappings. The parameter r extends the scope of admissible (that is, continuous to some extent) functions, the parameter q restricts the scope of admissible functions. It is demonstrated here that, in general there are much more r-continuous functions that continuous. However, the class of (q, r)-continuous functions can be much less than the class of continuous ones.

As we can see, (q, r)-continuity only decreases the scope of continuous functions in comparison with r-continuity. A natural question arises: Why do we need the concept of (q, r)-continuity? One of the reasons to introduce this concept is necessity to study and utilize functions and operators in discrete spaces (sets) developing differential and integral calculi in this setting [cf., (Burgin, 2012c)].

To conclude, it is necessary to remark that all results from this chapter are true not only for hyperseminormed vector spaces, but also for hypernormed, seminormed and normed vector spaces because hypernorms, seminorms and norms are special cases of hyperseminorms. In addition, many results formulated and proved only for operators are also true for functionals and functions.

KEYWORDS

- **(q, r)-continuity**
- **convergence**
- **discontinuity**
- **fuzzy continuity**
- **fuzzy convergence**
- **limit**
- **r-continuity**
- **r-limit**
- **sequence**
- **twofold fuzzy continuity**

CHAPTER 8

APPROXIMATELY LINEAR OPERATORS

In this chapter, we introduce and study approximately additive, approximately homogeneous and approximately linear operators in vector spaces exploring different types and kinds of approximate additivity, approximate homogeneity and approximate linearity, as well as relations between these types of operators. This approach is based on the principles of neoclassical analysis where ordinary structures of analysis, such as functions, sequences, series, and operators, are studied by means of fuzzy concepts: fuzzy limits, fuzzy continuity, fuzzy derivatives and fuzzy gradients (Burgin, 2008).

One of the first problems related to neoclassical analysis was considered by Ulam in 1940. He introduced approximate homomorphisms. Namely, an *approximate homomorphism* in a category of groups is a mapping f from a group A to a group B with a metric **d** such that $f(xy)$ is not necessarily equal to $f(x)f(y)$, but must be "close" to $f(x)f(y)$, i.e., the distance $\mathbf{d}(f(x)f(y), f(xy))$ has to be less that some small number δ for all x and y from A. It is possible to consider approximate homomorphisms in different categories, e.g., in categories of rings, of topological spaces, etc.). Examples of approximate homomorphisms in the category of topological spaces are fuzzy continuous mappings, while an example of approximate homomorphisms in the category of numerical series is fuzzy summation studied in neoclassical analysis.

Studying approximate homomorphisms, Ulam formulated the problem whether there was a strict homomorphism $g: A \to B$, which approximates any approximate homomorphism f in the sense that $f(x)$ is always "close" to $g(x)$. Hyers gave a partial solution to this problem (Hyers, 1941). His result and the concept of Ulam stability in general were further extended in various directions [cf., e.g., (Rassias, 1978; Forti, 1987, 1995; Gajda, 1991; Isac and Rassias, 1993; Găvruta, 1994; Borelli and Forti, 1995; Jung, 1996, 1997, 1998, 1998a)]. Recent results on approximate homomorphisms are presented in Kanovei and Reeken (2000), Rassias and Rassias (2003, 2005), and Rassias (2007).

Linear operators are homomorphisms of vector spaces, which are also abelian groups. Thus, approximately linear operators are approximate homomorphisms of vector spaces. In particular, approximately linear operators introduced and studied in this chapter (cf. Definition 8.8) are approximate homomorphisms of vector spaces in the sense of Ulam (1964) when numbers p and q are sufficiently small. Thus, in the terminology of this book, Ulam's problem is the question whether any approximately linear operator is functionally r-approximately linear (cf. Definition 8.17) for a sufficiently small number r.

Von Neumann formulated another classical problems related to neoclassical analysis in his formulation of the uncertainty principle. He asked whether two almost commuting operators are close to (are small perturbations of) commuting operators (von Neumann, 1929a). To make this problem exact, it is necessary to consider some operator norm $||\cdot||$ and introduce the following concepts, which belong to neoclassical analysis.

Two operators (matrices) A and B are called *almost commuting* if $||AB - BA|| < \delta$ for some small number δ.

Two operators (matrices) A and B are called *near commuting* if there are commuting operators (matrices) X and Y such that $||A - X|| + ||B - Y|| < \varepsilon$for some small number ε.

Rosenthal (1969) studied this problem for matrices with respect to the Hilbert-Schmidt norm, while Halmos (1976, 1977) explored it with respect to the conventional operator norm. Later many mathematicians studied this problem and its different forms [cf., e.g., (Davidson, 1985; Hastings, 2009; Hastings and Loring, 2005; Kachkovskiy and Safarov, 2016; Lin, 1997)]. In particular, it was proved that almost commuting matrices need not be nearly commuting (Choi, 1988; Exel and Loring, 1989).

Here we consider (approximate) linearity as a synthesis of (approximate) homogeneity and (approximate) additivity. Note that in the context of neoclassical analysis, being approximate is a kind of being fuzzy. Indeed, it is possible to discern three kinds of *fuzziness as a relation* between objects, systems or concepts:

- An object (system or concept) A is *approximately* another object (system or concept) B if it is possible to estimate the difference between these objects or in other words, how far they are from one another, and this difference (distance) is sufficiently small.

- An object (system or concept) A is *vaguely* another object (system or concept) B if A and B are similar but it is impossible to estimate the difference between these objects.
- An object (system or concept) A is *roughly* another object (system or concept) B if it is possible to estimate the difference between these objects or in other words, how far they are from one another, and this difference (distance) is not small.

We start with approximate homogeneity.

Let r be a non-negative real number and consider two hyperseminormed vector spaces L and M over the field of real numbers $\boldsymbol{R}$. We define the hyper-seminorm by $\|\cdot\|$ in both vector spaces.

Definition 8.1. a) An operator A: $L \to M$ is called *homogeneous*, also called *homogeneous of the first degree*, if for any number k from $\boldsymbol{R}$ and any element a from L, we have

$$A(ka) = kA(a)$$

b) An operator A: $L \to M$ is called *r-homogeneous* if for any number k from $\boldsymbol{R}$ and any element a from L, we have

$$\| A(ka) - kA(a)\| < r$$

c) The number $h_{A,a} = \inf \{r;\ A \text{ is } r\text{-homogeneous at } a\}$ is called a *homogeneity deviation* of the operator A.

d) An operator A: $L \to M$ is called *approximately homogeneous* if it is *r-homogeneous* for some non-negative real number r.

Note that although any linear operator is homogeneous, there are homogeneous operators, which are not linear.

Example 8.1. Let us consider the normed space $\boldsymbol{R}^2$ with the basis $\{e_1, e_2\}$ and the two-dimensional operator $A: = \boldsymbol{R}^2 \to \boldsymbol{R}^2$ defined by the following formulas:

$$A(ce_1) = 2ce_1$$

$$A(ce_2) = 3ce_2$$

$$A(ae_1 + be_2) = ae_1 + be_2, \text{ when } a \neq 0 \text{ and } b \neq 0$$

This operator A is homogeneous but not linear.

Remark 8.1. In mathematics, there are several other constructions, which are also called by the name a homogeneous operator [cf., e.g., (Tomšič, 1974; Bagchi and Misra, 2001; Koránye and Misra, 2008; Ioku et al., 2016)].

Remark 8.2. Although it is possible to consider similar kinds of homogeneity of higher degrees, e.g., approximate homogeneity of higher degrees, we do not study here homogeneity of higher degrees.

Definitions imply the following result.

Lemma 8.1. If an operator $A: L \to M$ is r-homogeneous, then it is q-homogeneous for any $q > r$.

Indeed, if

$$\| A(ka) - kA(a)\| < r$$

then

$$\| A(ka) - kA(a)\| < q$$

This result explains that if an operator has one homogeneity deviation, then it has many homogeneity deviations.

The following result demonstrates that fuzzy homogeneity of operators is a natural extension of the conventional homogeneity.

Let us assume that M is a hypernormed (normed) vector space.

Lemma 8.2. An operator $A: L \to M$ is homogeneous if and only if it is 0-homogeneous.

Proof is left as an exercise.

Lemma 8.3. If an operator $A: L \to M$ is r-homogeneous, then for any real number k and any element a from L, we have

$$|k| \, \|A(a)\| < \| A(ka) \| + r \tag{8.1}$$

and

$$\| A(ka) \| < |k| \, \|A(a)\| + r \tag{8.2}$$

Proof. By r-homogeneity and the properties of a hypernorm, we have

$$\| kA(a) - A(ka)\| = \| A(ka) - kA(a)\| < r$$

Besides, $kA(a) = kA(a) - A(ka) + A(ka)$. Thus, by the triangle inequality, we have

$$|k|\ \|A(a)\| = \|kA(a)\| = \|\ kA(a) - A(ka) + A(ka)\ \| \leq$$

$$\|\ kA(a) - A(ka)\ \| + \|\ A(ka)\ \| < \|\ A(ka)\ \| + r$$

This gives us inequality (8.1).

Inequality (8.2) is proved in a similar way.

Lemma is proved.

In particular, by Lemma 8.3, we have the following sequence of equalities and inequalities

$$\|\ A(\mathbf{0})\ \| = \|\ A(0 \cdot \mathbf{0})\ \| < 0 \cdot \|A(a)\| + r = r$$

It gives us the following result.

Corollary 8.1. If an operator A: $L \rightarrow M$ is r-homogeneous, then $\|\ A(\mathbf{0})\ \| < r$.

Let us consider sums of operators and products of operators and numbers.

Proposition 8.1. a) If an operator A: $L \rightarrow M$ is r-homogeneous and an operator B: $L \rightarrow M$ is q-homogeneous, then the operator $A + B$: $L \rightarrow M$ is $(r + q)$-homogeneous, i.e., the homogeneity deviation of the sum is the sum of homogeneity deviations.

b) If an operator A: $L \rightarrow M$ is r-homogeneous and k is a real number, then the operator kA: $L \rightarrow M$ is $(|k| \cdot r)$-homogeneous.

Proof. Let us consider an r-homogeneous operator A: $L \rightarrow M$ and a q-homogeneous operator B: $L \rightarrow M$.Then we have

$$\|\ (A + B)(ka) - k(A + B)(a)\| = \|\ (A(ka) + B(ka)) - (kA(a) + kB(a))\| =$$

$$\|\ (A(ka) - kA(a)) + (B(ka) - kB(a))\| \leq$$

$$\|\ A(ka) - kA(a)\| + \|B(ka) - kB(a)\| < r + q$$

It means that the operator $A + B$ is $(r + q)$-homogeneous.

Now let us treat the operator kA.

$$\|\ kA(ta) - ktA(a)\| = |k| \cdot \|\ A(ta) - tA(a)\| < |k| \cdot r$$

It means that the operator kA is $(|k| \cdot r)$-homogeneous.

Proposition is proved.

Corollary 8.2. If operators A: $L \to M$ and B: $L \to M$ are approximately homogeneous, then the operators $A + B$ and kA are also approximately homogeneous.

Corollary 8.3. If an operator A: $L \to M$ is r-homogeneous and an operator B: $L \to M$ is q-homogeneous, then the operator $dA + cB$: $L \to M$ is $(|d|r + |c|q)$-homogeneous for any real numbers d and c.

Corollary 8.4. If operators A: $L \to M$ and B: $L \to M$ are approximately homogeneous, then the operator $dA + cB$: $L \to M$ is also approximately homogeneous for any real numbers d and c.

Corollary 8.5. If functionals A: $L \to \mathbf{R}$ and B: $L \to \mathbf{R}$ are approximately homogeneous, then the operators $A + B$ and kA are also approximately homogeneous.

Corollary 8.6. If a functional A: $L \to \mathbf{R}$ is r-homogeneous and a functional B: $L \to M$ is q-homogeneous, then the functional $dA + cB$: $L \to M$ is $(|d|r + |c|q)$-homogeneous for any real numbers d and c.

Corollary 8.7. If functionals A: $L \to \mathbf{R}$ and B: $L \to \mathbf{R}$are approximately homogeneous, then the functional $dA + cB$: $L \to M$ is also approximately homogeneous for any real numbers d and c.

Taking two operators A: $L \to M$ and B: $H \to N$, it is possible to build their *Cartesian product* $A \times B$: $L \times H \to M \times N$ defining $(A \times B)(x, z) = (A(x), B(z))$ for all elements $x \in L$ and $z \in H$.

Let us explore approximate additivity of Cartesian products of operators assuming, at first, that L and H are hyperseminormed spaces, while N and M are seminormed spaces.

Proposition 8.2. If an operator A: $L \to M$ is r-homogeneous, an operator B: $H \to N$ is q-homogeneous, and the Cartesian product $M \times N$ has the Euclidean seminorm (cf. Chapter 3), then the Cartesian product $A \times B$ is t-homogeneous where $t = (r^2 + q^2)^{1/2}$.

Proof. Let us consider an r-homogeneous operator A: $L \to M$ and q-homogeneous operator B: $H \to N$. Then

$$\| A(kx) - kA(x) \| < r \text{ for any real number } k \text{ and any element } x \text{ from } L$$

and

$$\| B(kz) - kB(z) \| < q \text{ for any real number } k \text{ and any element } z \text{ from } H$$

Consequently,

$$|| (A \cdot B)(k(x, z)) - k(A \cdot B)(x, z) || = || (A \cdot B)(kx, kz) - k(A \cdot B)(x, z) || =$$

$$|| (A(kx), B(kz)) - k(A(x), B(z)) || = (|| A(kx) - kA(x)||^2 + ||B(kz) - kB(z)||^2)^{1/2} <$$

$$(r^2 + q^2)^{1/2}$$

It means that the Cartesian product $A \times B$ is t-homogeneous where $t = (r^2 + q^2)^{1/2}$.

Proposition is proved.

Corollary 8.8. If operators A: $L \rightarrow M$ and B: $H \rightarrow N$ are r-homogeneous, and the Cartesian product $M \times N$ has the Euclidean seminorm, then the Cartesian product $A \times B$ is $\sqrt{2}\, r$-homogeneous.

Now let us assume that all spaces L, H, N and M are hyperseminormed vector spaces.

Proposition 8.3. If an operator A: $L \rightarrow M$ is r-homogeneous, an operator B: $H \rightarrow N$ is q-homogeneous, and the Cartesian product $M \times N$ has the Manhattan hyperseminorm (cf. Chapter 3), then the Cartesian product $A \times B$ is t-homogeneous where $t = r + q$.

Proof. Let us consider an r-homogeneous operator A: $L \rightarrow M$ and q-homogeneous operator B: $H \rightarrow N$. Then

$$|| A(kx) - kA(x)|| < r \text{ for any real number } k \text{ and any element } x \text{ from } L$$

and

$$|| B(kz) - kB(z)|| < q \text{ for any real number } k \text{ and any element } z \text{ from } H$$

Consequently,

$$|| (A \cdot B)(k(x, z)) - k(A \cdot B)(x, z) || = || (A \cdot B)(kx, kz) - k(A \cdot B)(x, z) || =$$

$$|| (A(kx), B(kz)) - k(A(x), B(z)) || = || A(kx) - kA(x)|| + ||B(kz) - kB(z)|| < r + q$$

It means that the Cartesian product $A \times B$ is t-homogeneous where $t = r + q$.

Proposition is proved.

Corollary 8.9. If operators A: $L \rightarrow M$ and B: $H \rightarrow N$ are r-homogeneous, and the Cartesian product $M \times N$ has the Manhattan hyperseminorm, then the Cartesian product $A \times B$ is $2r$-homogeneous.

Proposition 8.4. If an operator A: $L \to M$ is r-homogeneous, an operator B: $H \to N$ is q-homogeneous, and the Cartesian product $M \times N$ has the Chebyshev hyperseminorm (cf. Chapter 3), then the Cartesian product $A \times B$ is t-homogeneous where $t = \max\{r, q\}$.

Proof. Let us consider an r-homogeneous operator A: $L \to M$ and q-homogeneous operator B: $H \to N$. Then

$$\| A(kx) - kA(x) \| < r \text{ for any real number } k \text{ and any element } x \text{ from } L$$

and

$$\| B(kz) - kB(z) \| < q \text{ for any real number } k \text{ and any element } z \text{ from } H$$

Consequently,

$$\| (A{\cdot}B)(k(x, z)) - k(A{\cdot}B)(x, z) \| = \| (A{\cdot}B)(kx, kz) - k(A{\cdot}B)(x, z) \| =$$

$$\| (A(kx), B(kz)) - k(A(x), B(z)) \| = \max\{\| A(kx) - kA(x)\|, \|B(kz) - kB(z)\|\} <$$

$$\max\{r, q\}.$$

It means that the Cartesian product $A \times B$ is t-homogeneous where $t = \max\{r, q\}$.

Proposition is proved.

Corollary 8.10. If operators A: $L \to M$ and B: $H \to N$ are r-homogeneous, and the Cartesian product $M \times N$ has the Chebyshev hyperseminorm, then the Cartesian product $A \times B$ is also r-homogeneous.

Corollary 8.11. a) If operators A: $L \to M$ and B: $H \to N$ are approximately homogeneous, and the Cartesian product $M \times N$ has the Chebyshev hyperseminorm or the Manhattan hyperseminorm, then the Cartesian product $A \times B$ is also approximately homogeneous.

b) If L and H are hyperseminormed spaces, N and M are seminormed spaces, operators A: $L \to M$ and B: $H \to N$ are approximately homogeneous, and the Cartesian product $M \times N$ has the Euclidean seminorm, then the Cartesian product $A \times B$ is also approximately homogeneous.

Let us assume that M is a hyperseminormed linear algebra. Then it is possible to define the product $A{\cdot}B$: $L \to M$ for any operators A: $L \to M$ and B: $L \to M$ by the formula $(A{\cdot}B)(x) = A(x){\cdot}B(x)$ and estimate the homogeneity of this product. However, it is possible that the product of approximately homogeneous operators is not approximately homogeneous.

Example 8.2. Let us take the one-dimensional operator $E(x) = x$ in the space $\boldsymbol{R}$. It is homogeneous and even linear. However, the operator (mapping) $(E{\cdot}E)(x) = x^2$ is not r-homogeneous for any number r from $\boldsymbol{R}^+$.

This example reflects the general case. Indeed, taking two (approximately) homogeneous operators $A: L \to M$ and $B: L \to M$ and considering their product $A{\cdot}B: L{\to}M$, we obtain

$$(A{\cdot}B)(kx) = A(kx){\cdot}B(kx) = (k{\cdot}A(x)){\cdot}(k{\cdot}B(x)) = k^2{\cdot}(A{\cdot}B)(x)$$

Thus, the product $A{\cdot}B$ is not homogeneous and not approximately homogeneous.

However, this brings us to the concept of homogeneity of higher degrees [cf., e.g., (Tomšič, 1974)]. Namely, an operator $A: L{\to}M$ is called *homogeneous of the degree n* if for any number k from $\boldsymbol{R}$ and any element a from L, we have

$$A(ka) = k^n \cdot A(a)$$

Homogeneity of degrees higher than 1 is studied elsewhere.

From global approximate homogeneity, we go to regional approximate homogeneity.

Definition 8.2. a) An operator $A: L{\to}M$ is called *homogeneous in* a set $X \subseteq L$ if for any number k from $\boldsymbol{R}$ and any element a from X such that ka also belongs to X, we have

$$A(ka) = kA(a)$$

b) An operator $A: L \to M$ is called *r-homogeneous in* a set $X \subseteq L$ if

$$\|A(kx) - kA(x)\| \leq r$$

for all $x \in X$ and any real number k such that kx also belongs to X.

c) The number $h_{A,X} = \inf \{r; A \text{ is } r\text{-homogeneous in } X\}$ is called a *homogeneity deviation* of the operator A *in the set* X.

d) An operator $A: L \to M$ is *approximately homogeneous in* a set $X \subseteq L$ if it is r-homogeneous in X for some non-negative real number r.

Note that a non-homogeneous operator can be homogeneous in each set from some partition of the space L as the following example demonstrates.

Example 8.3. Let us consider the following one-dimensional operator

$$A(x) = nx \text{ when } n - 1 \leq x < n \text{ where } n \in \boldsymbol{Z}$$

While $A(x)$ is homogeneous in each interval $[n - 1, n]$, it is not homogeneous and even not r-homogeneous for any number r. Indeed, taking $x = 1$ and $k = n > 0$, we have

$$\|A(kx) - kA(x)\| = |\,(n + 1)n - n \cdot n\,| = n$$

As n can be an arbitrary natural number, $\|A(kx) - kA(x)\|$ Cannot be less than some fixed real number. Thus, the operator A: $\boldsymbol{R} \to \boldsymbol{R}$ is not approximately homogeneous.

Lemma 8.4. If an operator A: $L \to M$ is r-homogeneous in a set X, then it is q-homogeneous in X for any $q > r$.

Proof is left as an exercise.

Let us assume that M is a hypernormed (normed) vector space.

Lemma 8.5. An operator A: $L \to M$ is homogeneous in a set X if and only if it is 0-homogeneous in X.

Proof is left as an exercise.

Proposition 8.5. a) If an operator A: $L \to M$ is r-homogeneous in a set X and an operator B: $L \to M$ is q-homogeneous in X, then the operator $A + B$: $L \to M$ is $(r + q)$-homogeneous in X, i.e., the homogeneity deviation of the sum is the sum of homogeneity deviations.

b) If an operator A: $L \to M$ is r-homogeneous in a set X and k is a real number, then the operator kA: $L \to M$ is $|k| \cdot r$-homogeneous in X.

Proof. Let us consider an r-homogeneous in X operator A: $L \to M$ and a q-homogeneous in X operator B: $L \to M$. Then taking a point a from X and a real number k such that ka also belongs to X, we have

$$\|\,(A + B)(ka) - k(A + B)(a)\| = \|\,(A(ka) + B(ka)) - (kA(a) + kB(a))\| =$$

$$\|\,(A(ka) - kA(a)) + (B(ka) + kB(a))\| \leq$$

$$\|\,A(ka) - kA(a)\| + \|B(ka) + kB(a)\| < r + q$$

It means that the operator $A + B$ is $(r + q)$-homogeneous in X.

Now let us treat the operator kA taking a point a from X and a real number t such that ta also belongs to X. Then we have

$$\| kA(ta) - ktA(a)\| = |k| \cdot \| A(ta) - tA(a)\| < |k| \cdot r$$

It means that the operator kA is $|k| \cdot r$-homogeneous in X.

Proposition is proved.

Corollary 8.12. If operators A: $L \to M$ and B: $L \to M$ are approximately homogeneous in a set X, then the operator $A + B$ and kA are also approximately homogeneous in X.

Corollary 8.13. If an operator A: $L \to M$ is r-homogeneous in a set X and an operator B: $L \to M$ is q-homogeneous, then the operator $dA + cB$: $L \to M$ is $(|d|r + |c|q)$-homogeneous in X for any numbers $d, c \in \mathbf{R} \setminus \{0\}$.

Corollary 8.14. If operators A: $L \to M$ and B: $L \to M$ are approximately homogeneous in a set X, then the operator $dA + cB$: $L \to M$ is also approximately homogeneous in X for any numbers $d, c \in \mathbf{R} \setminus \{0\}$.

Rassias studied approximately odd mappings, which are closely related to homogeneous operators (Rassias, 2007). An operator A: $L \to M$ is called *odd* if $A(x) = -A(-x)$ for any $x \in L$.

Let us consider hyperseminormed vector spaces L and M over a field $\boldsymbol{F}$ where $\boldsymbol{F}$ is either the field $\boldsymbol{R}$ of all real numbers or the field $\boldsymbol{C}$ of all complex numbers.

Definition 8.3. a) An operator A: $L \to M$ is called *r-odd* if it satisfies the following inequality

$$\|A(x) + A(-x)\| \leq r$$

for all $x \in L$.

b) An operator A: $L \to M$ is called *approximately odd* if it is r-odd for some non-negative real number r.

Taking $k = -1$ in the inequality for approximate homogeneity, we have the following result.

Lemma 8.6. Any r-odd operator is q-odd when $q > r$.

Proof is left as an exercise.

Lemma 8.7. Any r-homogeneous operator is r-odd.

Indeed, if A is an r-homogeneous operator, then

$$\|A(x) + A(-x)\| = \|A(-x) - (-1)A(x)\| \leq r$$

i.e., A is an r-odd operator.

Let us assume that M is a hypernormed vector space. Then the concept of an approximately odd mapping (operator) forms a natural extension of the concept of an odd mapping (operator).

Let us assume that M is a hypernormed (normed) vector space.

Lemma 8.8. Any operator $A: L \to M$ is odd if and only if it is 0-odd.

Indeed, if operator $A: L \to M$ is odd, then it is 0-odd because the equality $A(x) = -A(-x)$ implies the inequality $\|A(x) + A(-x)\| \leq 0$.

At the same time, if $\|A(x) + A(-x)\| \leq 0$, then $\|A(x) + A(-x)\| = 0$ and by properties of hypernorms, $A(x) + A(-x) = 0$ implying $A(x) = -A(-x)$. Consequently, the operator A is odd.

Proposition 8.6. a) If an operator A: $L \to M$ is r-odd and an operator B: $L \to M$ is q-odd, then the operator $A + B$: $L \to M$ is $(r + q)$-odd.

b) If an operator A: $L \to M$ is r-odd and k is a real number, then the operator kA: $L \to M$ is $|k|r$-odd.

Proof. Let us consider an r-odd operator A: $L \to M$ and a q-odd operator B: $L \to M$.Then we have

$$\| (A + B)(a) + (A + B)(-a)\| = \| A(a) + B(a) + A(-a) + B(-a))\| =$$

$$\| (A(a) + A(-a)) + (B(a) + B(-a))\| \leq$$

$$\| A(a) + A(-a)\| + \|B(a) + B(-a)\| < r + q$$

It means that the operator $A + B$ is $(r + q)$-odd.

Now let us treat the operator kA.

$$\| kA(a) + kA(-a)\| = |k|\cdot \| A(a) + A(-a)\| < |k|\cdot r$$

It means that the operator kA is $|k|\cdot r$-odd.

Proposition is proved.

Approximately odd operators are closely related to centered operators.

Definition 8.4. a) An operator A: $L \to M$ is called *centered* if it satisfies the following inequality

$$A(\mathbf{0}) = 0$$

b) An operator A: $L \to M$ is called *r-centered* if it satisfies the following inequality

$$\| A(\mathbf{0})\| \leq r$$

c) An operator A: $L \rightarrow M$ is called *approximately centered* if it is r-centered for some non-negative real number r.

Lemma 8.9. Any r-odd operator is r-centered.

Indeed,

$$\| A(\mathbf{0})\| = ||½A(\mathbf{0}) + ½A(-\mathbf{0})\| = ½ \| A(\mathbf{0}) + A(-\mathbf{0})\| \leq r$$

Proposition 8.7. Any r-homogeneous operator is 0-centered and centered when $||.||$ is a hypernorm.

Proof. Let us assume that A is an r-homogeneous operator but not centered. It means that $|| A(\mathbf{0})|| = t \neq 0$. By the properties of hyperseminorms $A(\mathbf{0}) \neq \mathbf{0}$. Then taking an arbitrary positive number k, we have

$$|| A(k\cdot\mathbf{0}) - k \cdot A(\mathbf{0})|| = || A(\mathbf{0}) - k \cdot A(\mathbf{0})|| = (k-1) \cdot || A(\mathbf{0})|| = t(k-1) > r$$

for sufficiently big number k.

Thus, if A is an r-homogeneous operator, then $|| A(\mathbf{0})|| = 0$, i.e., A is 0-centered.

When $||.||$ is a hypernorm, $|| A(\mathbf{0})|| = 0$ implies $A(\mathbf{0}) = 0$, i.e., A is centered.

Proposition is proved.

Let us assume that M is a hypernormed vector space. Then the concept of an approximately odd mapping (operator) forms a natural extension of the concept of an odd mapping (operator).

Let us assume that M is a hypernormed (normed) vector space.

Lemma 8.10. An operator A is 0-centered if and only if it is centered, i.e., $A(\mathbf{0}) = 0$.

Proof is left as an exercise.

Lemma 8.11. Any r-centered operator is q-centered when $q > r$.

Proof is left as an exercise.

Let q be a non-negative real number.

Definition 8.5. a) An operator A: $L \rightarrow M$ is called *additive* if for any elements a, b from L, we have

$$A(a+b) = A(a) - A(b)$$

b) An operator A: $L \rightarrow M$ is called *q-additive* if for any elements a, b from L, we have

$$\| A(a+b) - A(a) - A(b)\| < q$$

c) An operator A: $L \to M$ is called *strongly q-additive* if for any elements a, b from L, we have

$$\| A(a+b) - A(a) - A(b)\| < q$$

and

$$\| A(a-b) - (A(a) - A(b))\| < q$$

d) The number $q_{A,a} = \inf \{r; A \text{ is } r\text{-additive}\}$ is called the (*strong*) *additivity deviation* of the operator A.

e) An operator A: $L \to M$ is called *approximately additive* if it is r-additive for some real number r.

f) An operator A: $L \to M$ is called *strongly approximately additive* if it is strongly r-additive for some real number r.

g) An operator A: $L \to M$ is called *strongly additive* if for any elements a, b from L, we have

$$A(a+b) = A(a) + A(b)$$

and

$$A(a-b) = A(a) - A(b)$$

Note that any strongly r-additive operator is r-additive and any strongly additive operator is additive.

Example 8.4. An arbitrary constant operator C: $L \to M$ defined as $C(x) = c$ where x is any element from L and c is a fixed element from M is not additive but is strongly approximately additive. Indeed, we have

$$A(x+y) = c \neq A(x) + A(y) = 2c$$

but

$$\| A(x+y) - A(x) - A(y)\| = \| c \| < \| c \| + 1$$

and

$$\| A(a-b) - (A(a) - A(b))\| = \| c \| < \| c \| + 1.$$

Definitions imply the following result.

Lemma 8.12. If an operator A: $L \to M$ is (strongly) r-additive, then it is (strongly) q-additive for any $q > r$.

This result explains that if an operator has one homogeneity deviation, then it has many homogeneity deviations.

Corollary 8.15. If a functional A: $L \to \boldsymbol{R}$ is (strongly) r-additive, then it is (strongly) q-additive for any $q > r$.

The following result shows that fuzzy additivity of operators is a natural extension of the conventional additivity.

Let us assume that M is a hypernormed (normed) vector space.

Lemma 8.13. An operator A: $L \to M$ is additive if and only if it is 0-additive.

Proof is left as an exercise.

Corollary 8.16. If a functional A: $L \to \boldsymbol{R}$ is additive if and only if it is 0-additive.

Let us find additional properties of r-additive operators.

Proposition 8.8. a) If an operator A: $L \to M$ is r-additive, then for any number k and for any elements a and b from L, the following inequalities are true

$$\| A(a + b) \| < \| A(a) + A(b)\| + r \tag{8.3}$$

$$\| A(a) + A(b)\| < \| A(a + b) \| + r \tag{8.4}$$

b) If an operator A: $L \to M$ is strongly r-additive, then for any number k and for any elements a and b from L, (8.3), (8.4) and the following inequalities are true

$$\| A(a - b) \| < \| A(a) - A(b)\| + r \tag{8.5}$$

$$\| A(a) - A(b)\| < \| A(a - b) \| + r \tag{8.6}$$

Proof. By the properties of a norm, we have

$$\| kA(a) - A(ka)\| = \| A(ka) - kA(a)\| < r$$

Besides,

$$A(a + b) = A(a + b) - (A(a) + A(b)) + (A(a) + A(b))$$

Thus, by the triangle inequality, we have

$$\| A(a+b) \| \leq \| A(a+b) - (A(a) + A(b)) \| + \| A(a) + A(b)\| < \| A(a) + A(b)\| + r$$

This gives us inequality (8.3).

Inequality (8.4) is proved in a similar way.

We see that

$$A(a-b) = A(a-b) - (A(a) - A(b)) + (A(a) - A(b))$$

and thus,

$$\| A(a-b) \| \leq \| A(a-b) - (A(a) - A(b)) \| + \| A(a) - A(b)\| < \| A(a) - A(b)\| + r$$

This gives us inequality (8.5).

Inequality (8.6) is proved in a similar way.

Proposition is proved.

Lemma 8.14. Any r-additive operator is r-centered.

Indeed,

$$\| A(\mathbf{0})\| = \|A(\mathbf{0} + \mathbf{0}) + A(\mathbf{0}) - A(\mathbf{0})\| \leq r$$

Corollary 8.17. Any additive operator is centered.

Lemma 8.15. Any strongly additive operator is odd.

Indeed, if A is a strongly additive operator, then

$$A(a + (-b)) = A(a) + A(-b)$$

and

$$A(a-b) = A(a) - A(b)$$

As

$$A(a + (-b)) = A(a-b)$$

we have

$$A(-b) = -A(b)$$

Let us consider sums of operators and products of operators and numbers.

Proposition 8.9. a) If an operator A: $L \to M$ is (strongly) r-additive and an operator B: $L \to M$ is (strongly) q-additive, then the operator $A + B$: $L \to M$ is (strongly)$(r + q)$-additive.

b) If an operator $A: L \to M$ is (strongly) r-additive and k is a real number, then the operator $kA: L \to M$ is (strongly)($|k| \cdot r$)-additive.

Proof. a) Let us consider an r-additive operator $A: L \to M$ and q-additive operator $B: H \to N$. Then

$\| A(x + y) - A(x) - A(y)\| < r$ for arbitrary elements x and y from L

and

$\| B(z + u) - B(z) - B(u)\| < q$ for arbitrary elements z and u from H

Consequently, we have

$$\| (A + B)(x + y) - (A + B)(x) - (A + B)(y) \| =$$

$$\| A(x + y) - A(x) - A(y) + B(x + y) - B(x) - B(y) \| \leq$$

$$\| A(x + y) - A(x) - A(y)\| + \|B(x + y) - B(x) - B(y) \| < r + q$$

It means that the operator $A + B$ is $(r + q)$-additive.

Now let us treat strong approximate additivity, assuming that the operator A is strongly r-additive and the operator B is strongly q-additive. Then we also have

$$\| (A + B)(x - y) - (A + B)(x) + (A + B)(y) \| =$$

$$\| A(x - y) - A(x) + A(y) + B(x - y) - B(x) + B(y) \| \leq$$

$$\| A(x - y) - A(x) + A(y)\| + \|B(x - y) - B(x) + B(y) \| < r + q$$

Together with the first part of the proof, it means that the operator $A + B$ is strongly$(r + q)$-additive.

b) Let us consider an r-additive operator $A: L \to M$. Then

$\| A(x + y) - A(x) - A(y)\| < r$ for arbitrary elements x and y from L

Consequently, we have

$$\| kA(x + y) - kA(x) - kA(y)\| =$$

$$\|k(A(x + y) - A(x) - A(y))\| =$$

$$|k| \cdot ||A(x+y) - A(x) - A(y))|| < |k| \cdot r$$

It means that the operator kA: $L \to M$ is ($|k| \cdot r$)-additive.

Now let us treat strong approximate additivity, assuming that the operator A is strongly r-additive. Then we also have

$$|| kA(x-y) - kA(x) + kA(y)|| =$$

$$||k(A(x-y) - A(x) + A(y))|| =$$

$$|k| \cdot ||A(x-y) - A(x) + A(y))|| < |k| \cdot r$$

Together with the first part of the proof, it means that the operator kA: $L \to M$ is strongly ($|k| \cdot r$)-additive.

Proposition is proved.

Corollary 8.18. If operators A: $L \to M$ and B: $L \to M$ are approximately additive, then operators $A + B$ and kA are also approximately additive.

Corollary 8.19. If an operator A: $L \to M$ is (strongly) r-additive and an operator B: $L \to M$ is (strongly) q-additive, then the operator $dA + cB$: $L \to M$ is (strongly) ($|d|r + |c|q$)-additive for any numbers $d, c \in \boldsymbol{R} \backslash \{0\}$.

Corollary 8.20. If operators A: $L \to M$ and B: $L \to M$ are (strongly) approximately additive, then the operator $dA + cB$: $L \to M$ is also (strongly) approximately additive for any numbers $d, c \in \boldsymbol{R} \backslash \{0\}$.

Corollary 8.21. a) If a functional A: $L \to \boldsymbol{R}$ is (strongly) r-additive and a functional B: $L \to \boldsymbol{R}$ is (strongly) q-additive, then the functional $A + B$: $L \to \boldsymbol{R}$ is (strongly) ($r + q$)-additive.

b) If a functional A: $L \to M$ is (strongly) r-additive and k is a real number, then the functional kA: $L \to \boldsymbol{R}$ is (strongly) $|k| \cdot r$-additive.

Corollary 8.22. If functionals A: $L \to \boldsymbol{R}$ and B: $L \to \boldsymbol{R}$ are approximately additive, then functionals $A + B$ and kA are also approximately additive.

Corollary 8.23. If a functional A: $L \to \boldsymbol{R}$ is (strongly) r-additive and a functional B: $L \to \boldsymbol{R}$ is (strongly) q-additive, then the functional $dA + cB$: $L \to \boldsymbol{R}$ is (strongly) ($|d|r + |c|q$)-additive for any numbers $d, c \in \boldsymbol{R} \setminus \{0\}$.

Corollary 8.24. If functionals A: $L \to \boldsymbol{R}$ and B: $L \to \boldsymbol{R}$ are (strongly) approximately additive, then the functional $dA + cB$: $L \to \boldsymbol{R}$ is also (strongly) approximately additive for any numbers $d, c \in \boldsymbol{R} \setminus \{0\}$.

Now let us explore approximate additivity of Cartesian products of operators assuming, at first, that L and H are hyperseminormed spaces, while N and M are seminormed spaces.

Proposition 8.10. If an operator $A: L \to M$ is (strongly) r-additive, an operator $B: H \to N$ is (strongly) q-additive, and the Cartesian product $M \times N$ has the Euclidean seminorm, then the Cartesian product $A \times B$ is (strongly) t-additive where $t = (r^2 + q^2)^{1/2}$.

Proof. Let us consider an r-additive operator $A: L \to M$ and q-additive operator $B: H \to N$. Then

$$\| A(x + y) - A(x) - A(y)\| < r \text{ for arbitrary elements } x \text{ and } y \text{ from } L$$

and

$$\| B(z + u) - B(z) - B(u)\| < q \text{ for arbitrary elements } z \text{ and } u \text{ from } H$$

Consequently, we have

$$\| (A{\cdot}B)((x, z) + (y, u)) - (A{\cdot}B)(x, z) - (A{\cdot}B)(y, u) \| =$$

$$\| (A{\cdot}B)(x + y, z + u) - (A{\cdot}B)(x, z) - (A{\cdot}B)(y, u) \| =$$

$$\| (A(x + y), B(z + u)) - (A(x), B(z)) - (A(y), B(u)) \| =$$

$$\| (A(x + y) - A(x) - A(y)), (B(z + u) - B(z) - B(u)) \| =$$

$$(\| A(x + y) - A(x) - A(y) \|^2 + \|B(z + u) - B(z) - B(u) \|^2)^{1/2} < (r^2 + q^2)^{1/2}$$

It means that the Cartesian product $A \times B$ is t-additive where $t = (r^2 + q^2)^{1/2}$.

Now let us study strong approximate additivity, assuming that the operator A is strongly r-additive and the operator B is strongly q-additive. Then we have

$$\| A(x - y) - A(x) + A(y)\| < r \text{ for arbitrary elements } x \text{ and } y \text{ from } L$$

and

$$\| B(z - u) - B(z) + B(u)\| < q \text{ for arbitrary elements } z \text{ and } u \text{ from } H$$

Consequently, we have

$$\| (A{\cdot}B)((x, z) - (y, u)) - (A{\cdot}B)(x, z) + (A{\cdot}B)(y, u) \| =$$

$$\| (A{\cdot}B)(x - y, z - u) - (A{\cdot}B)(x, z) + (A{\cdot}B)(y, u) \| =$$

$$\| (A(x-y), B(z-u)) - (A(x), B(z)) + (A(y), B(u)) \| =$$

$$\| (A(x-y) - A(x) + A(y)), (B(z-u) - B(z) + B(u)) \| =$$

$$(\| A(x-y) - A(x) + A(y) \|^2 + \|B(z-u) - B(z) + B(u) \|^2)^{1/2} < (r^2 + q^2)^{1/2}$$

Together with the first part of the proof, it means that the Cartesian product $A \times B$ is strongly t-additive where $t = (r^2 + q^2)^{1/2}$.

Proposition is proved.

Corollary 8.25. If operators A: $L \to M$ and B: $H \to N$ are (strongly) r-additive and the Cartesian product $M \times N$ has the Euclidean seminorm, then the Cartesian product $A \times B$ is (strongly) $\sqrt{2} \cdot r$-additive.

Corollary 8.26. If operators A: $L \to M$ and B: $H \to N$ are (strongly) approximately additive and the Cartesian product $M \times N$ has the Euclidean seminorm, then the Cartesian product $A \times B$ is (strongly) approximately r-additive.

Let us assume that L, H, N and M are hyperseminormed spaces.

Proposition 8.11. If an operator A: $L \to M$ is (strongly) r-additive, an operator B: $H \to N$ is (strongly) q-additive, and the Cartesian product $M \times N$ has the Manhattan hyperseminorm (cf. Chapter 3), then the Cartesian product $A \times B$ is (strongly) t-additive where $t = r + q$.

Proof. Let us consider an r-additive operator A: $L \to M$ and q-additive operator B: $H \to N$. Then

$$\| A(x+y) - A(x) - A(y)\| < r \text{ for arbitrary elements } x \text{ and } y \text{ from } L$$

and

$$\| B(z+u) - B(z) - B(u)\| < q \text{ for arbitrary elements } z \text{ and } u \text{ from } H$$

Consequently,

$$\| (A \cdot B)((x, z) + (y, u)) - (A \cdot B)(x, z) - (A \cdot B)(y, u) \| =$$

$$\| (A \cdot B)(x + y, z + u) - (A \cdot B)(x, z) - (A \cdot B)(y, u) \| =$$

$$\| (A(x+y), B(z+u)) - (A(x), B(z)) - (A(y), B(u)) \| =$$

$$\| (A(x+y) - A(x) - A(y)), (B(z+u) - B(z) - B(u)) \| =$$

$$\| A(x+y) - A(x) - A(y) \| + \|B(z+u) - B(z) - B(u) \| < r + q$$

It means that the Cartesian product $A \times B$ is t-additive where $t = r + q$.

Now let us study strong approximate additivity, assuming that the operator A is strongly r-additive and the operator B is strongly q-additive. Then we have

$$\| A(x - y) - A(x) + A(y) \| < r \text{ for arbitrary elements } x \text{ and } y \text{ from } L$$

and

$$\| B(z - u) - B(z) + B(u) \| < q \text{ for arbitrary elements } z \text{ and } u \text{ from } H$$

Consequently, we have

$$\| (A \cdot B)((x, z) - (y, u)) - (A \cdot B)(x, z) + (A \cdot B)(y, u) \| =$$

$$\| (A \cdot B)(x - y, z - u) - (A \cdot B)(x, z) + (A \cdot B)(y, u) \| =$$

$$\| (A(x - y), B(z - u)) - (A(x), B(z)) + (A(y), B(u)) \| =$$

$$\| (A(x - y) - A(x) + A(y)), (B(z - u) - B(z) + B(u)) \| =$$

$$(\| A(x - y) - A(x) + A(y) \| + \|B(z - u) - B(z) + B(u) \| < r + q$$

Together with the first part of the proof, it means that the Cartesian product $A \times B$ is strongly t-additive where $t = r + q$.

Proposition is proved.

Corollary 8.27. If operators $A: L \to M$ and $B: H \to N$ are (strongly) r-additive and the Cartesian product $M \times N$ has the Manhattan hyperseminorm, then the Cartesian product $A \times B$ is (strongly) $2r$-additive.

Corollary 8.28. If operators $A: L \to M$ and $B: H \to N$ are (strongly) approximately additive and the Cartesian product $M \times N$ has the Manhattan hyperseminorm, then the Cartesian product $A \times B$ is (strongly) approximately additive.

Proposition 8.12. If an operator $A: L \to M$ is (strongly) r-additive, an operator $B: H \to N$ is (strongly) q-additive, and the Cartesian product $M \times N$ has the Chebyshev hyperseminorm (cf. Chapter 3), then the Cartesian product $A \times B$ is (strongly) t-additive where $t = \max \{r, q\}$.

Proof. Let us consider an r-additive operator $A: L \to M$ and q-additive operator $B: H \to N$. Then

$\| A(x + y) - A(x) - A(y)\| < r$ for arbitrary elements x and y from L

and

$\| B(z + u) - B(z) - B(u)\| < q$ for arbitrary elements z and u from H

Consequently,

$$\| (A\cdot B)((x, z) + (y, u)) - (A\cdot B)(x, z) - (A\cdot B)(y, u) \| =$$

$$\| (A\cdot B)(x + y, z + u) - (A\cdot B)(x, z) - (A\cdot B)(y, u) \| =$$

$$\| (A(x + y), B(z + u)) - (A(x), B(z)) - (A(y), B(u)) \| =$$

$$\| (A(x + y) - A(x) - A(y)), (B(z + u) - B(z) - B(u)) \| =$$

$$\max \{\| A(x + y) - A(x) - A(y) \|, \|B(z + u) - B(z) - B(u) \|\} < \max \{r, q\}$$

It means that the Cartesian product $A\times B$ is t-additive where $t = \max \{r, q\}$.

Now let us study strong approximate additivity, assuming that the operator A is strongly r-additive and the operator B is strongly q-additive. Then we have

$\| A(x - y) - A(x) + A(y)\| < r$ for arbitrary elements x and y from L

and

$\| B(z - u) - B(z) + B(u)\| < q$ for arbitrary elements z and u from H

Consequently, we have

$$\| (A\cdot B)((x, z) - (y, u)) - (A\cdot B)(x, z) + (A\cdot B)(y, u) \| =$$

$$\| (A\cdot B)(x - y, z - u) - (A\cdot B)(x, z) + (A\cdot B)(y, u) \| =$$

$$\| (A(x - y), B(z - u)) - (A(x), B(z)) + (A(y), B(u)) \| =$$

$$\| (A(x - y) - A(x) + A(y)), (B(z - u) - B(z) + B(u)) \| =$$

$$\max \{ \| A(x - y) - A(x) + A(y) \|, \|B(z - u) - B(z) + B(u) \|\} < \max \{r, q\}$$

Together with the first part of the proof, it means that the Cartesian product $A \times B$ is strongly t-additive where $t = \max\{r, q\}$.

Proposition is proved.

Corollary 8.29. If operators A: $L \to M$ and B: $H \to N$ are (strongly) r-additive, and the Cartesian product $M \times N$ has the Chebyshev norm, then the Cartesian product $A \times B$ is also (strongly) r-additive.

Corollary 8.30. If operators A: $L \to M$ and B: $H \to N$ are (strongly) approximately additive, and the Cartesian product $M \times N$ has the Chebyshev hyperseminorm, then the Cartesian product $A \times B$ is also (strongly) approximately additive.

Let us assume that M is a hyperseminormed linear algebra. Then it is possible to define the product $A \cdot B$: $L \to M$ for any operators A: $L \to M$ and B: $H \to N$ and estimate the homogeneity of this product. However, it is possible that the product of approximately additive operators is not approximately additive as Example 8.2 demonstrates.

In some cases, approximate additivity implies strong approximate additivity.

Proposition 8.13. If an operator A: $L \to M$ is r-additive and q-odd, then it is strongly$(r + q)$-additive.

Proof. Let us consider an r-additive and q-odd operator A: $L \to M$. To prove the necessary statement, we need to check the second inequality in Definition 8.5.c. Then we have

$$\| A(a-b) - (A(a) - A(b))\| = \| A(a-b) - A(a) + A(b)\| =$$

$$\| A(a-b) - A(a) + A(b) - A(-b) + A(-b)\| =$$

$$\| A(a+(-b)) - A(a) - A(-b) + A(b) + A(-b)\| \leq$$

$$\| A(a+(-b)) - A(a) - A(-b)\| + \|A(b) + A(-b)\| < r + q$$

Thus, by Lemma 8.4, the operator A is strongly$(r + q)$-additive.

Proposition is proved.

Corollary 8.31. If an operator A: $L \to M$ is r-additive and q-homogeneous, then it is strongly $(r + q)$-additive.

Corollary 8.32. If an operator A: $L \to M$ is r-additive and r-odd, then it is strongly $2r$-additive.

Corollary 8.33. If an operator A: $L \to M$ is r-additive and r-homogeneous, then it is strongly $2r$-additive.

Corollary 8.34. If an operator A: $L \to M$ is approximately additive and approximately odd, then it is strongly approximately additive.

From global approximate additivity, we go to *regional approximate additivity*.

Definition 8.6. a) An operator A: $L \to M$ is called *q-additive in a set* $X \subseteq L$ if for any elements a from X and b from L, we have

$$\| A(a+b) - A(a) - A(b) \| < q$$

b) An operator A: $L \to M$ is called *strongly q-additive in a set* $X \subseteq L$ if for any elements a from X and b from L, we have

$$\| A(a+b) - A(a) - A(b) \| < q$$

and

$$\| A(a-b) - (A(a) - A(b)) \| < q$$

c) The number $q_{A, a}$ = inf $\{r$; A is r-additive in $X\}$ is called the (*strong*) *additivity deviation* of the operator A *inX*.

d) An operator A: $L \to M$ is called *approximately additive in a set* $X \subseteq L$ if it is r-additive in X for some real number r.

e) An operator A: $L \to M$ is called *strongly approximately additive in a set* $X \subseteq L$ if it isstrongly r-additive in X for some real number r.

Note that a non-additive and even non-approximately additive operator can be additive in each set from some partition of the space L as the following example demonstrates.

Example 8.5. The operator $A(x)$ from Example 8.3 is additive in each interval $[n-1, n]$, but it is not additive and even not approximately additive in the whole space $L = \mathbf{R}$. Indeed, taking $y = 1$ and $x = n$, we have

$$\|A(x+y) - A(x) - A(y)\| = |(n+1)(n+2) - (n+1) - 2| = 2$$

As n can be an arbitrary natural number, $\|A(x+y) - A(x) - A(y)\|$ Cannot be less than some fixed real number. Thus, the operator A: $\mathbf{R} \to \mathbf{R}$ is not approximately additive.

Lemma 8.16. If an operator A: $L \to M$ is (strongly) r-additive in a set X, then it is (strongly) q-additive in X for any $q > r$.

Proof directly follows from definitions.

Let us assume that L and H are hyperseminormed spaces, while N and M are seminormed spaces.

Proposition 8.14. If an operator $A: L \to M$ is (strongly) r-additive in $X \subseteq L$, an operator $B: H \to N$ is (strongly) q-additive in $Y \subseteq H$, and the Cartesian product $M \times N$ has the Euclidean seminorm, then the Cartesian product $A \times B$ is (strongly) t-additive in $X \times Y$ where $t = (r^2 + q^2)^{1/2}$.

Proof is similar to the proof of Proposition 8.10.

Corollary 8.35. If operators $A: L \to M$ and $B: H \to N$ are (strongly) r-additive in $X \subseteq L$ and $Y \subseteq H$, respectively, while the Cartesian product $M \times N$ has the Euclidean seminorm, then the Cartesian product $A \times B$ is (strongly) ($\sqrt{2}\, r$)-additive in $X \times Y$.

Let us assume that all spaces L, H, N and M are hyperseminormed vector spaces.

Proposition 8.15. If an operator $A: L \to M$ is (strongly) r-additive in $X \subseteq L$, an operator $B: H \to N$ is (strongly) q-additive in $Y \subseteq H$, and the Cartesian product $M \times N$ has the Manhattan hyperseminorm, then the Cartesian product $A \times B$ is (strongly) t-additive in $X \times Y$ where $t = r + q$.

Proof is similar to the proof of Proposition 8.11.

Corollary 8.36. If operators $A: L \to M$ and $B: H \to N$ are (strongly) r-additive in $X \subseteq L$ and $Y \subseteq H$, respectively, while the Cartesian product $M \times N$ has the Manhattan hyperseminorm, then the Cartesian product $A \times B$ is (strongly) $2r$-additive in $X \times Y$.

Proposition 8.16. If an operator $A: L \to M$ is (strongly) r-additive in $X \subseteq L$, an operator $B: H \to N$ is (strongly) q-additive in $Y \subseteq H$, and the Cartesian product $M \times N$ has the Chebyshev hyperseminorm, then the Cartesian product $A \times B$ is (strongly) t-additive in $X \times Y$ where $t = \max \{r, q\}$.

Proof is similar to the proof of Proposition 8.12.

Corollary 8.37. If operators $A: L \to M$ and $B: H \to N$ are (strongly) r-additive in $X \subseteq L$ and $Y \subseteq H$, respectively, while the Cartesian product $M \times N$ has the Chebyshev hyperseminorm, then the Cartesian product $A \times B$ is also (strongly) r-additive in $X \times Y$.

Corollary 8.38. a) If operators $A: L \to M$ and $B: H \to N$ are (strongly) approximately additive in $X \subseteq L$ and $Y \subseteq H$, respectively, while the Cartesian product $M \times N$ has the Chebyshev hyperseminorm or the Manhattan hyperseminorm, then the Cartesian product $A \times B$ is also (strongly) approximately additive in $X \times Y$.

b) If L and H are hyperseminormed spaces, N and M are seminormed spaces, operators $A: L \to M$ and $B: H \to N$ are (strongly) approximately additive in $X \subseteq L$ and $Y \subseteq H$, respectively, while the Cartesian product $M \times N$ has the Euclidean seminorm, then the Cartesian product $A \times B$ is also (strongly) approximately additive in $X \times Y$.

Let us study properties of sums of operators and of products of operators and numbers.

Proposition 8.17. a) If an operator $A: L \to M$ is (strongly) r-additive in X and an operator $B: L \to M$ is (strongly) q-additive in X, then the operator $A + B: L \to M$ is (strongly) $(r + q)$-additive in X.

b) If an operator $A: L \to M$ is (strongly) r-additive in X and k is a real number, then the operator $kA: L \to M$ is (strongly) $|k| \cdot r$-additive in X.

Proof is similar to the proof of Proposition 8.9.

Corollary 8.39. If operators $A: L \to M$ and $B: L \to M$ are approximately additive in X, then operators $A + B$ and kA are also approximately additive in X.

Some authors studied another kind of approximate additivity [cf., e.g., (Rassias and Rassias, 2003; Rassias and Rassias, 2005; Rassias, 2007)]. We slightly generalize their concept.

Let us take $r \geq 0$ and $0 \leq p < 1$.

Definition 8.7. a) An operator $A: L \to M$ is called *calibrated* (r, p)-*additive* if the following condition is satisfied

$$\|A(x + y) - A(x) - A(y)\| \leq r \cdot (\|x\|^p + \|y\|^p)$$

for all $x, y \in L$.

b) An operator $A: L \to M$ is called *calibrated r-additive* if the following condition is satisfied

$$\|A(x + y) - A(x) - A(y)\| \leq r \cdot (\|x\| + \|y\|)$$

for all $x, y \in L$.

c) An operator $A: L \to M$ is *weakly* C-*approximately additive* if it is calibrated (r, p)-additive for some non-negative real number r and some $0 \leq p < 1$.

d) An operator $A: L \to M$ is C-*approximately additive* if it is calibrated r-additive for some non-negative real number r.

Lemma 8.17. a) If an operator A is calibrated (r, p)-additive and $r < q$, then A is calibrated (q, p)-additive.

b) If an operator A is calibrated r-additive and $r < q$, then A is calibrated q-additive.

Proof is left as an exercise.

Proposition 8.18. a) If an operator $A: L \to M$ is calibrated (r, p)-additive and an operator $B: L \to M$ is calibrated (q, p)-additive, then the operator $A + B: L \to M$ is calibrated $(r + q, p)$-additive.

b) If an operator $A: L \to M$ is calibrated (r, p)-additive and k is a real number, then the operator $kA: L \to M$ is calibrated $(|k| \cdot r, p)$-additive.

Proof. a) Let us consider a calibrated (r, p)-additive operator $A: L \to M$ and calibrated q-additive operator $B: H \to N$. Then

$$\|A(x + y) - A(x) - A(y)\| \leq r \cdot (\|x\|^p + \|y\|^p) \text{ for arbitrary elements } x \text{ and } y \text{ from } L$$

and

$$\|B(x + y) - B(x) - B(y)\| \leq q \cdot (\|x\|^p + \|y\|^p) \text{ for arbitrary elements } x \text{ and } y \text{ from } H$$

Consequently, we have

$$\| (A + B)(x + y) - (A + B)(x) - (A + B)(y) \| =$$

$$\| A(x + y) - A(x) - A(y) + B(x + y) - B(x) - B(y) \| \leq$$

$$\| A(x + y) - A(x) - A(y)\| + \|B(x + y) - B(x) - B(y) \| <$$

$$r \cdot (\|x\|^p + \|y\|^p) + q \cdot (\|x\|^p + \|y\|^p) = (r + q) \cdot (\|x\|^p + \|y\|^p)$$

It means that the operator $A + B: L \to M$ is calibrated $(r + q, p)$-additive.

b) Let us consider a calibrated (r, p)-additive operator $A: L \to M$. Then

$$\| A(x + y) - A(x) - A(y)\| < r \cdot (\|x\|^p + \|y\|^p) \text{ for arbitrary elements } x \text{ and } y \text{ from } L$$

Consequently, we have

$$\| kA(x + y) - kA(x) - kA(y)\| =$$

$$\|k(A(x + y) - A(x) - A(y))\| =$$

$$|k| \cdot \|A(x + y) - A(x) - A(y))\| < |k| \cdot r \cdot (\|x\|^p + \|y\|^p)$$

It means that the operator $kA: L \to M$ is calibrated $(|k|\cdot r, p)$-additive.

Proposition is proved.

Corollary 8.40. If operators $A: L \to M$ and $B: L \to M$ are weakly C-approximately additive, then the operators $A + B$ and kA are also weakly C-approximately additive.

Corollary 8.41. If functionals $A: L \to \mathbf{R}$ and $B: L \to \mathbf{R}$ are weakly C-approximately additive, then the functionals $A + B$ and kA are also weakly C-approximately additive.

Now let us study the situation when $p = r$.

Proposition 8.19. a) If an operator $A: L \to M$ is calibrated r-additive and an operator $B: L \to M$ is calibrated q-additive, then the operator $A + B: L \to M$ is calibrated $(r + q)$-additive.

b) If an operator $A: L \to M$ is calibrated r-additive and k is a real number, then the operator $kA: L \to M$ is calibrated $|k|\cdot r$-additive.

Proof is similar to the proof of Proposition 8.18.

Corollary 8.42. If operators $A: L \to M$ and $B: L \to M$ are C-approximately additive, then the operators $A + B$ and kA are also C-approximately additive.

Corollary 8.43. If functionals $A: L \to \mathbf{R}$ and $B: L \to \mathbf{R}$are C-approximately additive, then the functionals $A + B$ and kA are also C-approximately additive.

In analysis, mathematicians studied different classes of functions and operators. One of the important classes with many good properties is the class of Lipschitz functions (operators). We remind that an operator (mapping) $A: L \to M$ is called *Lipschitz* if there is a number c such that for any elements x and y from L

$$\| A(y) - A(x) \| < c \cdot \| y - x \|$$

Proposition 8.20. Any centered Lipschitz operator (mapping) $A: L \to M$ is C-approximately additive.

Proof. Let us consider a centered Lipschitz operator $A: L \to M$. Then there is a number c such that for any elements x and y from L

$$\| A(y) - A(x) \| < c \cdot \| y - x \|$$

By properties of hyperseminorms, we have

$$\| A(x+y) - A(x) - A(y) \| = \| A(x+y) - (A(x) + A(y)) \| \leq$$

$$\| A(x+y) \| + \| A(x) + A(y) \| \leq \| A(x+y) \| + \| A(x) \| + \| A(y) \|$$

$A(\mathbf{0}) = \mathbf{0}$ because A is a centered operator.

$$\| A(x+y) \| + \| A(x) \| + \| A(y) \| = \| A(x+y) - A(\mathbf{0}) \| + \| A(x) - A(\mathbf{0}) \| + \| A(y) - A(\mathbf{0}) \| <$$

$$c \cdot \| x + y - \mathbf{0} \| + c \cdot \| x - \mathbf{0} \| + c \cdot \| y - \mathbf{0} \| = c(\| x + y \| + \| x \| + \| y \|) \leq$$

$$c(\| x \| + \| y \| + \| x \| + \| y \|) = 2c(\| x \| + \| y \|)$$

It means that the operator A: $L \rightarrow M$ is calibrated $2c$-additive and thus, C-approximately additive.

Proposition is proved.

Additivity and homogeneity taken together give us linearity of operators.

Definition 8.8. a) An operator A: $L \rightarrow M$ is called *linear* if for all elements $x, y \in L$ and real numbers $a, b \in \boldsymbol{R}$, we have

$$A(ax + by) = aA(x) + bA(y)$$

b) An operator A: $L \rightarrow M$ is called (q, r)-*linear* if it is r-homogeneous and q-additive.

c) The number $p = \max \{ q_A, r_A \}$ where $q_A = \inf \{r;\ A$ is r-additive$\}$ and $h_A = \inf \{r;\ A$ is r-homogeneous$\}$ is called the *combined linearity deviation* of the operator A.

d) An operator A: $L \rightarrow M$ is called r-*linear* if it is both r-homogeneous and r-additive.

e) The number $l_A = \inf \{r;\ A$ is r-linear$\}$ is called a *linearity deviation* of the operator A.

f) An operator A: $\boldsymbol{L} \rightarrow \boldsymbol{M}$ is called *approximately linear* if it is (q, r)-linear for some numbers q and r.

Example 8.6. Let us consider the one-dimensional operator $A(a) = x + 0.05\sin x$. This operator A is 0.05-linear and continuous in $\boldsymbol{R}$.

Lemmas 8.4 and 8.7 imply the following result.

Let us assume that M is a hypernormed (normed) vector space.

Lemma 8.18. The following properties are equivalent:

1) An operator A: $L \rightarrow M$ is linear.
2) The operator A is 0-linear.
3) The operator A is (0, 0)-linear.

Propositions 8.8 and Lemmas 8.3 imply the following result.

Proposition 8.21. If an operator A: $L \rightarrow M$ is r-linear, then for any elements a and b from L, the following inequalities are true

$$|k| \cdot \|A(a)\| < \| A(ka) \| + r \tag{8.7}$$

$$\| A(ka) \| < |k| \, \|A(a)\| + r \tag{8.8}$$

$$\| A(a+b) \| < \| A(a) + A(b)\| + r \tag{8.9}$$

$$\| A(a) + A(b)\| < \| A(a+b) \| + r \tag{8.10}$$

$$\| A(a-b) \| < \| A(a) - A(b)\| + 2r \tag{8.11}$$

$$\| A(a) - A(b)\| < \| A(a-b) \| + 2r \tag{8.12}$$

Lemmas 8.3 and 8.6 imply the following results.

Corollary 8.44. If an operator A: $L \rightarrow M$ is (q, r)-linear, then it is (u, v)-linear for any $u > q$ and $v > r$.

Corollary 8.45. If an operator A: $L \rightarrow M$ is r-linear, then it is q-linear for any $q > r$.

Propositions 8.1 and 8.9 imply the following result.

Proposition 8.22. a) If an operator A: $L \rightarrow M$ is (r, t)-linear and an operator B: $L \rightarrow M$ is (q, p)-linear, then the operator $A + B$: $L \rightarrow M$ is $(r + q, t + p)$-linear.

b) If an operator A: $L \rightarrow M$ is (r, t)-linear and k is a real number, then the operator kA: $L \rightarrow M$ is $(|k|r, |k|t)$-linear.

Proof is left as an exercise.

Taking $r = t$ and $q = p$ in Proposition 8.22, we obtain the following result.

Corollary 8.46. a) If an operator A: $L \rightarrow M$ is r-linear and an operator B: $L \rightarrow M$ is q-linear, then the operator $A + B$: $L \rightarrow M$ is $(r + q)$-linear.

b) If an operator A: $L \rightarrow M$ is r-linear and k is a real number, then the operator kA: $L \rightarrow M$ is $(|k|r)$-linear.

Corollary 8.47. The combined linearity deviation of the sum is less than or equal to the sum of the combined linearity deviations.

Corollary 8.48. If an operator A: $L \rightarrow M$ is (r, t)-linear and an operator B: $L \rightarrow M$ is (q, p)-linear, then the operator $dA + cB$: $L \rightarrow M$ is $(|d|r + |c|q), |d|t + |c|p)$-linear for any numbers $d, c \in \boldsymbol{R} \setminus \{0\}$.

Corollary 8.49. If an operator A: $L \to M$ is r- linear and an operator B: $L \to M$ is q-linear, then the operator $dA + cB$: $L \to M$ is $(|d|r + |c|q)$-linear for any numbers $d, c \in \mathbf{R}\backslash\{0\}$.

Corollary 8.50. If operators A: $L \to M$ and B: $L \to M$ are approximately linear, then the operator $dA + cB$: $L \to M$ is also approximately linear for any numbers $d, c \in \mathbf{R}\backslash\{0\}$.

Now let us explore linearity of the Cartesian product of operators assuming, at first, that L and H are hyperseminormed spaces, while N and M are seminormed spaces.

Propositions 8.2 and 8.10 imply the following result.

Proposition 8.23. If an operator A: $L \to M$ is r-linear, an operator B: $H \to N$ is q-linear, and the Cartesian product $M \times N$ has the Euclidean seminorm, then the Cartesian product $A \times B$ is t-linear where $t = (r^2 + q^2)^{1/2}$.

Proof is left as an exercise.

Corollary 8.51. If operators A: $L \to M$ and B: $H \to N$ are r-linear, and the Cartesian product $M \times N$ has the Euclidean seminorm, then the Cartesian product $A \times B$ is $\sqrt{2}\, r$-linear.

Let us assume that L, H, N and M are hyperseminormed spaces. Then Propositions 8.3 and 8.11 imply the following result.

Proposition 8.24. If an operator A: $L \to M$ is r-linear, an operator B: $H \to N$ is q-linear, and the Cartesian product $M \times N$ has the Manhattan hyperseminorm, then the Cartesian product $A \times B$ is t-linear where $t = r + q$.

Proof is left as an exercise.

Corollary 8.52. If operators A: $L \to M$ and B: $H \to N$ are r-linear, and the Cartesian product $M \times N$ has the Manhattan hyperseminorm, then the Cartesian product $A \times B$ is $2r$-linear.

Propositions 8.4 and 8.12 imply the following result.

Proposition 8.25. If an operator A: $L \to M$ is r-linear, an operator B: $H \to N$ is q-linear, and the Cartesian product $M \times N$ has the Chebyshev hyperseminorm, then the Cartesian product $A \times B$ is t-linear where $t = \max\{r, q\}$.

Proof is left as an exercise.

Corollary 8.53. If operators A: $L \to M$ and B: $H \to N$ are r-linear, and the Cartesian product $M \times N$ has the Chebyshev hyperseminorm, then the Cartesian product $A \times B$ is also r-linear.

Corollary 8.54. a) If operators A: $L \to M$ and B: $H \to N$ are approximately linear, and the Cartesian product $M \times N$ has the Chebyshev hyperseminorm or the Manhattan hyperseminorm, then the Cartesian product $A \times B$ is also approximately linear.

b) If L and H are hyperseminormed spaces, N and M are seminormed spaces, operators A: $L \to M$ and B: $H \to N$ are approximately linear, and the Cartesian product $M \times N$ has the Euclidean seminorm, then the Cartesian product $A \times B$ is also approximately linear.

From global approximate linearity, we go to regional approximate linearity.

Definition 8.9. a) An operator A: $L \to M$ is called *(r, q)-linear in a set* $X \subseteq L$ if it is r-homogeneous in X and q-additive in X.

b) An operator A: $L \to M$ is called *r-linear in a set* $X \subseteq L$ if it is (r, r)-linear in X.

c) An operator A: $L \to M$ is *approximately linear in a set* $X \subseteq L$ if it is (r, q)-linear in X for some non-negative real numbers r and q.

d) The number $p = \max \{q_A, r_A\}$ where $q_A = \inf \{r;\ A \text{ is } r\text{-additive in } X\}$ and $h_A = \inf \{r;\ A \text{ is } r\text{-homogeneous in } X\}$ is called the *combined linearity deviation* of the operator A in X.

e) The number $l_A = \inf \{r;\ A \text{ is } r\text{-linear in } X\}$ is called the *linearity deviation* of the operator A in X.

Examples 8.3 and 8.5 show that a non-linear operator can be linear in each part from some partition of the space L. In the case of real functions, such a mapping is called a *piecewise linear function* as it is a function composed of straight-line sections. A more general example is a *piecewise-defined function*, pieces of which are affine functions.

Example 8.7. *Segmented linear regression*, also known as *piecewise linear regression* or *broken-stick regression*, is a method in statistics in which the domain of the independent variable x is partitioned into intervals and linear regression is applied in each interval (Fox, 2016). Segmented linear regression is also performed on multivariate data by partitioning the domains of different independent variables. If data modeled by a segmented linear regression form a function $y(x)$, then this function is approximately linear in intervals (in multidimensional rectangular prisms) of the used partition. Segmented regression is useful when the independent variables, clustered into different groups, exhibit different statistical relationships between the dependent and independent variables in these regions.

Moreover, segmentation may be true for approximately linear operators, namely, a non-linear operator can be linear in each set from some partition of the space L but not approximately linear (in L). However, when only one element of the partition has the infinite measure, this is not true.

Lemma 8.19. If an operator A: $L \to M$ is (r, q)-linear in a set X, then it is (l, h)-linear in X for any $l > r$ and $h > q$.

Proof is left as an exercise.

Propositions 8.5 and 8.17 imply the following result.

Proposition 8.26. a) If an operator A: $L \to M$ is (r, t)-linear in X and an operator B: $L \to M$ is (q, p)-linear in X, then the operator $A + B$: $L \to M$ is $(r + q, t + p)$-linear in X, i.e., the combined linearity deviation of the sum in X is the sum of the combined linearity deviations in X.

b) If an operator A: $L \to M$ is (r, t)-linear in X and k is a real number, then the operator kA: $L \to M$ is $(|k|r, |k|t)$-linear in X.

Proof is left as an exercise.

Taking $r = t$ and $q = p$ in Proposition 8.26, we obtain the following result.

Corollary 8.55. a) If an operator A: $L \to M$ is r-linear in X and an operator B: $L \to M$ is q-linear in X, then the operator $A + B$: $L \to M$ is $(r + q)$-linear in X.

b) If an operator A: $L \to M$ is r-linear and k is a real number, then the operator kA: $L \to M$ is $(|k|r)$-linear.

Corollary 8.56. If an operator A: $L \to M$ is (r, t)-linear in a subset X of the space L and an operator B: $L \to M$ is (q, p)-linear in X, then the operator $dA + cB$: $L \to M$ is $(|d|r + |c|q), |d|t + |c|p)$-linear in X for any numbers $d, c \in \boldsymbol{R} \setminus \{0\}$.

Corollary 8.57. If an operator A: $L \to M$ is r-linear in a set X and an operator B: $L \to M$ is q-linear in X, then the operator $dA + cB$: $L \to M$ is $(|d|r + |c|q)$-linear in X for any numbers $d, c \in \boldsymbol{R} \setminus \{0\}$.

Corollary 8.58. If operators A: $L \to M$ and B: $L \to M$ are approximately linear in a set X, then the operator $dA + cB$: $L \to M$ is also approximately linear in X for any numbers $d, c \in \boldsymbol{R} \setminus \{0\}$.

Now let us explore relative linearity of Cartesian products of operators assuming, at first, that L and H are hyperseminormed spaces, while N and M are seminormed spaces.

Propositions 8.2 and 8.10 imply the following result.

Proposition 8.27. If an operator A: $L \to M$ is r-linear in $X \subseteq L$, an operator B: $H \to N$ is q-linear in $Y \subseteq H$, and the Cartesian product $M \times N$ has the Euclidean seminorm, then the Cartesian product $A \times B$ is t-linear in $X \times Y$ where $t = (r^2 + q^2)^{1/2}$.

Proof is left as an exercise.

Corollary 8.59. If operators A: $L \to M$ and B: $H \to N$ are r-linear in $X \subseteq L$ and $Y \subseteq H$, respectively, while the Cartesian product $M \times N$ has the Euclidean seminorm, then the Cartesian product $A \times B$ is $\sqrt{2}\, r$-linear in $X \times Y$.

Let us assume that L, H, N and M are hyperseminormed spaces. Then Propositions 8.3 and 8.11 imply the following result.

Proposition 8.28. If an operator $A: L \to M$ is r-linear in $X \subseteq L$, an operator $B: H \to N$ is q-linear in $Y \subseteq H$, and the Cartesian product $M \times N$ has the Manhattan hyperseminorm, then the Cartesian product $A \times B$ is t-linear in $X \times Y$ where $t = r + q$.

Proof is left as an exercise.

Corollary 8.60. If operators $A: L \to M$ and $B: H \to N$ are r-linear in $X \subseteq L$ and $Y \subseteq H$, respectively, while the Cartesian product $M \times N$ has the Manhattan hyperseminorm, then the Cartesian product $A \times B$ is $2r$-linear in $X \times Y$.

Propositions 8.4 and 8.12 imply the following result.

Proposition 8.29. If an operator $A: L \to M$ is r-linear in $X \subseteq L$, an operator $B: H \to N$ is q-linear in $Y \subseteq H$, and the Cartesian product $M \times N$ has the Chebyshev hyperseminorm, then the Cartesian product $A \times B$ is t-linear in $X \times Y$ where $t = \max \{r, q\}$.

Proof is left as an exercise.

Corollary 8.61. If operators $A: L \to M$ and $B: H \to N$ are r-linear in $X \subseteq L$ and $Y \subseteq H$, respectively, while the Cartesian product $M \times N$ has the Chebyshev hyperseminorm, then the Cartesian product $A \times B$ is also r-linear in $X \times Y$.

Corollary 8.62. a) If operators $A: L \to M$ and $B: H \to N$ are approximately linear in $X \subseteq L$ and $Y \subseteq H$, respectively, while the Cartesian product $M \times N$ has the Chebyshev hyperseminorm or the Manhattan hyperseminorm, then the Cartesian product $A \times B$ is also approximately linear in $X \times Y$.

b) If L and H are hyperseminormed spaces, N and M are seminormed spaces, operators $A: L \to M$ and $B: H \to N$ are approximately linear in $X \subseteq L$ and $Y \subseteq H$, respectively, while the Cartesian product $M \times N$ has the Euclidean seminorm, then the Cartesian product $A \times B$ is also approximately linear in $X \times Y$.

Now let us look at some relations between approximate linearity, approximate additivity and fuzzy continuity. For instance, for linear operators, continuity at **0** implies continuity everywhere (Rudin, 1991). A similar statement is true for additive r-continuous in L operators.

Proposition 8.30. A strongly additive operator (mapping) $A: L \to M$ is r-continuous in L if and only if it is r-continuous at the point **0**.

Proof. Necessity. Any r-continuous in L operator is r-continuous at the point **0**.

Sufficiency. Let us consider an additive r-continuous at the point **0** operator $A: L \to M$ and a sequence $l = \{a_i \in L;\ i = 1, 2, 3, \dots\}$ that converges to a point a. Then the sequence $h = \{a - a_i \in L;\ i = 1, 2, 3, \dots\}$ converges to the point $\mathbf{0}$ in L. As the operator A is r-continuous at the point $\mathbf{0}$, the sequence $g = \{A(a - a_i) \in L;\ i = 1, 2, 3, \dots\}$ r-converges to the point $A(\mathbf{0})$. It means that the element $A(\mathbf{0})$ is an r-limit of the sequence g, i.e., $A(\mathbf{0}) = r\text{-}\lim_{i\to\infty} A(a - a_i)$, and by Definition 7.1, for any $\varepsilon \in \boldsymbol{R}^{++}$, there is a natural number n such that for any $i > n$, we have

$$\| A(a - a_i) - A(\mathbf{0})\| < r + \varepsilon$$

and as A is a strongly additive operator, we have

$$\| A(a) - A(a_i)\| = \| A(a - a_i)\| = \| A((a - a_i) - \mathbf{0}) \| =$$

$$\| A(a - a_i) - A(\mathbf{0})\| < r + \varepsilon$$

for any $i > n$. It means that $A(a) = q\text{-}\lim_{i\to\infty} A(a_i)$, i.e., the operator (mapping) A is r-continuous at x and thus, A is r-continuous in L.

Proposition is proved.

Corollary 8.63. A linear operator (mapping) A: $L \to M$ is r-continuous in L if and only if it is r-continuous at the point $\mathbf{0}$.

Corollary 8.64. A strongly additive operator (mapping) A: $L \to M$ is fuzzy continuous in L if and only if it is fuzzy continuous at the point $\mathbf{0}$.

Corollary 8.65. A linear operator (mapping) A: $L \to M$ is fuzzy continuous in L if and only if it is fuzzy continuous at the point $\mathbf{0}$.

Proposition 8.30 also gives us the following classical result (Dunford and Schwartz, 1958; Rudin, 1991; Kolmogorov and Fomin, 1999).

Corollary 8.66. A linear operator A: $L \to M$ is continuous at the point $\mathbf{0}$ from L if and only if A is continuous in L.

For approximately linear and approximately additive in L operators, we have a weaker result.

Proposition 8.31. If a strongly q-additive operator (mapping) A: $L \to M$ is r-continuous at the point $\mathbf{0}$ from L, then A is u-continuous in L where $u = q + r + \| A(\mathbf{0})\|$.

Proof. Let us consider strongly q-additive r-continuous at the point $\mathbf{0}$ operator A: $L \to M$ and a sequence $l = \{a_i \in L;\ i = 1, 2, 3, \dots\}$ that converges to a point a. Then the sequence $h = \{a - a_i \in L;\ i = 1, 2, 3, \dots\}$ converges to the point $\mathbf{0}$ in L. As the operator A is r-continuous at the point $\mathbf{0}$, the sequence $g = \{A(a - a_i) \in L;\ i = 1, 2, 3, \dots\}$ r-converges to the point $A(\mathbf{0})$. It means that the element $A(\mathbf{0})$ is an r-limit of the sequence g, i.e., $A(\mathbf{0}) = r\text{-}\lim_{i\to\infty} A(a - a_i)$, and by Definition 7.1, for any $\varepsilon \in \boldsymbol{R}^{++}$, there is $n > 0$ such that for any $i > n$, we have

$$\| A(a - a_i) - A(\mathbf{0})\| < r + \varepsilon$$

Consequently, by properties of hyperseminorms, we have

$$\| A(a - a_i) \| = \| A(a - a_i) - A(\mathbf{0}) + A(\mathbf{0})\| \leq$$

$$\| A(a - a_i) - A(\mathbf{0})\| + \|A(\mathbf{0})\| < r + t + \varepsilon$$

for any $i > n$, where $t = \| A(\mathbf{0})\|$.

Therefore, for any $i > n$,

$$\| A(a) - A(a_i)\| = \| A(a) - A(a_i) - A(a - a_i) + A(a - a_i) \| \leq$$

$$\| A(a) - A(a_i) - A(a - a_i) \| + \|A(a - a_i)\| < r + t + q + \varepsilon = p + \varepsilon$$

where $p = r + t + q$. It means that $A(a) = p\text{-}\lim_{i\to\infty} A(a_i)$, i.e., the operator (mapping) A is p-continuous at x and thus, A is p-continuous in L as $p \geq r$.

Proposition is proved.

Corollary 8.67. If an approximately strongly additive operator (mapping) A: $L\to M$ is fuzzy continuous at the point $\mathbf{0}$ from L, then A is fuzzy continuous in L.

Corollary 8.68. If a q-linear operator (mapping) A: $L\to M$ is r-continuous at the point $\mathbf{0}$ from L, then A is u-continuous in L where $u = q + r + \| A(\mathbf{0})\|$.

Corollary 8.69. If an approximately linear operator (mapping) A: $L\to M$ is fuzzy continuous at the point $\mathbf{0}$ from L, then A is fuzzy continuous in L.

The following example shows that Proposition 8.31 and Corollary 8.68 give exact estimates of the continuity defect for approximately linear operators as there are such operators for which it is impossible to improve these estimates.

Example 8.8. Let us consider the one-dimensional operator A defined by the following formula

$$A(x) = \begin{cases} 3x & \text{if } x \leq 1 \\ 3x + 0.1 & \text{if } x > 1 \end{cases}$$

This operator is (0.1)-linear, (0.1)-continuous at the point 1 and continuous at all other points. In particular, A is continuous at 0. This shows that for approximately linear operators, continuity at 0 does imply continuity at other points. Besides, in contrast to linear operators, the defect of continuity of approximately linear operators can be different at different points.

Operator spaces often have hypernorms, hyperseminorms, seminorms and norms, which are constructed using seminorms and norms in vector spaces. For instance, for a bounded operator $D: L \to M$ from a seminormed (normed) vector space L into a seminormed (normed) vector space M, the conventional operator seminorm (the *unit supremum seminorm*) is given by the formula

$$||D||_1 = \sup \{||D(x)||; x \in L, ||x|| \leq 1\}$$

Another operator seminorm (the *standard supremum seminorm*) is given by the formula

$$||D||_2 = \sup \{||D(x)||; x \in L, ||x|| = 1\}$$

There is also an operator norm (the *relational supremum norm*), which is frequently used in functional analysis and is given by the formula

$$||D||_3 = \sup \{||D(x)||/||x||; x \in L, ||x|| \neq 0\}$$

When D is a linear operator, the unit supremum seminorm coincides with the relational supremum norm, as well as with the standard supremum seminorm.

The norm $||.||_3$ is defined only for operators with the bounded ratio $||D(x)||/||x||$. For unbounded operators, the value $\sup\{||D(x)||/||x||; x \in l\}$ tends to infinity for some sequences l in L. In this case, it is possible to define the operator hypernorm using the following construction.

At first, we build the cumulative cover $\{T_n = \{x; x \in L, 1/n < ||x|| < n\}; n = 1, 2, 3, \ldots\}$ of the vector space L. Then if the operator D is bounded in all sets T_n, we define

$$a_n = ||D||^{(n)} = \sup \{||D(x)||/||x||; x \in T_n\}$$

and

$$||D||^{(-)} = \mathrm{Hn}(a_n)_{n\in\omega}$$

When $||.||$ is a norm in M, the function $||.||^{(\infty)}$ is a hypernorm in the space of all operators from L into M. In a general case, $||.||^{(\infty)}$ is a hyperseminorm, which is called the *relational supremum hyperseminorm*. Note that this hyperseminorm is defined for all operators from L into M that are bounded in all sets T_n. As a result, the space of all such operators becomes a hyperseminormed or hypernormed vector space. This essentially expands the space of operators where it is possible to develop a theory similar to the theory of normed spaces, such Banach spaces or Hilbert spaces.

For instance, taking the one-dimensional operator Sq: $\boldsymbol{R} \to \boldsymbol{R}$ such that $Sq(x) = x^2$ and the absolute value as the norm in $\boldsymbol{R}$, we see that

$$||Sq(x)||/||x|| = |x|$$

for all x and this value grows with x. So,

$$||Sq||_1 = ||Sq||_2 = 1$$

while $||Sq||_3$ does not exist or if we allow infinite values $||Sq||_3 = \infty$. However, the second option does not allow discerning different operators with infinite norm. For instance, all three operators $Cb(x) = x^3$, $Sq(x) = x^2$ and $Sqr(x) = x^2 + x$ will have the same infinite norm while they are essentially different. Besides, using these infinite norms, we will not be able to find that the difference $Sqr - Sq$ has the finite hypernorm equal to 1.

In contrast to this, the hypernorm $||.||^{(\infty)}$ is defined for all three operators:

$$||Sq||^{(\infty)} = \mathrm{Hn}(n)_{n \in \omega}$$

$$||Cb||^{(\infty)} = \mathrm{Hn}(n^2)_{n \in \omega}$$

$$||\ Sqr\ ||^{(\infty)} = \mathrm{Hn}(n)_{n \in \omega} + 1$$

and

$$||\ Sqr - Sq\ ||^{(\infty)} = \mathrm{Hn}(n)_{n \in \omega} + 1$$

Operator norms and hypernorms bring us to another kind of approximate homogeneity, additivity and linearity.

Let us assume that $||.||$ is an operator hyperseminorm (seminorm) [cf., e.g., (Rudin, 1991)]. For instance, taking a hyperseminorm (seminorm) q in M and an element a from L, we can define a hyperseminorm (seminorm) of an arbitrary operator A: $L \to M$ by the following formula

$$\|A\|_a = q(A(a))$$

Definition 8.10. a) An operator $A: L \to M$ is called *functionally r-homogeneous* if there is a homogeneous operator $B: L \to M$ such that

$$\|A - B\| < r$$

b) The number $fh_A = \inf \{r; A \text{ is functionally } r\text{-homogeneous}\}$ is called a *functional homogeneity deviation* of the operator A.

c) An operator $A: L \to M$ is F-*approximately homogeneous* if it is functionally r-homogeneous for some non-negative real number r.

As $\|.\|$ is a hypernorm (norm), we have the following result.

Lemma 8.20. An operator $A: L \to M$ is functionally 0-homogeneous if and only if A is homogeneous.

Proof is left as an exercise.

Lemma 8.21. If an operator $A: L\to M$ is functionally r-homogeneous, then it is functionally l- homogeneous for any $l > r$.

Proof is left as an exercise.

Proposition 8.32. a) If an operator $A: L\to M$ is functionally r-homogeneous and an operator $C: L\to M$ is functionally q-homogeneous, then the operator $A + C: L\to M$ is functionally $(r + q)$-homogeneous, i.e., the functional homogeneity deviation of the sum is the sum of the functional homogeneity deviations.

b) If an operator $A: L\to M$ is functionally r-homogeneous and k is a real number, then the operator $kA: L\to M$ is functionally $|k|\cdot r$-homogeneous.

Proof. a) Let us consider a functionally r-homogeneous operator $A: L\to M$ and a functionally q-homogeneous operator $B: L\to M$. Then we have

$$\|A - B\| < r$$

and

$$\|C - D\| < r$$

for some homogeneous operators $B: L \to M$ and $D: L \to M$. By Proposition 8.1.a and Lemma 8.2, the sum $B + D$ is a homogeneous operator. Thus, we have

$$\|(A + C) - (B + D)\| = \|(A - B) + (C - D)\| \leq$$

$$\|A - B\| + \|C - D\| < r + q$$

It means that the operator $A + C$ is functionally $(r + q)$-homogeneous.

2) Now let us deal with the operator kA. Note that if the operator B is homogeneous, then by Proposition 8.1.b and Lemma 8.2, the operator kB is homogeneous. Thus, we have

$$\| kA - kB \| = |k| \cdot \|A - B\| < |k| \cdot r$$

It means that the operator kA is functionally $|k|\cdot r$-homogeneous.

Proposition is proved.

Corollary 8.70. If operators $A: L \to M$ and $B: L \to M$ are F-approximately homogeneous, then the operator $A + B$ and kA are also F-approximately homogeneous.

From global relative homogeneity, we come to regional relative homogeneity.

Definition 8.11. a) An operator $A: L \to M$ is called *functionally r-homogeneous in* a set X if there is a homogeneous in X operator $B: L \to M$ such that

$$\|A - B\| < r$$

b) An operator $A: L \to M$ is F-*approximately homogeneous in* a set X if it is functionally r-homogeneous in X for some non-negative real number r.

A functionally non-homogeneous operator can be functionally homogeneous in some set. For instance, any operator $A: L \to M$, which is defined at a point a from L, is functionally homogeneous in the set $X = \{a\}$ because there is always a linear, and thus, homogeneous operator $B: L \to M$ such that $A(a) = B(a)$.

Lemma 8.22. If an operator $A: L \to M$ is functionally r-homogeneous in a set X, then it is functionally l- homogeneous in X for any $l > r$.

Let us consider functional homogeneity of sums of operators and products of operators and numbers.

Proposition 8.33. a) If an operator $A: L \to M$ is functionally r-homogeneous in a set X and an operator $C: L \to M$ is functionally q-homogeneous in X, then the operator $A + C: L \to M$ is functionally $(r + q)$-homogeneous in X.

b) If an operator $A: L \to M$ is functionally r-homogeneous in X and k is a real number, then the operator $kA: L \to M$ is functionally $|k|\cdot r$-homogeneous in X.

Proof is similar to the proof of Proposition 8.5.

Corollary 8.71. If operators A: $L \to M$ and B: $L \to M$ are F-approximately homogeneous in X, then the operators $A + B$ and kA are also F-approximately homogeneous in X.

There is one more type of homogeneity.

Definition 8.12. a) An operator A: $L \to M$ is called *pointwise r-homogeneous* if there is a homogeneous operator B: $L \to M$ such that

$$\|A(x) - B(x)\| < r \text{ for all elements } x \text{ from } L$$

b) An operator A: $L \to M$ is P-*approximately homogeneous* if it is pointwise r-homogeneous for some non-negative real number r.

c) The number $ph_A = \inf \{r; A \text{ is pointwise } r\text{-homogeneous}\}$ is called the *pointwise r-homogeneity deviation* of the operator A.

For instance, the function sin x is pointwise 1-homogeneous.

Let us assume that $\|.\|$ is a hypernorm (norm) in M.

Lemma 8.23. An operator A: $L \to M$ is pointwise 0-homogeneous if and only if A is homogeneous.

Proof is left as an exercise.

Proposition 8.34. a) If an operator A: $L \to M$ is pointwise r-homogeneous and an operator C: $L \to M$ is pointwise q-homogeneous, then the operator $A + C$: $L \to M$ is pointwise $(r + q)$-homogeneous.

b) If an operator A: $L \to M$ is pointwise r-homogeneous and k is a real number, then the operator kA: $L \to M$ is pointwise $|k| \cdot r$-homogeneous.

Proof. a) Let us consider a pointwise r-homogeneous operator A: $L \to M$ and a pointwise q-homogeneous operator B: $L \to M$. Then we have

$$\|A(x) - B(x)\| < r$$

and

$$\|C(x) - D(x)\| < r$$

for all x from L and some homogeneous operators B: $L \to M$ and D: $L \to M$. By Proposition 8.1.a and Lemma 8.2, the sum $B + D$ is a homogeneous operator. Thus we have

$$\|(A(x) + C(x)) - (B(x) + D(x))\| = \|(A(x) - B(x)) + (C(x) - D(x))\| \leq$$

$$\|A(x) - B(x)\| + \|C(x) - D(x)\| < r + q$$

It means that the operator $A + C$ is pointwise$(r + q)$-homogeneous.

2) Now let us treat the operator kA. Note that if the operator B is pointwise homogeneous, then by Proposition 8.1.b and Lemma 8.2, the operator kB is pointwise homogeneous. Thus, we have

$$\| kA(x) - kB(x)\| = |k| \cdot \|A(x) - B(x)\| < |k|\cdot r$$

It means that the operator kA is pointwise $|k|\cdot r$-homogeneous.

Proposition is proved.

Corollary 8.72. If operators A: $L \to M$ and B: $L \to M$ are P-approximately homogeneous, then the operators $A + B$ and kA are also P-approximately homogeneous.

There are definite relations between homogeneity deviations.

Lemma 8.24. If an operator A is pointwise r-homogeneous and $r < q$, then A is pointwise q-homogeneous.

Proof is left as an exercise.

Proposition 8.35. Any pointwise r-homogeneous operator A: $L \to M$ is $2r$-homogeneous.

Proof. Let us consider a pointwise r-homogeneous operator A: $L \to M$. Then for some homogeneous operator B: $L \to M$ and some non-negative real number r, we have

$$\|A(x) - B(x)\| < r \text{ for all } x \text{ from } L$$

This allows us to get the following sequence of equalities and inequalities based on properties of hyperseminorms:

$$\|A(kx) - kA(x)\| = \| A(kx) - kA(x) - B(kx) + kB(x)\| =$$

$$\|(A(kx) - B(kx)) + (kB(x) - kA(x))\| \leq$$

$$\| A(kx) - B(kx)\| + \| kB(x) - kA(x)\| \leq r + r = 2r$$

It means that operator A: $L \to M$ is $2r$-homogeneous.

Proposition is proved.

Proposition 8.35 implies the following result.

Corollary 8.73. Any P-approximately homogeneous operator A: $L \to M$ is approximately homogeneous.

Let us take an operator A: $L \to M$ where L and M are normed spaces and consider the *standard supremum operator seminorm* defined by the following formula

$$\|A\| = \sup \{ \|A(x)\|;\ x \in L,\ \|x\| = 1\}$$

Proposition 8.36. If a bounded operator $A: L \to M$ is pointwise r-homogeneous, then A is functionally r-homogeneous with respect to the standard supremum seminorm.

Indeed, if

$$\|A(x) - B(x)\| < r$$

for some homogeneous operator B and all x from L, then

$$\|A(x) - B(x)\| < r$$

for all x from L with $\|x\| = 1$. Consequently,

$$\|A - B\| = \sup \{ \|A(x) - B(x)\|;\ x \in L,\ \|x\| = 1\} < r$$

It means that operator A is functionally r-homogeneous with respect to the standard supremum seminorm.

Corollary 8.74. Any bounded P-approximately homogeneous operator $A: L \to M$ is F-approximately homogeneous.

Naturally, we come to the regional pointwise homogeneity.

Definition 8.13. a) An operator $A: L \to M$ is called *pointwise r-homogeneous in* a set $X \subseteq L$ if there is a homogeneous operator $B: L \to M$ such that

$$\|A(x) - B(x)\| < r \text{ for all } x \text{ from } X$$

b) The number $ph_{A,\,X} = \inf \{r;\ A \text{ is pointwise } r\text{-homogeneous in } X\}$ is called the *pointwise homogeneity deviation* of the operator A in a set X.

c) An operator $A: L \to M$ is P-*approximately homogeneous in* a set $X \subseteq L$ if it is pointwise r-homogeneous in X for some non-negative real number r.

A pointwise non-homogeneous operator can be pointwise homogeneous in some subset of the whole space. For instance, any operator $A: L \to M$, which is defined at a point a from L, is pointwise homogeneous in the set $X = \{a\}$ because there is always a linear, and thus, homogeneous operator $B: L \to M$ such that $A(a) = B(a)$.

Lemma 8.25. If an operator $A: L \to M$ is pointwise r-homogeneous in a set X, then it is pointwise l-homogeneous in X for any $l > r$.

Proof is left as an exercise.

Let us assume that $||.||$ is a hypernorm (norm) in M.

Lemma 8.26. An operator A: $L \to M$ is pointwise 0-homogeneous in X if and only if A is homogeneous in X.

Proof is left as an exercise.

Let us consider sums of operators and products of operators and numbers.

Proposition 8.37. a) If an operator A: $L \to M$ is pointwise r-homogeneous in X and an operator C: $L \to M$ is pointwise q-homogeneous in X, then the operator $A + C$: $L \to M$ is pointwise$(r + q)$-homogeneous in X.

b) If an operator A: $L \to M$ is pointwise r-homogeneous in X and k is a real number, then the operator kA: $L \to M$ is pointwise $|k| \cdot r$-homogeneous in X.

Proof is similar to the proof of Proposition 8.34.

Corollary 8.75. If operators A: $L \to M$ and B: $L \to M$ are P-approximately homogeneous in X, then the operators $A + B$ and kA are also P-approximately homogeneous in X.

There are definite relations between homogeneity deviations.

Proposition 8.38. Any pointwise r-homogeneous in X operator A: $L \to M$ is $2r$-homogeneous in X.

Proof is similar to the proof of Proposition 8.35.

Proposition 8.38 implies the following result.

Corollary 8.76. Any P-approximately homogeneous in X operator A: $L \to M$ is approximately homogeneous in X.

Let us study relations between pointwise homogeneity and functional homogeneity.

Proposition 8.39. If a bounded in X operator A is pointwise r-homogeneous in X, then A is functionally r-homogeneous with respect to the standard supremum seminorm.

Proof is similar to the proof of Proposition 8.36.

Corollary 8.77. Any bounded P-approximately homogeneous in X operator A: $L \to M$ is F-approximately homogeneous in X.

Now we define approximate additivity by comparing a given operator with an additive operator.

Let us assume that $||.||$ is an operator hyperseminorm (seminorm).

Definition 8.14. a) An operator A: $L \to M$ is called *functionally r-additive* if there is an additive operator B: $L \to M$ such that

$$||A - B|| < r$$

b) The number $fa_A = \inf \{r;\ A \text{ is functionally } r\text{-additive}\}$ is called the *functional additivity deviation* of operator A.

c) An operator $A: L \to M$ is F-*approximately additive* if it is functionally r-additive for some non-negative real number r.

Example 8.9. The one-dimensional operator (function) $D(x) = \sin x$ is functionally 1-additive if the norm $||A|| = \sup \{ |A(x)|; x \in \boldsymbol{R}\}$.

Let us assume that M is a hypernormed (normed) vector space.

Lemma 8.27. An operator $A: L \to M$ is functionally 0-additive if and only if A is additive.

Indeed, if $A: L \to M$ is functionally 0-additive, then

$$||A - B|| < 0$$

for some additive operator B and because $||.||$ is a hypernorm (norm), we have

$$||A - B|| = 0$$

Consequently, $A = B$.

Lemma 8.28. If an operator A is functionally r-additive and $r < q$, then A is functionally q-additive.

Proof is left as an exercise.

Let us consider the popular operator norm, namely, the relational supremum norm

$$||A|| = \sup \{(||A(x)||/||x||); x \in L\}$$

It is defined for bounded operators (Rudin, 1991).

Proposition 8.40. If a bounded operator A is functionally r-additive with respect to the relational supremum norm, then A is calibrated $2r$-additive.

Proof. Let us consider a functionally r-additive operator $A: L \to M$. Then

$$||A(x) - B(x)\,||/||x|| \leq ||A - B||$$

and

$$||A(x) - B(x)\,|| \leq ||A - B||\cdot||x|| < r \cdot ||x||$$

for some additive operator $B: L \to M$ and any $x \in L$. Consequently, as B is an additive operator, we have

$$||A(x + y) - A(x) - A(y)|| = ||A(x + y) - A(x) - A(y) - B(x + y) + B(x) - B(y)|| =$$

$$||(A(x + y) - B(x + y)) + (B(x) - A(x)) + (B(y) - A(y))|| \leq$$

$$||A(x + y) - B(x + y)|| + ||B(x) - A(x)|| + ||B(y) - A(y)|| <$$

$$r \cdot(||x + y|| + ||x|| + ||y||) \leq r \cdot(||x|| + ||y|| + ||x|| + ||y||) = 2\, r \cdot(||x|| + ||y||)$$

$$\text{i.e., } ||A(x + y) - A(x) - A(y)|| < 2\, r \cdot(||x|| + ||y||).$$

It means that the operator A is calibrated $2r$-additive.

Proposition is proved.

Corollary 8.78. Any F-approximately additive operator A: $L \to M$ is C-approximately additive.

It is possible to extend these results to the relational supremum hypernorm. Let us consider the cumulative cover $\{T_n = \{x; x \in L, 1/n < ||x|| < n\}; n = 1, 2, 3, \ldots\}$ of the vector space L.

Proposition 8.41. If a bounded in all sets T_n operator A is functionally r-additive with respect to the relational supremum hypernorm, then A is calibrated $2r$-additive.

Proof is similar to the proof of Proposition 8.40.

Let us consider sums of operators.

Proposition 8.42. a) If an operator A: $L \to M$ is functionally r-additive and an operator C: $L \to M$ is functionally q-additive, then the operator $A + C$: $L \to M$ is functionally $(r + q)$-additive, i.e., the functional additivity deviation of the sum is the sum of functional additivity deviations.

b) If an operator A: $L \to M$ is functionally r-additive and k is a real number, then the operator kA: $L \to M$ is functionally $|k| \cdot r$-additive.

Proof. a) Let us consider a functionally r-additive operator A: $L \to M$ and a functionally q-additive operator B: $L \to M$. Then we have

$$||A - B|| < r$$

and

$$||C - D|| < r$$

for some additive operators B: $L \to M$ and D: $L \to M$. By Proposition 8.9.a and Lemma 8.13, the sum $B + D$ is an additive operator. Thus, we have

$$||\,(A + C) - (B + D)\,|| = ||\,(A - B) + (C - D)|| \leq$$

$$||\,A - B\,|| + ||\,C - D\,|| < r + q$$

It means that the operator $A + C$ is functionally $(r + q)$-additive.

2) Now let us treat the operator kA. Note that if the operator B is additive, then by Proposition 8.9.b and Lemma 8.13, the operator kB is additive. Thus, we have

$$\| kA - kB \| = |k| \cdot \|A - B\| < |k| \cdot r$$

It means that the operator kA is functionally $|k| \cdot r$-additive.

Proposition is proved.

Corollary 8.79. If operators $A: L \to M$ and $B: L \to M$ are F-approximately additive, then the operator $A + B$ and kA are also F-approximately additive.

Definition 8.15. a) An operator $A: L \to M$ is called *functionally r-additive in* a set X if there is an additive in X operator $B: L \to M$ such that

$$\|A - B\| < r$$

b) The number $fa_{A,X} = \inf \{r;\ A \text{ is functionally } r\text{-additive in } X\}$ is called the *functional additivity deviation* of operator A in X.

c) An operator $A: L \to M$ is F-*approximately additive* in X if it is r-additive in X for some non-negative real number r.

Let us assume that M is a hypernormed (normed) vector space.

Lemma 8.29. An operator $A: L \to M$ is functionally 0-additive in X if and only if A is additive in X.

Proof is left as an exercise.

Lemma 8.30. If an operator A is functionally r-additive in X and $r < q$, then A is functionally q-additive in X.

Proof is left as an exercise.

Proposition 8.43. If a bounded operator A is functionally r-additive in X with respect to the relational supremum norm, then A is calibrated $2r$-additive in X.

Proof is similar to the proof of Proposition 8.40.

Corollary 8.80. Any F-approximately additive in X operator $A: L \to M$ is C-approximately additive in X.

It is possible to extend these results to the relational supremum hypernorm. Let us consider the cumulative cover $\{T_n = \{x;\ x \in L,\ 1/n < \|x\| < n\};\ n = 1, 2, 3, \ldots\}$ of the vector space L.

Proposition 8.44. If a bounded in all sets $T_n \cap X$ operator A is functionally r-additive in X with respect to the relational supremum hypernorm, then A is calibrated $2r$-additive in X.

Proof is similar to the proof of Proposition 8.40.

Let us consider sums of operators and their products with numbers.

Proposition 8.45. a) If an operator $A: L \to M$ is functionally r-additive in X and an operator $C: L \to M$ is functionally q-additive in X, then the operator $A + C: L \to M$ is functionally $(r + q)$-additive in X.

b) If an operator $A: L \to M$ is functionally r-additive in X and k is a real number, then the operator $kA: L \to M$ is functionally $|k| \cdot r$-additive in X.

Proof is similar to the proof of Proposition 8.42.

Corollary 8.81. If operators $A: L \to M$ and $B: L \to M$ are F-approximately additive in X, then the operator $A + B$ and kA are also F-approximately additive in X.

There is one more type of approximate additivity.

Definition 8.16. a) An operator $A: L \to M$ is called *pointwise r-additive* if there is an additive operator $B: L \to M$ such that

$$\|A(x) - B(x)\| < r \text{ for all } x \text{ from } L$$

b) The number $pa_A = \inf \{r; A \text{ is pointwise } r\text{-additive}\}$ is called the *pointwise additivity deviation* of the operator A.

c) An operator $A: L \to M$ is P-*approximately additive* if it is pointwise r-additive for some non-negative real number r.

Note that pointwise approximate additivity does not coincide with functional approximate additivity. It is possible that an operator A is functionally 0-additive but not pointwise r-additive for any positive real number r as the following example demonstrates.

Example 8.10. Let us take the one-dimensional operator $Sq(x) = x^2$. It is functionally 0-additive and functionally 0-linear because $\| Sq - E \|_2 = 0$ where $\|.\|_2$ is the standard supremum seminorm and $E(x) = x$ is a linear operator (function). However, the value $\|Sq(x) - E(x)\|$ tends to infinity when x tends to infinity and thus, Sq is not pointwise r-additive for any positive real number r.

Let us assume that $\|.\|$ is a hypernorm (norm) in M.

Lemma 8.31. An operator $A: L \to M$ is pointwise 0-additive if and only if A is additive.

Proof is left as an exercise.

Lemma 8.32. If an operator A is pointwise r-additive and $r < q$, then A is pointwise q-additive.

Proof is left as an exercise.

Proposition 8.46. Any pointwise r-additive operator A: $L \to M$ is $3r$-additive.

Proof. Let us consider a pointwise r-additive operator A: $L \to M$. Then for some additive operator B: $L \to M$ and some non-negative real number r, we have

$$\|A(x) - B(x)\| < r \text{ for all } x \text{ from } L$$

This allows us to get the following sequence of equalities and inequalities:

$$\|A(x+y) - A(x) - A(y)\| = \|A(x+y) - A(x) - A(y) - B(x+y) + B(x) + B(y)\| =$$

$$\|(A(x+y) - B(x+y)) + (B(x) - A(x)) + (B(y) - A(y))\| \leq$$

$$\|A(x+y) - B(x+y)\| + \|B(x) - A(x)\| + \|B(y) - A(y)\| \leq r + r + r = 3r$$

It means that operator A: $L \to M$ is $3r$-additive.

Proposition is proved.

Proposition 8.46 implies the following result.

Proposition 8.47. Any P-approximately additive operator A: $L \to M$ is approximately additive.

Let us consider the *standard supremum seminorm operator norm*

$$\|A\| = \sup \{ \|A(x)\|; x \in L, \|x\| = 1\}$$

Proposition 8.48. If a bounded with respect to the standard supremum seminorm operator norm operator A is pointwise r-additive, then A is functionally r-additive with respect to the standard supremum seminorm.

Indeed, if

$$\|A(x) - B(x)\| < r$$

for some additive operator B and all x from L, then

$$\|A(x) - B(x)\| < r$$

for all x from L with $\|x\| = 1$. Consequently,

$$\|A - B\| = \sup \{ \|A(x) - B(x)\|; x \in L, \|x\| = 1\} < r$$

It means that the A is functionally r-additive with respect to the unit supremum seminorm.

Corollary 8.82. Any bounded P-approximately additive operator $A: L \to M$ is F-approximately additive.

Let us consider sums of operators.

Proposition 8.49. a) If an operator $A: L \to M$ is pointwise r-additive and an operator $C: L \to M$ is pointwise q-additive, then the operator $A + C: L \to M$ is pointwise$(r + q)$-additive, i.e., the pointwise additivity deviation of the sum is the sum of pointwise additivity deviations.

b) If an operator $A: L \to M$ is pointwise r-additive and k is a real number, then the operator $kA: L \to M$ is pointwise $|k|\cdot r$-additive.

Proof. a) Let us consider a pointwise r-additive operator $A: L \to M$ and a pointwise q-additive operator $B: L \to M$. Then we have

$$\|A(x) - B(x)\| < r$$

and

$$\|C(x) - D(x)\| < r$$

for all x from L and some additive operators $B: L \to M$ and $D: L \to M$. By Proposition 8.1.a and Lemma 8.2, the sum $B + D$ is a additive operator. Thus, we have

$$\|(A(x) + C(x)) - (B(x) + D(x))\| = \|(A(x) - B(x)) + (C(x) - D(x))\| \leq$$

$$\|A(x) - B(x)\| + \|C(x) - D(x)\| < r + q$$

It means that the operator $A + C$ is pointwise$(r + q)$-additive.

2) Now let us treat the operator kA. Note that if the operator B is additive, then by Proposition 8.1.b and Lemma 8.2, the operator kB is additive. Thus, we have

$$\|kA(x) - kB(x)\| = |k|\cdot \|A(x) - B(x)\| < |k|\cdot r$$

It means that the operator kA is pointwise $|k|\cdot r$-additive.

Proposition is proved.

Corollary 8.83. If operators $A: L \to M$ and $B: L \to M$ are P-approximately additive, then the operators $A + B$ and kA are also P-approximately additive.

Proposition 8.49 implies the following result.

Proposition 8.50. a) If a functional A: $L \rightarrow \boldsymbol{R}$ is pointwise r-additive and functional C: $L \rightarrow \boldsymbol{R}$ is pointwise q-additive, then the functional $A + C$: $L \rightarrow \boldsymbol{R}$ is pointwise $(r + q)$-additive, i.e., the pointwise additivity deviation of the sum is the sum of pointwise additivity deviations.

b) If a functional A: $L \rightarrow \boldsymbol{R}$ is pointwise r-additive and k is a real number, then the functional kA: $L \rightarrow \boldsymbol{R}$ is pointwise $|k| \cdot r$-additive.

Corollary 8.84. If functionals A: $L \rightarrow \boldsymbol{R}$ and B: $L \rightarrow \boldsymbol{R}$ are P-approximately additive, then the functionals $A + B$ and kA are also P-approximately additive.

From global approximate additivity, we go to regional approximate additivity.

Definition 8.17. a) An operator A: $L \rightarrow M$ is called *pointwise r-additive in* a set $X \subseteq L$ if there is an additive operator B: $L \rightarrow M$ such that

$$\|A(x) - B(x)\| < r \text{ for all } x \text{ from } X$$

b) An operator A: $L \rightarrow M$ is P-*approximately additive in* a set $X \subseteq L$ if it is pointwise r-additive in X for some non-negative real number r.

Let us study properties of pointwise r-additive in a set X operators assuming, at first, that M is a hypernormed (normed) vector space.

Lemma 8.32. An operator A: $L \rightarrow M$ is pointwise 0-additive in a set X if and only if A is additive in a set X.

Proof is left as an exercise.

Let us assume that M and L are hyperseminormed vector spaces.

Lemma 8.34. If an operator A: $L \rightarrow M$ is pointwise r-additive in a set X and $r < q$, then A is pointwise q-additive in X.

Proof is left as an exercise.

Proposition 8.51. a) If an operator A: $L \rightarrow M$ is pointwise r-additive in X and an operator C: $L \rightarrow M$ is pointwise q-additive in X, then the operator $A + C$: $L \rightarrow M$ is pointwise $(r + q)$-additive in X, i.e., the pointwise additivity deviation in X of the sum is the sum of pointwise additivity deviations in X.

b) If an operator A: $L \rightarrow M$ is pointwise r-additive in X and k is a real number, then the operator kA: $L \rightarrow M$ is pointwise $|k| \cdot r$-additive in X.

Proof is similar to the proof of Proposition 8.49.

Corollary 8.85. If operators A: $L \rightarrow M$ and B: $L \rightarrow M$ are P-approximately additive in X, then the operators $A + B$ and kA are also P-approximately additive in X.

Now we define functional approximate linearity by comparing a given operator with a linear operator or by combining properties of approximate homogeneity and approximate additivity.

Let us assume that ||.|| is an operator hypernorm (norm).

Definition 8.18. a) An operator A: $L \rightarrow M$ is called *functionally* (r, q)-*linear* if it is functionally r-homogeneous and functionally q-additive.

b) An operator A: $L \rightarrow M$ is called *functionally r-linear* if there is a linear operator B: $L \rightarrow M$ such that

$$||A - B|| < r$$

c) An operator A: $L \rightarrow M$ is F-*approximately linear* if it is functionally r-linear for some non-negative real number r.

F-approximate linearity means that the operator is not far from some linear operator in the operator hypernorm (norm).

Example 8.11. *Linear regression* is an approach in statistics to modeling a possible relationship between a scalar dependent variable y and one or more explanatory variables denoted x (Draper and Smith, 1998). The case of one explanatory variable x is called *simple linear regression*. In the case of than one explanatory variable, the technique is called *multiple linear regression*. In linear regression, data are modeled using linear predictor functions, and unknown model parameters are estimated from the data. Such models are called *linear models*. If data modeled by a linear regression form a function $y(x)$, then this function is F-approximately linear.

Let us assume that M is a hypernormed (normed) vector space.

Lemma 8.29. An operator A: $L \rightarrow M$ is functionally 0-linear if and only if A is linear.

Proof is similar to the proof of Lemma 8.25.

Lemma 8.30. A functionally r-linear operator A is functionally (r, r)-linear.

Proof is left as an exercise.

Lemma 8.31. If an operator A is functionally (r, p)-linear, $r < q$ and $p < t$, then A is functionally (q, t)-linear.

Proof is left as an exercise.

Corollary 8.86. If an operator A is functionally r-linear and $r < q$, then A is functionally q-linear.

Proposition 8.37 and 8.29 imply the following result.

Proposition 8.52. a) If an operator A: $L \rightarrow M$ is functionally (r, p)-linear and an operator C: $L \rightarrow M$ is functionally (q, t)-linear, then the operator $A + C$: $L \rightarrow M$ is functionally $(r + q, p + t)$-linear.

b) If an operator A: $L \to M$ is functionally (r, p)-linear and k is a real number, then the operator kA: $L \to M$ is functionally $(|k| \cdot r, |k| \cdot p)$-linear.

Corollary 8.87. a) If a functional A: $L \to \boldsymbol{R}$ is functionally (r, p)-linear and functional C: $L \to \boldsymbol{R}$ is functionally (q, t)-linear, then the functional $A + C$: $L \to \boldsymbol{R}$ is functionally $(r + q, p + t)$-linear.

b) If a functional A: $L \to \boldsymbol{R}$ is functionally (r, p)-linear and k is a real number, then the functional kA: $L \to \boldsymbol{R}$ is functionally $(|k| \cdot r, |k| \cdot p)$-linear.

Proposition 8.53. a) If an operator A: $L \to M$ is functionally r-linear and an operator C: $L \to M$ is functionally q-linear, then the operator $A + C$: $L \to M$ is functionally $(r + q)$-linear.

b) If an operator A: $L \to M$ is functionally r-linear and k is a real number, then the operator kA: $L \to M$ is functionally $|k| \cdot r$-linear.

Proof is similar to the proof of Proposition 8.37.

Corollary 8.88. If operators A: $L \to M$ and B: $L \to M$ are F-approximately linear, then the operators $A + B$ and kA are also F-approximately linear.

Corollary 8.89. a) If a functional A: $L \to \boldsymbol{R}$ is functionally r-linear and functional C: $L \to \boldsymbol{R}$ is functionally q-linear, then the functional $A + C$: $L \to \boldsymbol{R}$ is functionally $(r + q)$-linear.

b) If a functional A: $L \to \boldsymbol{R}$ is functionally r-linear and k is a real number, then the functional kA: $L \to \boldsymbol{R}$ is functionally $|k| \cdot r$-linear.

Corollary 8.90. If functionals A: $L \to \boldsymbol{R}$ and B: $L \to \boldsymbol{R}$ are F-approximately linear, then the functionals $A + B$ and kA are also F-approximately linear.

Now we come to the pointwise approximate linearity.

Definition 8.19. a) An operator A: $L \to M$ is called *pointwise* (q, r)*-linear* if it is pointwise r-homogeneous and pointwise q-additive.

b) The number $t = \max \{ ph_A, pa_A \}$ where $pa_A = \inf \{r;\ A$ is pointwise r-additive$\}$ and $ph_A = \inf \{r;\ A$ is pointwise r-homogeneous$\}$ is called the *combined pointwise linearity deviation* of the operator A.

c) An operator A: $L \to M$ is called *pointwise r-linear* if there is a linear operator B: $L \to M$ such that

$$\|A(x) - B(x)\| < r \text{ for all } x \text{ from } L$$

d) The number $l_A = \inf \{r;\ A$ is pointwise r-linear in $X\}$ is called the *pointwise linearity deviation* of the operator A.

e) An operator A: $L \to M$ is P-*approximately linear* if it is pointwise r-linear for some non-negative real number r.

Let us consider some examples.

Example 8.12. Let us consider the one-dimensional operator $A(x) = \lfloor x \rfloor$. It is pointwise 1-linear and functionally 1-linear because $|A(x) - B(x)| < 1$ for the one-dimensional linear operator $B(x) = x$ and all x from $\boldsymbol{R}$.

Example 8.13. Let us consider the space $\boldsymbol{R}$ and the one-dimensional operator $A(x) = x\cdot \sin\pi x$ in this space. Let us take the constant operator $0(x) = 0$ for all numbers x from $\boldsymbol{R}$.

The operator $A(x)$ is functionally 2-linear because

$$\|A(x) - 0(x)\| = \|A(x)\| = \sup\,\{|x\cdot \sin\pi x|/|x|;\, x \in \boldsymbol{R}\} =$$

$$= \sup\,\{|x| \cdot |\sin\pi x|/|x|;\, x \in \boldsymbol{R}\} = \sup\,\{|\sin\pi x|;\, x \in \boldsymbol{R}\} < 2$$

However, there is no number k such that $A(x)$ is a pointwise k-linear operator.

Indeed, let us assume that $A(x)$ is a pointwise k-linear operator. Then there is a linear operator C: $\boldsymbol{R} \rightarrow \boldsymbol{R}$ such that

$$\|A(x) - C(x)\| < k \text{ for all } x \text{ from } \boldsymbol{R}$$

The operator C has the form $C(x) = ax + b$. Then we have

$$|\, x \cdot \sin\pi x - ax - b| < k \text{ for all } x \text{ from } \boldsymbol{R}$$

Consequently,

$$|\, x \cdot \sin\pi x - ax - b| = |\, x \cdot (\sin\pi x - a) - b|$$

There are three possibilities for a: $a < 1$, $a > 1$ and $a = 1$.

If $a < 1$, then $\sin\pi x - a > 0$ when $x = ½, 5/2, 9/2, \ldots, (4n + 1)/2, \ldots$ Thus, for all these values of x, the value $x\cdot(\sin\pi x - a)$ grows without limits and as a result, the absolute value $|\, x\cdot(\sin\pi x - a) - b|$ becomes larger than any k.

If $a \geq 1$, then $\sin\pi x - a < 0$ when $x = 3/2, 7/2, 11/2, \ldots, (4n - 1)/2, \ldots$ Thus, for all these values of x, the value $-x\cdot(\sin\pi x - a)$ grows without limits and as a result, the absolute value $|x\cdot(\sin\pi x - a) - b\,| = |-x\cdot(\sin\pi x - a) + b\,|$ becomes larger than any k.

Therefore $A(x)$ cannot be a pointwise k-linear operator for any number k.

By the same token, $A(x)$ is not a pointwise k-homogeneous operator for any number k and not a pointwise k-additive operator for any number k.

In addition, we have

$$\|A(kx) - kA(x)\| = |kx \cdot \sin\pi kx - kx \cdot \sin\pi x| = |kx| \cdot |\sin\pi kx - \sin\pi x|$$

Taking $x = 1$, we see that $\sin\pi x = 0$ and $\sin\pi k = 1$ for all $k = ½, 5/2, 9/2, \ldots, (4n + 1)/2, \ldots$ Consequently the value

$$\|A(k \cdot 1) - kA(1)\| = |k| \cdot |\sin\pi k|$$

grows without limits. As a result, $A(x)$ is not a r-homogeneous operator for any number r.

Let us look whether the operator A is r-additive for some real number r. Suppose, A is such an operator. Then

$$\|A(x + y) - A(x) - A(y)\| = |(x + y) \cdot \sin\pi(x + y) - x \cdot \sin\pi x - y \cdot \sin\pi y| \leq r$$

for all $x, y \in \boldsymbol{R}$. Consequently, taking $y = 1$, we have

$$|(x + y) \cdot \sin\pi(x + y) - x \cdot \sin\pi x - y \cdot \sin\pi y| = |(x + 1) \cdot \sin\pi(x + 1) - x \cdot \sin\pi x|$$

Then taking $x = 3/2, 7/2, 11/2, \ldots, (4n - 1)/2, \ldots$, we obtain the sequence

$$l = \{4n;\ n = 1, 2, 3, \ldots\}$$

as

$$|(((4n - 1)/2) + 1) \cdot \sin\pi(((4n - 1)/2) + 1) - ((4n - 1)/2) \cdot \sin\pi((4n - 1)/2)| =$$

$$|((4n + 1)/2) \cdot \sin\pi((4n + 1)/2) - ((4n - 1)/2) \cdot \sin\pi((4n - 1)/2)| =$$

$$|((4n + 1)/2) - ((4n - 1)/2) \cdot (-1)| = 4n$$

As the sequence l grows without limits, $A(x)$ is not a r-additive operator for any number r.

Consequently, $A(x)$ is not a r-linear operator for any number r.

Let us study properties of pointwise r-linear operators assuming, at first, that M is a hypernormed (normed) vector space.

Lemma 8.32. An operator $A: L \to M$ is pointwise 0-linear if and only if A is linear.

Proof is left as an exercise.

Let us assume that M is a hyperseminormed vector space.

Lemma 8.33. A pointwise r-linear operator A is pointwise (r, r)-linear.

Proof is left as an exercise.

Lemma 8.34. If an operator A is pointwise r-linear and $r < q$, then A is pointwise q-linear.

Proof is left as an exercise.

Let us study relations between approximate linearity and P-approximate linearity.

Propositions 8.34 and 8.49 give us the following result.

Theorem 8.3. Any pointwise(r, q)-linear operator A: $L \to M$ is ($2r$, $3q$)-linear.

Theorem 8.3 and Lemmas 8.33 and 8.34 imply the following result.

Corollary 8.91. Any pointwise r-linear operator A: $L \to M$ is $3r$-linear.

Corollary 8.92. Any P-approximately linear operator A: $L \to M$ is approximately linear.

Let us take the standard supremum norm of operators, which is defined for bounded operators in the following way:

$$\|A\| = \sup \{\|A(x)\|; x \in L, \|x\| - 1 \}$$

Theorem 8.4. Any pointwise(r, q)-linear bounded operator A: $L \to M$ is functionally (r, q)-linear with respect to the standard supremum norm in the space of operators.

Indeed, by Proposition 8.36, A is functionally r-homogeneous with respect to the standard supremum norm in the space of operators, and by Proposition 8.48, A is functionally q-additive with respect to the standard supremum norm in the space of operators. Thus, A is functionally (r, q)-linear with respect to the standard supremum norm in the space of operators.

Corollary 8.93. Any bounded pointwise r-linear operator A: $L \to M$ is functionally r-linear.

Corollary 8.94. Any bounded P-approximately linear operator A: $L \to M$ is F-approximately linear.

A combination of regional pointwise homogeneity and regional pointwise additivity give us regional pointwise linearity.

Definition 8.20. a) An operator A: $L \to M$ is called *pointwise*(r, q)*-linear in* a set $X \subseteq L$ if it is pointwise r-homogeneous in X and pointwise q-additive in X.

b) An operator A: $L \to M$ is called *pointwise r-linear in a set* $X \subseteq L$ if there is a linear operator B: $L \to M$ such that

$$\|A(x) - B(x)\| < r \text{ for all } x \text{ from } X$$

c) An operator $A: L \to M$ is P-*approximately linear in a set* $X \subseteq L$ if it is pointwise r-linear in X for some non-negative real number r.

Let us assume that M is a hypernormed (normed) vector space.

Lemma 8.22. An operator $A: L \to M$ is pointwise 0-linear in X if and only if A is linear in X.

Proof is left as an exercise.

Lemma 8.23. If an operator A is pointwise r-linear in X and $r < q$, then A is pointwise q-linear in X.

Proof is left as an exercise.

Propositions 8.37 and 8.51 give us the following result.

Proposition 8.54. a) If an operator $A: L \to M$ is pointwise (r, p)-linear in a set X and an operator $C: L \to M$ is pointwise (q, t)-linear in X, then the operator $A + C: L \to M$ is pointwise $(r + q, p + t)$-linear in X.

b) If an operator $A: L \to M$ is pointwise (r, p)-linear in X and k is a real number, then the operator $kA: L \to M$ is pointwise $(|k| \cdot r, |k| \cdot p)$-linear in X.

Corollary 8.95. a) If an operator $A: L \to M$ is pointwise r-linear in X and an operator $C: L \to M$ is pointwise q-linear in X, then the operator $A + C: L \to M$ is pointwise $(r + q)$-linear in X.

b) If an operator $A: L \to M$ is pointwise r-linear in X and k is a real number, then the operator $kA: L \to M$ is pointwise $|k| \cdot r$-linear in X.

Corollary 8.96. If operators $A: L \to M$ and $B: L \to M$ are P-approximately linear in X, then the operators $A + B$ and kA are also P-approximately linear in X.

Corollary 8.97. The combined pointwise linearity deviation of the sum is less than or equal to the sum of the combined pointwise linearity deviations.

Corollary 8.98. a) If a functional $A: L \to \boldsymbol{R}$ is pointwise(r, p)-linear in a set X and functional $C: L \to \boldsymbol{R}$ is pointwise (q, t)-linear in X, then the functional $A + C: L \to \boldsymbol{R}$ is pointwise $(r + q, p + t)$-linear in X.

b) If a functional $A: L \to \boldsymbol{R}$ is pointwise (r, p)-linear in X and k is a real number, then the functional $kA: L \to \boldsymbol{R}$ is pointwise $(|k| \cdot r, |k| \cdot p)$-linear in X.

Corollary 8.99. a) If a functional $A: L \to \boldsymbol{R}$ is pointwise r-linear in X and a functional $C: L \to \boldsymbol{R}$ is pointwise q-linear in X, then the functional $A + C: L \to \boldsymbol{R}$ is pointwise $(r + q)$-linear in X.

b) If a functional $A: L \to \boldsymbol{R}$ is pointwise r-linear in X and k is a real number, then the functional $kA: L \to M$ is pointwise $|k| \cdot r$-linear in X.

Corollary 8.100. If functionals $A: L \to \boldsymbol{R}$ and $B: L \to \boldsymbol{R}$ are P-approximately linear in X, then the functionals $A + B$ and kA are also P-approximately linear in X.

Let us consider one more type of approximate linearity.

Definition 8.21. a) An operator A: $L \to M$ is called *integrally r-linear* if

$$\|A(kx + hy) - kA(x) - hA(y)\| \leq r$$

for all $x, y \in L$ and all real numbers k and h.

b) The number $l_A = \inf \{r;\ A$ is integrally r-linear $\}$ is called the *integral linearity deviation* of the operator A.

c) An operator A: $L \to M$ is I-*approximately linear* if it is integrally r-linear for some non-negative real number r.

d) The number $p = \max \{q_A, r_A\}$ where $q_A = \inf \{r;\ A$ is r-additive in $X\}$ and $h_A = \inf \{r;\ A$ is r-homogeneous in $X\}$ is called the *combined linearity deviation* of the operator A in X.

e) The number $l_A = \inf \{r; A$ is r-linear in $X\}$ is called a *linearity deviation* of the operator A in X.

Integral linearity is defined independently from integral homogeneity and integral additivity comprising the latter as particular cases.

Example 8.14. The one-dimensional operator A: $\boldsymbol{R} \to \boldsymbol{R}$ defined by the formula $A(x) = nx + \sin x$ is integrally 2-linear.

Let us assume that M is a hypernormed (normed) vector space.

Lemma 8.35. An operator A: $L \to M$ is integrally 0-linear if and only if A is linear.

Proof is left as an exercise.

Lemma 8.36. If an operator A is integrally r-linear and $r < q$, then A is integrally q-linear.

Proof is left as an exercise.

Proposition 8.55. Any integrally r-linear operator A: $L \to M$ is r-linear.

Proof. Let us consider an integrally r-linear operator A: $L \to M$. Then for the real number r, all $x, y \in L$ and all real numbers k and h, we have

$$\|A(kx + hy) - kA(x) - hA(y)\| \leq r$$

Taking $k = h = 1$, we have the identity of approximate additivity

$$\|A(x + y) - A(x) - A(y)\| \leq r$$

In addition,

$$\|A(kx) - kA(x)\| = \|A(\tfrac{1}{2}kx + \tfrac{1}{2}kx) - \tfrac{1}{2}Af(x) - \tfrac{1}{2}Af(x)\| \leq r$$

Thus, operator A is integrally r-linear.

Proposition is proved.

Proposition 8.56. Any (q, r)-linear operator A: $L \rightarrow M$ is integrally$(r + 2q)$-linear.

Proof. Let us consider an approximately linear operator A: $L \rightarrow M$. Then for some non-negative real numbers r and q, all $x, y \in L$ and all real numbers k, we have

$$\|f(x + y) - f(x) - f(y)\| \leq r$$

and

$$\|f(kx) - kf(x)\| \leq q$$

Consequently,

$$\|A(kx + hy) - kA(x) - hA(y)\| =$$

$$\|A(kx + hy) - A(kx) - A(hy) + A(kx) - A(hy) - kA(x) - hA(y)\| \leq$$

$$\|A(kx + hy) - A(kx) - A(hy)\| + \|A(kx) - kA(x)\| + \| A(hy) - hA(y)\| \leq r + q + q = t$$

Thus, operator A is integrally t-linear and therefore, I-approximately linear.

Proposition is proved.

Propositions 8.55 and 8.56 give us the following result.

Corollary 8.101. An operator A: $L \rightarrow M$ is I-approximately linear if and only if A is approximately linear.

Let us study properties of sums of operators.

Proposition 8.57. a) If an operator A: $L \rightarrow M$ is integrally r-linear and an operator B: $L \rightarrow M$ is integrally q-linear, then the operator $A + B$: $L \rightarrow M$ is integrally $(r + q)$-linear.

b) If an operator A: $L \rightarrow M$ is integrally r-linear and t is a real number, then the operator tA: $L \rightarrow M$ is integrally $|t| \cdot r$-linear.

Proof. a) Let us consider an integrally r-linear operator A: $L \rightarrow M$ and integrally q-linear operator B: $H \rightarrow N$. Then

$$\| A(kx + hy) - A(kx) - A(hy)\| < r$$

for arbitrary elements x and y from L and all real numbers k and h and

$$|| B(kx + hy) - B(kx) - B(hy)|| < q$$

for arbitrary elements z and u from H and all real numbers k and h.

Consequently, we have

$$|| (A + B)(kx + hy) - (A + B)(kx) - (A + B)(hy) || =$$

$$|| A(kx + hy) - A(kx) - A(hy) + B(kx + hy) - B(kx) - B(hy)|| \leq$$

$$|| A(kx + hy) - A(kx) - A(hy) || + ||B(kx + hy) - B(kx) - B(hy)|| < r + q$$

It means that the operator $A + B$: $L \rightarrow M$ is integrally $(r + q)$-linear.

b) Let us consider an integrally r-linear operator A: $L \rightarrow M$. Then $|| A(kx + hy) - A(kx) - A(hy)|| < r$ for arbitrary elements x and y from L and all real numbers k and h

Consequently, for a real number t, we have

$$|| tA(kx + hy) - tA(kx) - tA(hy)|| =$$

$$||t(A(kx + hy) - A(kx) - A(hy)) || =$$

$$|t| \cdot ||A(kx + hy) - A(kx) - A(hy) || < |t| \cdot r$$

It means that the operator tA: $L \rightarrow M$ is integrally $|t| \cdot r$-linear.

Proposition is proved.

Corollary 8.102. If operators A: $L \rightarrow M$ and B: $L \rightarrow M$ are integrally approximately linear, then the operators $A + B$ and kA are also integrally approximately linear.

Definition 8.22. a) An operator A: $L \rightarrow M$ is called *integrally r-linear in* a set $X \subseteq L$ if

$$||A(kx + hy) - kA(x) - hA(y)|| \leq r$$

for all $x, y \in X$ and all real numbers k and h.

b) An operator A: $L \rightarrow M$ is I-*approximately linear in* a set $X \subseteq L$ if it is integrally r-linear in X for some non-negative real number r.

Lemma 8.37. If an operator A is integrally r-linear in X and $r < q$, then A is integrally q-linear in X.

Proof is left as an exercise.

Let us assume that M is a hypernormed (normed) vector space.

Lemma 8.38. An operator $A: L \to M$ is integrally 0-linear in X if and only if A is linear in X.

Proof is left as an exercise.

Let us find relations between linearity and integral linearity.

Proposition 8.58. Any integrally r-linear in X operator $A: L \to M$ is r-linear in X.

Proof is similar to the proof of Proposition 8.55.

The inverse relation is also valid.

Proposition 8.59. Any (q, r)-linear in X operator $A: L \to M$ is integrally$(r + 2q)$-linear in X.

Proof is similar to the proof of Proposition 8.56.

Application of Propositions 8.58 and 8.59 gives us the following result.

Corollary 8.103. An operator $A: L \to M$ is I-approximately linear if and only if A is approximately linear.

Let us consider sums of operators and products of operators and numbers.

Proposition 8.60. a) If an operator $A: L \to M$ is integrally r-linear in X and an operator $B: L \to M$ is integrally q-linear in X, then the operator $A + B: L \to M$ is integrally $(r + q)$-linear in X.

b) If an operator $A: L \to M$ is integrally r-linear in X and t is a real number, then the operator $tA: L \to M$ is integrally $|t| \cdot r$-linear in X.

Proof is similar to the proof of Proposition 8.57.

Corollary 8.104. If operators $A: L \to M$ and $B: L \to M$ are I-approximately linear in X, then the operators $A + B$ and kA are also I-approximately linear in X.

There are also other concepts of approximately linear functionals and operators. For instance, Šemrl (1988) studied ε-approximately linear functionals. Namely, let us take a normed vector space L and a real number $\varepsilon > 0$.

Definition 8.23. A functional f on L is called *ε-approximately linear* if

$$|f(\lambda x + \mu y) - \lambda f(x) - \mu f(y)| \leq \varepsilon(|\lambda| \, \|x\| + |\mu| \, \|y\|)$$

for all scalars λ, μ and all vectors $x, y \in L$.

Šemrl (1998) obtained the following property of ε-approximately linear functionals.

Theorem 8.8. (Šemrl, 1998). If H is a real or complex Hilbert space and a ε-approximately linear functional f on L is bounded, then there exists a continuous linear functional g on H such that $|f(x) - g(x)| \leq 37\varepsilon\|x\|$ for every $x \in H$.

This type of approximate linearity for functionals is generalized in the following definitions.

Let us take $r \geq 0$ and $0 \leq p \leq 1$.

Definition 8.24. a) An operator A: $L \to M$ is called *calibrated* (r, p)-*linear* if

$$\|A(ax + by) - aA(x) - bA(y)\| \leq r(|a|\ \|x\|^p + |b|\ \|y\|^p)$$

for all real numbers a, b and all vectors x and y from L.

b) An operator A: $L \to M$ is called *calibrated r-linear* if

$$\|A(ax + by) - aA(x) - bA(y)\| \leq r(|a|\ \|x\| + |b|\ \|y\|)$$

for all real numbers a, b and all vectors x and y from L.

c) An operator A: $L \to M$ is called C-*approximately linear* if it is calibrated r-linear for some number r.

d) An operator A: $L \to M$ is called *weakly* C-*approximately linear* if it is calibrated (r, p)-linear for some numbers r and $p < 1$.

Lemma 8.16. a) If an operator A is calibrated r-linear and $r < q$, then A is calibrated q-linear.

b) If an operator A is calibrated (r, p)-linear and $r < q$, then A is calibrated (q, p)-linear.

Proof is left as an exercise.

Proposition 8.61. a) If an operator A: $L \to M$ is calibrated r-linear and an operator B: $L \to M$ is calibrated q-linear, then the operator $A + B$: $L \to M$ is calibrated $(r + q)$-linear.

b) If an operator A: $L \to M$ is calibrated r-linear and k is a real number, then the operator kA: $L \to M$ is calibrated $(|k| \cdot r)$-linear.

Proof. a) Let us consider a calibrated r-linear operator A: $L \to M$ and calibrated q-linear operator B: $L \to M$. Then we have

$$\|A(ax + by) - aA(x) - bA(y)\| \leq r \cdot (|a|\ \|x\| + |b|\ \|y\|)$$

and

$$\|B(ax + by) - aB(x) - bB(y)\| \leq q \cdot (|a|\ \|x\| + |b|\ \|y\|)$$

for all real numbers a, b and arbitrary elements x and y from L.

Consequently, we have

$$\| (A + B)(ax + by) - a(A + B)(x) - b(A + B)(y) \| =$$

$$\| A(ax + by) - aA(x) - bA(y) + B(ax + by) - aB(x) - bB(y)\| \leq$$

$$\| A(ax + by) - aA(x) - bA(y)\| + \| B(ax + by) - aB(x) - bB(y)\| <$$

$$r\cdot(|a|\ \|x\| + |b|\ \|y\|) + q\cdot(|a|\ \|x\| + |b|\ \|y\|) = (r + q)\cdot(|a|\ \|x\| + |b|\ \|y\|)$$

It means that the operator $A + B$: $L \to M$ is calibrated $(r + q)$-linear.

b) Let us consider a calibrated r-additive operator A: $L \to M$. Then

$$\| A(ax + by) - aA(x) - bA(y)\| < r\cdot(|a|\ \|x\| + |b|\ \|y\|)$$

for arbitrary elements x and y from L. Consequently, we have

$$\| kA(ax + by) - kaA(x) - kbA(y)\| =$$

$$\|k(A(ax + by) - aA(x) - bA(y))\| =$$

$$|k| \cdot \|A(ax + by) - aA(x) - bA(y))\| < |k|\cdot r\cdot(|a|\ \|x\| + |b|\ \|y\|)$$

It means that the operator kA: $L \to M$ is calibrated $(|k|\cdot r)$-linear.

Proposition is proved.

Corollary 8.105. If operators A: $L \to M$ and B: $L \to M$ are C-approximately linear, then the operators $A + B$ and kA are also C-approximately linear.

Corollary 8.106. a) If a functional A: $L \to \boldsymbol{R}$ is calibrated r-linear and a functional B: $L \to \boldsymbol{R}$ is calibrated q-linear, then the functional $A + B$: $L \to \boldsymbol{R}$ is calibrated $(r + q)$-linear.

b) If a functional A: $L \to \boldsymbol{R}$ is calibrated r-linear and k is a real number, then the functional kA: $L \to \boldsymbol{R}$ is calibrated $(|k|\cdot r)$-linear.

From global calibrated approximate linearity, we go to regional calibrated approximate linearity.

Definition 8.25. a) An operator A: $L \to M$ is called *calibrated r-linear in* a set $X \subseteq L$ if

$$\|A(\lambda x + \mu y) - \lambda A(x) - \mu A(y)\| \leq \varepsilon(|\lambda|\ \|x\| + |\mu|\ \|y\|)$$

for all real numbers λ, μ and all vectors x and y from X.

b) An operator A: $L \to M$ is called C-*approximately linear in* a set $X \subseteq L$ if it is calibrated r-linear in X for some number r.

Lemma 8.16. If an operator A is calibrated r-linear in X and $r < q$, then A is calibrated q-linear in X.

Proof is left as an exercise.

Proposition 8.62. a) If an operator A: $L \rightarrow M$ is calibrated r-linear in X and an operator B: $L \rightarrow M$ is calibrated q-linear in X, then the operator $A + B$: $L \rightarrow M$ is calibrated $(r + q)$-linear in X.

b) If an operator A: $L \rightarrow M$ is calibrated r-linear in X and k is a real number, then the operator kA: $L \rightarrow M$ is calibrated $(|k| \cdot r)$-linear in X.

Proof. a) Let us consider a calibrated r-linear in X operator A: $L \rightarrow M$ and calibrated q-linear in X operator B: $L \rightarrow M$. Then we have

$$\|A(ax + by) - aA(x) - bA(y)\| \leq r \cdot (|a|\ \|x\| + |b|\ \|y\|)$$

and

$$\|B(ax + by) - aB(x) - bB(y)\| \leq q \cdot (|a|\ \|x\| + |b|\ \|y\|)$$

for all real numbers a, b and arbitrary elements x and y from X. Consequently, we have

$$\| (A + B)(ax + by) - a(A + B)(x) - b(A + B)(y) \| =$$

$$\| A(ax + by) - aA(x) - bA(y) + B(ax + by) - aB(x) - bB(y)\| \leq$$

$$\| A(ax + by) - aA(x) - bA(y)\| + \| B(ax + by) - aB(x) - bB(y)\| <$$

$$r \cdot (|a|\ \|x\| + |b|\ \|y\|) + q \cdot (|a|\ \|x\| + |b|\ \|y\|) = (r + q) \cdot (|a|\ \|x\| + |b|\ \|y\|)$$

It means that the operator $A + B$: $L \rightarrow M$ is calibrated $(r + q)$-linear in X.

b) Let us consider a calibrated r-additive in X operator A: $L \rightarrow M$. Then

$$\| A(ax + by) - aA(x) - bA(y)\| < r \cdot (|a|\ \|x\| + |b|\ \|y\|)$$

for arbitrary elements x and y from X. Consequently, we have

$$\| kA(ax + by) - kaA(x) - kbA(y)\| =$$

$$\|k(A(ax + by) - aA(x) - bA(y))\| =$$

$$|k| \cdot \|A(ax + by) - aA(x) - bA(y))\| < |k| \cdot r \cdot (|a|\ \|x\| + |b|\ \|y\|)$$

It means that the operator kA: $L \to M$ is calibrated ($|k| \cdot r$)-linear in X.

Proposition is proved.

Corollary 8.107. If operators A: $L \to M$ and B: $L \to M$ are C-approximately linear in X, then the operators $A + B$ and kA are also C-approximately linear in X.

To conclude, it is necessary to remark that all results from this chapter proved for hyperseminormed vector spaces are also true for hypernormed, seminormed and normed vector spaces because hypernorms, seminorms and norms are special cases of hyperseminorms. In addition, many results formulated and proved only for operators are also true for functionals and functions.

KEYWORDS

- **approximate additivity**
- **approximate homogeneity**
- **approximate linearity**
- **approximately additive**
- **approximately homogeneous**
- **approximately linear**
- **C-approximately linear**
- **calibrated *r*-linear**

CHAPTER 9

FROM BOUNDEDNESS TO MULTIBOUNDEDNESS

In this chapter, we explore the idea of boundedness in multihyperseminormed, multihypernormed, multiseminormed and multinormed vector spaces, demonstrating a possibility of a variety of different explications, conceptualizations and formalizations of this idea.

Let us consider hyperseminormed vector space L over a field $\boldsymbol{F}$ with a hyperseminorm $||.||$ where $\boldsymbol{F}$ is either the field $\boldsymbol{R}$ of all real numbers or the field $\boldsymbol{C}$ of all complex numbers.

Taking a positive number p from $\boldsymbol{F}$, we define different types of bounded sets in hyperseminormed vector spaces.

Definition 9.1. a) A set $X \subseteq L$ is called *totally p-bounded* if $||x - y|| < p$ for any elements x and y from X.

b) The number p is called a *measure of total boundedness* of the set X.

c) A set $X \subseteq L$ is called *totally bounded* if it is totally p-bounded for some positive number p.

If $L = \boldsymbol{R}$, then the set $X = \{x;\ 0 < x < 1\}$ is totally 1-bounded, while the set $X = \{\ x;\ 0 < x\}$ is totally unbounded.

Lemma 9.1. If a set $X \subseteq L$ is totally p-bounded, then it is totally q-bounded for any $q > p$.

Proof is left as an exercise.

Let us find what relations and operations preserve boundedness of sets and to what extent.

Proposition 9.1.(a) If X, $Y \subseteq L$, X is totally p-bounded, Y is totally q-bounded and $X \cap Y \neq \varnothing$, then the union $X \bigcup Y$ is totally $(p + q)$-bounded.

(b) If $X \subseteq Y \subseteq L$ and Y is totally q-bounded, then X is totally q-bounded.

(c) If $X, Y \subseteq L$, X is totally p-bounded, and Y is totally q-bounded, then the sum $X + Y$ is totally $(p + q)$-bounded.

(d) If $X \subseteq L$ and X is totally q-bounded, then its closure CX is totally q-bounded.

Proof. (a) Let us consider a totally p-bounded set X, totally q-bounded set Y, an element x and z from X, and elements y and u from Y. By the initial conditions, there is an element z, which belongs both to X and to Y. As a result, we have

$$\|(x - y)\| = \|(x - z) + (z - y)\| < \|(x - z)\| + \|(y - z)\| < p + q$$

It means that the union $X \bigcup Y$ is totally $(p + q)$-bounded.

Part (a) is proved.

Part (b) follows from definitions.

(c) Let us consider a totally p-bounded set X, totally q-bounded set Y, elements x and z from X, and elements y and u from Y. Then $x + y$ and $z + u$ belong to $X + Y$ and we have

$$\|(x + y) - (z + u)\| = \|(x - z) + (y - u)\| < \|(x - z)\| + \|(y - u)\| < p + q$$

It means that the sum $X + Y$ is totally $(p + q)$-bounded.

(d) Let us consider a totally q-bounded set X and elements y and u from closure CX of X. As any hyperseminormed vector space is a metric space (cf. Chapter 3), there are two sequences $l = \{x_i; x_i \in X, i = 1, 2, 3, \ldots\}$ that converges to u and $h = \{y_i; y_i \in X, i = 1, 2, 3, \ldots\}$ that converges to y. Then for any $\varepsilon > 0$, there is a number n such that $\|(x_i - u)\| < \varepsilon$ when $i > n$, and there is a number m such that $\|(x_i - u)\| < \varepsilon$ when $i > m$. Taking $t = \max\{m, n\}$, for all $i > t$, we have

$$\|y - u\| = \|(y - y_i) + (y_i - x_i) + (x_i - u)\| \leq \|(y - y_i)\| + \|(y_i - x_i)\| + \|(x_i - u)\|$$
$$< q + 2\varepsilon$$

As ε is an arbitrary small positive number, $\|(y - u)\| < q$. Thus, the set CX is totally q-bounded.

Proposition is proved.

Corollary 9.1. (a) If $X, Y \subseteq L$, sets X and Y are totally bounded, and $X \cap Y \neq \varnothing$, then the union $X \bigcup Y$ is totally bounded.

(b) If $X \subseteq Y \subseteq L$, while the set Y is totally bounded, then X is totally bounded.

(c) If X, $Y \subseteq L$, X and Y are totally bounded, then the sum $X + Y$ is totally $(p + q)$-bounded.

(d) If $X \subseteq L$ and X is totally bounded, then its closure CX is totally bounded.

Remark 9.1. Condition $X \cap Y \neq \varnothing$ is essential in Proposition 9.1.a and Corollary 9.1.a as the following example demonstrates.

Example 9.1. It is proved (cf. Chapter 3) that the set of all real hypernumbers $\boldsymbol{R}_\omega$ is a hypernormed vector space. Let us consider sets $X = \{\mathbf{0}\}$ and $Y = \{\alpha = \mathrm{Hn}(i)_{i\in\omega}\}$. Then the union $X \cup Y = \{\mathbf{0}, \alpha\}$ is not totally bounded because the hypernorm $\|\alpha - \mathbf{0}\| = \|\alpha\|$ is larger than any real number (cf. Chapter 2).

Corollary 9.2. If X, $Y \subseteq L$, X is totally p-bounded, then the intersection $X \cap Y$ is totally p-bounded.

Corollary 9.1.b implies the following result.

Corollary 9.3. If X, $Y \subseteq L$, X is totally bounded, then the intersection $X \cap Y$ is totally bounded.

Let us consider the Cartesian product $L_1 \times L_2 \times L_3 \times \ldots \times L_n$ with the Manhattan hyperseminorm of hyperseminormed vector spaces $L_1, L_2, L_3, \ldots, L_n$ and sets $X_1 \subseteq L_1, X_2 \subseteq L_2, X_3 \subseteq L_3, \ldots, X_n \subseteq L_n$.

Proposition 9.2. If each set X_i is totally p_i-bounded ($i = 1, 2, 3, \ldots, n$), then the Cartesian product of these sets $X_1 \times X_2 \times X_3 \times \ldots \times X_n$ is totally p-bounded in the Cartesian product $L_1 \times L_2 \times L_3 \times \ldots \times L_n$ of hyperseminormed vector spaces where $p = p_1 + p_2 + p_3 + \ldots + p_n$.

Proof. 1. Let us consider the Cartesian product $L_1 \times L_2 \times L_3 \times \ldots \times L_n$ of hyperseminormed vector spaces with the Euclidean norm, p_i-bounded sets $X_1 \subseteq L_1, X_2 \subseteq L_2, X_3 \subseteq L_3, \ldots, X_n \subseteq L_n$, their Cartesian product of these sets $X = X_1 \times X_2 \times X_3 \times \ldots \times X_n$ and two elements x and z from X.

By the definition of the Manhattan hyperseminorm (cf. Chapter 3), we have

$$\|x - z\| = \|(x_1, x_2, x_3, \ldots, x_n) - (z_1, z_2, z_3, \ldots, z_n)\| = \|(x_1 - z_1, x_2 - z_2, x_3 - z_3, \ldots, x_n - z_n)\| =$$

$$\|x_1 - z_1\| + \|x_2 - z_2\| + \|x_3 - z_3\| + \ldots + \|x_n - z_n)\| < p_1 + p_2 + p_3 + \ldots + p_n = p$$

Proposition is proved.

Let us consider the Cartesian product $L_1 \times L_2 \times L_3 \times \ldots \times L_n$ with the Euclidean seminorm of seminormed vector spaces $L_1, L_2, L_3, \ldots, L_n$ and sets $X_1 \subseteq L_1, X_2 \subseteq L_2, X_3 \subseteq L_3, \ldots, X_n \subseteq L_n$.

Proposition 9.3. If each set X_i is totally p_i-bounded ($i = 1, 2, 3, \ldots, n$), then the Cartesian product of these sets $X_1 \times X_2 \times X_3 \times \ldots \times X_n$ is totally t-bounded in the Cartesian product $L_1 \times L_2 \times L_3 \times \ldots \times L_n$ of hyperseminormed vector spaces where $t = (p_1^2 + p_2^2 + p_3^2 + \ldots + p_n^2)^{1/2}$.

Proof. 1. Let us consider the Cartesian product $L_1 \times L_2 \times L_3 \times \ldots \times L_n$ of hyperseminormed vector spaces with the Euclidean seminorm, p_i-bounded sets $X_1 \subseteq L_1$, $X_2 \subseteq L_2$, $X_3 \subseteq L_3$, …, $X_n \subseteq L_n$, their Cartesian product of these sets $X = X_1 \times X_2 \times X_3 \times \ldots \times X_n$ and two elements x and z from X.

By the definition of the Euclidean norm (cf. Chapter 3), we have

$$\|x - z\| = \|(x_1, x_2, x_3, \ldots, x_n) - (z_1, z_2, z_3, \ldots, z_n)\| = \|(x_1 - z_1, x_2 - z_2, x_3 - z_3, \ldots, x_n - z_n)\| =$$

$$(\|x_1 - z_1\|^2 + \|x_2 - z_2\|^2 + \|x_3 - z_3\|^2 + \ldots + \| x_n - z_n) \|^2)^{1/2} < (p_1^2 + p_2^2 + p_3^2 + \ldots + p_n^2)^{1/2} = t$$

Proposition is proved.

Let us consider the Cartesian product $L_1 \times L_2 \times L_3 \times \ldots \times L_n$ with the Chebyshev hyperseminorm of hyperseminormed vector spaces L_1, L_2, L_3, …, L_n and sets $X_1 \subseteq L_1$, $X_2 \subseteq L_2$, $X_3 \subseteq L_3$, …, $X_n \subseteq L_n$.

Proposition 9.4. If each set X_i is totally p_i-bounded ($i = 1, 2, 3, \ldots, n$), then the Cartesian product of these sets $X_1 \times X_2 \times X_3 \times \ldots \times X_n$ is totally q-bounded in the Cartesian product $L_1 \times L_2 \times L_3 \times \ldots \times L_n$ of hyperseminormed vector spaces where $q = \max \{p_1, p_2, p_3, \ldots, p_n\}$.

Proof. 1. Let us consider the Cartesian product $L_1 \times L_2 \times L_3 \times \ldots \times L_n$ of hyperseminormed vector spaces with the Euclidean norm, p_i-bounded sets $X_1 \subseteq L_1$, $X_2 \subseteq L_2$, $X_3 \subseteq L_3$, …, $X_n \subseteq L_n$, their Cartesian product of these sets $X = X_1 \times X_2 \times X_3 \times \ldots \times X_n$ and two elements x and z from X.

By the definition of the Chebyshev hyperseminorm (cf. Chapter 3), we have

$$\|x - z\| = \|(x_1, x_2, x_3, \ldots, x_n) - (z_1, z_2, z_3, \ldots, z_n)\| = \|(x_1 - z_1, x_2 - z_2, x_3 - z_3, \ldots, x_n - z_n)\| =$$

$$\max \{\|x_1 - z_1\|, \|x_2 - z_2\|, \|x_3 - z_3\|, \ldots \| x_n - z_n) \|\} < \max \{p_1, p_2, p_3, \ldots, p_n\} = q$$

Proposition is proved.

Let us take an element a from X.

Definition 9.2. a) A set $X \subseteq L$ is called *totally p-bounded at a point a* if $\|x - a\| < p$ for any element x from X.

b) The number p is called a *measure of total boundedness* of the set X *at the point a.*

c) A set $X \subseteq L$ is called *totally bounded at a point a* if it is totally p-bounded at a for some positive number p.

If $L = \boldsymbol{R}$, then the set $X = \{x;\ 0 < x < 1\}$ is totally 1-bounded, while the set $Z = \{x;\ 0 < x\}$ is totally unbounded.

Lemma 9.2. If a set $X \subseteq L$ is totally p-bounded at a point a, then it is totally q-bounded for any $q > p$ at a.

Proof is left as an exercise.

Proposition 9.4 implies the following results.

Corollary 9.4. The Cartesian product of totally bounded at a point sets is totally bounded at a point in the Cartesian product $L_1 \times L_2 \times L_3 \times \ldots \times L_n$ of hyperseminormed vector spaces if it has either the Manhattan hyperseminorm or the Chebyshev hyperseminorm.

Corollary 9.5. The Cartesian product of totally bounded at a point sets is totally bounded at a point in the Cartesian product $L_1 \times L_2 \times L_3 \times \ldots \times L_n$ of seminormed vector spaces if it has the Euclidean seminorm.

Proposition 9.5.(a) If X, $Y \subseteq L$, X is totally p-bounded at a point a and Y is totally q-bounded at a, then the union $X \bigcup Y$ is totally t-bounded at a and at b where $t = \max\{k, h\}$.

(b) If X, $Y \subseteq L$, X is totally p-bounded at a point a, Y is totally q-bounded at b and $\|a - b\| = r$, then the union $X \bigcup Y$ is totally $(t + r)$-bounded at a and at b where $t = \max\{k, h\}$. (b) If $X \subseteq Y \subseteq L$ and Y is totally q-bounded at a, then X is totally q-bounded at a.

(c) If X, $Y \subseteq L$, X is totally p-bounded at a point a and Y is totally q-bounded at a point b, then the sum $X + Y$ is totally $(p + q)$-bounded at the point $a + b$.

(d) If $X \subseteq L$ and X is totally q-bounded at a point a, then its closure CX is totally q-bounded at a point a.

Proof. The statement (a) follows from definitions because any element from $X \bigcup Y$ belongs either to A or to B.

(b) Let us consider a totally p-bounded at a set X and totally q-bounded at b set Y. By the initial conditions, if an element x belongs to X, then we have

$$\|(x - b)\| = \|(x - a) + (a - b)\| < \|x - a\| + \|b - a\| < p + r$$

It means that the set X is totally $(p + r)$-bounded at b. By Lemma 9.2, the set X is totally $(t + r)$-bounded at b as $p + r \leq t + r$.

In a similar way, if an element y belongs to Y, then we have

$$\|(y - a)\| = \|(y - b) + (b - a)\| < \|y - b\| + \|b - a\| < q + r$$

It means that the set Y is totally $(q + r)$-bounded at a. By Lemma 9.2, the set Y is totally $(t + r)$-bounded at a as $q + r \leq t + r$. Thus, by the statement (a), the set $X \bigcup Y$ is totally $(t + r)$-bounded at b and at a.

(c) Let us consider a totally p-bounded at a set X, a totally q-bounded at b set Y, elements x from X and y from Y. Then $x + y$ belongs to $X + Y$, and we have

$$\|(x + y) - (a + b)\| = \|(x - a) + (y - b)\| < \|x - a\| + \|y - b\| < p + q$$

(d) Let us consider a totally q-bounded at a point a set X and an element y from the closure CX of X. As any hyperseminormed vector space is a metric space (cf. Chapter 3), there is a sequence $h = \{y_i; y_i \in X, i = 1, 2, 3, \ldots\}$ that converges to y. Then for any $\varepsilon > 0$, there is a number m such that $\|(x_i - u)\| < \varepsilon$ when $i > m$. As a result, for $i > m$, we have

$$\|y - a\| = \|(y - y_i) + (y_i - a)\| \leq \|(y - y_i)\| + \|(y_i - a)\| < q + \varepsilon$$

As ε is an arbitrary small positive number, $\|y - a\| < q$. Thus, the set CX is totally q-bounded at the point a.

Proposition is proved.

Corollary 9.6. (a) If X, $Y \subseteq L$, while X and Y are totally bounded at a point a, then the union $X \bigcup Y$ is totally bounded at a.

(b) If $X \subseteq Y \subseteq L$ and Y is totally bounded at a, then X is totally bounded at a.

(c) If X, $Y \subseteq L$, while X and Y are totally bounded at a, then the sum $X + Y$ is totally bounded at a.

(d) If $X \subseteq L$ and X is totally bounded at a point a, then its closure CX is totally bounded at a point a.

Corollary 9.7. If X, $Y \subseteq L$ and X is totally p-bounded at a, then the intersection $X \cap Y$ is totally p-bounded at a.

Corollary 9.6.b implies the following result.

Corollary 9.8. If X, $Y \subseteq L$ and X is totally bounded at a point a, then the intersection $X \cap Y$ is totally bounded at a.

Corollary 9.9. (a) If X, $Y \subseteq L$, X is totally p-bounded at a point a, Y is totally q-bounded at a, then the union $X \bigcup Y$ is totally t-bounded at a where $t = \max \{k, h\}$.

(b) If X, $Y \subseteq L$, X is totally p-bounded at a point a and Y is totally q-bounded at a point a, then the sum $X + Y$ is totally $(p + q)$-bounded at the point $a + b$.

Proposition 9.6. Any totally p-bounded at some point set X is totally $2p$-bounded.

Proof. Let us consider a totally p-bounded at a point a set X and elements x and y from X. Then, we have

$$\|(x - y)\| = \|(x - a) + (a - y)\| < \|x - a\| + \|y - a\| < p + p = 2p$$

It means that the set X is totally $2p$-bounded at a.

Proposition is proved.

Corollary 9.10. Any totally bounded at a point a set X is totally bounded.

Definitions imply the following result.

Proposition 9.7. Any totally p-bounded set X is totally p-bounded at any of its points.

Proof is left as an exercise.

Corollary 9.11. Any totally bounded X is totally bounded at any of its points.

Corollary 9.12. A set X is totally bounded if and only if it is totally bounded at any of its points.

Now let us study another kind of boundedness.

Let us consider a positive number k from $\boldsymbol{F}$, and the neighborhood $O_1(\mathbf{0}) = \{x; \|x\| < 1\}$ of $\mathbf{0}$.

Definition 9.3. a) A set $X \subseteq L$ is called *k-bounded* if $X \subseteq k \cdot O_1(\mathbf{0})$.

b) The number k is called a *measure of boundedness* of the set X.

c) A set $X \subseteq L$ is called *bounded* if it is k-bounded for some positive number k.

Example 9.3. If $L = \boldsymbol{R}$, then the set $X = \{x; 0 < x < 3\}$ is 3-bounded, while the set $Z = \{ x; 0 < x\}$ is unbounded.

Let us consider the neighborhood $O_h(\mathbf{0}) = \{x; \|x\| < h\}$ of $\mathbf{0}$.

Lemma 9.3. If $X \subseteq k \cdot O_h(\mathbf{0})$ for some positive real numbers h and k, then the set X is (kh)-bounded.

Indeed, if $x \in k \cdot O_h(\mathbf{0})$, then $x = ky$ where $y \in O_h(\mathbf{0})$, and consequently,

$$||x|| = ||ky|| = k \cdot ||y|| < kh$$

as $||y|| < h$. Thus, $X \subseteq kh \cdot O_1(\mathbf{0})$.

Remark 9.2. In seminormed vector spaces, neighborhoods $O_h(\mathbf{0})$ form a local base at $\mathbf{0}$ (Rudin, 1991). However, this is not true for hyperseminormed and hypernormed vector spaces as the following example demonstrates.

Example 9.2. Let us consider the hypernormed vector space $\boldsymbol{R}_\omega$ (cf. Example 3.8) with the topology υ generated by the sets $O_\beta x = \{y \in L; q(x - y) < \beta\}$ and the hypernumber $\beta = \text{Hn}(b_i)_{i\in\omega}$ where $b_i = 1$ for $i = 2, 4, \ldots, 2n, \ldots$ and $b_i = 0$ for $i = 1, 3, \ldots, 2n - 1, \ldots$.

Then open in the topology υ (cf. Chapter 3) set $O_\beta\mathbf{0}$ does not contain any set $O_k x$ where k is a positive real number, and thus, neighborhoods $O_h(\mathbf{0})$ do not form a local base at $\mathbf{0}$ in the topology υ.

Lemma 9.4. A set $X \subseteq L$ is bounded if and only if $X \subseteq k \cdot O_h(\mathbf{0})$ for some positive real numbers h and k where $O_h(\mathbf{0}) = \{x; ||x|| < h\}$.

Proof. Sufficiency. Indeed, by Lemma 9.3, if $X \subseteq k \cdot O_h(\mathbf{0})$, then the set X is bounded.

Necessity. If the set X is bounded, then by definition, $X \subseteq k \cdot O_1(\mathbf{0})$ and we can take $h = 1$.

Lemma is proved.

As the neighborhoods $O_h(\mathbf{0})$ form a local base of $\mathbf{0}$ in the vector space L, this result brings us to the conventional concept of boundedness in topological vector spaces (Rudin, 1991).

Definition 9.4. A subset X of a topological vector space L is called *(essentially) bounded* if X for any neighborhood U of $\mathbf{0}$, there is a number k such that $X \subseteq kU$.

Lemma 9.5. Essential boundedness and boundedness coincide in topological vector spaces.

Proof is left as an exercise.

It is possible to extend this concept to semitopological vector spaces.

Definition 9.5. A subset X of a semitopological vector space L is called *essentially bounded* if X for any neighborhood U of $\mathbf{0}$, there is a number k such that $X \subseteq kU$.

Note that there is an essential difference between essential boundedness in semitopological vector spaces and essential boundedness in topological vector spaces. For instance, any compact set is essentially bounded in topological vector spaces (Rudin, 1991) but this is not true for semitopological vector spaces.

Nevertheless, there are definite relations between essential boundedness and boundedness in hyperseminormed vector spaces.

Lemma 9.6. Essential boundedness implies boundedness in semitopological vector spaces.

Proof. Taking any essentially bounded set X, we see that it is bounded because $O_h(\mathbf{0})$ is a neighborhood of $\mathbf{0}$ for any positive real number h.

Proposition 9.8. If the neighborhoods $O_h(\mathbf{0})$ form a local base of $\mathbf{0}$ in the vector space L, then essential boundedness and boundedness coincide in L.

Proof. As by Lemma 9.6, essential boundedness implies boundedness, we need to prove only the opposite implication.

Let us consider a bounded set X and neighborhood U of $\mathbf{0}$. As the neighborhoods $O_h(\mathbf{0})$ form a local base of $\mathbf{0}$, there is a neighborhood $O_h(\mathbf{0}) \subseteq U$. As X is bounded, there is a number k such that $X \subseteq O_h(\mathbf{0})$. Consequently, $X \subseteq kU$.

Proposition is proved.

Let us find relations between boundedness and total boundedness. As the set $O_1(\mathbf{0})$ is totally 2-bounded, we have the following result.

Proposition 9.9. Any k-bounded set X is totally $2k$-bounded.

Proof is left as an exercise.

Corollary 9.13. Any bounded set X is totally bounded.

However, the converse of Corollary 9.13 is not true as the following example demonstrates.

Example 9.3. As it is proved by Burgin (2012), the set of all real hypernumbers $\boldsymbol{R}_\omega$ is a hypernormed space where the hypernorm $||\cdot||$ is defined by the following formula:

If α is a real hypernumber, i.e., $\alpha = \mathrm{Hn}(a_i)_{i\in\omega}$ with $a_i \in \boldsymbol{R}$ for all $i \in \omega$, then $||\, \alpha \,|| = \mathrm{Hn}(|a_i|)_{i\in\omega}$

In this space, the set $X = \{\alpha = \mathrm{Hn}(i)_{i\in\omega}, \beta = \mathrm{Hn}(i + 1)_{i\in\omega}\}$ is totally bounded because $||\alpha - \beta\,|| = 1$ but is not bounded because it is not a subset of any set $k \cdot O_1(\mathbf{0})$ as both its elements are larger than any real number.

Nonetheless, for normed and seminormed vector spaces, boundedness and total boundedness coincide.

Lemma 9.7. If a set $X \subseteq L$ is k-bounded, then it is h-bounded for any $h > k$.

Proof is left as an exercise.

Proposition 9.10.(a) If X, $Y \subseteq L$, X is k-bounded and Y is h-bounded, then the union $X \cup Y$ is t-bounded where $t = \max \{k, h\}$.

(b) If $X \subseteq Y \subseteq L$ and Y is k-bounded, then X is k-bounded.

(c) If $X, Y \subseteq L$, X is k-bounded and Y is h-bounded, then the sum $X + Y$ is $(k + h)$-bounded.

(d) If $X \subseteq L$ and X is k-bounded, then its closure CX is $(k + \varepsilon)$-bounded for any positive real number ε.

Proof. (a) Let us consider a k-bounded set X and a h-bounded set Y. By definition, all elements from X, belong to $k \cdot \mathrm{O}_1(\mathbf{0})$ and all elements from Y belong to $h \cdot \mathrm{O}_1(\mathbf{0})$. Taking an element x from X and an element y from Y, we have $\|x\| < p$ and $\|y\| < q$. Thus, $\|z\| < t$ for an arbitrary element z from X. Consequently, z belongs to $t \cdot \mathrm{O}_1(\mathbf{0})$ and the set $X \bigcup Y$ is t-bounded.

Part (b) follows from definitions.

(c) Let us consider a k-bounded set X, a h-bounded set Y, and elements x from X and y from Y. Then the element $x + y$ belongs to the set $X + Y$. By definition, all elements from X, belong to $k \cdot \mathrm{O}_1(\mathbf{0})$ and all elements from Y belong to $h \cdot \mathrm{O}_1(\mathbf{0})$. Thus, x belong to $k \cdot \mathrm{O}_1(\mathbf{0})$ and y belong to $h \cdot \mathrm{O}_1(\mathbf{0})$, i.e., $\|x\| < k$ and $\|y\| < h$. Then we have

$$\| x + y\| < \|x\| + \|y\| < k + h$$

Thus, the sum $X + Y$ is $(k + h)$-bounded.

(d) Let us consider a k-bounded set X and elements y and u from closure CX of X. As any hyperseminormed vector space is a metric space (cf. Chapter 3), there is a sequence $h = \{y_i; y_i \in X, i = 1, 2, 3, \ldots\}$ that converges to y. Then all elements y_i belong to $k \cdot \mathrm{O}_1(\mathbf{0})$. Consequently, y belongs to the closure $\mathrm{C}(k \cdot \mathrm{O}_1(\mathbf{0}))$ of the set $k \cdot \mathrm{O}_1(\mathbf{0})$ and thus, y belongs to $(k + \varepsilon) \cdot \mathrm{O}_1(\mathbf{0})$ for any positive real number ε. By Lemma 9.3, the set CX is $(k + \varepsilon)$-bounded.

Proposition is proved.

Corollary 9.14.(a) If $X, Y \subseteq L$, while X and Y are bounded, then the union $X \bigcup Y$ is bounded.

(b) If $X \subseteq Y \subseteq L$ and Y is bounded, then X is bounded.

(c) If $X, Y \subseteq L$, while X and Y are bounded, then the sum $X + Y$ is bounded.

(d) If $X \subseteq L$ and X is bounded then its closure CX is totally bounded.

Corollary 9.15. If $X, Y \subseteq L$, X is k-bounded, then the intersection $X \cap Y$ is k-bounded.

Corollary 9.14.b implies the following result.

Corollary 9.16. If $X, Y \subseteq L$, X is bounded, then the intersection $X \cap Y$ is bounded.

Properties of totally bounded sets bring us to the concept of bornology and bornological spaces.

The concept of bornological space is similar to the concept of topological space, while bornology is similar to topology (Hogbe-Nlend, 1972; Beer and Levi, 2009; Rodráguez-López and Sánchez-Granero, 2011).

Namely, a *bornology* on a set X is a collection $\boldsymbol{B} = \{ B_i;\ i \in I\}$ of subsets of X such that

1. $\boldsymbol{B}$ covers X, i.e., $X = \bigcup_{B_i \boldsymbol{B}} B_i$.
2. $\boldsymbol{B}$ is stable under inclusions, i.e., if $A \in \boldsymbol{B}$ and $A' \subseteq A$, then $A' \in \boldsymbol{B}$.
3. $\boldsymbol{B}$ is stable under finite unions, i.e., if $B_1, \ldots, B_n \in \boldsymbol{B}$, then $\bigcup_{i=1}^{n} B_i \in \boldsymbol{B}$.

Elements of the collection B are usually called *bounded sets*. However, it is necessary to understand difference between this usage of the term "bounded" and traditional uses. The pair (X, B) is called a *bornological set* or a *bornological space*.

As any one-element set from L is a bounded set, we have the following result.

Proposition 9.11. All bounded sets in a hyperseminormed vector space L form a bornology on the space $V = \{x;\ x \in L,\ \|x\| \text{ is finite}\}$.

Proof is left as an exercise.

In turn, Corollary 9.14 implies the following result.

Proposition 9.12. All bounded sets in a seminormed vector space L form a bornology on L.

Proof is left as an exercise.

Let us consider a positive number k from $\boldsymbol{F}$, an element x from L and the neighborhood $\mathrm{O}_1(x) = \{z;\ \|x - z\| < 1\}$ of x.

Definition 9.6. a) A set $X \subseteq L$ is called *k-bounded at x* if $X \subseteq k \cdot \mathrm{O}_1(x)$.

b) The number k is called a *measure of boundedness* of the set X *at a point x*.

c) A set $X \subseteq L$ is called *bounded at x* if it is k-bounded at x for some positive number k from $\boldsymbol{F}$.

If $L = \boldsymbol{R}_\omega$, then the set $X = \{x;\ 0 < \alpha - |x| < 3\}$ where $\alpha = \mathrm{Hn}(i)_{i\in\omega}$ is 3-bounded at α, while the set $Z = \{\alpha + x;\ 0 < x\}$ is unbounded at any hypernumber from $\boldsymbol{R}_\omega$.

Lemma 9.8. If $X \subseteq k \cdot \mathrm{O}_h(kx)$ for some positive real numbers h and k, then the set X is (kh)-bounded at the point kx where $\mathrm{O}_h(u) = \{z;\ \|u - z\| < h\}$.

Indeed, if $z \in k \cdot \mathrm{O}_h(kx)$, then $z = ky$ where $y \in \mathrm{O}_h(x)$, and consequently,

$$\|kx - z\| = \|kx - ky\| = k \cdot \|x - y\| < kh$$

as $\|x - y\| < h$. Thus, $X \subseteq kh \cdot \mathrm{O}_1(kx)$.

However, the set X from Lemma 9.8 can be unbounded at the point x as the following example demonstrates.

Example 9.4. It is proved (cf. Chapter 3) that the set of all real hypernumbers $\boldsymbol{R}_\omega$ is a hypernormed vector space. Then the set $X = \{z = \mathrm{Hn}(5i)_{i\in\omega}\}$ is the subset of the set $5\cdot \mathrm{O}_h(5x)$ where $x = \mathrm{Hn}(i)_{i\in\omega}$ (cf. Chapter 2). However, there is no positive real number k such that X is k-bounded at the point x because

$$\|z - x\| = \mathrm{Hn}(5i - i)_{i\in\omega} = \mathrm{Hn}(4i)_{i\in\omega}$$

is larger than any positive real number k.

Lemma 9.9. A set $X \subseteq L$ is bounded at x if and only if $X \subseteq k \cdot \mathrm{O}_h(kx)$ for some positive real numbers h and k.

Proof. Sufficiency. Indeed, by Lemma 9.8, if $X \subseteq k \cdot \mathrm{O}_h(kx)$, then the set X is bounded at the point x.

Necessity. If the set X is bounded at the point x, then by definition, $X \subseteq k \cdot \mathrm{O}_1(x)$ for some positive real number k. Taking an element z from X, we have $z = ky$ with $y \in \mathrm{O}_1(x)$. By definition, $\|x - y\| < 1$. Consequently, $k\|x - y\| < k$. Thus, by properties of hyperseminorm, we have

$$\|kx - z\| = \|kx - ky\| = k\|x - y\| < k$$

Consequently, $z \in k \cdot \mathrm{O}_h(kx)$, and we can take $h = k$.

Lemma is proved.

As the set $\mathrm{O}_1(x)$ is totally 2-bounded, we have the following result.

Proposition 9.13. For an element x from L, any k-bounded at x set X is totally $2k$-bounded.

Proof is left as an exercise.

Corollary 9.17. Any bounded at x set X is totally bounded.

Proposition 9.14. Any totally k-bounded set X is k-bounded at any of its points.

Proof is left as an exercise.

Corollary 9.18. Any totally bounded set X is bounded at any of its points.

This and Corollaries 9.4 and 9.15 give us the following result

Corollary 9.19. The Cartesian product of bounded at a point sets is bounded at a point in the Cartesian product $L_1 \times L_2 \times L_3 \times \ldots \times L_n$ of hyperseminormed vector spaces if it has either the Manhattan hyperseminorm or the Chebyshev hyperseminorm.

Corollary 9.20. The Cartesian product of bounded at a point sets is bounded at a point in the Cartesian product $L_1 \times L_2 \times L_3 \times \ldots \times L_n$ of seminormed vector spaces if it has the Euclidean seminorm.

There are definite relations between measures of boundedness.

Lemma 9.10. If a set $X \subseteq L$ is k-bounded at x, then it is h-bounded at x for any $h > k$.

Proof is left as an exercise.

Proposition 9.15. (a) If $X, Y \subseteq L$, X is k-bounded at a point x, Y is h-bounded at y and $||x - y|| = r$, then the union $X \bigcup Y$ is $(t + r)$-bounded at x and at y where $t = \max \{k, h\}$.

(b) If $X \subseteq Y \subseteq L$ and Y is k-bounded at x, then X is k-bounded at x.

(c) If $X, Y \subseteq L$, X is k-bounded at a point x and Y is h-bounded at a point y, then the sum $X + Y$ is $(k + h)$-bounded at the point $x + y$.

(d) If $X \subseteq L$ and X is k-bounded at a point a, then its closure $\mathrm{C}X$ is k-bounded at a point a.

Proof. (a) Let us consider a k-bounded at x set X and an h-bounded at x set Y. By definition, all elements from X, belong to $k \cdot \mathrm{O}_1(x)$ and all elements from Y belong to $h \cdot \mathrm{O}_1(y)$. Taking an element z from X, we have

$$||x - z|| < k$$

and

$$||y - z|| = ||y - x + x - z|| \leq ||y - x\,|| + ||x - z|| < k + r$$

In a similar way, taking an element u from Y, we have

$$||y - u|| < h$$

and

$$||x - u|| = ||x - y + y - u|| \leq ||x - y\,|| + ||y - u|| < h + r$$

Consequently, z and u belong to $(t + r) \cdot \mathrm{O}_1(x)$ and to $(t + r) \cdot \mathrm{O}_1(y)$, where $t = \max \{k, h\}$. Thus, the set $X \bigcup Y$ is $(t + r)$-bounded.

Part (b) follows from definitions.

(c) Let us consider a k-bounded at x set X, h-bounded at y set Y, element z from X, and element u from Y. Then $x + y$ and $z + u$ belong to $X + Y$ and we have

$$||(\,x + y) - (z + u)|| = ||(\,x - z) + (y - u)|| < ||(\,x - z)|| + ||(y - u)|| < k + h$$

It means that the element $z + u$ belongs to $(k + h) \cdot O_1(x + y)$ and consequently, the set $X + Y$ is $(k + h)$-bounded.

(d) Let us consider a k-bounded at x set X and an element y from the closure CX of X. As any hyperseminormed vector space is a metric space (cf. Chapter 3), there is a sequence $h = \{y_i; y_i \in X, i = 1, 2, 3, \ldots\}$ that converges to y. Then all elements y_i belong to $k \cdot O_1(x)$. Consequently, y belongs to the closure $C(k \cdot O_1(x))$ of the set $k \cdot O_1(x)$ and thus, y belongs to $(k + \varepsilon) \cdot O_1(x)$ for any positive real number ε. By Lemma 9.3, the set CX is $(k + \varepsilon)$-bounded at x.

Proposition is proved.

Corollary 9.21. (a) If $X, Y \subseteq L$, while X and Y are k-bounded at a point x, then the union $X \bigcup Y$ is t-bounded at x where $t = \max \{k, h\}$.

(b) If $X, Y \subseteq L$, while X and Y are k-bounded at a point x, then the sum $X + Y$ is $(k + h)$-bounded at the point $x + x$.

Corollary 9.22.(a) If $X, Y \subseteq L$, X is bounded at a point x, Y is bounded at y, then the union $X \bigcup Y$ is bounded at x and at y.

(b) If $X \subseteq Y \subseteq L$ and Y is bounded at a point x, then X is bounded at x.

(c) If $X, Y \subseteq L$, X is k-bounded at a point x, Y is h-bounded at y, then the sum $X + Y$ is bounded at a point $x + y$.

(d) If $X \subseteq L$ and X is k-bounded at a point x, then its closure CX is k-bounded at a point x.

Corollary 9.23. (a) If $X, Y \subseteq L$, while X and Y are bounded at a point x, then the union $X \bigcup Y$ is bounded at x.

(b) If $X, Y \subseteq L$, while X and Y are bounded at a point x, then the sum $X + Y$ is bounded at a point $x + x$.

Corollary 9.24. If $X, Y \subseteq L$, X is k-bounded at x, then the intersection $X \cap Y$ is k-bounded at x.

Corollary 9.22.b implies the following result.

Corollary 9.25. If $X, Y \subseteq L$, X is bounded at x, then the intersection $X \cap Y$ is bounded at x.

There is one more natural concept of boundedness.

Definition 9.7. a) A set $X \subseteq L$ is called *relatively k-bounded* if it is k-bounded at some point x.

b) The number k is called a *measure of relative boundedness* of the set X.

c) A set $X \subseteq L$ is called *relatively bounded at* x if it is relatively k-bounded for some positive number k.

If $L = \boldsymbol{R}_\omega$, then the set $X = \{x; 0 < \{\alpha - |x| < 1000\}$ where $\{\alpha = \text{Hn}(i)_{i \in \omega}$ is relatively bounded, while the set $X = \{\{\alpha + x; 0 < x\}$ is relatively unbounded.

Lemma 9.11. A set $X \subseteq L$ is relatively bounded if and only if $X \subseteq k \cdot O_h(x)$ for some element x from L and some positive real numbers h and k.

Proof is left as an exercise.

Proposition 9.12 implies the following result.

Proposition 9.16. Any relatively k-bounded set X is totally $2k$-bounded.

Proof is left as an exercise.

Corollary 9.26. Any bounded at x set X is totally bounded.

Corollary 9.27. Any relatively bounded set X is totally bounded.

Corollary 9.16 implies the following result.

Proposition 9.17. Any totally k-bounded set X is relatively k-bounded.

Proof is left as an exercise.

Corollary 9.28. Any totally bounded set X is relatively bounded.

Lemma 9.12. If a set $X \subseteq L$ is relatively k-bounded, then it is relatively h-bounded for any $h > k$.

Proof is left as an exercise.

Proposition 9.18. (a) If $X \subseteq Y \subseteq L$ and Y is relatively k-bounded, then X is relatively k-bounded.

(b) If X, $Y \subseteq L$, X is relatively k-bounded and Y is relatively h-bounded, then the sum $X + Y$ is relatively $(k + h)$-bounded.

(c) If $X \subseteq L$ and X is relatively k-bounded, then its closure CX is relatively $(k + \varepsilon)$-bounded.

Proof. Part (a) directly follows from definitions.

(b) Let us consider a relatively k-bounded set X and a relatively h-bounded set Y. By definition, X is relatively k-bounded at some x and Y is relatively h-bounded at some y. By Theorem 9.6, if X, $Y \subseteq L$, X is k-bounded at a point x and Y is h-bounded at a point y, then the sum $X + Y$ is $(k + h)$-bounded at the point $x + y$. It means that $X + Y$ is relatively $(k + h)$-bounded.

(c) Let us consider a relatively k-bounded set X and an element y from closure CX of X. By definition, the set X is relatively k-bounded at some point x. By Proposition 9.11, the set CX is $(k + \varepsilon)$-bounded at x, and thus, it is relatively $(k + \varepsilon)$-bounded.

Proposition is proved.

Corollary 9.29. (a) If $X \subseteq Y \subseteq L$ and Y is relatively bounded, then X is relatively bounded.

(b) If X, $Y \subseteq L$, while X and Y are relatively bounded, then the sum $X + Y$ is relatively bounded.

(c) If $X \subseteq L$ and X is relatively bounded, then its closure CX is relatively bounded.

Corollary 9.30. If $X, Y \subseteq L$, X is relatively k-bounded, then the intersection $X \cap Y$ is relatively k-bounded.

Corollary 9.29.a implies the following result.

Corollary 9.31. If $X, Y \subseteq L$ and X is relatively bounded, then the intersection $X \cap Y$ is relatively bounded.

Remark 9.3. In contrast to this, a similar statement for the union $X \cup Y$ is not true in general as Example 9.1 demonstrates. Indeed, it describes sets X and Y, the union $X \cup Y$ of which is not totally bounded. At the same time, by Corollary 9.27, any relatively bounded set is totally bounded. So, the union $X \cup Y$ is not relatively bounded.

To make boundedness preserved by finite unions, we introduce more flexible conditions of boundedness.

Definition 9.8. a) A set $X \subseteq L$ is called *k-multibounded* if $X = \bigcup_{i=1}^{n} X_i$ and all sets X_i are k-bounded at some points x_i.

b) The number k is called a *measure of multiboundedness* of the set X.

c) A set $X \subseteq L$ is called *multibounded* if $X = \bigcup_{i=1}^{n} X_i$ and all sets X_i are bounded at some points x_i.

If $L = \boldsymbol{R}_\omega$, then the set $X = \{[0, 3], \alpha\}$ where $\{\alpha = \mathrm{Hn}(i)_{i\in\omega}$ is 3-multibounded, while the set $X = \{\{\alpha + x;\ 0 < x\}$ is not multibounded.

Lemma 9.13. a) Any relatively k-bounded set X is k-multibounded.

b) Any totally k-bounded set X is k-multibounded.

c) Any k-bounded set X is k-multibounded.

Proof is left as an exercise.

Lemma 9.14. If a set $X \subseteq L$ is k-multibounded, then it is h-multibounded for any $h > k$.

Proof is left as an exercise.

Proposition 9.19. (a) If $X, Y \subseteq L$, X is k-multibounded and Y is h-multibounded, then the union $X \cup Y$ is t-multibounded where $t = \max\{k, h\}$.

(b) If $X \subseteq Y \subseteq L$ and Y is k-multibounded, then X is k-multibounded.

(c) If $X, Y \subseteq L$, X is k-multibounded and Y is h-multibounded, then the sum $X + Y$ is $(k + h)$-multibounded.

(d) If $X \subseteq L$ and X is k-multibounded, then its closure CX is $(k + \varepsilon)$-multibounded.

Proof. (a) Let us consider a k-multibounded set X and an h-multibounded set Y. By definition, $X = \bigcup_{i=1}^{n} X_i$ and all sets X_i are k-bounded at some points x_i, while $Y = \bigcup_{j=1}^{m} Y_j$ and all sets Y_j are h-bounded at some points y_j. Then $X \cup Y = \bigcup_{r=1}^{n+m} Z_r$ where $Z_r = X_r$ when $r < n + 1$ and $Z_r = Y_r$ when $m + 1 >$

$r > n$. As all sets Z_r are t-bounded at some points z_r where $t = \max\{k, h\}$, the union $X \bigcup Y$ is t-multibounded.

Part (b) directly follows from definitions.

(c) Let us consider a k-multibounded set X and an h-multibounded set Y. By definition, $X = \bigcup_{i=1}^{n} X_i$ and all sets X_i are k-bounded at some points x_i, while $Y = \bigcup_{j=1}^{m} Y_j$ and all sets Y_j are h-bounded at some points y_j. Then $X + Y = \bigcup_{r=1}^{nm} Z_r$ where $Z_r = X_i + Y_j$ for some i and j. By Proposition 9.11.c, all sets Z_r are $(k + h)$-bounded at some points z_r. Consequently, the sum $X + Y$ is $(k + h)$-multibounded.

(d) Let us consider a k-multibounded set X. By definition, $X = \bigcup_{i=1}^{n} X_i$ and all sets X_i are k-bounded at some points x_i ($i = 1, 2, 3, \ldots, n$). By Proposition 9.17, the closure CX_i of each set X_i is $(k + \varepsilon)$-bounded at the point x_i ($i = 1, 2, 3, \ldots, n$). Besides, the closure of a union of a finite number of sets is the union of the closures of these sets (Kuratowski, 1966), i.e., $CX = \bigcup_{i=1}^{n} CX_i$. Thus, by definition, the set CX is $(k + \varepsilon)$-multibounded.

Proposition is proved.

Corollary 9.32. (a) If $X \subseteq Y \subseteq L$ and Y is multibounded, then X is multibounded.

(b) If $X, Y \subseteq L$, while X and Y are multibounded, then the sum $X + Y$ is multibounded.

Corollary 9.33. If $X, Y \subseteq L$, X is k-multibounded, then the intersection $X \cap Y$ is k-multibounded.

Corollary 9.32.a implies the following result.

Corollary 9.34. If $X, Y \subseteq L$, X is multibounded, then the intersection $X \cap Y$ is multibounded.

Let us consider the Cartesian product $L_1 \times L_2 \times L_3 \times \ldots \times L_n$ with the Manhattan hyperseminorm with respect to the hyperseminormed vector spaces $L_1, L_2, L_3, \ldots, L_n$ and sets $X_1 \subseteq L_1, X_2 \subseteq L_2, X_3 \subseteq L_3, \ldots, X_n \subseteq L_n$.

Proposition 9.20. If each set X_i is p_i-multibounded ($i = 1, 2, 3, \ldots, n$), then the Cartesian product of these sets $X_1 \times X_2 \times X_3 \times \ldots \times X_n$ is p-multibounded in the Cartesian product $L_1 \times L_2 \times L_3 \times \ldots \times L_n$ of hyperseminormed vector spaces where $p = 2(p_1 + p_2 + p_3 + \ldots + p_n)$.

Proof. Let us consider the Cartesian product $L_1 \times L_2 \times L_3 \times \ldots \times L_n$ of hyperseminormed vector spaces with the Manhattan hyperseminorm, p_i-multibounded sets $X_1 \subseteq L_1, X_2 \subseteq L_2, X_3 \subseteq L_3, \ldots, X_n \subseteq L_n$, and their Cartesian product of these sets $X = X_1 \times X_2 \times X_3 \times \ldots \times X_n$. By definition, for all $i = 1, 2, 3, \ldots, n$, we have

$$X_i = \bigcup_{i=1}^{n} X_{ij}$$

and all sets X_{ij} are p_i-bounded at some points x_{ij} ($j = 1, 2, 3, \ldots, n_i$). By Proposition 9.12, all sets X_{ij} are totally $2p_i$-bounded at the points x_{ij} ($j = 1, 2, 3, \ldots, n_i$).

At the same time,

$$X_1 \times X_2 \times X_3 \times \ldots \times X_n = \bigcup X_{1j_1} \times X_{2j_2} \times X_{3j_3} \times \ldots \times X_{nj_n}$$

By Proposition 9.2, all Cartesian products $X_{1j_1} \times X_{2j_2} \times X_{3j_3} \times \ldots \times X_{nj_n}$ are totally p-bounded where

$$p = 2(p_1 + p_2 + p_3 + \ldots + p_n)$$

By Proposition 9.13, all sets $X_{1j_1} \times X_{2j_2} \times X_{3j_3} \times \ldots \times X_{nj_n}$ are p-bounded. Thus, the Cartesian product $X_1 \times X_2 \times X_3 \times \ldots \times X_n$ is p-multibounded in the Cartesian product $L_1 \times L_2 \times L_3 \times \ldots \times L_n$.

Proposition is proved.

Let us consider the Cartesian product $L_1 \times L_2 \times L_3 \times \ldots \times L_n$ with the Euclidean seminorm with respect to the seminormed vector spaces L_1, L_2, L_3, …, L_n and sets $X_1 \subseteq L_1$, $X_2 \subseteq L_2$, $X_3 \subseteq L_3$, …, $X_n \subseteq L_n$.

Proposition 9.21. If each set X_i is p_i-multibounded ($i = 1, 2, 3, \ldots, n$), then the Cartesian product of these sets $X_1 \times X_2 \times X_3 \times \ldots \times X_n$ is t-multibounded in the Cartesian product $L_1 \times L_2 \times L_3 \times \ldots \times L_n$ of hyperseminormed vector spaces where $t = (p_1^2 + p_2^2 + p_3^2 + \ldots + p_n^2)^{1/2}$.

Proof. Let us consider the Cartesian product $L_1 \times L_2 \times L_3 \times \ldots \times L_n$ of seminormed vector spaces with the Euclidean seminorm, p_i-multibounded sets $X_1 \subseteq L_1$, $X_2 \subseteq L_2$, $X_3 \subseteq L_3$, …, $X_n \subseteq L_n$, and their Cartesian product of these sets $X = X_1 \times X_2 \times X_3 \times \ldots \times X_n$. By definition, for all $i = 1, 2, 3, \ldots, n$, we have

$$X_i = \bigcup_{i=1}^{n} X_{ij}$$

and all sets X_{ij} are p_i-bounded at some points x_{ij} ($j = 1, 2, 3, \ldots, n_i$). By Proposition 9.12, all sets X_{ij} are totally $2p_i$-bounded at the points x_{ij} ($j = 1, 2, 3, \ldots, n_i$).

At the same time,

$$X_1 \times X_2 \times X_3 \times \ldots \times X_n = \bigcup X_{1j_1} \times X_{2j_2} \times X_{3j_3} \times \ldots \times X_{nj_n}$$

By Proposition 9.3, all Cartesian products $X_{1j_1} \times X_{2j_2} \times X_{3j_3} \times \ldots \times X_{nj_n}$ are totally p-bounded where

$$p = ((2p_1)^2 + (2p_2)^2 + (2p_3)^2 + \ldots + (2p_n^2))^{1/2} = 2(p_1^2 + p_2^2 + p_3^2 + \ldots + p_n^2)^{1/2}$$

By Proposition 9.13, all sets $X_{1j_1} \times X_{2j_2} \times X_{3j_3} \times \ldots \times X_{nj_n}$ are p-bounded. Thus, the Cartesian product $X_1 \times X_2 \times X_3 \times \ldots \times X_n$ is p-multibounded in the Cartesian product $L_1 \times L_2 \times L_3 \times \ldots \times L_n$.

Proposition is proved.

Let us consider the Cartesian product $L_1 \times L_2 \times L_3 \times \ldots \times L_n$ with the Chebyshev hyperseminorm with respect to the hyperseminormed vector spaces $L_1, L_2, L_3, \ldots, L_n$ and sets $X_1 \subseteq L_1, X_2 \subseteq L_2, X_3 \subseteq L_3, \ldots, X_n \subseteq L_n$.

Proposition 9.22. If each set X_i is p_i-multibounded ($i = 1, 2, 3, \ldots, n$), then the Cartesian product of these sets $X_1 \times X_2 \times X_3 \times \ldots \times X_n$ is q-multibounded in the Cartesian product $L_1 \times L_2 \times L_3 \times \ldots \times L_n$ of hyperseminormed vector spaces where $q = \max \{p_1, p_2, p_3, \ldots, p_n\}$.

Proof is similar to the proof of Proposition 9.20.

Corollary 9.35. The Cartesian product of multibounded sets is multibounded in the Cartesian product $L_1 \times L_2 \times L_3 \times \ldots \times L_n$ of hyperseminormed vector spaces if it has either the Manhattan hyperseminorm or the Chebyshev hyperseminorm.

Corollary 9.36. The Cartesian product of multibounded sets is multibounded in the Cartesian product $L_1 \times L_2 \times L_3 \times \ldots \times L_n$ of seminormed vector spaces if it has the Euclidean seminorm.

Theorem 9.1. Any compact subset K of a complete hyperseminormed space L is multibounded.

Proof. Let us consider a compact set $X \subseteq L$ and the neighborhood $O_1(\mathbf{0}) = \{x; \|x\| < 1\}$ of $\mathbf{0}$. Then the union of all open sets $x + O_1(\mathbf{0})$ with x from X covers X. By properties of compact sets, there is a finite number of elements $x_1, x_2, x_3, \ldots, x_n$ from X such that the sets $x_1 + O_1(\mathbf{0})$, $x_2 + O_1(\mathbf{0})$, $x_3 + O_1(\mathbf{0})$, $\ldots$, $x_n + O_1(\mathbf{0})$ cover X.

We define $X_i = X \cap (x_i + O_1(\mathbf{0}))$, $i = 1, 2, 3, \ldots, n$. By Definition 9.4, each set X_i is bounded at the point x_i ($i = 1, 2, 3, \ldots, n$). Besides, $X = \bigcup_{i=1}^{n} X_i$ because the sets $x_1 + O_1(\mathbf{0})$, $x_2 + O_1(\mathbf{0})$, $x_3 + O_1(\mathbf{0})$, $\ldots$, $x_n + O_1(\mathbf{0})$ cover X. Thus, by Definition 9.6, the set X is multibounded.

Theorem is proved.

Corollary 9.37. Any compact subset X of a hypernormed vector space L is multibounded.

As any seminorm is a hyperseminorm, we have the following result.

Corollary 9.38. Any compact subset X of a seminormed vector space L is multibounded.

As any norm is a seminorm, we have the following result.

Corollary 9.39. Any compact subset X of a normed vector space L is multibounded.

Remark 9.4. A compact subset K of a hyperseminormed vector space L is not necessarily totally bounded, as well as not necessarily bounded, as the following example demonstrates.

Example 9.5. As it is proved by Burgin (2012), the set of all real hypernumbers $\boldsymbol{R}_\omega$ is a hypernormed space where the hypernorm $\|\cdot\|$ is defined by the following formula:

If α is a real hypernumber, i.e., $\alpha = \text{Hn}(a_i)_{i\in\omega}$ with $a_i \in \boldsymbol{R}$ for all $i\in\omega$, then $\| \alpha \| = \text{Hn}(|a_i|)_{i\in\omega}$

Let us consider the set $K = \{0, 1, \alpha = \text{Hn}(n)_{n\in\omega}\}$. This set is compact but not bounded in $\boldsymbol{R}_\omega$ because $\| \alpha - \mathbf{0} \| = \|\alpha\| = \text{Hn}(|n|)_{n\in\omega} = \text{Hn}(n)_{n\in\omega}$ is larger than any real number.

There are other definitions of boundedness used in the literature. For instance, Narici and Beckenstein study the following concept (Narici and Beckenstein, 1985).

Definition 9.9. A set $X \subseteq L$ is called *precompact* (or *totally bounded* in terminology of Narici and Beckenstein) if for any neighborhood V of $\mathbf{0}$, there is a finite number of elements $x_1, x_2, x_3, \ldots, x_n$ from X such that the sets $x_1 + V, x_2 + V, x_3 + V, \ldots, x_n + V$ cover X.

Let us consider a hyperseminormed vector space L.

Lemma 9.15. Any compact set $X \subseteq L$ is precompact.

Indeed, let us take a neighborhood V of $\mathbf{0}$ and a compact set $X \subseteq L$. Then the set $P = \{ x + V; x \in L\}$ is an open cover of X. By the definition of a compact set (cf. Appendix), there is a finite subset $x_1 + V, x_2 + V, x_3 + V, \ldots, x_n + V$ of P, which is also an open cover of X. Consequently, X is a precompact set.

Let us consider a locally compact hyperseminormed vector space L.

Theorem 9.2. A closed set $X \subseteq L$ is precompact if and only if it is multibounded.

Proof. Necessity. Let us consider a precompact set $X \subseteq L$ and the neighborhood $\text{O}_1(\mathbf{0}) = \{x; \|x\| < 1\}$ of $\mathbf{0}$. By Definition 9.9, there is a finite number of elements $x_1, x_2, x_3, \ldots, x_n$ from X such that the sets $x_1 + \mathrm{O}_1(\mathbf{0}), x_2 + \mathrm{O}_1(\mathbf{0}), x_3 + \mathrm{O}_1(\mathbf{0}), \ldots, x_n + \mathrm{O}_1(\mathbf{0})$ cover X.

We define $X_i = X \cap (x_i + \mathrm{O}_1(\mathbf{0}))$, $i = 1, 2, 3, \ldots, n$. By Definition 9.4, each set X_i is bounded at the point $x_i (i = 1, 2, 3, \ldots, n)$. Besides, $X = \bigcup_{i=1}^{n} X_i$ because the sets $x_1 + \mathrm{O}_1(\mathbf{0}), x_2 + \mathrm{O}_1(\mathbf{0}), x_3 + \mathrm{O}_1(\mathbf{0}), \ldots, x_n + \mathrm{O}_1(\mathbf{0})$ cover X. By Definition 9.8, the set X is multibounded. Then by Proposition 9.13, the closure CX of X is multibounded.

Sufficiency. Let us assume that the closure CX of a set $X \subseteq L$ is multibounded. By Definition 9.8, $CX = \bigcup_{i=1}^{n} X_i$ and each set X_i is bounded at the point x_i, i.e. $X_i \subseteq k_i \cdot \mathrm{O}_1(x)$ for some positive real number $k_i (i = 1, 2, 3, \ldots, n)$. By Lemma 5.7, each $\mathrm{C}(k_i \cdot \mathrm{O}_1(x))$ is a compact space. By properties of closure, $\mathrm{C}X_i \subseteq \mathrm{C}(k_i \cdot \mathrm{O}_1(x))$.

As a closed subset of a compact space, each $\mathrm{C}X_i$ is also a compact space (Kuratowski, 1966). Thus, it is possible to cover each $\mathrm{C}X_i$ by a finite number of the sets $x_1 + V, x_2 + V, x_3 + V, \ldots, x_n + V$ where elements $x_1, x_2, x_3, \ldots, x_n$ belong to $\mathrm{C}X_i$ and V is an arbitrary neighborhood of $\mathbf{0}$ $(i = 1, 2, 3, \ldots, n)$. As $CX = \bigcup_{i=1}^{n} X_i$, all these sets $x_t + V$ cover CX and by Definition 9.9, the set $X = CX$ is precompact.

Theorem is proved.

Proposition 9.23. If subsets X_i $(i = 1, 2, 3, \ldots, n)$ of a hyperseminormed vector space L are precompact, then their union $X = \bigcup_{i=1}^{n} X_i$ is also precompact.

Proof. Let us consider precompact subsets X_i $(i = 1, 2, 3, \ldots, n)$ of a hyperseminormed vector space L and their union $X = \bigcup_{i=1}^{n} X_i$. By definition, for all $i = 1, 2, 3, \ldots, n$, we have

$$X_i = \bigcup_{j=1}^{n_i} X_{ij}$$

and all sets X_{ij} are p_i-bounded at some points x_{ij} $(j = 1, 2, 3, \ldots, n_i)$. Consequently,

$$X = \bigcup_{i=1}^{n} {}_{j=1}^{n_i} X_{ij}$$

and all sets X_{ij} are p_i-bounded at some points x_{ij} $(j = 1, 2, 3, \ldots, n_i; i = 1, 2, 3, \ldots, n)$. Thus, their union X is a precompact set.

Proposition is proved.

Corollary 9.40. If subsets X_i $(i = 1, 2, 3, \ldots, n)$ of a hypernormed vector space L are precompact, then their union is also precompact.

Let us consider the Cartesian product $L_1 \times L_2 \times L_3 \times \dots \times L_n$ with the Euclidean seminorm with respect to the seminormed vector spaces L_1, L_2, L_3, …, L_n and sets $X_1 \subseteq L_1$, $X_2 \subseteq L_2$, $X_3 \subseteq L_3$, …, $X_n \subseteq L_n$.

Proposition 9.24. If subsets X_i of seminormed vector spaces $L_i (i = 1, 2, 3, \dots, n)$ are precompact, then their Cartesian product $X_1 \times X_2 \times X_3 \times \dots \times X_n$ with the Euclidean seminorm is also precompact in the Cartesian product $L_1 \times L_2 \times L_3 \times \dots \times L_n$ with the Euclidean seminorm.

Proof. Let us consider the Cartesian product $L_1 \times L_2 \times L_3 \times \dots \times L_n$ of seminormed vector spaces, p_i-multibounded sets $X_1 \subseteq L_1$, $X_2 \subseteq L_2$, $X_3 \subseteq L_3$, …, $X_n \subseteq L_n$, and their Cartesian product of these sets $X = X_1 \times X_2 \times X_3 \times \dots \times X_n$. By Theorem 9.2, all sets $X_1, X_2, X_3, \dots, X_n$ are p_i-multibounded for some numbers $p_1, p_2, p_3, \dots, p_n$.

By definition, for all $i = 1, 2, 3, \dots, n$, we have

$$X_i = \bigcup_{j=1}^{n_i} X_{ij}$$

and all sets X_{ij} are p_i-bounded at some points x_{ij} ($j = 1, 2, 3, \dots, n_i$). By Proposition 9.12, all sets X_{ij} are totally $2p_i$-bounded at the points x_{ij} ($j = 1, 2, 3, \dots, n_i$).

At the same time,

$$X_1 \times X_2 \times X_3 \times \dots \times X_n = \bigcup (X_{1j_1} \times X_{2j_2} \times X_{3j_3} \times \dots \times X_{nj_n})$$

By Proposition 9.3, all Cartesian products $X_{1j_1} \times X_{2j_2} \times X_{3j_3} \times \dots \times X_{nj_n}$ are totally p-bounded where

$$p = ((2p_1)^2 + (2p_2)^2 + (2p_3)^2 + \dots + (2p_n{}^2))^{1/2} = 2^{1/2}(p_1{}^2 + p_2{}^2 + p_3{}^2 + \dots + p_n{}^2)^{1/2}$$

By Proposition 9.13, all sets $X_{1j_1} \times X_{2j_2} \times X_{3j_3} \times \dots \times X_{nj_n}$ are p-bounded. Therefore, the Cartesian product $X_1 \times X_2 \times X_3 \times \dots \times X_n$ is p-multibounded and thus, multibounded in the Cartesian product $L_1 \times L_2 \times L_3 \times \dots \times L_n$. Then by Theorem 9.2, the Cartesian product $X_1 \times X_2 \times X_3 \times \dots \times X_n$ is precompact in the Cartesian product $L_1 \times L_2 \times L_3 \times \dots \times L_n$.

Proposition is proved.

Corollary 9.41. If subsets X_i of normed vector spaces $L_i (i = 1, 2, 3, \dots, n)$ are precompact, then their Cartesian product $X_1 \times X_2 \times X_3 \times \dots \times X_n$ with the Euclidean seminorm is also precompact in the Cartesian product $L_1 \times L_2 \times L_3 \times \dots \times L_n$ with the Euclidean seminorm.

Let us consider the Cartesian product $L_1 \times L_2 \times L_3 \times \ldots \times L_n$ with the Manhattan hyperseminorm with respect to the hyperseminormed vector spaces $L_1, L_2, L_3, \ldots, L_n$ and sets $X_1 \subseteq L_1, X_2 \subseteq L_2, X_3 \subseteq L_3, \ldots, X_n \subseteq L_n$.

Proposition 9.25. If subsets X_i of hyperseminormed vector spaces L_i (i = 1, 2, 3, …, n) are precompact, then their Cartesian product $X_1 \times X_2 \times X_3 \times \ldots \times X_n$ with the Manhattan hyperseminorm is also precompact in the Cartesian product $L_1 \times L_2 \times L_3 \times \ldots \times L_n$ with the Manhattan hyperseminorm.

Proof. Let us consider the Cartesian product $L_1 \times L_2 \times L_3 \times \ldots \times L_n$ of seminormed vector spaces, p_i-multibounded sets $X_1 \subseteq L_1, X_2 \subseteq L_2, X_3 \subseteq L_3, \ldots, X_n \subseteq L_n$, and their Cartesian product of these sets $X = X_1 \times X_2 \times X_3 \times \ldots \times X_n$. By Theorem 9.2, all sets $X_1, X_2, X_3, \ldots, X_n$ are p_i-multibounded for some numbers $p_1, p_2, p_3, \ldots, p_n$.

By definition, for all i = 1, 2, 3,…, n, we have

$$X_i = \bigcup_{j=1}^{n_i} X_{ij}$$

and all sets X_{ij} are p_i-bounded at some points x_{ij} (j = 1, 2, 3,…, n_i). By Proposition 9.12, all sets X_{ij} are totally $2p_i$-bounded at the points x_{ij} (j = 1, 2, 3, …, n_i).

At the same time,

$$X_1 \times X_2 \times X_3 \times \ldots \times X_n = \bigcup (X_{1j_1} \times X_{2j_2} \times X_{3j_3} \times \ldots \times X_{nj_n})$$

By Proposition 9.2, all Cartesian products $X_{1j_1} \times X_{2j_2} \times X_{3j_3} \times \ldots \times X_{nj_n}$ are totally p-bounded where

$$p = 2p_1 + 2p_2 + 2p_3 + \ldots + 2p_n.$$

By Proposition 9.13, all sets $X_{1j_1} \times X_{2j_2} \times X_{3j_3} \times \ldots \times X_{nj_n}$ are p-bounded. Therefore, the Cartesian product $X_1 \times X_2 \times X_3 \times \ldots \times X_n$ is p-multibounded and thus, multibounded in the Cartesian product $L_1 \times L_2 \times L_3 \times \ldots \times L_n$. Then by Theorem 9.2, the Cartesian product $X_1 \times X_2 \times X_3 \times \ldots \times X_n$ is precompact in the Cartesian product $L_1 \times L_2 \times L_3 \times \ldots \times L_n$.

Proposition is proved.

Corollary 9.42. If subsets X_i of hypernormed vector spaces L_i (i = 1, 2, 3, …, n) are precompact, then their Cartesian product $X_1 \times X_2 \times X_3 \times \ldots \times X_n$ with the Manhattan hyperseminorm is also precompact in the Cartesian product $L_1 \times L_2 \times L_3 \times \ldots \times L_n$ with the Manhattan hyperseminorm.

Let us consider the Cartesian product $L_1 \times L_2 \times L_3 \times \ldots \times L_n$ with the Chebyshev hyperseminorm with respect to the hyperseminormed vector spaces $L_1, L_2, L_3, \ldots, L_n$ and sets $X_1 \subseteq L_1, X_2 \subseteq L_2, X_3 \subseteq L_3, \ldots, X_n \subseteq L_n$.

Proposition 9.26. If subsets X_i of hyperseminormed vector spaces L_i ($i = 1, 2, 3, \ldots, n$) are precompact, then their Cartesian product $X_1 \times X_2 \times X_3 \times \ldots \times X_n$ with the Chebyshev hyperseminorm is also precompact in the Cartesian product $L_1 \times L_2 \times L_3 \times \ldots \times L_n$ with the Chebyshev hyperseminorm.

Proof is the similar to the proof of Proposition 9.25.

Corollary 9.43. If subsets X_i of hypernormed vector spaces L_i ($i = 1, 2, 3, \ldots, n$) are precompact, then their Cartesian product $X_1 \times X_2 \times X_3 \times \ldots \times X_n$ with the Chebyshev hyperseminorm is also precompact in the Cartesian product $L_1 \times L_2 \times L_3 \times \ldots \times L_n$ with the Chebyshev hyperseminorm.

Proposition 9.27. Any precompact subset X of a hyperseminormed vector space L is multibounded.

Proof is the same as the proof of necessity in Theorem 9.2.

Corollary 9.44. Any precompact subset X of a hypernormed vector space L is multibounded.

Corollary 9.45. Any precompact subset X of a seminormed vector space L is multibounded.

Corollary 9.46. Any precompact subset X of a normed vector space L is multibounded.

Remark 9.5. In topological vector spaces, any precompact set X is essentially bounded (Narici and Beckenstein, 1985: Theorem 7.1.1). For semitopological vector spaces, this result is not valid as the following example demonstrates.

Example 9.6. As it is proved by Burgin (2012), the set of all real hypernumbers $\boldsymbol{R}_\omega$ is a hypernormed space and thus, a semitopological vector space (cf. Theorem 5.5).

Let us consider the set $K = \{0, 1, \alpha = \mathrm{Hn}(n)_{n\in\omega}\}$. This set is precompact but not bounded and not essentially bounded in $\boldsymbol{R}_\omega$ because $\| \alpha - \mathbf{0} \| = \|\alpha\| = \mathrm{Hn}(|n|)_{n\in\omega} = \mathrm{Hn}(n)_{n\in\omega}$ is larger than any real number, while by Proposition 9.8, essential boundedness and boundedness coincide in $\boldsymbol{R}_\omega$.

There are also other concepts of boundedness. For instance, Narici and Beckenstein study the following concept of boundedness (Narici and Beckenstein, 1985).

Definition 9.10. A set $X \subseteq L$ is called *Cauchy bounded* if for any neighborhood V of $\mathbf{0}$ and any denumerable subset E of X, there are two distinct elements x and z from E such that the difference $x - z$ belongs to V.

Proposition 9.28. A subset X of a hyperseminormed vector space L is precompact if and only if it is Cauchy bounded.

Proof. Necessity. Let us take an infinite subset E of a precompact subset X of L and a neighborhood V of $\mathbf{0}$. As by Theorem 5.4, L is a semitopological vector space, it is possible to assume that there is a neighborhood U of $\mathbf{0}$ such that $U - U \subseteq V$ (cf. Lemma 5.2).

By the definition of a precompact set, for the neighborhood U of $\mathbf{0}$, there is a finite number of elements $x_1, x_2, x_3, \ldots, x_n$ from X such that the sets $x_1 + U$, $x_2 + U$, $x_3 + U$, …, $x_n + U$ cover X. As the set E is infinite, at least, one of the sets, say $x_j + U$, contains two distinct elements x and z from E. Then we have

$$x - y \in (x_j + U) - (x_j + U) = U - U \subseteq V$$

So, the difference $x - z$ belongs to V and subset X of L is Cauchy bounded.

Sufficiency. Let us assume that a set $X \subseteq L$ is not precompact. Then there is a symmetric neighborhood V of $\mathbf{0}$ such that no finite number of the sets $x + V$ covers X. As a result, there is an element z_1 from X that does not belong to V. By the same token, there is an element z_2 from X that does not belong to $z_1 + V$. Continuing this process and applying mathematical induction, we obtain a sequence $l = \{z_i; i = 1, 2, 3, \ldots\}$ of elements from X such that for any natural number n, the element z_n does not belong to the union$_{i=1}^{n-1}$ $(z_i + V)$.

Then the difference of any two elements from l does not belong to V. Indeed, as the neighborhood V is symmetric, $V = -V$ and taking $i < j$, we see that each membership relation $z_j - z_{i\in} V$ or $z_i - z_j \in V$ implies $z_j \in z_i + V$. This contradicts the choice of the elements z_i and shows that X is not Cauchy bounded. Consequently, if a subset X of L is Cauchy bounded, then X is precompact.

Theorem is proved.

Corollary 9.47. A subset X of a hypernormed vector space L is precompact if and only if it is Cauchy bounded.

Corollary 9.48. A subset X of a seminormed vector space L is precompact if and only if it is Cauchy bounded.

Corollary 9.49. A subset X of a normed vector space L is precompact if and only if it is Cauchy bounded.

Corollary 9.50. Any Cauchy bounded subset X of a hyperseminormed (hypernormed) vector space L is multibounded.

KEYWORDS

- **Cartesian product**
- **Cauchy bounded**
- **closure**
- **compact**
- **multibounded**
- **neighborhood**
- **operator**
- **precompact**
- ***q*-bounded set**
- **vector space**

CHAPTER 10

BOUNDEDNESS AND FUZZY CONTINUITY

As we have found, an important property of operators and functionals in hyperseminormed vector spaces is boundedness. We continue our exploration of this property and it relation to continuity. However, to expand the scope of applications, continuity is changed here to a more general concept of fuzzy continuity, while we use the standard concepts of boundedness studied in Chapter 6. At first, we treat total boundedness of sets in hyperseminormed vector spaces deriving the concept of a bounded operator between hyperseminormed vector spaces. Note that, as a rule, results derived for hyperseminormed vector spaces remain true for hypernormed, seminormed and normed vector spaces.

Let us consider hyperseminormed vector spaces L and M over a field $\boldsymbol{F}$ where $\boldsymbol{F}$ is either the field $\boldsymbol{R}$ of all real numbers or the field $\boldsymbol{C}$ of all complex numbers. Both hyperseminorms in L and M are denoted by the same symbol $\|.\|$.

Definition 10.1.

a) An operator (mapping) $A: L \to M$ is called (p, q)-*bounded at* a point a from the space L if for any element b from L, the inequality $\|b - a\| < p$ implies the inequality $\| A(b) - A(a)\| < q$.

b) An operator (mapping) $A: L \to M$ is called *bounded at* a point a from L if it is (p, q)-bounded at a for some positive numbers p and q.

c) An operator (mapping) $A: L \to M$ is called *totally bounded at* a point a from the space L if for any number p, there is a number q such that for any element b from L, the inequality $\| b - a \| < p$ implies the inequality $\| A(b) - A(a)\| < q$.

d) An operator (mapping) $A: L \to M$ is called *p-bounded at* a point a from the space L if it is (p, q)-bounded at a for some positive number q.
e) An operator (mapping) $A: L \to M$ is called *q-cobounded at* a point a from the space L if it is (p, q)-bounded at a for some positive numbers p and q.

Example 10.1. Let us take $L = M = \boldsymbol{R}$ and consider the following operator, which is called the Heaviside function

$$A(x) = \begin{cases} 1 & \text{if } x \geq 0 \\ 0 & \text{if } x < 0 \end{cases}$$

This operator A is (1, 2)-bounded and totally bounded at $\mathbf{0}$ because for any point b from $\boldsymbol{R}$, the inequality $||A(b) - A(\mathbf{0})|| < 2$ is valid, but A is not (0.1, 0.5)-bounded at $\mathbf{0}$.

Example 10.2. Let us take $L = M = \boldsymbol{R}$ and consider the one-dimensional operator (mapping) tg x. This operator is (0.1, 1)-bounded and 0.2-bounded at $\mathbf{0}$ but it is not (4, 10)-bounded, not 4-bounded and not totally bounded at $\mathbf{0}$.

Lemma 10.1. If $t \leq p$, $q \leq u$ and an operator (mapping) $A: L \to M$ is (p, q)-bounded at a point a, then A is (t, u)-bounded at a.

Proof. Let us consider a (p, q)-bounded at a point a from the space L operator $A: L \to M$. Then taking an arbitrary point b from the space L, $t \leq p$ and $u \geq q$, we have

$$||b - a|| < t \Rightarrow ||b - a|| < p \Rightarrow || A(b) - A(a)|| < q \Rightarrow ||A(b) - A(a)|| < u$$

Thus, the operator (mapping) A is (t, u)-bounded at a.

Lemma is proved.

Corollary 10.1. If $t < p$ and an operator (mapping) $A: L \to M$ is (p, q)-bounded at a point a, then A is (t, q)-bounded at a.

Corollary 10.2. If $q < u$ and an operator (mapping) $A: L \to M$ is (p, q)-bounded at a point a, then A is (p, u)-bounded at a.

Corollary 10.3. If $t < p$ and an operator (mapping) $A: L \to M$ is p-bounded at a point a, then A is t-bounded at a.

Corollary 10.4. If $q < u$ and an operator (mapping) $A: L \to M$ is q-cobounded at a point a, then A is u-cobounded at a.

Definitions imply the following results.

Lemma 10.2. Any totally bounded at a point a operator A: $L \to M$ is bounded at a, q-cobounded at a for some positive real number q and p-bounded at a for any positive real number p.

Proof is left as an exercise.

Lemma 10.3. Any bounded at a point a operator A: $L \to M$ is q-cobounded at a for some positive real number q and p-bounded at a for some positive real number p.

Proof is left as an exercise.

Lemma 10.4. a) Any p-bounded at a point a operator A: $L \to M$ is bounded at a.

b) Any q-cobounded at a point a operator A: $L \to M$ is bounded at a.

Proof is left as an exercise.

Lemma 10.5. An operator (mapping) A: $L \to M$ is (p, q)-bounded at a point a if and only if A maps any totally p-bounded at the point a set into a totally q-bounded at the point $A(a)$ set.

Proof is left as an exercise.

Lemma 10.6. An operator (mapping) A: $L \to M$ is q-cobounded at a point a if and only if for any number $p \leq t$, there is a positive real number q such that A maps any totally t-bounded at the point a set into a totally q-bounded at the point $A(a)$ set.

Proof is left as an exercise.

Lemma 10.7. An operator (mapping) A: $L \to M$ is totally bounded at a point a if and only if A maps any totally bounded at the point a set into a totally bounded at the point $A(a)$ set.

Proof is left as an exercise.

Lemma 10.8. An operator (mapping) A: $L \to M$ is totally bounded at a point a if and only if for any number p, there is a number q such that A is (p, q)-bounded at the point a.

Proof is left as an exercise.

These results allow us to define exact measures of boundedness of an operator at a point.

Lower boundedness $\text{Bound}_p(A, a)$ of an operator A at a point a shows the lower limit of the measure of total boundedness for the image of totally p – bounded set. It is defined by the formula

$$\text{Bound}_p(A, a) = \inf \{q;\ A \text{ is } (p, q)\text{-bounded at } a\}$$

Upper boundedness $\text{Bound}^q(A, a)$ of an operator A at a point a shows the upper limit of the measure of total boundedness for a set that can be mapped by A into totally q – bounded set. It is defined by the formula

$$\text{Bound}^q(A, a) = \sup \{p; A \text{ is } (p, q)\text{-bounded at } a\}$$

Lemma 10.9. The lower boundedness $\text{Bound}_p(A, a)$ of an operator A at a point a is defined if and only if A is (p, q)-bounded at a for some number $q \geq \text{Bound}_p(A, a)$.

Proof is left as an exercise.

Lemma 10.10. Theupper boundedness $\text{Bound}^q(A, a)$ of an operator A at a point a is defined if and only if A is (p, q)-bounded at a for some number $p \leq \text{Bound}^q(A, a)$.

Proof is left as an exercise.

Proposition 10.1. If the lower boundedness $\text{Bound}_p(A, a)$ is defined and $r < p$, then the lower boundedness $\text{Bound}_r(A, a)$ is also defined and $\text{Bound}_r(A, a) \leq \text{Bound}_p(A, a)$.

Indeed, by Lemma 10.7, if an operator (mapping) A: $L \rightarrow M$ is (p, q)-bounded at a point a, then A is (r, q)-bounded at a. Thus, we have

$$\text{Bound}_r(A, a) = \inf \{q; A \text{ is } (r, q)\text{-bounded at } a\} \leq \inf \{q; A \text{ is } (p, q)\text{-bounded at } a\} = \text{Bound}_p(A, a)$$

Proposition 10.2. If the upper boundedness $\text{Bound}^q(A, a)$ is defined and $q < t$, then the upper boundedness $\text{Bound}^t(A, a)$ is also defined and $\text{Bound}^q(A, a) \leq \text{Bound}^t(A, a)$.

Indeed, by Lemma 10.7, if an operator (mapping) A: $L \rightarrow M$ is (p, q)-bounded at a point a, then A is (p, t)-bounded at a. Thus, we have

$$\text{Bound}^q(A, a) = \sup \{p; A \text{ is } (p, q)\text{-bounded at } a\} \leq \sup \{p; A \text{ is } (p, t)\text{-bounded at } a\} = \text{Bound}^t(A, a)$$

If A: $L \rightarrow M$ is operator (mapping) of hyperseminormed vector spaces L and M, it is possible to build a *shift*(also called *translation*) of the operator A taking an element u from M and defining the operator $A + u$ by the following formula:

$$(A + u)(x) = A(x) + u$$

Proposition 10.3. An operator (mapping) $A: L \to M$ is (p, q)-bounded at a point a if and only if any its shift $A + u$ is (p, q)-bounded at the point a.

Proof. Necessity. Let us take a (p, q)-bounded at a point a operator (mapping) $A: L \to M$. It means that for any element b from L, the inequality $\| b - a \| < p$ implies the inequality $\| A(b) - A(a)\| < q$. Then for the same element b, we have

$$\| (A(b) + u) - (A(a) + u) \| = \| A(b) - A(a) \| < q$$

Thus, the operator $A + u$ is (p, q)-bounded at the point a.

Sufficiency. As we have proved that (p, q)-boundedness of an operator implies(p, q)-boundedness of any its shift, (p, q)-boundedness of the operator $A + u$ implies(p, q)-boundedness of the operator A because A is a shift of $A + u$.

Proposition is proved.

Corollary 10.5. An operator (mapping) $A: L \to M$ is p-bounded at a point a if and only if any its shift $A + u$ is p-bounded at the point a.

Corollary 10.6. An operator (mapping) $A: L \to M$ is q-cobounded at a point a if and only if any its shift $A + u$ is q-cobounded at the point a.

Corollary 10.7. An operator (mapping) $A: L \to M$ is bounded at a point a if and only if any its shift $A + u$ is bounded at the point a.

Corollary 10.8. An operator (mapping) $A: L \to M$ is totally bounded at a point a if and only if any its shift $A + u$ is totally bounded at the point a.

Proposition 10.4. a) If $B = A + u$ and the lower boundedness $\text{Bound}_p(A, a)$ is defined, then the lower boundedness $\text{Bound}_p(B, a)$ is also defined and $\text{Bound}_p(A, a) = \text{Bound}_p(B, a)$.

b) If $B = A + u$ and the upper boundedness $\text{Bound}^q(A, a)$ is defined, then the upper boundedness $\text{Bound}^q(B, a)$ is also defined and $\text{Bound}^q(A, a) = \text{Bound}^q(B, a)$.

Proof is left as an exercise.

Now let us study boundedness of compositions of operators.

Proposition 10.5. If an operator (mapping) $A: L \to M$ is (p, q)-bounded at a point a from the space L, an operator (mapping) $B: M \to N$ is (r, t)-bounded at the point $A(a)$ from the space M and $q \leq r$, then their composition BA: $L \to N$ is (p, t)-bounded at the point a.

Proof. Let us consider a (p, q)-bounded at a point a from the space L operator (mapping) $A: L \to M$ and an (r, t)-bounded at the point $A(a)$ from the space M operator (mapping) B: $M \to N$ assuming $q \leq r$. Then taking an element b from L such that $\| b - a \| < p$, we have

$$\| A(b) - A(a) \| < q \leq r$$

Consequently,

$$\| BA(b) - BA(a) \| < t$$

Thus, the composition BA: $L \to N$ is (p, t)-bounded at a point a.

Proposition is proved.

Corollary 10.9. If an operator (mapping) A: $L \to M$ is totally bounded at a point a from the space L and an operator (mapping) B: $M \to N$ is totally bounded at the point $A(a)$ from the space M, then their composition BA: $L \to N$ is totally bounded at the point a.

Corollary 10.10. If an operator (mapping) A: $L \to M$ is q-cobounded at a point a from the space L, an operator (mapping) B: $M \to N$ is r-bounded at the point $A(a)$ from the space M and $q \leq r$, then their composition BA: $L \to N$ is (p, t)-bounded inside X.

Let us consider operators (mappings) A: $L \to M$ and B: $M \to N$.

Proposition 10.6. If the lower boundedness $\mathrm{Bound}_p(A, a)$ is defined, the lower boundedness $\mathrm{Bound}_r(B, A(a))$ is defined and $\mathrm{Bound}_p(A, a) < r$, then the lower boundedness $\mathrm{Bound}_r(BA, a)$ is also defined and $\mathrm{Bound}_r(BA, a) \leq \mathrm{Bound}_r(B, A(a))$.

Proof. If the lower boundedness $\mathrm{Bound}_p(A, a)$ is defined, then by Lemma 10.9, A is a (p, q)-bounded at a operator for some number q. As $\mathrm{Bound}_p(A, a) < r$, it is possible to assume that $q < r$. If the lower boundedness $\mathrm{Bound}_r(B, A(a))$ is defined, then by Lemma 10.9, B is a (r, t)-bounded at $A(a)$ operator for some number t. Then by Proposition 10.3, the composition BA: $L \to N$ is (p, t)-bounded at the point a.

Consequently, by Lemma 10.9, the lower boundedness $\mathrm{Bound}_r(BA, a)$ is defined and by Proposition 10.3, $\mathrm{Bound}_r(BA, a) \leq \mathrm{Bound}_r(B, A(a))$.

Proposition is proved.

Proposition 10.7. If the upper boundedness $\mathrm{Bound}^q(A, a)$ is defined, the upper boundedness $\mathrm{Bound}^t(B, A(a))$ is defined and $q < \mathrm{Bound}^t(B, A(a))$, then the upper boundedness $\mathrm{Bound}^q(BA, a)$ is also defined and $\mathrm{Bound}^q(BA, a) \geq \mathrm{Bound}^t(B, A(a))$.

Proof is similar to the proof of Proposition 10.4.

Now let us study boundedness of sums of operators.

Proposition 10.8. If an operator (mapping) $A: L \to M$ is (p, q)-bounded at a point a from the space L and an operator (mapping) $B: L \to M$ is (r, t)-bounded at a, then their sum $A + B: L \to M$ is (u, v)-bounded at the point a where $u = \min\{p, r\}$ and $v = q + t$.

Proof. Let us consider a (p, q)-bounded at a point a from the space L operator $A: L \to M$ and an (r, t)-bounded at a operator $B: L \to M$. Then taking an element b from L such that $\| b - a \| < u$, we have

$$\| b - a \| < u \leq p$$

and

$$\| b - a \| < u \leq r$$

Consequently,

$$\| A(b) - A(a)\| < q$$

and

$$\| B(b) - B(a)\| < t$$

Thus,

$$\| (A + B)(b) - (A + B)(a)\| = \| (A(b) + B(b)) - (A(a) + B(a))\| =$$

$$\| (A(b) - A(a)) + (B(b) - B(a))\| \leq$$

$$\| A(b) - A(a)\| + \|B(b) - B(a)\| < q + t$$

It means that the operator $A + B: L \to M$ is (u, v)-bounded at the point a.

Proposition is proved.

Corollary 10.11. If operators (mappings) $A: L \to M$ and $B: L \to M$ are bounded at a point a from the space L, then their sum $A + B: L \to M$ is also bounded at a.

Corollary 10.12. If operators (mappings) $A: L \to M$ and $B: L \to M$ are totally bounded at a point a from the space L, then their sum $A + B: L \to M$ is also totally bounded at a.

Let us consider operators (mappings) $A: L \to M$ and $B: L \to M$.

Proposition 10.9. If the lower boundedness $\mathrm{Bound}_p(A, a)$ and the lower boundedness $\mathrm{Bound}_p(B, a)$ are defined, then the lower boundedness $\mathrm{Bound}_p(B + A, a)$ is also defined and the following inequalities are valid:

(1) $\text{Bound}_p(A, a) \leq \text{Bound}_p(B + A, a)$;
(2) $\text{Bound}_p(B, a) \leq \text{Bound}_p(B + A, a)$;
(3) $\text{Bound}_p(B + A, a) \leq \text{Bound}_p(A, a) + \text{Bound}_p(B, a)$.

Proof. If the lower boundedness $\text{Bound}_p(A, a)$ is defined, then by Lemma 10.9, A is a (p, q)-bounded at a operator for some number q. If the lower boundedness $\text{Bound}_p(B, A(a))$ is defined, then by Lemma 10.9, B is a (p, t)-bounded at $A(a)$ operator for some number t. Then by Proposition 10.8, the sum $A + B: L \to M$ is (p, v)-bounded at the point a where $v = q + t$.

Consequently, by Lemma 10.9, the lower boundedness $\text{Bound}_p(B + A, a)$ is defined and by Proposition 10.8, $\text{Bound}_p(A, a) \leq \text{Bound}_p(B + A, a)$, $\text{Bound}_p(B, a) \leq \text{Bound}_p(B + A, a)$ and $\text{Bound}_p(B + A, a) \leq \text{Bound}_p(A, a) + \text{Bound}_p(B, a)$.

Proposition is proved.

Proposition 10.10. If the upper boundedness $\text{Bound}^q(A, a)$ is defined and the lower boundedness $\text{Bound}^q(B, a)$ is defined, then the lower boundedness $\text{Bound}^q(B + A, a)$ is also defined and the following inequalities are valid:

(1) $\text{Bound}^q(B + A, a) \leq \text{Bound}_p(A, a)$
(2) $\text{Bound}^q(B + A, a) \leq \text{Bound}_p(B, a)$

Proof is similar to the proof of Proposition 10.7.

Let us assume that M is a hyperseminormed linear algebra and L is a hyperseminormed vector space. In this case, it is possible to define the product $A \cdot B: L \to M$ for any operators $A: L \to M$ and $B: L \to M$ and estimate the boundedness of this product.

Proposition 10.11. If an operator (mapping) $A: L \to M$ is (p, q)-bounded at a point a from the space L, an operator (mapping) $B: L \to M$ is (r, t)-bounded at a, and the hyperseminorms of both operators at a are finite, i.e., $\|A(a)\| < w$ and $\|B(a)\| < s$ for some positive real numbers w and s, then their product $A \cdot B: L \to M$ is $(u, v(v + s + w))$-bounded at the point a where $u = \min\{p, r\}$ and $v = \max\{q, t\}$.

Proof. Let us consider a (p, q)-bounded at a point a operator (mapping) $A: L \to M$ and an (r, t)-bounded at a operator (mapping) $B: L \to M$. Then for any element b from L, the inequality $\| b - a \| < p$ implies the inequality $\| A(b) - A(a)\| < q$ and the inequality $\| b - a \| < p$ implies the inequality $\| B(b) - B(a)\| < t$. Taking $u = \min\{p, r\}$, we obtain that the inequality $\| b - a \| < u$ implies the inequality $\| A(b) - A(a)\| < q$ and $\| B(b) - B(a)\| < t$. Consequently, the inequality $\| b - a \| < u$ implies the inequality $\|A(b) - A(a)\| < v$ and $\| B(b) - B(a)\| < v$ where $v = \max\{q, t\}$.

Let us denote $\|A(a)\|$ by w and $\|B(a)\|$ by s. Then assuming $\| b - a \| < u$, by the properties of a hyperseminorm in hyperseminormed linear algebras, we have

$$\|B(b)\| = \|B(b) - B(a) + B(a)\| \leq \|B(b) - B(a)\| + \|B(a)\| < v + s$$

and

$$\| (A \cdot B)(b) - (A \cdot B)(a)\| = \| A(b) \cdot B(b) - A(a) \cdot B(a)\| =$$

$$\| A(b) \cdot B(b) - A(a) \cdot B(b) + A(a) \cdot B(b) - A(a) \cdot B(a)\| \leq$$

$$\| A(b) \cdot B(b) - A(a) \cdot B(b) \| + \|A(a) \cdot B(b) - A(a) \cdot B(a)\| \leq$$

$$\| (A(b) - A(a)) \cdot B(b) \| + \|A(a) \cdot (B(b) - B(a))\| \leq$$

$$\| A(b) - A(a)\| \cdot \|B(b)\| + \|A(a)\| \cdot \|B(b) - B(a)\| <$$

$$v(v + s) + wv = v(v + s + w)$$

It means that the operator $A \cdot B$: $L \to M$ is $(u, v(v + s + w))$-bounded at the point a.

Proposition is proved.

Note that the hyperseminorm or the hypernorm of an element in a vector space can be finite even when it is not a real number. For instance, it is possible to define a hyperseminorm in the set of real functions such that the hyperseminorm of the function $\sin x$ will be equal to the proper hypernumber defined by the sequence (1, 0, 1, 0, 1, …).

Corollary 10.13. If operators (mappings) A: $L \to M$ and B: $L \to M$ are bounded at a point a from the space L and the hyperseminorms of both operators at a are finite, then their product $A \cdot B$: $L \to M$ is also bounded at a.

Corollary 10.14. If operators (mappings) A: $L \to M$ and B: $L \to M$ are totally bounded at a point a from the space L and the hyperseminorms of both operators at a are finite, then their product $A \cdot B$: $L \to M$ is also totally bounded at a.

Corollary 10.15. If a functional A: $L \to \boldsymbol{R}$ is (p, q)-bounded at a point a from the space L, a functional B: $L \to \boldsymbol{R}$ is (r, t)-bounded at a, and the hyperseminorms of both functionals at the point a are finite, then their product $A \cdot B$: $L \to \boldsymbol{R}$ is $(u, v(v + s + w))$-bounded at the point a where $\|A(a)\| = w$, $\|B(a)\| = s$, $u = \min \{p, r\}$ and $v = \max \{q, t\}$.

Corollary 10.16. If functionals A: $L \to \boldsymbol{R}$ and B: $L \to \boldsymbol{R}$ are bounded at a point a from the space L and the hyperseminorms of both functionals at a are finite, then their product $A \cdot B$: $L \to \boldsymbol{R}$ is also bounded at a.

Corollary 10.17. If functionals A: $L \to \boldsymbol{R}$ and B: $L \to \boldsymbol{R}$ are totally bounded at a point a from the space L and the hyperseminorms of both functionals at a are finite, then their product $A \cdot B$: $L \to \boldsymbol{R}$ is also totally bounded at a.

Corollary 10.18. If a function A: $\boldsymbol{R} \to \boldsymbol{R}$ is (p, q)-bounded at a point a from the space $\boldsymbol{R}$, a function B: $\boldsymbol{R} \to \boldsymbol{R}$ is (r, t)-bounded at a, and the hyperseminorms of both functions at a are finite, then their product $A \cdot B$: $\boldsymbol{R} \to \boldsymbol{R}$ is $(u, v(v + s + w))$-bounded at the point a where $\|A(a)\| = w$, $\|B(a)\| = s$, $u = \min \{p, r\}$ and $v = \max \{q, t\}$.

Corollary 10.19. If functions A: $\boldsymbol{R} \to \boldsymbol{R}$ and B: $\boldsymbol{R} \to \boldsymbol{R}$ are bounded at a point a from the space $\boldsymbol{R}$ and the hyperseminorms of both functions at a are finite, then their product $A \cdot B$: $\boldsymbol{R} \to \boldsymbol{R}$ is also bounded at a.

Corollary 10.20. If functions A: $\boldsymbol{R} \to \boldsymbol{R}$ and B: $\boldsymbol{R} \to \boldsymbol{R}$ are totally bounded at a point a from the space $\boldsymbol{R}$ and the hyperseminorms of both functions at a are finite, then their product $A \cdot B$: $\boldsymbol{R} \to \boldsymbol{R}$ is also totally bounded at a.

Condition that the hyperseminorms of both operators at a are finite is essential as the following example demonstrates.

Example 10.3. Let us take the one-dimensional vector space $L = \boldsymbol{R}$ and the two-dimensional vector space $M = V\{\boldsymbol{R}, \alpha\}$ generated by elements 1 and $\alpha = (i)_{i \in \omega}$ in the vector space $\boldsymbol{R}_\omega$. We take the point 1 in L and define operators A: $L \to M$ and B: $L \to M$ by the following rules:

$$A(x) = x \text{ for all } x \text{ from } L$$

$$B(x) = \alpha \text{ for all } x \neq 0 \text{ from } L \text{ and } B(0) = 0$$

By definition, the operator A is (p, p)-bounded at 1 for any $p > 0$ and the operator B is (p, q)-bounded at 1 for any $p, q > 0$. However, for any b from L, if $\| b - 1 \| = k > 0$, then

$$\| (A \cdot B)(b) - (A \cdot B)(1)\| = \| (A(b) \cdot B(b) - A(1) \cdot B(1)\| = \| (A(b) \cdot \alpha - 1 \cdot \alpha\| =$$

$$\|\alpha\| \cdot \| (A(b) - 1\| = k\alpha$$

because $B(a) = B(b) = \alpha$. According to the order in the space of hypernumbers (cf., Chapter 2), the hypernumber $k\alpha$ is larger than any natural number q. Therefore, the product $A \cdot B$: $L \to M$ is not (p, q)-bounded at 1 for any positive real numbers p and q.

However, in some cases, the condition that the hyperseminorms of both operators at a are finite can be automatically true.

Let us assume M is a seminormed linear algebra and L is a seminormed vector space.

Corollary 10.21. If an operator (mapping) A: $L \to M$ is (p, q)-bounded at a point a from the space L and an operator (mapping) B: $L \to M$ is (r, t)-bounded at a, then their product $A \cdot B$: $L \to M$ is (u, $v(v + s + w)$)-bounded at the point a where $\|A(a)\| = w$, $\|B(a)\| = s$, $u = \min \{p, r\}$ and $v = \max \{q, t\}$.

Corollary 10.22. If operators (mappings) A: $L \to M$ and B: $L \to M$ are bounded at a point a from the space L, then their product $A \cdot B$: $L \to M$ is also bounded at a.

Corollary 10.23. If operators (mappings) A: $L \to M$ and B: $L \to M$ are totally bounded at a point a from the space L, then their product $A \cdot B$: $L \to M$ is also totally bounded at a.

Corollary 10.24. If a functional A: $L \to \mathbf{R}$ is (p, q)-bounded at a point a from the space L and a functional B: $L \to \mathbf{R}$ is (r, t)-bounded at a, then their product $A \cdot B$: $L \to \mathbf{R}$ is (u, $v(v + s + w)$)-bounded at the point a where $\|A(a)\| = w$, $\|B(a)\| = s$, $u = \min \{p, r\}$ and $v = \max \{q, t\}$.

Corollary 10.25. If functionals A: $L \to \mathbf{R}$ and B: $L \to \mathbf{R}$ are bounded at a point a from the space L, then their product $A \cdot B$: $L \to \mathbf{R}$ is also bounded at a.

Corollary 10.26. If functionals A: $L \to \mathbf{R}$ and B: $L \to \mathbf{R}$ are totally bounded at a point a from the space L, then their product $A \cdot B$: $L \to \mathbf{R}$ is also totally bounded at a.

Thus, we can see how results for operators imply results for functionals and functions, while results for hyperseminorms imply results for hypernorms, seminorms and norms.

Let us consider hyperseminormed vector spaces L and M.

Proposition 10.12. If an operator (mapping) A: $L \to M$ is (p, q)-bounded at a point a from the space L, then the operator (mapping) dA: $L \to M$ is (p, $|d|q$)-bounded at a for any number $d \in \mathbf{R} \backslash \{0\}$.

Indeed, the inequality

$$\| b - a \| < p$$

implies the inequality

$$\| dA(b) - dA(a)\| = \| d\,(A(b) - A(a))\| = |d| \cdot \| A(b) - dA(a)\| < |d| \cdot q$$

for any point b from L.

Corollary 10.27. If an operator $A: L \to M$ is bounded at a point a from the space L, then the operator dA is also bounded at a for any number $d \in \boldsymbol{R} \setminus \{0\}$.

Corollary 10.28. If an operator $A: L \to M$ is totally bounded at a point a from the space L, then the operator dA is also totally bounded at a for any number $d \in \boldsymbol{R} \setminus \{0\}$.

Corollary 10.29. If a functional $A: L \to \boldsymbol{R}$ is (p, q)-bounded at a point a from the space L, then the functional $dA: L \to \boldsymbol{R}$ is $(p, |d|q)$-bounded at a for any number $d \in \boldsymbol{R} \setminus \{0\}$.

Corollary 10.30. If a functional $A: L \to \boldsymbol{R}$ is bounded at a point a from the space L, then the functional dA is also bounded at a for any number $d \in \boldsymbol{R} \setminus \{0\}$.

Corollary 10.31. If a functional $A: L \to \boldsymbol{R}$ is totally bounded at a point a from the space L, then the functional dA is also totally bounded at a for any number $d \in \boldsymbol{R} \setminus \{0\}$.

Corollary 10.32. For any operator A,

$\text{Bound}_p(kA, a) \leq |k| \ \text{Bound}_p(A, a)$

Proposition 10.8 and 10.12 give us the following results.

Corollary 10.33. If an operator (mapping) $A: L \to M$ is (p, q)-bounded at a point a from the space L and an operator (mapping) $B: L \to M$ is (r, t)-bounded at a, then the operator $dA + cB: L \to M$ is (u, v)-bounded at the point a where $u = \min \{p, r\}$ and $v = |d|q + |c|t$ for any numbers $d, c \in \boldsymbol{R} \setminus \{0\}$.

Corollary 10.34. If operators (mappings) $A: L \to M$ and $B: L \to M$ are bounded at a point a from the space L, then the operator $dA + cB: L \to M$ is also bounded at a for any numbers $d, c \in \boldsymbol{R} \setminus \{0\}$.

Corollary 10.35. If operators (mappings) $A: L \to M$ and $B: L \to M$ are totally bounded at a point a from the space L, then the operator $dA + cB: L \to M$ is also totally bounded at a for any numbers $d, c \in \boldsymbol{R} \setminus \{0\}$.

Corollary 10.36. If a functional $A: L \to \boldsymbol{R}$ is (p, q)-bounded at a point a from the space L and a functional $B: L \to \boldsymbol{R}$ is (r, t)-bounded at a, then the functional $dA + cB: L \to \boldsymbol{R}$ is (u, v)-bounded at the point a where $u = \min \{p, r\}$ and $v = |d|q + |c|t$ for any numbers $d, c \in \boldsymbol{R} \setminus \{0\}$.

Corollary 10.37. If functionals $A: L \to \boldsymbol{R}$ and $B: L \to \boldsymbol{R}$ are bounded at a point a from the space L, then the functional $dA + cB: L \to \boldsymbol{R}$ is also bounded at a for any numbers $d, c \in \boldsymbol{R} \setminus \{0\}$.

Corollary 10.38. If functional $A: L \to \boldsymbol{R}$ and $B: L \to \boldsymbol{R}$ are totally bounded at a point a from the space L, then the functional $dA + cB: L \to \boldsymbol{R}$ is also totally bounded at a for any numbers $d, c \in \boldsymbol{R} \setminus \{0\}$.

These results show that linear combinations of operators (functionals or functions) preserves many properties of their constituents.

It is possible to introduce natural characteristics of boundedness.

Proposition 10.13. If the lower boundedness $\text{Bound}_p(A, a)$ is defined, then the lower boundedness $\text{Bound}_p(dA, a)$ is also defined for any number $d \in \boldsymbol{R} \setminus \{0\}$.

Proof is similar to the proof of Proposition 10.9.

Proposition 10.14. If the upper boundedness $\text{Bound}^q(A, a)$ is defined, then the lower boundedness $\text{Bound}^q(dA, a)$ is also defined for any number $d \in \boldsymbol{R} \setminus \{0\}$.

Proof is similar to the proof of Proposition 10.10.

Now let us study operators with additional properties.

Lemma 10.11. A centered operator (mapping) $A: L \to M$ is (p, q)-bounded at the point $\mathbf{0}$ if and only if the inequality $||x|| < p$ implies the inequality $||A(x)|| < q$ for any element x from the space L.

Proof. Necessity. Let us consider a centered operator $A: L \to M$ that is (p, q)-bounded at the point $\mathbf{0}$. Then $A(\mathbf{0}) = \mathbf{0}$ and

$$||\mathbf{0} - x|| = ||x|| < p$$

implies

$$|| Ax || = || \mathbf{0} - A(x)|| = || A(\mathbf{0}) - A(x)|| < q$$

Sufficiency. If the inequality $||x|| < p$ implies the inequality $|| A(x)|| < q$, then the inequality $|| x - \mathbf{0} || < p$ implies the inequality $||A(x) - A(\mathbf{0})|| < q$ because for a centered operator A, we have $A(\mathbf{0}) = \mathbf{0}$.

Lemma is proved.

Corollary 10.39. A homogeneous operator (mapping) $A: L \to M$ is (p, q)-bounded at the point $\mathbf{0}$ if and only if the inequality $||x|| < p$ implies the inequality $|| A(x)|| < q$ for any element x from L.

Proof. Necessity. Let us consider a homogeneous operator $A: L \to M$ that is (p, q)-bounded at the point $\mathbf{0}$. It means that for any element x from L, the inequality $||x|| < p$ implies the inequality $|| A(x)|| < q$ because

$$||A(\mathbf{0})|| = ||A(0 \cdot \mathbf{0})|| = || 0 \cdot A(\mathbf{0})|| = 0 \cdot ||A(\mathbf{0})|| = 0$$

and as M is a hypernormed vector space, $A(\mathbf{0}) = \mathbf{0}$.

Sufficiency. If the inequality $||x|| < p$ implies the inequality $||A(x)|| < q$, then the inequality $||x - \mathbf{0}|| < p$ implies the inequality $||A(x) - A(\mathbf{0})|| < q$ because for a homogeneous operator A, we have $A(\mathbf{0}) = \mathbf{0}$.

Corollary is proved.

Corollary 10.40. A linear operator (mapping) A: $L \to M$ is (p, q)-bounded at the point $\mathbf{0}$ if and only if the inequality $||x|| < p$ implies the inequality $||A(x)|| < q$ for any element x from L.

Thus, it is possible to define boundedness of linear operators using property from Corollary 10.40.

Additive operators have additional properties of boundedness.

Proposition 10.15. An additive operator (mapping) A: $L \to M$ is (p, q)-bounded at the point $\mathbf{0}$ if and only if A is (p, q)-bounded at all points from the space L.

Proof. Sufficiency directly follows from definitions.

Necessity. Let us consider an additive operator A: $L \to M(p, q)$-bounded at the point $\mathbf{0}$. It means that for any element x from L, the inequality $||x|| < p$ implies the inequality $|| A(x)|| < q$ because by Corollary 8.12, the operator A is centered.

Then taking $x = b - a$, we obtain for any elements a and b from L, the inequality $|| b - a || < p$ implies the inequality $|| A(a) - A(b)|| = || A(b - a)|| < q$. It means that the operator A is (p, q)-bounded at the point a.

Proposition is proved.

Let us assume that M is a hypernormed vector space.

Corollary 10.41. An additive operator (mapping) A: $L \to M$ is p-bounded at the point $\mathbf{0}$ if and only if A is p-bounded at all points from the space L.

Corollary 10.42. A linear operator (mapping) A: $L \to M$ is (p, q)-bounded at the point $\mathbf{0}$ if and only if A is (p, q)-bounded at all points from the space L.

It is possible to extend these results to approximately linear operators.

Let us assume that M is a hypernormed vector space.

Proposition 10.16. If an (r, t)-linear operator (mapping) A: $L \to M$ is (p, q)-bounded at the point $\mathbf{0}$, then A is $(p, q + r + t)$-bounded at all points from the space L.

Proof. Let us consider an (r, t)-linear operator A: $L \to M$ that is (p, q)-bounded at the point $\mathbf{0}$ and take a point a from the space L. Then for an arbitrary element b from L such that the inequality $|| b - a || < p$ is true, we see that by Lemma 10.11, (p, q)-boundedness of A at the point $\mathbf{0}$ implies

$$||A(b + (-a)) || = ||A(b - a)) || < q$$

In addition, we have

$$\| A(b) - A(a)\| = \| A(b) + A(-a) - A(-a) - A(a)\| =$$

$$\| -A(b + (-a)) + A(b) + A(-a) + A(b + (-a)) - A(-a) - A(a)\| \leq$$

$$\| -A(b + (-a)) + A(b) + A(-a) \| + \|A(b + (-a)) \| + \| -A(-a) - A(a)\| =$$

$$\| A(b + (-a)) - A(b) - A(-a) \| + \|A(b + (-a)) \| + \|A(-a) + A(a)\| < r + q + t$$

It means that A is $(p, q + r + t)$-bounded at all points from the space L. Proposition is proved.

Corollary 10.43. If an approximately linear operator (mapping) $A: L \to M$ is bounded at the point **0**, then A is bounded at all points from the space L.

Now let us study boundedness of Cartesian products of operators considering operators in seminormed vector spaces.

Proposition 10.17. If L, M, H and N are seminormed vector spaces, an operator $A: L \to M$ is (p, q)-bounded at a point a from L, an operator $B: H \to N$ is (r, t)-bounded at a point b from H, and the Cartesian product $M \times N$ has the Euclidean seminorm (cf. Chapter 3), then the Cartesian product $A \times B$ is (u, v)-bounded at the point (a, b) from $L \times N$ where $u = \min \{p, r\}$ and $v = (q^2 + t^2)^{1/2}$.

Proof. Let us consider a (p, q)-bounded at a point a operator $A: L \to M$ and(r, t)-bounded at a point b operator $B: H \to N$. Then the inequality

$$\| a - x \| < p$$

implies the inequality

$$\| A(a) - A(x)\| < q$$

and the inequality

$$\| b - y \| < r$$

implies the inequality

$$\| A(a) - A(x)\| < t$$

Then we have

$$\| (A \cdot B)((a, b)) - (A \cdot B)(x, y) \| = \| (A(a), B(b)) - (A(x), B(y)) \| =$$

$$(\| A(a) - A(x)\|^2 + \|B(b) - B(y)\|^2)^{1/2} < (r^2 + q^2)^{1/2}$$

when

$$\| a - x \| < u$$

and

$$\| b - y \| < u$$

because $u \leq p$ and $u \leq r$. It means that the Cartesian product $A \times B$ is (u, v)-bounded at the point (a, b).

Proposition is proved.

Corollary 10.44. If L, M, H and N are seminormed vector spaces, an operator A: $L \to M$ is (p, q)-bounded at a point a from L, an operator B: $H \to N$ is (p, q)-bounded at a point b from H, and the Cartesian product $M \times N$ has the Euclidean seminorm, then the Cartesian product $A \times B$ is $(p, \sqrt{2}q)$-bounded at the point (a, b) from $L \times N$.

Now let us study boundedness of Cartesian products of operators considering operators in hyperseminormed vector spaces.

Proposition 10.18. If an operator A: $L \to M$ is (p, q)-bounded at a point a from L, an operator B: $H \to N$ is (r, t)-bounded at a point b from H, and the Cartesian product $M \times N$ has the Manhattan hyperseminorm (cf. Chapter 3), then the Cartesian product $A \times B$ is (u, v)-bounded at the point (a, b) from $L \times N$ where $u = \min \{p, r\}$ and $v = q + t$.

Proof. Let us consider a (p, q)-bounded at a point a operator A: $L \to M$ and(r, t)-bounded at a point b operator B: $H \to N$. Then the inequality

$$\| a - x \| < p$$

implies the inequality

$$\| A(a) - A(x) \| < q$$

and the inequality

$$\| b - y \| < r$$

implies the inequality

$$\| A(a) - A(x) \| < t$$

Then we have

$$\| (A \cdot B)((a, b)) - (A \cdot B)(x, y) \| = \| (A(a), B(b)) - (A(x), B(y)) \| =$$

$$\| A(a) - A(x) \| + \| B(b) - B(y) \| < r + q = v$$

when

$$\| a - x \| < u$$

and

$$\| b - y \| < u$$

because $u \leq p$ and $u \leq r$. It means that the Cartesian product $A \times B$ is (u, v)-bounded at the point (a, b).

Proposition is proved.

Corollary 10.45. If an operator A: $L \to M$ is (p, q)-bounded at a point a from L, an operator B: $H \to N$ is (p, q)-bounded at a point b from H, and the Cartesian product $M \times N$ has the Manhattan hyperseminorm, then the Cartesian product $A \times B$ is $(p, 2q)$-bounded at the point (a, b) from $L \times N$.

Proposition 10.19. If an operator A: $L \to M$ is (p, q)-bounded at a point a from L, an operator B: $H \to N$ is (r, t)-bounded at a point b from H, and the Cartesian product $M \times N$ has the Chebyshev hyperseminorm (cf. Chapter 3), then the Cartesian product $A \times B$ is (u, v)-bounded at the point (a, b) from $L \times N$ where $u = \min \{p, r\}$ and $v = \max \{q + t\}$.

Proof. Let us consider a (p, q)-bounded at a point a operator A: $L \to M$ and(r, t)-bounded at a point b operator B: $H \to N$. Then the inequality

$$\| a - x \| < p$$

implies the inequality

$$\| A(a) - A(x) \| < q$$

and the inequality

$$\| b - y \| < r$$

implies the inequality

$$\| A(a) - A(x) \| < t$$

Then we have

$$\| (A \cdot B)((a, b)) - (A \cdot B)(x, y) \| = \| (A(a), B(b)) - (A(x), B(y)) \| =$$

$$\max \{\| A(a) - A(x) \|, \|B(b) - B(y)\| < \max \{q, t\} = v$$

when

$$\| a - x \| < u$$

and

$$\| b - y \| < u$$

because $u \leq p$ and $u \leq r$. It means that the Cartesian product $A \times B$ is (u, v)-bounded at the point (a, b).

Proposition is proved.

Corollary 10.46. If an operator $A: L \to M$ is (p, q)-bounded at a point a from L, an operator $B: H \to N$ is (p, q)-bounded at a point b from H, and the Cartesian product $M \times N$ has the Chebyshev hyperseminorm, then the Cartesian product $A \times B$ is also (p, q)-bounded at the point (a, b) from $L \times N$.

Corollary 10.47. a) If an operator $A: L \to M$ is (totally) bounded at a point a from L, an operator $B: H \to N$ is (totally) bounded at a point b from H, and the Cartesian product $M \times N$ has the Chebyshev hyperseminorm or the Manhattan hyperseminorm, then the Cartesian product $A \times B$ is (totally) bounded at the point (a, b) from $L \times N$.

b) If L, M, H and N are seminormed vector spaces, an operator $A: L \to M$ is (totally) bounded at a point a from L, an operator $B: H \to N$ is (totally) bounded at a point b from H, and the Cartesian product $M \times N$ has the Euclidean seminorm, then the Cartesian product $A \times B$ is (totally) bounded at the point (a, b) from $L \times N$.

From local boundedness, we come to regional boundedness.

Definition 10.2. a) An operator (mapping) $A: L \to M$ is called (p, q)-*bounded inside* a set $X \subseteq L$ if for any elements a, b from X, the inequality $\| b - a \| < p$ implies the inequality $\| A(b) - A(a) \| < q$.

b) An operator (mapping) $A: L \to M$ is called *bounded inside* a set X if it is (p, q)-bounded inside a set X for some positive numbers p and q.

c) An operator (mapping) $A: L \to M$ is called *p-bounded inside* a set X if it is (p, q)-bounded inside X for some positive number q.

d) An operator (mapping) $A: L \to M$ is called *q-cobounded inside* a set X if it is (p, q)-bounded inside X for some positive numbers p and q.

e) An operator (mapping) $A: L \to M$ is called *totally bounded inside* a set $X \subseteq L$ if for any number p and any elements a, b from X, there is a number q such that inequality $\| b - a \| < p$ implies the inequality $\| A(b) - A(a) \| < q$.

f) An operator (mapping) $A: L \to M$ is called *uniformly totally bounded inside a set* $X \subseteq L$ if for any number p, there is a number q such that for any elements a, b from X, inequality $\| b - a \| < p$ implies the inequality $\| A(b) - A(a) \| < q$.

Example 10.4. Let us take $L = M = \mathbf{R}$ and consider the one-dimensional operator (mapping) $C(x) = x^2$. This operator is (1, 1)-bounded inside the interval [0, 1] but not inside the interval [10, 11].

It is possible that an operator is totally bounded at all points of X but is not uniformly totally bounded inside X. It is possible that a totally bounded operator inside X is not uniformly totally bounded inside X as the following example demonstrates.

Example 10.5. Let us consider the one-dimensional operator C: $\mathbf{R}\to\mathbf{R}$ with $C(x) = x^2$. The operator C is totally bounded at all points of the interval $[7, \infty)$ but is not uniformly totally bounded inside $[7, \infty)$. Besides, C is totally bounded inside $[7, \infty)$ but is not uniformly totally bounded inside $[7, \infty)$.

Lemma 10.1 implies the following results.

Lemma 10.12. a) Any p-bounded inside a set X operator (mapping) A: $L\to M$ is t-bounded inside X for any number $t < p$.

b) Any uniformly totally bounded inside a set X operator (mapping) A: $L\to M$ is totally bounded inside X and p-bounded inside X for any number p.

Proof is left as an exercise.

Corollary 10.48. a) Any p-bounded inside a set X functional A: $L\to\mathbf{R}$ is t-bounded inside X for any number $t < p$.

b) Any uniformly totally bounded inside a set X functional A: $L\to\mathbf{R}$ is totally bounded inside X and p-bounded inside X for any number p.

Lemma 10.13. Any totally bounded inside the set X operator (mapping) A: $L\to M$ is p-bounded inside X for some number p.

Proof is left as an exercise.

Corollary 10.49. Any totally bounded inside the set X functional A: $L\to\mathbf{R}$ is p-bounded inside X for some number p.

Lemma 10.14. If $t \leq p$, $q \leq u$ and an operator (mapping) A: $L\to M$ is (p, q)-bounded inside a set X, then A is (t, u)-bounded inside the set X.

Proof is left as an exercise.

Corollary 10.50. If $t < p$ and an operator (mapping) A: $L\to M$ is (p, q)-bounded inside a set X, then A is (t, q)-bounded inside X.

Corollary 10.51. If $q < u$ and an operator (mapping) A: $L\to M$ is (p, q)-bounded inside a set X, then A is (p, u)-bounded inside X.

Corollary 10.52. If $t \leq p$, $q \leq u$ and a functional A: $L\to\mathbf{R}$ is (p, q)-bounded inside a set X, then A is (t, u)-bounded inside the set X.

Boundedness of an operator implies boundedness of all its shifts.

Proposition 10.20. An operator (mapping) A: $L\to M$ is (p, q)-bounded inside a set X if and only if any its shift $A + u$ is (p, q)-bounded inside X.

Proof is similar to the proof of Proposition 10.3.

Corollary 10.53. An operator (mapping) $A: L \to M$ is p-bounded inside X if and only if any its shift $A + u$ is p-bounded inside X.

Corollary 10.54. An operator (mapping) $A: L \to M$ is q-cobounded inside X if and only if any its shift $A + u$ is q-cobounded inside X.

Corollary 10.55. An operator (mapping) $A: L \to M$ is bounded inside X if and only if any its shift $A + u$ is bounded inside X.

Corollary 10.56. An operator (mapping) $A: L \to M$ is totally bounded inside X if and only if any its shift $A + u$ is totally bounded inside X.

Approximate boundedness at a point is intrinsically related to approximate boundedness in a set.

Proposition 10.21. A (p, q)-bounded at a point x operator (mapping) $A: L \to M$ is $(p, 2q)$-bounded inside some closed neighborhood of x.

Proof. Let us consider a (p, q)-bounded at a point x operator $A: L \to M$ and the neighborhood $O_{\frac{1}{2}p}(x) = \{z; \|z - x\| < \frac{1}{2} p\}$ of the point x. Then by properties of hyperseminorm, the hyperseminorm of the difference between any points a and b from $O_{\frac{1}{2}p}(x)$ is less than p. Indeed, by the properties of a hyperseminorm, we have

$$\| a - b \| = \| a - x + x - b \| \leq \| a - x \| + \| x - b\| < \tfrac{1}{2} p + \tfrac{1}{2} p = p$$

Consequently, the hyperseminorm of the difference between their images $A(a)$ and $A(b)$ of the points a and b from $O_{\frac{1}{2}p}(x)$ is less than q. Indeed, by the properties of a hyperseminorm, we have

$$\| A(a) - A(b) \| = \| A(a) - A(x) + A(x) - A(b)\| \leq \| A(a) - A(x) \| + \| A(x) - A(b)\| < q + q = 2q$$

It means that the operator (mapping) $A: L \to M$ is $(p, 2q)$-bounded inside the neighborhood $O_{\frac{1}{2}p}(x)$ and thus inside the closed neighborhood $O_{\frac{1}{3}p}(x) = \{z; \|z - x\| \leq \frac{1}{3} p\}$ of the point x.

Proposition is proved.

Proposition 10.22. A (p, q)-bounded inside the neighborhood $O_r(x)$ of some point x operator (mapping) $A: L \to M$ is (p, q)-bounded at the point x if $p \leq r$.

Indeed, if for some point b, we have $\| b - x \| < p$, then $b \in O_r(x)$ because $p \leq r$.Thus, $\| A(b) - A(x)\| < q$ because A is (p, q)-bounded inside $O_p(x)$. As b is an arbitrary point from L, the operator A is (p, q)-bounded at the point x.

Corollary 10.57. An operator (mapping) A: $L \rightarrow M$ is bounded at x if and only if it is bounded inside some closed neighborhood of x.

Proposition 10.22 is not true for smaller neighborhoods as the following example demonstrates.

Example 10.6. Let us consider the one-dimensional operator A: $\boldsymbol{R} \rightarrow \boldsymbol{R}$ defined by the following formula

$$A(x) = \begin{cases} 1/(x-1) & \text{if } x > 1 \\ 0 & \text{if } 0 \leq x \leq 1 \end{cases}$$

The operator A is (p, q)-bounded for any $p, q > 0$ inside the interval [0, 1] and in particular, A is (2, 1)-bounded inside the interval [0, 1]. However, A is not (2, 1)-bounded at the point 0 because, for example, $\|1.2 - 1\| < 2$ but $\| A(1.2) - A(1)\| = 5 > 1$.

Even more, Example 10.6 shows that it is possible that an operator A is (p, q)-bounded inside a set but it is not (p, q)-bounded at any point of this set. Indeed, the A is (2, 1)-bounded inside the interval [0, 1]. However, A is not (2, 1)-bounded at any point b of this interval because, for example, $\|1.2 - b\| < 2$ for any from [0, 1] but $\| A(1.2) - A(b)\| = 5 > 1$.

Lemma 10.15. An operator (mapping) A: $L \rightarrow M$ is totally bounded inside a set X if and only if for any number p, there is a number q such that A is (p, q)-bounded inside the set X.

Proof is left as an exercise.

Now let us study boundedness of compositions of operators.

Proposition 10.23. If an operator (mapping) A: $L \rightarrow M$ is (p, q)-bounded inside a set $X \subseteq L$, an operator (mapping) B: $M \rightarrow N$ is (r, t)-bounded inside the set $A(X)$ from the space M and $q \leq r$, then their composition BA: $L \rightarrow N$ is (p, t)-bounded inside X.

Proof. Let us consider a (p, q)-bounded inside a set $X \subseteq L$ operator (mapping) A: $L \rightarrow M$ and an (r, t)-bounded inside the set $A(X)$ from the space M operator (mapping) B: $M \rightarrow N$ assuming $q \leq r$. Then taking elements a, b from X such that $\| b - a \| < p$, we have

$$\|A(b) - A(a)\| < q \leq r$$

Consequently,

$$\| BA(b) - BA(a)\| < t$$

Thus, the composition BA: $L\to N$ is (p, t)-bounded inside X.

Proposition is proved.

Corollary 10.58. If an operator (mapping) A: $L\to M$ is totally bounded inside X and an operator (mapping) B: $M\to N$ is totally bounded inside $A(X)$ from the space M, then their composition BA: $L\to N$ is totally bounded inside X.

Corollary 10.59. If an operator (mapping) A: $L\to M$ is q-cobounded inside a set $X\subseteq L$, an operator (mapping) B: $M\to N$ is r-bounded inside the set $A(X)$ from the space M and $q\leq r$, then their composition BA: $L\to N$ is (p, t)-bounded inside X.

Now let us study boundedness of sums of operators.

Proposition 10.24. If an operator (mapping) A: $L\to M$ is (p, q)-bounded inside a set $X\subseteq L$ and an operator (mapping) B: $L\to M$ is (r, t)-bounded inside X, then their sum $A + B$: $L\to M$ is (u, v)-bounded inside X where $u = \min\{p, r\}$ and $v = q + t$.

Proof. Let us consider a (p, q)-bounded inside X operator A: $L\to M$ and an (r, t)-bounded inside X operator B: $L\to M$. Then taking elements a, b from L such that $\| b - a \| < u$, we have

$$\| b - a \| < u \leq p$$

and

$$\| b - a \| < u \leq r$$

Consequently,

$$\| A(b) - A(a)\| < q$$

and

$$\| B(b) - B(a)\| < t$$

Thus,

$$\| (A + B)(b) - (A + B)(a)\| = \| (A(b) + B(b)) - (A(a) + B(a))\| =$$

$$\| (A(b) - A(a)) + (B(b) - B(a))\| \leq$$

$$\| A(b) - A(a)\| + \|B(b) - B(a)\| < q + t$$

It means that the operator $A + B$: $L\to M$ is (u, v)-bounded inside X.

Proposition is proved.

Corollary 10.60. If operators (mappings) A: $L \to M$ and B: $L \to M$ are bounded inside X, then their sum $A + B$: $L \to M$ is also bounded inside X.

Corollary 10.61. If operators (mappings) A: $L \to M$ and B: $L \to M$ are totally bounded inside X, then their sum $A + B$: $L \to M$ is also totally bounded inside X.

Corollary 10.62. If a functional A: $L \to \boldsymbol{R}$ is (p, q)-bounded inside a set $X \subseteq L$ and a functional B: $L \to \boldsymbol{R}$ is (r, t)-bounded inside X, then their sum $A + B$: $L \to \boldsymbol{R}$ is (u, v)-bounded inside X where $u = \min \{p, r\}$ and $v = q + t$.

Corollary 10.63. If functionals A: $L \to \boldsymbol{R}$ and B: $L \to \boldsymbol{R}$ are bounded inside X, then their sum $A + B$: $L \to \boldsymbol{R}$ is also bounded inside X.

Corollary 10.64. If functionals A: $L \to \boldsymbol{R}$ and B: $L \to \boldsymbol{R}$ are totally bounded inside X, then their sum $A + B$: $L \to \boldsymbol{R}$ is also totally bounded inside X.

Now let us study boundedness of products of operators assuming that M is a hyperseminormed linear algebra.

Proposition 10.25. If a set $X \subseteq L$ is l-bounded, $\mathbf{0} \in X$, a centered operator (mapping) A: $L \to M$ is (p, q)-bounded inside X with $2l < p$ and a centered operator (mapping) B: $L \to M$ is (r, t)-bounded inside X with $2l < r$, then their product $A \cdot B$: $L \to M$ is $(u, 2v^2)$-bounded inside X where $v = \max \{q, t\}$ and $u = \min \{p, r\}$.

Proof. Let us take an l-bounded in L set X. By Proposition 9.9, X is totally $2l$-bounded.

Now let us consider a centered (p, q)-bounded inside X operator (mapping) A: $L \to M$ and a centered (r, t)-bounded inside X operator (mapping) B: $L \to M$. Then for any elements a and b from X, the inequality $\| b - a \| < p$ implies the inequality $\| A(b) - A(a) \| < q$ and the inequality $\| b - a \| < p$ implies the inequality $\| B(b) - B(a) \| < t$. Taking $u = \min \{p, r\}$, we obtain that by Corollary 10.1, the inequality $\| b - a \| < u$ implies the inequality $\| A(b) - A(a) \| < q$ and $\| B(b) - B(a) \| < t$. Consequently, the inequality $\| b - a \| < u$ implies the inequality $\| A(b) - A(a) \| < v$ and $\| B(b) - B(a) \| < v$ where $v = \max \{q, t\}$.

In particular, we have $\| B(\mathbf{0}) \| = 0$, $\| A(\mathbf{0}) \| = 0$ and

$$\| B(b) \| = \| B(b) - B(\mathbf{0}) \| < v$$

because $\| b - \mathbf{0} \| < 2l < u$ and

$$\| A(a) \| = \| A(a) - A(\mathbf{0}) \| < v$$

because $\| a - \mathbf{0} \| < 2l < u$.

In addition, by the properties of a hyperseminorm in hyperseminormed linear algebras, we have

$$\| (A \cdot B)(b) - (A \cdot B)(a)\| = \| A(b) \cdot B(b) - A(a) \cdot B(a)\| =$$

$$\| A(b) \cdot B(b) - A(a) \cdot B(b) + A(a) \cdot B(b) - A(a) \cdot B(a)\| \leq$$

$$\| A(b) \cdot B(b) - A(a) \cdot B(b) \| + \|A(a) \cdot B(b) - A(a) \cdot B(a)\| \leq$$

$$\| (A(b) - A(a)) \cdot B(b) \| + \|A(a) \cdot (B(b) - B(a))\| \leq$$

$$\| A(b) - A(a)\| \cdot \|B(b) \| + \|A(a) \| \cdot \|B(b) - B(a)\| <$$

$$vv + vv = 2v^2$$

It means that the operator $A \cdot B$: $L \to M$ is $(u, 2v^2)$-bounded inside X.

Proposition is proved.

Note that the condition $\mathbf{0} \in X$ is essential in Proposition 10.25 as the following example demonstrates.

Example 10.7. Let us take the one-dimensional vector space $L = \boldsymbol{R}$ and the two-dimensional vector space $M = V\{\boldsymbol{R}, \alpha\}$ generated by elements 1 and $\alpha = (i)_{i \in \omega}$ in the vector space $\boldsymbol{R}_\omega$. We take $X = [1, 2]$ and define operators A: $L \to M$ and B: $L \to M$ by the following rules:

$$A(x) = x \text{ for all } x \text{ from } L$$

$$B(x) = \alpha \text{ for all } x \neq 0 \text{ from } L \text{ and } B(0) = 0$$

By definition, the operator A is (p, p)-bounded inside X for any $p > 0$ and the operator B is (p, q)-bounded inside X for any $p, q > 0$. However, for any a and b from X, if $\| b - a \| = k > 0$, then

$$\| (A \cdot B)(b) - (A \cdot B)(a)\| = \| (A(b) \cdot B(b) - A(a) \cdot B(a)\| = \| (A(b) \cdot \alpha - A(a) \cdot \alpha\| =$$

$$\|\alpha\| \cdot \| (A(b) - A(a)\| = k\alpha$$

because $B(a) = B(b) = \alpha$. According to the order in the space of hypernumbers (cf., Chapter 2), the hypernumber $k\alpha$ is larger than any natural number q. Therefore, the product $A \cdot B$: $L \to M$ is not (p, q)-bounded inside X for any positive real numbers p and q.

Note that for any $p, q > 0$, the operator B is not (p, q)-bounded inside the interval $[0, 1]$ or in the whole space $L = \boldsymbol{R}$.

Proposition 10.26. If an operator (mapping) A: $L \to M$ is (p, q)-bounded inside X, then the operator (mapping) dA: $L \to M$ is $(p, |d|q)$-bounded inside X for any number $d \in \boldsymbol{R} \setminus \{0\}$.

Indeed, the inequality

$$\| b - a \| < p$$

implies the inequality

$$\| dA(b) - dA(a)\| = \| d\,(A(b) - A(a))\| = |d| \cdot \| A(b) - dA(a)\| < |d| \cdot q$$

for any point b from L.

Corollary 10.65. If an operator A: $L \to M$ is bounded inside X, then the operator dA is also bounded inside X for any number $d \in \boldsymbol{R} \setminus \{0\}$.

Corollary 10.66. If an operator A: $L \to M$ is totally bounded inside X, then the operator dA is also totally bounded inside X for any number $d \in \boldsymbol{R} \setminus \{0\}$.

Proposition 10.24 and 10.26 give us the following results.

Corollary 10.67. If an operator (mapping) A: $L \to M$ is (p, q)-bounded inside X and an operator (mapping) B: $L \to M$ is (r, t)-bounded inside X, then the operator $dA + cB$: $L \to M$ is (u, v)-bounded inside X where $u = \min \{p, r\}$ and $v = |d|q + |c|t$ for any numbers $d, c \in \boldsymbol{R} \setminus \{0\}$.

Corollary 10.68. If operators (mappings) A: $L \to M$ and B: $L \to M$ are bounded inside X, then the operator $dA + cB$: $L \to M$ is also bounded inside X for any numbers $d, c \in \boldsymbol{R} \setminus \{0\}$.

Corollary 10.69. If operators (mappings) A: $L \to M$ and B: $L \to M$ are totally bounded inside X, then the operator $dA + cB$: $L \to M$ is also totally bounded inside X for any numbers $d, c \in \boldsymbol{R} \setminus \{0\}$.

Now we go from boundedness inside to the stronger concept of boundedness in.

Definition 10.3. a) An operator (mapping) A: $L \to M$ is called (p, q)-*bounded in* a set $X \subseteq L$ if it is (p, q)-bounded at any point a from X.

b) An operator (mapping) A: $L \to M$ is called *bounded in* a set X if at any point a from X, it is (p, q)-bounded for some numbers $p, q > 0$.

c) An operator (mapping) A: $L \to M$ is called p-*bounded in* a set $X \subseteq L$ if it is p-bounded at any point a from X.

d) An operator (mapping) A: $L \rightarrow M$ is called *q-cobounded in* a set X if it is q-cobounded atany point a from X.

e) An operator (mapping) A: $L \rightarrow M$ is called *uniformly totally bounded in* a set $X \subseteq L$ if for any number p, there is a number q such that A is (p, q)-bounded at any point a from X.

f) An operator (mapping) A: $L \rightarrow M$ is called *totally bounded in* a set $X \subseteq L$ if for any number p and any point a from X, there is a number q such that A is (p, q)-bounded at a.

It is possible that an operator is totally bounded at all points of X but is not uniformly totally bounded in X. It is possible that a totally bounded operator in X is not uniformly totally bounded in X as the following example demonstrates.

Example 10.8. Let us consider the one-dimensional operator C: $\boldsymbol{R} \rightarrow \boldsymbol{R}$ with $C(x) = x^2$. The operator C is totally bounded at all points of the interval $[7, \infty)$ but is not uniformly totally bounded in $[7, \infty)$. Besides, C is totally bounded in $[7, \infty)$ but is not uniformly totally bounded in $[7, \infty)$.

Boundedness *in* is stronger than boundedness *inside*.

Lemma 10.16. If an operator (mapping) A: $L \rightarrow M$ is (p, q)-bounded in a set X, then A is (p, q)-bounded inside X.

Proof is left as an exercise.

The converse is not true in a general case as the following example demonstrates.

Example 10.9. Let us consider the one-dimensional operator A: $\boldsymbol{R} \rightarrow \boldsymbol{R}$ defined by the following formula

$$A(x) = \begin{cases} 1/x & \text{if } x < 0 \\ 1 & \text{if } 0 \leq x \end{cases}$$

The operator A is (p, q)-bounded for any $p, q > 0$ inside the interval $[0, 1]$ but it is not (r, t)-bounded in the interval $[0, 1]$ for any positive real numbers r and t.

However, in some cases, boundedness *in* and boundedness *inside* coincide.

Definition 10.4. A subset X of a hyperseminormed vector space L is called an *h-component* of L if $\|a - b)\| > h$ for any point a from X and any point b from $L \backslash X$.

Example 10.10. Let us consider $L = \boldsymbol{R}_\omega$ and the hypernumber $\alpha = \mathrm{Hn}(i)_{i\in\omega}$. As it is demonstrated in Chapter 3, L is a hypernormed vector space over the field $\boldsymbol{R}$. Taking the set

$X = \{\alpha + x;\ x$ is a finite hypernumber x from $\boldsymbol{R}_\omega\}$,

we see that for any positive real number h, X is an h-component X of a hyperseminormed vector space L because a hypernumber $\beta = \mathrm{Hn}(b_i)_{i\in\omega}$ does not belong to X if and only if the difference $\beta - \delta$ is infinite for any hypernumber δ from X. The norm $\|\beta - \delta\|$ of the difference $\beta - \delta$ is also infinite.

Proposition 10.27. An operator (mapping) $A: L\to M$ is (p, q)-bounded in an h-component X of a hyperseminormed vector space L with $p \leq h$ if and only if A is (p, q)-bounded inside X.

Indeed, if $p \leq h$ and X is h-component L, then for any element b from L and some element a from X, the inequality $\| b - a \| < p$ is possible only if b also belongs to X.

Lemma 10.17. a) Any p-bounded in a set X operator (mapping) $A: L\to M$ is t-bounded in X for any number $t < p$.

b) Any uniformly totally bounded in a set X operator (mapping) $A: L\to M$ is totally bounded in X and p-bounded in X for any number p.

Proof. Let us consider a p-bounded in a subset X of the space L operator $A: L\to M$. By definition, the operator A is (p, q)-bounded in X for some number q. Then taking an arbitrary point b from the space L, an arbitrary point a from X and a positive number $t \leq p$, we have

$$\| b - a \| < t \Rightarrow \| b - a \| < p \Rightarrow \| A(b) - A(a)\| < q$$

Thus, the operator (mapping) A is (t, q)-bounded in X. Therefore, by definition, the operator A is t-bounded in X.

Part (b) directly follows from definitions.

Lemma is proved.

Lemma 10.18. Any totally bounded in the set X operator (mapping) A: $L\to M$ is p-bounded in X for some number p.

Proof is left as an exercise.

Lemma 10.19. If $t \leq p$, $q \leq u$ and an operator (mapping) $A: L\to M$ is (p, q)-bounded in a set X, then A is (t, u)-bounded in the set X.

Proof is left as an exercise.

Corollary 10.70. If $t < p$ and an operator (mapping) $A: L\to M$ is (p, q)-bounded in a set X, then A is (t, q)-bounded in X.

Corollary 10.71. If $q < u$ and an operator (mapping) A: $L \to M$ is (p, q)-bounded in a set X, then A is (p, u)-bounded in X.

Proposition 10.3 implies the following result.

Proposition 10.28. An operator (mapping) A: $L \to M$ is (p, q)-bounded in a set X if and only if any its shift $A + u$ is (p, q)-bounded in X.

Proof is similar to the proof of Proposition 10.3.

Corollary 10.72. An operator (mapping) A: $L \to M$ is p-bounded in X if and only if any its shift $A + u$ is p-bounded in X.

Corollary 10.73. An operator (mapping) A: $L \to M$ is q-cobounded in X if and only if any its shift $A + u$ is q-cobounded in X.

Corollary 10.74. An operator (mapping) A: $L \to M$ is bounded in X if and only if any its shift $A + u$ is bounded in X.

Corollary 10.75. An operator (mapping) A: $L \to M$ is totally bounded in X if and only if any its shift $A + u$ is totally bounded in X.

Now let us study boundedness of compositions of operators.

Proposition 10.29. If an operator (mapping) A: $L \to M$ is (p, q)-bounded in a set $X \subseteq L$, an operator (mapping) B: $M \to N$ is (r, t)-bounded in the set $A(X)$ from the space M and $q \leq r$, then their composition BA: $L \to N$ is (p, t)-bounded in X.

Proof. Let us consider a (p, q)-bounded in a set $X \subseteq L$ operator (mapping) A: $L \to M$ and an (r, t)-bounded in the set $A(X)$ from the space M operator (mapping) B: $M \to N$ assuming $q \leq r$. Then taking elements a from X and b from L such that $\| b - a \| < p$, we have

$$\| A(b) - A(a) \| < q \leq r$$

Consequently,

$$\| BA(b) - BA(a) \| < t$$

Thus, the composition BA: $L \to N$ is (p, t)-bounded in X.

Proposition is proved.

Corollary 10.76. If an operator (mapping) A: $L \to M$ is totally bounded in X and an operator (mapping) B: $M \to N$ is totally bounded in $A(X)$ from the space M, then their composition BA: $L \to N$ is totally bounded in X.

Corollary 10.77. If an operator (mapping) A: $L \to M$ is q-cobounded in a set $X \subseteq L$, an operator (mapping) B: $M \to N$ is r-bounded in the set $A(X)$ from the space M and $q \leq r$, then their composition BA: $L \to N$ is (p, t)-bounded in X.

Now let us study boundedness of sums of operators.

Proposition 10.30. If an operator (mapping) $A: L \to M$ is (p, q)-bounded in a set $X \subseteq L$ and an operator (mapping) $B: L \to M$ is (r, t)-bounded in X, then their sum $A + B: L \to M$ is (u, v)-bounded in X where $u = \min \{p, r\}$ and $v = q + t$.

Proof. Let us consider a (p, q)-bounded in X operator $A: L \to M$ and an (r, t)-bounded in X operator $B: L \to M$. Then taking elements a from X and b from L such that $\| b - a \| < u$, we have

$$\| b - a \| < u \leq p$$

and

$$\| b - a \| < u \leq r$$

Consequently,

$$\| A(b) - A(a)\| < q$$

and

$$\| B(b) - B(a)\| < t$$

Thus,

$$\| (A + B)(b) - (A + B)(a)\| = \| (A(b) + B(b)) - (A(a) + B(a))\| =$$

$$\| (A(b) - A(a)) + (B(b) - B(a))\| \leq$$

$$\| A(b) - A(a)\| + \|B(b) - B(a)\| < q + t$$

It means that the operator $A + B: L \to M$ is (u, v)-bounded in X.

Proposition is proved.

Corollary 10.78. If operators (mappings) $A: L \to M$ and $B: L \to M$ are bounded in X, then their sum $A + B: L \to M$ is also bounded in X.

Corollary 10.79. If operators (mappings) $A: L \to M$ and $B: L \to M$ are totally bounded in X, then their sum $A + B: L \to M$ is also totally bounded in X.

Lemma 10.20. An operator (mapping) $A: L \to M$ is totally bounded in a set X if and only if for any number p, there is a number q such that A is (p, q)-bounded in the set X.

Proof is left as an exercise.

Now let us study boundedness of products of operators assuming that M is a hyperseminormed linear algebra.

Proposition 10.31. If a set $X \subseteq L$ is l-bounded, $\mathbf{0} \in X$, a centered operator (mapping) $A: L \to M$ is (p, q)-bounded in X with $2l < p$ and a centered operator (mapping) $B: L \to M$ is (r, t)-bounded in X with $2l < r$, then their product $A \cdot B: L \to M$ is $2v^2$-bounded in X where $v = \max \{q, t\}$.

Proof. Let us take an l-bounded in L set X. By Proposition 9.9, X is totally $2l$-bounded.

Now let us consider a centered (p, q)-bounded in X operator (mapping) $A: L \to M$ and a centered (r, t)-bounded in X operator (mapping) $B: L \to M$. Then for any elements a from X and b from L, the inequality $\| b - a \| < p$ implies the inequality $\| A(b) - A(a) \| < q$ and the inequality $\| b - a \| < p$ implies the inequality $\| B(b) - B(a) \| < t$. Taking $u = \min \{p, r\}$, we obtain that by Corollary 10.1, the inequality $\| b - a \| < u$ implies the inequality $\| A(b) - A(a) \| < q$ and $\| B(b) - B(a) \| < t$. Consequently, the inequality $\| b - a \| < u$ implies the inequality $\| A(b) - A(a) \| < v$ and $\| B(b) - B(a) \| < v$ where $v = \max \{q, t\}$.

In particular, we have $\| B(\mathbf{0}) \| = 0$, $\| A(\mathbf{0}) \| = 0$ and

$$\| B(b) \| = \| B(b) - B(\mathbf{0}) \| < v$$

because $\| b - \mathbf{0} \| < 2l < u$ and

$$\| A(a) \| = \| A(a) - A(\mathbf{0}) \| < v$$

because $\| a - \mathbf{0} \| < 2l < u$.

In addition, by the properties of a hyperseminorm in hyperseminormed linear algebras, we have

$$\| (A \cdot B)(b) - (A \cdot B)(a) \| = \| A(b) \cdot B(b) - A(a) \cdot B(a) \| =$$

$$\| A(b) \cdot B(b) - A(a) \cdot B(b) + A(a) \cdot B(b) - A(a) \cdot B(a) \| \leq$$

$$\| A(b) \cdot B(b) - A(a) \cdot B(b) \| + \| A(a) \cdot B(b) - A(a) \cdot B(a) \| \leq$$

$$\| (A(b) - A(a)) \cdot B(b) \| + \| A(a) \cdot (B(b) - B(a)) \| \leq$$

$$\| A(b) - A(a) \| \cdot \| B(b) \| + \| A(a) \| \cdot \| B(b) - B(a) \| <$$

$$vv + vv = 2v^2$$

It means that the operator $A \cdot B$: $L \rightarrow M$ is $2v^2$-bounded in X.

Proposition is proved.

Note that condition $\mathbf{0} \in X$ is essential in Proposition 10.31 as the following example demonstrates.

Example 10.11. In Example 10.7, operators A: $L \rightarrow M$ and B: $L \rightarrow M$ are defined in such a way that A is (p, p)-bounded in $X = [1, 2]$ for any $p > 0$ and the operator B is (p, q)-bounded in X for any $p, q > 0$. At the same time, their product $A \cdot B$: $L \rightarrow M$ is not (p, q)-bounded inside X and thus, in X for any positive real numbers p and q.

Proposition 10.32. If an operator (mapping) A: $L \rightarrow M$ is (p, q)-bounded in X, then the operator (mapping) dA: $L \rightarrow M$ is $(p, |d|q)$-bounded in X for any number $d \in \boldsymbol{R} \setminus \{0\}$.

Indeed, the inequality

$$\| b - a \| < p$$

implies the inequality

$$\| dA(b) - dA(a) \| = \| d\,(A(b) - A(a)) \| = |d| \cdot \| A(b) - dA(a) \| < |d| \cdot q$$

for any point b from L.

Corollary 10.80. If an operator A: $L \rightarrow M$ is bounded in X, then the operator dA is also bounded in X for any number $d \in \boldsymbol{R} \setminus \{0\}$.

Corollary 10.81. If an operator A: $L \rightarrow M$ is totally bounded in X, then the operator dA is also totally bounded in X for any number $d \in \boldsymbol{R} \setminus \{0\}$.

Propositions 10.30 and 10.32 give us the following results.

Corollary 10.82. If an operator (mapping) A: $L \rightarrow M$ is (p, q)-bounded in X and an operator (mapping) B: $L \rightarrow M$ is (r, t)-bounded in X, then the operator $dA + cB$: $L \rightarrow M$ is (u, v)-bounded in X where $u = \min \{p, r\}$ and $v = |d|q + |c|t$ for any numbers $d, c \in \boldsymbol{R} \setminus \{0\}$.

Corollary 10.83. If operators (mappings) A: $L \rightarrow M$ and B: $L \rightarrow M$ are bounded in X, then the operator $dA + cB$: $L \rightarrow M$ is also bounded in X for any numbers $d, c \in \boldsymbol{R} \setminus \{0\}$.

Corollary 10.84. If operators (mappings) A: $L \rightarrow M$ and B: $L \rightarrow M$ are totally bounded in X, then the operator $dA + cB$: $L \rightarrow M$ is also totally bounded in X for any numbers $d, c \in \boldsymbol{R} \setminus \{0\}$.

From absolute boundedness, we come to relative boundedness.

Definition 10.5. a) An operator (mapping) A: $L \rightarrow M$ is called *(p, q)-bounded at* a point *arelative to* a set X from the space L if for any element b from X, the inequality $\| b - a \| < p$ implies the inequality $\| A(b) - A(a) \| < q$.

b) An operator (mapping) A: $L \rightarrow M$ is called *bounded at* a point *a relative to* a set X from L if it is (p, q)-bounded at a relative to a set X for some positive numbers p and q.

c) An operator (mapping) A: $L \rightarrow M$ is called *p-bounded at* a point *a relative to* a set X from the space L if it is (p, q)-bounded at a relative to a set X for some positive number q.

d) An operator (mapping) A: $L \rightarrow M$ is called *q-cobounded at* a point *a relative to* a set X if it is (p, q)-bounded at a relative to a set X for some positive number p.

e) An operator (mapping) A: $L \rightarrow M$ is called *uniformly totally bounded at* a point *a relative to* a set X from the space L if for any number p, there is a number q such that for any element b from X, inequality $\| b - a \| < p$ implies the inequality $\| A(b) - A(a) \| < q$.

f) An operator (mapping) A: $L \rightarrow M$ is called *totally bounded at* a point *a relative to* a set X if for any number p and any element b from X, there is a number q such that A is (p, q)-bounded at a relative to a set X.

These concepts allow selection of boundedness domains for unbounded operators and functionals.

Lemma 10.21. If an operator A: $L \rightarrow M$ is (p, q)-bounded (bounded, p-bounded or totally bounded) at a point x from L, then it is (p, q)-bounded (bounded, p-bounded or totally bounded) at the point x relative to any set $X \subseteq L$.

Proof is left as an exercise.

However, a converse result is not true as the following example demonstrates.

Example 10.12. Let us take $L = M = \boldsymbol{R}$ and define the following operator

$$A(x) = \begin{cases} 0 & \text{if } x \leq 0 \\ 1/x & \text{if } x > 0 \end{cases}$$

The operator A is (p, q)-bounded at the point 0 relative to the set $(-\infty, 0]$ for any $p, q > 0$ but it not (p, q)-bounded at the point 0 for any $p, q > 0$.

Lemma 10.22. If $t \leq p$, $q \leq u$ and an operator (mapping) A: $L \rightarrow M$ is (p, q)-bounded at a point a relative to a set $X \subseteq L$, then A is (t, u)-bounded at a relative to X.

Proof is left as an exercise.

Corollary 10.85. If $t < p$ and an operator (mapping) A: $L \to M$ is (p, q)-bounded at a point a relative to X, then A is (t, q)-bounded at a relative to X.

Corollary 10.86. If $q < u$ and an operator (mapping) A: $L \to M$ is (p, q)-bounded at a point a relative to X, then A is (p, u)-bounded at a relative to X.

Lemma 10.23. If $t < p$ and an operator (mapping) A: $L \to M$ is p-bounded at a point a relative to X, then A is t-bounded at a relative to X.

Proof is left as an exercise.

Proposition 10.33. An operator (mapping) A: $L \to M$ is (p, q)-bounded at a point a relative to X if and only if any its shift $A + u$ is (p, q)-bounded at the point a relative to X.

Proof is similar to the proof of Proposition 10.3.

Corollary 10.87. An operator (mapping) A: $L \to M$ is p-bounded at a point a relative to X if and only if any its shift $A + u$ is p-bounded at the point a relative to X.

Corollary 10.88. An operator (mapping) A: $L \to M$ is q-cobounded at a point a relative to X if and only if any its shift $A + u$ is q-cobounded at the point a relative to X.

Corollary 10.89. An operator (mapping) A: $L \to M$ is bounded at a point a relative to X if and only if any its shift $A + u$ is bounded at the point a relative to X.

Corollary 10.90. An operator (mapping) A: $L \to M$ is totally bounded at a point a relative to X if and only if any its shift $A + u$ is totally bounded at the point a relative to X.

It is possible to show that *boundedness inside* is a special case of *relative boundedness*.

Proposition 10.34. An operator (mapping) A: $L \to M$ is (p, q)-bounded at all points from a set X relative to X if and only if it is (p, q)-bounded inside X.

Proof. Necessity. Let us consider a subset X of a hyperseminormed vector space L and an operator (mapping) A: $L \to M$ is (p, q)-bounded at all points from a set X relative to X. It means that for any elements a and b from X, the inequality $\| b - a \| < p$ implies the inequality $\| A(b) - A(a) \| < q$. Thus, A is (p, q)-bounded inside X.

Sufficiency. Let us consider a (p, q)-bounded inside X operator (mapping) A: $L \to M$. It means that for any elements a and b from X, the inequality $\| b - a \| < p$ implies the inequality $\| A(b) - A(a) \| < q$. Thus, A is (p, q)-bounded at all points from a set X relative to X.

Proposition is proved.

Corollary 10.91. An operator (mapping) $A: L \to M$ is bounded at all points from a set X relative to X if and only if it is bounded inside X.

Corollary 10.92. An operator (mapping) $A: L \to M$ is p-bounded at all points from a set X relative to X if and only if it is p-bounded inside X.

Corollary 10.93. An operator (mapping) $A: L \to M$ is q-cobounded at all points from a set X relative to X if and only if it is q-cobounded inside X.

Corollary 10.94. A functional $A: L \to \boldsymbol{R}$ is (p, q)-bounded at all points from a set X relative to X if and only if it is (p, q)-bounded inside X.

Corollary 10.95. A functional $A: L \to \boldsymbol{R}$ is bounded at all points from a set X relative to X if and only if it is bounded inside X.

Corollary 10.96. A functional $A: L \to \boldsymbol{R}$ is p-bounded at all points from a set X relative to X if and only if it is p-bounded inside X.

Corollary 10.97. A functional $A: L \to \boldsymbol{R}$ is q-cobounded at all points from a set X relative to X if and only if it is q-cobounded inside X.

It is also possible to show that *boundedness in* is a special case of *relative boundedness*.

Proposition 10.35. An operator (mapping) $A: L \to M$ is (p, q)-bounded at all points from X relative to the space L if and only if it is (p, q)-bounded in X.

Proof. Necessity. Let us consider a subset X of a hyperseminormed vector space L and an operator (mapping) $A: L \to M$ is (p, q)-bounded at all points from a set X relative to L. It means that for any elements a from X and b from L, the inequality $\| b - a \| < p$ implies the inequality $\| A(b) - A(a) \| < q$. Thus, A is (p, q)-bounded in X.

Sufficiency. Let us consider a(p, q)-bounded inside X operator (mapping) $A: L \to M$. It means that for any elements a from X and b from L, the inequality $\| b - a \| < p$ implies the inequality $\| A(b) - A(a) \| < q$. Thus, A is (p, q)-bounded at all points from a set X relative to L.

Proposition is proved.

Corollary 10.98. An operator (mapping) $A: L \to M$ is bounded at all points from a set X relative to L if and only if it is bounded in X.

Corollary 10.99. An operator (mapping) $A: L \to M$ is p-bounded at all points from a set X relative to L if and only if it is p-bounded in X.

Corollary 10.100. An operator (mapping) $A: L \to M$ is q-cobounded at all points from a set X relative to L if and only if it is q-cobounded in X.

Corollary 10.101. A functional $A: L \to \boldsymbol{R}$ is (p, q)-bounded at all points from a set X relative to L if and only if it is (p, q)-bounded in X.

Corollary 10.102. A functional $A: L \to \boldsymbol{R}$ is bounded at all points from a set X relative to L if and only if it is bounded in X.

Corollary 10.103. A functional A: $L \to \boldsymbol{R}$ is p-bounded at all points from a set X relative to L if and only if it is p-bounded in X.

Corollary 10.104. A functional A: $L \to \boldsymbol{R}$ is q-cobounded at all points from a set X relative to L if and only if it is q-cobounded in X.

From regional boundedness, we come to global boundedness.

Definition 10.6. a) An operator (mapping) A: $L \to M$ is called (p, q)-*bounded* if for any number p and any elements a, b from L, the inequality $\| b - a \| < p$ implies the inequality $\| A(b) - A(a) \| < q$.

b) An operator (mapping) A: $L \to M$ is called *bounded* if it is (p, q)-bounded for some positive numbers p and q.

c) An operator (mapping) A: $L \to M$ is called *p-bounded* if it is (p, q)-bounded for some positive number q.

d) An operator (mapping) A: $L \to M$ is called *q-cobounded* if it is (p, q)-bounded for some positive number p.

e) An operator (mapping) A: $L \to M$ is called *totally bounded* if for any number p and any elements a, b from L, there is a number q such that inequality $\| b - a \| < p$ implies the inequality $\| A(b) - A(a) \| < q$.

f) An operator (mapping) A: $L \to M$ is called *uniformly totally bounded* if for any number p, there is a number q such that for any elements a, b from L, inequality $\| b - a \| < p$ implies the inequality $\| A(b) - A(a) \| < q$.

These concepts do not coincide because Examples 10.1–10.10 show that:

1) An operator can be totally bounded at all points of but be not uniformly totally bounded.
2) A totally bounded operator can be not uniformly totally bounded.

Definitions imply the following result.

Proposition 10.36. a) An operator (mapping) A: $L \to M$ is (p, q)-bounded if and only if it is (p, q)-bounded at any point a from L.

b) An operator (mapping) A: $L \to M$ is bounded if and only if it is bounded at any point a from L.

c) An operator (mapping) A: $L \to M$ is p-bounded if and only if it is p-bounded at any point a from L.

d) An operator (mapping) A: $L \to M$ is uniformly p-bounded if and only if for any number $t \leq p$, there is a number q such that A is (p, q)-bounded at any point a from L.

e) An operator (mapping) A: $L \to M$ is uniformly totally bounded if for any number p, there is a number q such that A is (p, q)-bounded at any point a from L.

Informally, Proposition 10.36 means that any type of boundedness inside L coincides with the corresponding type of boundedness in L.

Proposition 10.36 and Corollary 10.104 imply the following result.

Lemma 10.24. a) If $t < p$ and an operator (mapping) A: $L \to M$ is (p, q)-bounded, then A is (t, q)-bounded.

b) If $q < u$ and an operator (mapping) A: $L \to M$ is (p, q)-bounded, then A is (p, u)-bounded.

Proposition 10.36 implies the following results.

Corollary 10.105. An additive operator (mapping) A: $L \to M$ is bounded at the point **0** if and only if it is bounded.

Corollary 10.106. An additive operator (mapping) A: $L \to M$ is totally bounded at the point **0** if and only if it is totally bounded.

Let us take a non-negative real number r.

Lemma 10.25. An strongly r-additive operator A: $L \to M$ is uniformly totally bounded if and only if it satisfies the following Condition (B):

for any number p, there is a number q such that for any element a from L, the inequality $\|a\| < p$ implies the inequality $\|A(a)\| < t$

Proof. Necessity. Let us take a bounded strongly r-additive operator A: $L \to M$ and an element a from L such that $\|a\| < p$. Then the inequality $\|a - \mathbf{0}\| = \|a\| < p$ implies by Definition 10.6, the inequality $\|A(a) - A(\mathbf{0})\| < q$ for some number q. Consequently, by Proposition 8.8,

$$\|A(a)\| = \|A(a - \mathbf{0})\| < \|A(a) - A(\mathbf{0})\| + r < q + r$$

Necessity is proved as we can take $t = q + r$.

Sufficiency. Let us assume that for an r-linear operator A, Condition (B) is satisfied. Taking $a, b \in L$ such that $\|a - b\| < p$, by Proposition 8.8, we have

$$\|A(a) - A(b)\| < \|A(a - b)\| + r$$

At the same time, by Condition (B),

$$\|A(a)\| = \|A(a + \mathbf{0})\| < t$$

Thus,

$$\|A(a) - A(b)\| < q = t + r$$

Lemma is proved.

Corollary 10.107. An r-linear functional A: $L \to \boldsymbol{R}$ is uniformly totally bounded if and only if it satisfies Condition (B).

Corollary 10.108. An r-linear operator A: $L \to M$ is uniformly totally bounded if and only if it satisfies Condition (B):

Corollary 10.109. An r-linear functional A: $L \to \boldsymbol{R}$ is uniformly totally bounded if and only if it satisfies Condition (B).

Remark 10.1. An operator (mapping) A: $L \to M$ can be bounded at all points from L but unbounded in L as the following example demonstrates.

Example 10.13. The set $\boldsymbol{R}$ of all real numbers is a one-dimensional normed vector space (Rudin, 1991). Let us consider the operator (mapping) A: $\boldsymbol{R} \to \boldsymbol{R}$ that assigns x^2 to any element x from $\boldsymbol{R}$. Then for any point a from $\boldsymbol{R}$, we have $|||A(a) - A(b)|| < q = (a + p)^2 - a^2 = 2ap + p^2$ when $|| b - a || < p$. Thus, A is bounded at all points from $\boldsymbol{R}$. At the same time, A is not bounded because the number a can be larger than any given number and consequently, q can grow without limits.

It means that local boundedness does not imply global boundedness.

However, taking exact characteristics of boundedness, we see that local boundedness can imply global boundedness. Namely, definitions imply the following result.

Proposition 10.37. An operator (mapping) A: $L \to M$ is (p, q)-bounded at all points from L if and only if it is (p, q)-bounded in L.

Proof is left as an exercise.

Lemma 10.26. An operator (mapping) A: $L \to M$ is uniformly totally bounded if and only if for any number p, there is a number q such that A is (p, q)-bounded.

Proof is left as an exercise.

Proposition 10.8 gives us the following result.

Corollary 10.110. If an (r, t)-linear operator (mapping) A: $L \to M$ is (p, q)-bounded at the point $\mathbf{0}$, then A is $(p, q + r + t)$-bounded.

In analysis, mathematicians studied different classes of functions and operators. One of the important classes with many good properties is the class of Lipschitz functions (operators). We remind that an operator (mapping) A: $L \to M$ is called *Lipschitz* if there is a number c such that for any elements a and b from L

$$|| A(b) - A(a)|| < c \cdot || b - a ||$$

Proposition 10.38. Any Lipschitz operator (mapping) A: $L \to M$ is totally bounded.

Indeed, if $\| b - a \| < p$, then $\| A(b) - A(a)\| < cp$ for any elements a, b from L.

Theorem 10.1. a) A (p, q)-linear operator A: $L \to M$ is bounded at the point **0** if and only if A is bounded.

b) A (p, q)-linear operator A: $L \to M$ is totally bounded at the point **0** if and only if A is uniformly totally bounded.

Proof. Sufficiency directly follows from definitions.

Necessity. (a) Let us consider a (p, q)-linear operator A: $L \to M$ that is bounded at the point **0**. It means that A is (p, q)-bounded at **0** for some positive numbers p and q. Taking arbitrary elements a and b from L such that the inequality $\| b - a \| < p$ is true, we see that by Lemma 10.11, (p, q)-boundedness of A at the point **0** implies

$$\|A(b + (-a)) \| = \|A(b - a)) \| < q$$

In addition, we have

$$\| A(b) - A(a)\| = \| A(b) + A(-a) - A(-a) - A(a)\| =$$

$$\| - A(b + (-a)) + A(b) + A(-a) + A(b + (-a)) - A(-a) - A(a)\| \leq$$

$$\| - A(b + (-a)) + A(b) + A(-a) \| + \|A(b + (-a)) \| + \| - A(-a) - A(a)\| =$$

$$\| A(b + (-a)) - A(b) - A(-a) \| + \|A(b + (-a)) \| + \|A(-a) + A(a)\| < r + q + t$$

It means that A is $(p, q + r + t)$-bounded at the point a. As a is an arbitrary point from L, A is $(p, q + r + t)$-bounded at all points from the space L. By definition, A is a bounded operator.

(b) Now let us consider a (p, q)-linear operator A: $L \to M$ that is totally bounded at the point **0**. It means that for any positive real number p, there is a number q such that A is (p, q)-bounded at **0**. Then as it is proved above, A is $(p, q + r + t)$-bounded at all points from the space L. As positive real number p is arbitrary, A is a uniformly totally bounded operator.

Theorem is proved.

Theorem 10.1 and Proposition 10.38 imply the following results.

Corollary 10.111. A linear operator A: $L \to M$ is (totally) bounded at the point **0** if and only if A is (uniformly totally) bounded.

Let us assume $\mathbf{0} \in X \subseteq L$.

Corollary 10.112. A (p, q)-linear operator A: $L \rightarrow M$ is (totally) bounded at the point **0** if and only if A is (uniformly totally) bounded in X.

Linearity of the operator A is essential in Theorem 10.1 and its corollaries as the following example demonstrates.

Example 10.14. Let us take $L = M = \boldsymbol{R}$ and consider the one-dimensional operator (mapping) $1/(x - 1)$. This operator is $(0.1, 1)$-bounded and thus, bounded at the point **0**. However, A is not bounded at the point 1 and thus, it is not bounded.

Let us investigate boundedness of Cartesian products of operators in seminormed vector spaces.

Proposition 10.39. If L, M, H and N are seminormed vector spaces, an operator A: $L \rightarrow M$ is (p, q)-bounded, an operator B: $H \rightarrow N$ is (r, t)-bounded, and the Cartesian product $M \times N$ has the Euclidean seminorm, then the Cartesian product $A \times B$ is (u, v)-bounded in $L \times N$ where $u = \min \{p, r\}$ and $v = (q^2 + t^2)^{1/2}$.

Proof. Let us consider a (p, q)-bounded operator A: $L \rightarrow M$ and(r, t)-bounded operator B: $H \rightarrow N$. Then the inequality

$$\| a - x \| < p$$

implies the inequality

$$\| A(a) - A(x) \| < q$$

and the inequality

$$\| b - y \| < r$$

implies the inequality

$$\| A(a) - A(x) \| < t$$

Then we have

$$\| (A \cdot B)((a, b)) - (A \cdot B)(x, y) \| = \| (A(a), B(b)) - (A(x), B(y)) \| =$$

$$(\| A(a) - A(x) \|^2 + \| B(b) - B(y) \|^2)^{1/2} < (r^2 + q^2)^{1/2}$$

when

$$\| a - x \| < u$$

and

$$\| b - y \| < u$$

because $u \leq p$ and $u \leq r$. It means that the Cartesian product $A \times B$ is (u, v)-bounded because a and x are arbitrary points from L and b and y are arbitrary points from H.

Proposition is proved.

Corollary 10.113. If L, M, H and N are seminormed vector spaces, an operator A: $L \rightarrow M$ is (p, q)-bounded, an operator B: $H \rightarrow N$ is (p, q)-bounded, and the Cartesian product $M \times N$ has the Euclidean seminorm, then the Cartesian product $A \times B$ is $(p, \sqrt{2}q)$-bounded.

Let us investigate boundedness of Cartesian products of operators in hyperseminormed vector spaces.

Proposition 10.40. If an operator A: $L \rightarrow M$ is (p, q)-bounded, an operator B: $H \rightarrow N$ is (r, t)-bounded, and the Cartesian product $M \times N$ has the Manhattan hyperseminorm, then the Cartesian product $A \times B$ is (u, v)-bounded where $u = \min \{p, r\}$ and $v = q + t$.

Proof. Let us consider a (p, q)-bounded operator A: $L \rightarrow M$ and(r, t)-bounded operator B: $H \rightarrow N$. Then the inequality

$$\| a - x \| < p$$

implies the inequality

$$\| A(a) - A(x) \| < q$$

and the inequality

$$\| b - y \| < r$$

implies the inequality

$$\| A(a) - A(x) \| < t$$

Then we have

$$\| (A \cdot B)((a, b)) - (A \cdot B)(x, y) \| = \| (A(a), B(b)) - (A(x), B(y)) \| =$$

$$\| A(a) - A(x) \| + \| B(b) - B(y) \| < r + q = v$$

when

$$\| a - x \| < u$$

and

$$\| b - y \| < u$$

because $u \leq p$ and $u \leq r$. It means that the Cartesian product $A \times B$ is (u, v)-bounded because a and x are arbitrary points from L and b and y are arbitrary points from H.

Proposition is proved.

Corollary 10.114. If an operator $A: L \to M$ is (p, q)-bounded, an operator $B: H \to N$ is (p, q)-bounded, and the Cartesian product $M \times N$ has the Manhattan hyperseminorm, then the Cartesian product $A \times B$ is $(p, 2q)$-bounded.

Proposition 10.41. If an operator $A: L \to M$ is (p, q)-bounded, an operator $B: H \to N$ is (r, t)-bounded, and the Cartesian product $M \times N$ has the Chebyshev hyperseminorm, then the Cartesian product $A \times B$ is (u, v)-bounded where $u = \min \{p, r\}$ and $v = \max \{q + t\}$.

Proof. Let us consider a (p, q)-bounded operator $A: L \to M$ and(r, t)-bounded operator $B: H \to N$. Then the inequality

$$\| a - x \| < p$$

implies the inequality

$$\| A(a) - A(x) \| < q$$

and the inequality

$$\| b - y \| < r$$

implies the inequality

$$\| A(a) - A(x) \| < t$$

Then we have

$$\| (A \cdot B)((a, b)) - (A \cdot B)(x, y) \| = \| (A(a), B(b)) - (A(x), B(y)) \| =$$

$$\max \{\| A(a) - A(x) \|, \|B(b) - B(y)\| < \max \{q, t\} = v$$

when

$$\| a - x \| < u$$

and

$$\| b - y \| < u$$

because $u \leq p$ and $u \leq r$. It means that the Cartesian product $A \times B$ is (u, v)-bounded because a and x are arbitrary points from L and b and y are arbitrary points from H.

Proposition is proved.

Corollary 10.115. If an operator A: $L \rightarrow M$ is (p, q)-bounded, an operator B: $H \rightarrow N$ is (p, q)-bounded and the Cartesian product $M \times N$ has the Chebyshev hyperseminorm, then the Cartesian product $A \times B$ is also (p, q)-bounded.

Corollary 10.116. a) If an operator A: $L \rightarrow M$ is (totally or uniformly totally) bounded, an operator B: $H \rightarrow N$ is (totally or uniformly totally) bounded, and the Cartesian product $M \times N$ has the Chebyshev hyperseminorm or the Manhattan hyperseminorm, then the Cartesian product $A \times B$ is (totally or uniformly totally) bounded.

b) If L, M, H and N are seminormed vector spaces, an operator A: $L \rightarrow M$ is (totally or uniformly totally) bounded, an operator B: $H \rightarrow N$ is (totally or uniformly totally) bounded from H, and the Cartesian product $M \times N$ has the Euclidean seminorm, then the Cartesian product $A \times B$ is (totally or uniformly totally) bounded.

Proposition 10.3 implies the following result.

Proposition 10.42. An operator (mapping) A: $L \rightarrow M$ is (p, q)-bounded if and only if any its shift $A + u$ is (p, q)-bounded.

Proof is similar to the proof of Proposition 10.3.

Corollary 10.117. An operator (mapping) A: $L \rightarrow M$ is p-bounded if and only if any its shift $A + u$ is p-bounded.

Corollary 10.118. An operator (mapping) A: $L \rightarrow M$ is q-cobounded if and only if any its shift $A + u$ is q-cobounded.

Corollary 10.119. An operator (mapping) A: $L \rightarrow M$ is bounded if and only if any its shift $A + u$ is bounded.

Corollary 10.120. An operator (mapping) A: $L \rightarrow M$ is totally bounded if and only if any its shift $A + u$ is totally bounded.

As we have seen in Chapter 6, boundedness is closely related to continuity. Let us study relations between fuzzy continuity and boundedness taking real numbers r, q and p, such that $1 > r > p$. For instance, numbers $r = 1/3$ and $p = 1/7$ satisfy these conditions.

Theorem 10.2. A (p, q)-linear operator A: $L \rightarrow M$ is r-continuous if and only if it is uniformly totally bounded.

Proof. Necessity. Let us take a (p, q)-linear operator $A: L \to M$ and assume that it is r-continuous, but not uniformly totally bounded. It means (cf. Lemma 10.10) that there are a number a and a sequence $l = \{c_i \in L; i = 1, 2, 3, \ldots\}$ such that $\| c_i \| < a$, but $\| A(c_i) \| > i$ for all $i = 1, 2, 3, \ldots$.

Let us put $d_i = (1/i)c_i$ for all $i = 1, 2, 3, \ldots$. By properties of hyperseminorms, we have

$$\| d_i \| = \| (1/i)c_i \| = (1/i) \| c_i \|$$

As $\| c_i \| < a$ for all i, the sequence $l = \{d_i \in L; i = 1, 2, 3, \ldots\}$ converges to $\mathbf{0}$.

Now let us find some inequalities for the number q using properties of hyperseminorms and the initial conditions of the theorem.

$$\| A((1/i)c_i)\| = \| A((1/i)c_i) - (1/i)\cdot A(c_i) + (1/i)\cdot A(c_i)\| \leq$$

$$\| A((1/i)c_i) - (1/i)\cdot A(c_i)\| + \|(1/i)\cdot A(c_i)\| < q + \|(1/i)\cdot A(c_i)\|$$

This implies

$$\| A((1/i)c_i)\| - \|(1/i)\cdot A(c_i)\| < q$$

As an operator A is functionally (p, q)-linear, it is also q-homogeneous. Thus, we have

$$\|A(\mathbf{0})\| = \|A(0\cdot\mathbf{0}) + 0\cdot A(\mathbf{0})\| < q$$

and

$$\| A(d_i) - A(\mathbf{0})\| \geq \| A(d_i)\| - \|A(\mathbf{0})\| > \|A(d_i)\| - q = \| A((1/i)c_i)\| - q >$$

$$q + \|(1/i)\cdot A(c_i)\| - q = \|(1/i)\cdot A(c_i)\| = (1/i)\cdot \|A(c_i)\| > 1$$

because $\|A(c_i)\| > i$ for all $i = 1, 2, 3, \ldots$

As by the initial conditions of the theorem, we have $r < 1$, the point $A(\mathbf{0})$ is not an r-limit of the sequence $\{A(d_i) \in L; i = 1, 2, 3, \ldots\}$. Consequently, the operator A is not r-continuous. Then by *reductio adabsurdum*, if the operator A is r-continuous, then it is uniformly totally bounded.

Sufficiency. Let us take a (p, q)-linear operator $A: L \to M$ that it is not r-continuous. It means (cf. Lemma 10.10) that there is a sequence $l = \{c_i \in L; i = 1, 2, 3, \ldots\}$ such that $a = \lim_{i\to\infty} c_i$, but $A(a) \neq r\text{-}\lim_{i\to\infty} A(c_i)$. In this case, it is possible to assume (Burgin, 2008) that $\| A(a) - A(c_i)\| > r$ for all $i = 1, 2, 3, \ldots$ because if this is not true, we can select a subsequence of l with this property.

As $a = \lim_{i\to\infty} c_{i,}$ there is a sequence $\{n_{k;} k = 1, 2, 3, \ldots\}$ of natural numbers such that $\| a - c_i \| < 1/k$ when $i > n_k$. Consequently, if $u_i = k(a - c_i)$ for all $n_{k+1} \geq i > n_k$ and $k = 1, 2, 3, \ldots$, then

$$\| u_i \| = \| k(a - c_i)\| = k\| a - c_i \| < k(1/k) = 1$$

Thus, the set $\{u_i; i = 1, 2, 3, \ldots\}$ is bounded.

Now let us find some inequalities for the numbers p and q using properties of hyperseminorms and the initial conditions of the theorem.

$$\| A(a) - A(c_i) \| = \| A(a) - A(c_i) - A(a - c_i) + A(a - c_i) \| \leq$$

$$\|A(a) - A(c_i) - A(a - c_i)\| + \| A(a - c_i) \| < p + \| A(a - c_i) \|$$

and consequently,

$$\| A(a - c_i)\| > \| A(a) - A(c_i) \| - p$$

In addition,

$$\| A(k(a - c_i))\| = \| A(k(a - c_i)) - k{\cdot}A((a - c_i) + k{\cdot}A((a - c_i)\| \leq$$

$$\| A(k(a - c_i)) - k{\cdot}A((a - c_i)\| + \| k{\cdot}A((a - c_i)\| < q + \| k{\cdot}A((a - c_i)\|$$

and consequently,

$$k{\cdot} \|A((a - c_i)\| = \| k{\cdot}A((a - c_i)\| > \| A(k(a - c_i))\| - q$$

Besides,

$$\| k{\cdot}A(a - c_i)\| = \| k{\cdot}A(a - c_i - A(k(a - c_i)) + A(k(a - c_i))\| \leq$$

$$\| A(k(a - c_i)) - k{\cdot}A((a - c_i)\| + \| A(k(a - c_i))\| < q + \| A(k(a - c_i))\|$$

and consequently,

$$\| A(k(a - c_i))\| > \| k{\cdot}A(a - c_i)\| - q = k{\cdot} \| A(a - c_i)\| - q >$$

$$k{\cdot}(\| A(a) - A(c_i) \| - p) - q > k{\cdot}(r - \varepsilon - p) - q$$

for any $\varepsilon > 0$ because $A(a)$ is an r-limit of the sequence $\{ A(c_i); i = 1, 2, 3, \ldots\}$.

As $r > p$, there is $\varepsilon > 0$ such that $r - \varepsilon > p$. Therefore, $r - \varepsilon - p > 0$ and thus, $k(r - p - q)$ tends to infinity, when k tends to infinity. Consequently, the set $\{A(u_i); i = 1, 2, 3, \ldots\}$ is not bounded. This implies that the operator A is not uniformly totally bounded. Then by *reductio ad absurdum*, if the operator A is uniformly totally bounded, then it is r-continuous.

Theorem is proved.

Note that number r in the theorem is an arbitrary number that satisfies the necessary conditions.

Corollary 10.121. If $1 - 2q > r > 2q$, then a q-linear operator A: $L \rightarrow M$ is r-continuous if and only if it is bounded.

Note that any q for which Corollary 10.121 is true has to be less than 0.25, while any r for which Corollary 10.121 is true can be close to 1 but is always less than 1.

It is possible to apply Theorem 10.2 to check fuzzy continuity for different numbers r. For instance, we have the following result.

Corollary 10.122. A 0.1-linear operator A: $L \rightarrow M$ is 0.5-continuous if and only if it is bounded.

Theorem 10.2 also gives us the following classical result (Dunford and Schwartz, 1958; Rudin, 1991; Kolmogorov and Fomin, 1999).

Corollary 10.123. A linear operator is continuous if and only if it is bounded.

However, for linear operators, Theorem 10.2 does not give new results as Theorem 10.3 demonstrates.

Let us take a real number r such that $1 > r > 0$.

Theorem 10.3. A linear r-continuous operator A: $L \rightarrow M$ is continuous.

Proof. If A is a linear r-continuous operator, then by Theorem 10.2, it is uniformly totally bounded and thus, conventionally bounded. By the properties of conventionally bounded linear operators proved in Chapter 6 for hyperseminormed vector spaces (cf. Corollary 6.59) and, for example, in Rudin (1991), for seminormed vector spaces, the operator A is continuous.

Theorem is proved.

Corollary 10.124. A linear r-continuous functional A: $L \rightarrow M$ is continuous.

Thus, Theorem 10.3 and Corollary 10.124 show that for linear operators and functionals, fuzzy continuity coincides with continuity when the continuity defect is sufficiently small, i.e., when it is less than one.

Now let us explore connections between fuzzy continuity and boundedness at a point for general operators. The Local Weierstrass Theorem states that a continuous at a point function is bounded at this point (Burgin, 2008). It gives only a sufficient condition of boundedness of a function at a point. Examples show that this is not a necessary condition for boundedness. Coming to the realm of fuzzy continuous operators, we are able to make the Weierstrass' result complete. The following theorem attains completion of the Local Weierstrass Theorem, giving a criterion of the operator boundedness at a point.

Let us consider two hyperseminormed vector spaces L and M.

Theorem 10.4. An operator A: $L \rightarrow M$ is fuzzy continuous at a point a from L if and only if it is defined and bounded at this point.

Proof. Necessity. Let us take an operator A: $L \rightarrow M$ and assume that it is r-continuous at a point a, but is not bounded at this point. It means that for any real numbers p, $q > 0$, the operator A is not (p, q)-bounded at a. In particular, A is not $(1/n, n)$-bounded at a. Consequently, there is a sequence of points $l = \{c_n; \in n = 1, 2, 3, \ldots\}$ such that $\| c_n - a \| < 1/n$ but $\| A(c_n) - A(a)\| > n$. By construction, the sequence l converges to a.

The operator $A(x)$ is fuzzy continuous at the point a, i.e., it is r-continuous at the point a for some positive real number r. Thus, by Definition 7.3, for any $\varepsilon > 0$, there is $\delta > 0$ such that the inequality $\| a - x \| < \delta$ implies the inequality $\| A(x) - A(a)\| < r + \varepsilon$. This means that the hypernumbers $\| A(c_n) - A(a)\|$ have to be bounded by some positive number k. This is not true. So, this contradiction shows that the operator A has to be bounded at the point a when it is fuzzy continuous at a point a.

Sufficiency. Let us consider a bounded at a point a operator A: $L \rightarrow M$. It means that there are numbers p and q such that for any element b from L, the inequality $\| b - a \| < p$ implies the inequality $\| A(b) - A(a)\| < q$.

This allows us to conclude that for any $\varepsilon > 0$, there is $\delta > 0$ such that if $\delta < p$, then the inequality $\| a - x \| < \delta$ implies the inequality $\| A(x) - A(a)\| < q + \varepsilon$. By Definition 7.3, the operator A is q-continuous at the point a and thus, fuzzy continuous at a point a.

Theorem is proved.

As any continuous at a point operator (function) is fuzzy continuous at this point (cf. Chapter 7), Theorem 10.4 implies the following classical result.

Corollary 10.125. A continuous at a point functional is bounded at this point.

Corollary 10.126. A continuous at a point functional is bounded at this point.

It is possible to ask a question whether it is possible to prove similar results for boundedness and fuzzy continuity inside any subset of the whole space or in the whole space. However, for operators defined in L or inside an open subset of L, Theorem 10.4 is not true in general because even a continuous operator can be unbounded. For instance, we can take functions (one-dimensional operators) x^2 in $L = \boldsymbol{R}$ or $\tan x$ in the interval $X = (-\pi/2, \pi/2)$. They are continuous but unbounded. Nevertheless, we can show that boundedness imply fuzzy continuity inside any subset of the whole space.

Let $X \subseteq L$.

Theorem 10.5. Any bounded in X operator $A: X \to M$ is fuzzy continuous in X.

Proof. Let us consider a bounded in X operator $A: L \to M$. It means that for any point a from X, there are numbers p and q such that for any element b from L, the inequality $\| b - a \| < p$ implies the inequality $\| A(b) - A(a)\| < q$.

This allows us to conclude that for any $\varepsilon > 0$, there is $\delta > 0$ such that if $\delta < p$, then the inequality $\| a - x \| < \delta$ implies the inequality $\| A(x) - A(a)\| < q + \varepsilon$. By Definition 7.3, the operator A is q-continuous at the point a and thus, fuzzy continuous at a point a. As a is an arbitrary point from X, the operator A is fuzzy continuous in X.

Theorem is proved.

Corollary 10.127. Any bounded in X function is fuzzy continuous in X.

Corollary 10.128. Any bounded in X functional is fuzzy continuous in X.

Theorem 10.6. Any bounded inside X operator $A: X \to M$ is fuzzy continuous inside X.

Proof is similar to the proof of Theorem 10.5.

Corollary 10.129. Any bounded inside X function is fuzzy continuous inside X.

Corollary 10.130. Any bounded inside X functional is fuzzy continuous inside X.

When $X = L$, Theorem 10.6 gives us the following result.

Theorem 10.7. Any bounded operator $A: L \to M$ is locally fuzzy continuous.

Note that as we have proved, boundedness of operators implies their local fuzzy continuity but the converse is not always true.

Corollary 10.131. Any bounded in X function is locally fuzzy continuous.

Corollary 10.132. Any bounded in X functional is locally fuzzy continuous. The first Weierstrass Theorem (Ross, 1996) gives a sufficient condition for boundedness of a function in a closed interval. Burgin proved a completion of the first Weierstrass Theorem, giving a criterion, i.e., necessary and sufficient conditions, for boundedness of a functionin a closed interval (Burgin, 2008). Here we extend this result to the spaces of operators defined on compact subsets of complete hyperseminormed spaces.

Completeness is an important property of normed spaces [cf., e.g., (Rudin, 1991)]. It has its natural counterpart in hypernormed and hyperseminormed spaces.

Definition 10.7. A hyperseminormed space L is called *complete* if any Cauchy sequence (cf. Chapter 3) in it converges.

Let us consider a compact subset K of a complete hyperseminormed space L.

Theorem 10.8. An operator $A: K \to M$ is fuzzy continuous inside K if and only if it is bounded inside K.

Proof. Necessity. Let us take an operator $A: L \to M$ and assume that it is fuzzy continuous inside K but is not bounded inside K. The first assumption means that A is r-continuous inside K for some $r \geq 0$. The second assumption means that for any real numbers $p, q > 0$, the operator A is not (p, q)-bounded inside K. In particular, A is not $(1/n, n)$-bounded inside K. Consequently, there is a sequence of points $l = \{c_n; \in n = 1, 2, 3, \ldots\}$ and a sequence of points $h = \{d_n; \in n = 1, 2, 3, \ldots\}$ such that for all $n = 1, 2, 3, \ldots$, $\| c_n - d_n \| < 1/n$ but $\| A(c_n) - A(d_n) \| > n$. In a compact set, any sequence contains a convergent subsequence (Kuratowski, 1966). Thus, it is possible to assume that the sequence l itself converges to some element a from K.

If $\lim_{i\to\infty} c_i = a$, then $\lim_{i\to\infty} \| c_i - a \| = 0$. At the same time, we have

$$\| d_i - a \| = \| d_i - c_i + c_i - a \| \leq \| d_i - c_i\| + \| c_i - a \|$$

By construction, $\lim_{i\to\infty} \| c_i - d_i\| = 0$. Consequently, $\lim_{i\to\infty} \| d_i - a \| = 0$. Besides,

$$\| A(c_n) - A(a) \| + \| A(a) - A(d_n) \| \geq \| A(c_n) - A(d_n) \| > n$$

Then either $\| A(c_n) - A(a) \| \geq n/2$ or $\| A(d_n) - A(a) \| \geq n/2$.

Thus, either there is an infinite sequence $u = \{c_{n_i}; \in i = 1, 2, 3, \ldots\}$ such that $\| A(c_{n_i}) - A(a) \| \geq n/2$ for all $i = 1, 2, 3, \ldots$ or there is an infinite sequence $v = \{d_{n_i}; \in i = 1, 2, 3, \ldots\}$ such that $\| A(d_{n_i}) - A(a) \| \geq n/2$ for all $i = 1, 2, 3, \ldots$.

In the first case, $\lim_{i\to\infty} c_{n_i} = a$, as u is a subsequence of the sequence l but $A(a)$ is not a q – limit of $A(c_{n_i})$ for any q. Consequently, A is not q-continuous inside X for any q. Thus, the operator $A(x)$ is not fuzzy continuous inside X.

The second case is treated in a similar way.

Sufficiency follows from Theorem 10.6.

Theorem is proved.

Corollary 10.133. A functional F: $K \to \boldsymbol{R}$ is fuzzy continuous inside a compact set K if and only if it is bounded inside K.

Corollary 10.134. A function f: $\boldsymbol{R} \to \boldsymbol{R}$ is fuzzy continuous inside a compact set K if and only if it is bounded inside K.

Such a classical result as the first Weierstrass Theorem [cf., (Ross, 1996)] is a direct corollary of Theorem 10.8 because any continuous function is fuzzy continuous (Burgin, 2008).

However, not every fuzzy continuous in X operator is bounded when X is not a compact space. For instance, the function (one-dimensional operator) $f(x) = 1/x$ is continuous and thus, fuzzy continuous in the open interval (0, 1) but it is not bounded. In particular, it means that in general, fuzzy continuous operators do not coincide with bounded operators.

For 2-fuzzy continuous inside compact sets operators, it is possible to strengthen the previous theorem.

Theorem 10.9. If an operator A: $L \to M$ is 2-fuzzy continuous inside a compact set K, then it is uniformly totally bounded inside K.

Proof. Let us take an operator A: $L \to M$ and assume that it is 2-fuzzy continuous inside K but is not uniformly totally bounded inside K. The first assumption means that A is (t, r)-continuous inside K for some $t, r \geq 0$. The second assumption means that for some real number p, there are no numbers q such that the operator A is not (p, q)-bounded inside K.

In particular, A is not (p, np)-bounded inside K for all $n = 1, 2, 3, \ldots$ Consequently, there are two sequences $l = \{c_n; n = 1, 2, 3, \ldots\}$ and $h = \{d_n; n = 1, 2, 3, \ldots\}$ of points from K such that for all $n = 1, 2, 3, \ldots$, $\| c_n - d_n \| < p$ but $\| A(c_n) - A(d_n) \| > n$.

In a compact set, any sequence contains a convergent subsequence (Kuratowski, 1966). Thus, it is possible to assume that the sequence l itself converges to some element a from K.

If $\lim_{i\to\infty} c_i = a$, then $\lim_{i\to\infty} \| c_i - a \| = 0$. At the same time, we have

$$\| d_i - a \| = \| d_i - c_i + c_i - a \| \leq \| d_i - c_i \| + \| c_i - a \|$$

By construction, $0 = p\text{-}\lim_{i\to\infty} \| c_i - d_i \|$ because for all $n = 1, 2, 3, \ldots$, $\| c_n - d_n \| < p$. Consequently, $0 = p\text{-}\lim_{i\to\infty} \| d_i - a \|$ and $a = p\text{-}\lim_{i\to\infty} \| d_i - a \|$.

Besides,

$$\| A(c_n) - A(a) \| + \| A(a) - A(d_n) \| \geq \| A(c_n) - A(d_n) \| > n$$

Then either $\| A(c_n) - A(a) \| \geq n/2$ or $\| A(d_n) - A(a) \| \geq n/2$.

Thus, either there is an infinite sequence $u = \{c_{n_i}; i = 1, 2, 3, \ldots\}$ such that $\| A(c_{n_i}) - A(a) \| \geq n/2$ for all $i = 1, 2, 3, \ldots$ or there is an infinite sequence $v = \{d_{n_i}; i = 1, 2, 3, \ldots\}$ such that $\| A(d_{n_i}) - A(a) \| \geq n/2$ for all $i = 1, 2, 3, \ldots$.

In the first case, $a = p\text{-}\lim_{i\to\infty} c_{n_i}$, as u is a subsequence of the sequence l and the equality $\lim_{i\to\infty} c_{n_i} = a$ implies the equality $a = p\text{-}\lim_{i\to\infty} c_{n_i}$, (Burgin, 2008). However, $A(a)$ is not a (p, q) – limit of $A(c_{n_i})$ for any q. Consequently, A is not (p, q) – continuous inside X for any q. Thus, the operator $A(x)$ is not 2-fuzzy continuous inside X.

In the second case, $a = p\text{-}\lim_{i\to\infty} d_{n_i}$, as u is a subsequence of the sequence l (Burgin, 2008). However, $A(a)$ is not a (p, q) – limit of $A(d_{n_i})$ for any q. Consequently, A is not (p, q)-continuous inside X for any q. Thus, the operator $A(x)$ is not 2-fuzzy continuous inside X. This contradiction with the made assumption concludes the proof.

Theorem is proved.

Corollary 10.135. If a functional $F: K \to \boldsymbol{R}$ is 2-fuzzy continuous inside a compact set K, then it is uniformly totally bounded inside K.

Corollary 10.136. If a function $f: \boldsymbol{R} \to \boldsymbol{R}$ is 2-fuzzy continuous inside a compact set K, then it is uniformly totally bounded inside K.

KEYWORDS

- **(*p*, *q*)-bounded**
- **2-fuzzy continuous**
- **bounded operator**
- **cobounded**
- **compact**
- **continuous**
- **fuzzy continuous**
- ***q*-bounded**
- **uniformly bounded operator**
- **uniformly totally bounded operator**

CHAPTER 11

CONCLUSION AND DIRECTIONS FOR FUTURE RESEARCH

The main results of this book are:

- Further development of the theory of hypernumbers and extrafunctions.
- Presentation and exploration of new theoretical tools, such as hypernorms, hyperseminorms and hypermetrics, for study of diverse mathematical structures, such as hypernormed vector spaces, hyperseminormed vector spaces, polyhypernormed vector spaces, polyhyperseminormed vector spaces, hyperpseudometric vector spaces, and hypermetric vector spaces.
- Presentation and study of the novel mathematical structures called semitopological vector spaces, as well as operators in these spaces and their mappings.
- Introduction and study of various types of fuzzy continuity of operators in normed, seminormed, hypernormed and hyperseminormed vector spaces
- Introduction and investigation of various types of approximate additivity, approximate homogeneity and approximate linearity of operators in normed, seminormed, hypernormed and hyperseminormed vector spaces
- Introduction and study of various types of boundedness of operators in hypernormed, polyhyperseminormed and hyperseminormed vector spaces, as well as establishing relations between boundedness and fuzzy continuity

Many classical theorems become corollaries of the results obtained in this book. In addition, constructions and outcomes from this book open innovative opportunities for a variety of practical fields, for example, when operator equations, which include differential and integral equations, are solved and these solutions are applied to different practical problems [cf., e.g., (Burgin and Dantsker, 1995, 2015; Burgin, et al, 2012)].

Obtained results also explicate new directions for further research and development. Let us list some of them.

At the beginning of this book, we have developed the theory of hypernumbers in a much more general framework than it was done before, namely, taking as basic construct a normed vector space and not only real or complex numbers (Chapter 2). This makes it possible to build generalized hypernumbers called P-hypernumbers, different properties of which are obtained bringing us to the following algebraic, analytic and topological problems.

Problem 1. Study algebraic structures in spaces of P-hypernumbers defining such operations as multiplication or division when it is possible.

There are also analytic operations performed with real and complex numbers, which would be useful to extend to P-hypernumbers.

Problem 2. Introduce and study analytic operations with P-hypernumbers such as taking the limit of a sequence or finding the sum of a series.

There are also topological structures in spaces of real and complex numbers, which would be useful to extend to P-hypernumbers.

Problem 3. Study topology in spaces of P-hypernumbers.

In many important cases, it is possible to define topology in vector spaces using norms, metrics, pseudometrics and seminorms. These topologies have many good qualities. We can ask similar questions about hypernorms, hypermetrics, hyperpseudometrics and hyperseminorms.

Problem 4. Study hypernorms, hypermetrics, hyperpseudometrics and hyperseminorms in spaces of P-hypernumbers delineating topologies defined by these characteristics.

Moreover, it is possible to extend the concepts of these characteristics.

Problem 5. Study hypernorms, hypermetrics, hyperpseudometrics and hyperseminorms that take values in spaces of P-hypernumbers.

Hypernumbers are intrinsically related to the theory of extrafunctions, which is also extended (in Chapter 4) to hyperfunctionals and hyperoperators, which include extrafunctions as a particular case. In particular, the spaces of generalized distributions are also enlarged demonstrating that they include conventional distributions of Schwartz, tempered distributions, ultradistributions and many other generalized functions.

As we know, there are different approaches to construction of distributions [cf., e.g., (Sobolev, 1936; Schwartz, 1945; Mikusinski, 1948; Halperin, 1952; König, 1953; Temple, 1953, 1955; Sikorski, 1954; Korevar, 1955; Gelfand and Vilenkin, 1964; Synowiec, 1983)], as well as to building other generalized functions [cf., e.g., (Fisher, 1969; Berg, 1978; Colombeau, 1986; Todorov, 1987)]. The most popular are the functional approach originally introduced by Sobolev (1936) and independently introduced and fully developed by Schwartz (1945), and the sequential approach elaborated by Mikusinski (1948). In the functional approach, distributions are defined as linear functionals; while in the sequential approach, distributions are defined as classes of equivalent fundamental function sequences. In the theory of extrafunctions, generalized distributions are constructed as special cases of distributions in the context of the sequential approach. Thus, it would be interesting to explore other possibilities

Problem 6. Use other approaches and in particular, the functional approach for building generalized distributions and studying their properties.

The classical theory of distributions builds, studies and applies distributions defined for real and complex numbers because this is the framework of the classical analysis, as well as of applications of mathematics in physics and engineering. The same is true for the theory of hypernumbers and extrafunctions. However, in recent years, mathematicians have also introduced and investigated p-adic distributions [cf., e.g., (Schneider and Teitelbaum, 2002; Albeverio et al., 2010)]. Thus, we have a worthy of note mathematical problems.

Problem 7. Introduce and study p-adic hypernumbers.

This brings us to the next problem.

Problem 8. Introduce and explore extrafunctions, hyperfunctionals and hyperoperators in the context of p-adic hypernumbers and p-adic norms.

Mathematicians have also studied expansions, asymptotic expansions and quasiasymptotic expansion of distributions [cf., e.g., (Korevar, 1959; Walter, 1965; Pilipović, 1991; Nikolić-Despotović and Pilipović, 1988, 1993; Estrada and Kanwal, 1994, 2002)].

Problem 9. Define and study expansions, asymptotic expansions and quasiasymptotic expansion of extrafunctions, hyperfunctionals and hyperoperators.

In this book, normed and seminormed vector spaces, which play an indispensable role in functional analysis, modern physics and other sciences, are studied as particular cases of hypernormed and hyperseminormed vector spaces (Chapter 3). This shift to more general structures allows extension of application of topological methods in mathematics and beyond.

Problem 10. Study what kinds of topology it is possible to define based on systems of seminorms, hypernorms or hyperseminorms.

In Chapter 3, we also studied several generalizations of norms, seminorms, hypernorms and hyperseminorms such as quasi-norms, pseudonorms and quasi-seminorms. However, there are other generalizations of norms. For instance, mathematicians studied fuzzy norms and fuzzy-normed linear spaces [cf., e.g., (Bag and Samanta, 2003; Chang and Mordeson, 1994; Felbin, 1992; Krishna and Sarma, 1994; Rhie et al., 1997)]. This brings us to the following problem.

Problem 11. Introduce and study fuzzy hypernorms, fuzzy hyperseminorms, fuzzy-hypernormed vector spaces and fuzzy-hyperseminormed vector spaces.

Besides, mathematicians also studied q-norms (Litvak, 1998).

A mapping $q: L \to \boldsymbol{R}$ is called a *q-norm* on a real vector space L if it satisfies the following conditions:

N1. $q(x) = 0$ if and only if $x = \mathbf{0}$.

N2. $q(ax) = |a| \cdot q(x)$ for any x from L and any number a from $\boldsymbol{R}$.

Nq3. $q(x + y)^q \leq q(x)^q + q(y)^q$ for any x and y from L

Problem 12. Introduce and study q-hypernorms, q-hyperseminorms, q-hypernormed vector spaces and q-hyperseminormed vector spaces.

One more important mathematical structure studied in this book is a semitopological vector space (Chapter 5). Semitopological vector spaces are more general than conventional topological vector spaces, which have been very useful for solving many problems in functional analysis. Semitopological vector spaces have even a larger scope of applications in comparison with topological vector spaces.

There are several important problems in the basic theory of semitopological vector spaces.

Problem 13. Study topology in semitopological vector spaces.

Problem 14. Study operations with semitopological vector spaces.

Problem 15. Explore applications of semitopological vector spaces.

In this book, it is demonstrated that hyperseminormed and hypernormed vector spaces are semitopological vector spaces because hypernorms and hyperseminorms define topological structures in vector spaces. For instance, we have also proved in this book (Theorem 5.5) that any polyhyperseminormed vector space is a semitopological vector space giving a positive answer to Problem 11 from Burgin (2013).

In addition, hypermetrics and hyperpseudometrics are introduced and it is demonstrated that hyperseminorms induce hyperpseudometrics, while hypernorms induce hypermetrics. Sufficient and necessary conditions for a hyperpseudometric (hypermetric) to be induced by a hyperseminorm (hypernorm) are found (Theorems 3.5, 3.6, 3.8 and 3.9).

At the same time, it is known that pseudometrics and metrics also define topological structures [cf., e.g., (Kuratowski, 1968)]. Thus, we have the following problem.

Problem 16. Study what kinds of topology it is possible to define with hyperpseudometrics and hypermetrics.

It is also demonstrated [cf. (Rudin, 1991)] that systems of seminorms characterize locally convex spaces but there are topological vector spaces topology in which is not defined by systems of seminorms. It is possible to ask if the same is true for semitopological vector spaces. Namely, we have the following problems.

Problem 17. Is the topology in a semitopological vector space always defined by a system of seminorms?

Problem 18. Is the topology in a semitopological vector space always defined by a system of hyperseminorms?

These two problems are related to the next two problems.

Problem 19. Find topological characteristics that allow one to define topology in a semitopological vector space by a system of seminorms.

Problem 20. Find topological characteristics that allow one to define topology in a semitopological vector space by a system of hyperseminorms.

It is also possible to generalize the concepts of hypernorms and hyperseminorms using spaces of P-hypernumbers instead of ordinary hypernumbers.

Problem 21. Study relations between semitopological vector spaces and hypernormed (hyperseminormed) vector space with hypernorms and hyperseminorms taking values in spaces of P-hypernumbers.

The traditional approach builds a topology in a poly(hyper)seminormed and poly(hyper)normed vector space using all (hyper)seminorms or (hyper)norms. However, it is possible to use each (hyper)seminorm or (hyper)norm for building a separate topology in the space. This brings us to polytopological spaces. For instance, with two norms, it is possible to build bitopological spaces studied by different authors [cf., (Künzi, 2009)].

Problem 22. Study polytopological vector spaces.

Topological vector spaces provide an efficient context for the development of integration [cf., (Choquet, 1969; Edwards and Wayment, 1970; Shuchat, 1972; Kurzweil, 2000)].

Problem 23. Study integration in semitopological (polyhyperseminormed and polyhypernormed) vector spaces.

At the same time, a new area of integration and hyperintegration in bundles with a hyperspace base has been developed by Burgin (2010, 2012, 2015) where construction of the hyperspace utilizes seminorms. This approach provides a base for developing the theory of extrafunction spaces in an abstract setting of algebraic systems and topological spaces or even in abstract categories, where integration plays an important role. Therefore, we naturally come to the following problem.

Problem 24. Study integration and hyperintegration in bundles with a hyperspace base where the hyperspace is built by means of hyperseminorms.

It is also possible to define norms and seminorms with values not only in number or hypernumber spaces but in more general spaces, e.g., in operator spaces.

Problem 25. Study vector spaces that have norms or/and seminorms with values in general spaces, e.g., in normed fields or seminormed rings.

Here we studied functionals and operators in hypernormed and hyperseminormed vector spaces. At the same time, functionals and operators are special cases of hyperfunctionals and hyperoperators in same way as functions are special cases of distributions, which in turn, are special cases of extrafunctions. Bounded, additive and linear functionals and operators occupy important place in functional analysis. This brings us to the following problems.

Problem 26. Study bounded hyperfunctionals and hyperoperators in polyhyperseminormed (semitopological) vector spaces.

Problem 27. Study (fuzzy) additive hyperfunctionals and hyperoperators in polyhyperseminormed (semitopological) vector spaces.

Problem 28. Study (fuzzy) linear hyperfunctionals and hyperoperators in polyhyperseminormed (semitopological) vector spaces.

Many laws of physics and other natural sciences essentially depend on the scale in which physical structures and processes are studied. To be able to represent this feature of physics in a mathematical setting, scalable topological spaces have been introduced and studied (Burgin, 2004, 2005, 2006). Scalable topological spaces are topological spaces enriched with a scale, which regulates precision with which topological spaces and their mapping are treated and applied.

Problem 29. Define and study scalable topological and scalable semitopological vector spaces.

In this book, boundedness and continuity are defined relative to systems of hyperseminorms or hypernorms. Utilization of different sets of hyperseminorms impacts the strength of corresponding topologies, namely, the larger is the set Q of hyperseminorms (hypernorms), the weaker topology it defines. In such a way, we obtain a definite scalability of spaces (Burgin, 2004, 2006) with systems of hyperseminorms (hypernorms), coming to the following problem.

Problem 30. Study scalability of topological and semitopological vector spaces defined by systems of hyperseminorms and hypernorms.

From the perspective of applications, we also suggest the following important problem.

Problem 31. Study approximately linear operator equations and in particular, approximately linear fuzzy differential and integral equations.

A bornology on a vector space L is a family B of subsets of L that is covering L, hereditary under inclusion, stable under finite unions, vector addition, scalar multiplication, and the formation of balanced hulls (Hogbe-Nlend, 1977; Beer and Levi, 2009). A bornological vector space is a locally convex topological vector space in which it is possible to recover topology from its bornology in a natural way (Meyer, 2004, 2004a).

In functional analysis, bornological vector spaces provide an efficient setting for solving many problems in noncommutative geometry and representation theory (Meyer, 2004, 2004a). In addition, bornological vector spaces allow researchers to extend applications of functional analysis and to strengthen its results (Schaefer, 1970; Hogbe-Nlend, 1972). Results and constructions from this book open new opportunities for further research in the area of bornological vector spaces.

Problem 32. Define and study semibornological vector spaces.

As in the case of topological vector spaces, scalability is an important property of bornological vector spaces.

Problem 33. Define and study scalable bornological vector spaces.

APPENDIX

1. **General Concepts and Structures**

$N = \{1, 2, 3, \ldots\}$ is the set of all natural numbers;

ω is the sequence of all natural numbers;

$\varnothing$ is the *empty set*, i.e., the set that has no elements.

$\boldsymbol{R}$ is the set of all real numbers;

$\boldsymbol{R}^+$ is the set of all non-negative real numbers;

$\boldsymbol{R}^{++}$ is the set of all positive real numbers;

$\boldsymbol{R}^-$ is the set of all non-positive real numbers;

$\boldsymbol{R}^{--}$ is the set of all negative real numbers;

$\boldsymbol{C}$ is the set of all complex numbers;

If a is a real number, then $|a|$ or $||a||$ denotes its absolute value or modulus;

If a is a complex number, then $||a||$ or $|a|$ denotes its magnitude or modulus;

If b is a vector from $\boldsymbol{R}^n$, then $||a||$ denotes its modulus;

In general, a sequence of elements a_i is denoted either by $\{a_i(x); i = 1, 2, 3, \ldots\}$ or by $\{a_i(x); i \in \omega\}$ or by $(a_i)_{i \in \omega}$;

$F(X, Y)$ is the set of all mappings (functions) from a set X into a set Y;

$F(\boldsymbol{R})$ or $F(\boldsymbol{R}, \boldsymbol{R})$ is the space of all real functions;

$C(\boldsymbol{R})$ or $C(\boldsymbol{R}, \boldsymbol{R})$ is the space of all continuous real functions;

$F[a, b]$ is the space of all real functions defined in the interval $[a, b]$.

$C[a, b]$ is the space of all continuous real functions defined in the interval $[a, b]$.

If X is a set (class), then $r \in X$ means that r belongs to X or r is a member of X.

If X and Y are sets (classes), then $Y \subseteq X$ means that Y is a *subset* (subclass) of X, i.e., Y is a set such that all elements of Y belong to X, and X is a *superset* of Y. A subset is *proper* if it does coincide with the whole set.

The *union* $Y \cup X$ of two sets (classes) Y and X is the set (class) that consists of all elements from Y and from X. The union $Y \cup X$ is called disjoint if $Y \cap X = \varnothing$.

The *intersection* $Y \cap X$ of two sets (classes) Y and X is the set (class) that consists of all elements that belong both to Y and to X.

The *union* $\bigcup_{i\in I} X_i$ of sets (classes) X_i is the set (class) that consists of all elements from all sets (classes) X_i, $i\in I$.

The *intersection* $\bigcap_{i\in I} X_i$ of sets (classes) X_i is the set (class) that consists of all elements that belong to each set (class) X_i, $i\in I$.

The *difference* $Y \setminus X$ of two sets (classes) Y and X is the set (class) that consists of all elements that belong to Y but does not belong to X.

The *symmetric difference* $Y \Delta X$ of two sets (classes) Y and X is equal to $(Y \setminus X) \bigcup (Y \setminus X)$.

If X is a set, then 2^X is the *power set* of X, which consists of all subsets of X. The *power set* of X is also denoted by PX or by βX.

If X and Y are sets (classes), then $X\times Y = \{(x, y); x \in X, y \in Y\}$ is the *Cartesian product* of X and Y, in other words, $X\times Y$ is the set (class) of all pairs (x, y), in which x belongs to X and y belongs to Y.

If $\{X_i; i \in I\}$ is a collection of sets indexed by a set I, then the Cartesian product $\prod_{i\in I} X_i$ of the sets in X is defined as

$$\prod_{i\in I} X_i = \{f: I \to \bigcup_{i\in I} X_i;\ \forall i\ (f(i) \in X_i)\}$$

It is the set of all functions defined on the index set I such that the value of each function f at a particular index i is an element from the set X_i. Even if each of the sets X_i is nonempty, their Cartesian product may be empty if the axiom of choice is not assumed because the axiom of choice is equivalent to the statement that every such product is nonempty.

The function $\pi_i: \prod_{i\in I} X_i \to X_i$ defined by $\pi_i(f) = f(i)$ is called the *canonical i-th projection mapping* or simply, the *i*-th *projection*.

Y^X is the set of all mappings from X into Y.

$$X^n = \underbrace{X \times X \times \ldots X \times X}_{n}$$

Elements of the set X^n have the form $(x_1, x_2, \ldots, x_n)$ with all $x_i \in X$ and are called n-tuples, or simply, tuples.

A fundamental structure of mathematics is *function*. However, functions are special kinds of binary relations between two sets.

A set $\boldsymbol{F}$ of subsets of a set X is called a filter on X if it satisfies the following conditions:

1. $\boldsymbol{F} \neq \varnothing$ and $\varnothing \notin \boldsymbol{F}$.
2. If $A \in \boldsymbol{F}$ and $A \subseteq B \subseteq X$, then $A \in \boldsymbol{F}$.
3. If $A, B \in \boldsymbol{F}$, then $A \cap B \in \boldsymbol{F}$.

A *binary relation* T between sets X and Y, also called *correspondence* from X to Y, is a subset of the direct product $X{\times}Y$. The set X is called the *domain* of T ($X = \mathrm{Dom}(T)$) and Y is called the *codomain* of T ($Y = \mathrm{Codom}(T)$). The *range* of the relation T is $\mathrm{Rg}(T) = \{y;\ \exists\, x \in X\,((x, y) \in T)\}$. The *domain of definition* also called the *definability domain* of the relation T is $\mathrm{DDom}(T) = \{x;\ \exists\, y \in Y\,((x, y) \in T)\}$. If $(x, y) \in T$, then one says that the elements x and y are in relation T, and one also writes $T(x, y)$.

The image $T(x)$ of an element x from X is the set $\{y;\ (x, y) \in T\}$ and the coimage $T^{-1}(y)$ of an element y from Y is the set $\{x;\ (x, y) \in T\}$.

Binary relations are also called *multivalued functions* (mappings or maps).

Taking binary relations $T \subseteq X{\times}Y$ and $R \subseteq Y{\times}Z$, it is possible to build a new binary relation $RT \subseteq X{\times}Z$ that is called the (*sequential*) *composition* or *superposition* of binary relations T and R and is defined as

$R \circ T = \{(x, z);\ x \in X,\ z \in Z;$ where $(x, y) \in T$ and $(y, z) \in R$ for some $y \in Y\}$.

A *preorder*, also called *quasiorder*, on a set (class) X is a binary relation Q on X that satisfies the following axioms:

O1. Q is *reflexive*, i.e., $x\mathrm{Q}x$ for all x from X.

O2. Q is *transitive*, i.e., $x\mathrm{Q}y$ and $y\mathrm{Q}z$ imply $x\mathrm{Q}z$ for all $x, y, z \in X$.

A *partial order* is a preorder that satisfies the following additional axiom:

O3. Q is *antisymmetric*, i.e., $x\mathrm{Q}y$ and $y\mathrm{Q}x$ imply $x = y$ for all $x, y \in X$.

A *strict* also called *sharp partial order* is a preorder that is not reflexive, is transitive and satisfies the following additional axiom:

O4. Q is *asymmetric*, i.e., only one relation $x\mathrm{Q}y$ or $y\mathrm{Q}x$ is true for all $x, y \in X$.

A *linear* or *total order* is a strict partial order that satisfies the following additional axiom:

O5. We have either $x\mathrm{Q}y$ or $y\mathrm{Q}x$ for all $x, y \in X$.

A set (class) X is *well-ordered* if there is a partial order on X such that any its non-empty subset has the least element. Such a partial order is called *well-ordering*.

An *equivalence* on a set (class) X is a binary relation Q on X that is reflexive, transitive and satisfies the following additional axiom:

O6. Q is *symmetric*, i.e., $x\mathrm{Q}y$ implies $y\mathrm{Q}x$ for all x and y from X.

If we have an equivalence σ on a set X, this set is a disjoint union of classes of the equivalence σ where each class consists of equivalent elements from

X and there are no equivalent elements in different classes. The set of these equivalence classes is called the quotient set of X and it is denoted by X/σ.

A *tolerance relation* is a binary relation that is reflexive and symmetric.

Traditionally, a *function* (also called a *mapping* or *map* or *total function* or *total mapping* or *everywhere defined function*) f from X to Y is defined as a binary relation between sets X and Y in which there are no elements from X which are corresponded to more than one element from Y and to any element from X, some element from Y is corresponded. At the same time, the function f is also denoted by $f: X \to Y$ or by $f(x)$. In the latter formula, x is a variable and not a definite element from X. The *support*, or *carrier*, of a function f is the closure of the set where $f(x) \neq 0$. Usually the element $f(a)$ is called the *image* of the element a and denotes the value of f on the element a from X. The *coimage* $f^{-1}(y)$ of an element y from Y is the set $\{x; f(x) = y\}$.

However, the traditional definition does not include all kinds of functions and their representations.

There are three basic forms of function representation (definition):

1. (The *set-theoretical*, e.g., *table*, *representation*) A function f is given as a subset R_f of the direct product $X \times Y$ such that the first element if each pair from R_f uniquely defines the second element in this pair, e.g., in a form of a table or of a list of pairs (x, y) where the first element x is taken from X, while the second element y is the image $f(x)$ of the first one. The set R_f is called the *graph* of the function f. When X and Y are sets of points in a geometrical space, e.g., their elements are real numbers, the graph of the function f is called the *geometrical graph* of f.
2. (The *analytic representation*) A function f is described by a formula, i.e., a relevant expression in a mathematical language, e.g., $f(x) = \sin(e^{x + \cos x})$.
3. (The *algorithmic representation*) A function f is given as an algorithm that computes $f(x)$ given x.

$f(x) \equiv a$ means that the function $f(x)$ is equal to a at all points where $f(x)$ is defined.

A function (mapping) f from X to Y is an *injection* if the equality $f(x) = f(y)$ implies the equality $x = y$ for any elements x and y from X, i.e., different elements from X are mapped into different elements from Y.

A function (mapping) f from X to Y is a *projection* also called *surjection* if for any y from Y there is x from X such that $f(x) = y$.

A function (mapping) f from X to Y is a *bijection* if it is both a projection and injection.

A function (mapping) f from X to Y is an *inclusion* if the equality $f(x) = x$ holds for any element x from X.

Two important concepts of mathematics are the domain and range of a function. However, there is some ambiguity for the first of them. Namely, there are two distinct meanings in current mathematical usage for this concept. In the majority of mathematical areas, including the calculus and analysis, the term "domain of f" is used for the set of all values x such that $f(x)$ is defined. However, some mathematicians (in particular, category theorists), consider the domain of a function $f: X \to Y$ to be X, irrespective of whether $f(x)$ is defined for all x in X. To eliminate this ambiguity, we suggest the following terminology consistent with the current practice in mathematics.

$F(X, Y)$ is the set of all mappings from X into Y. $PF(X, Y)$ is the set of all partial mappings from X into Y. Naturally, $F(X, Y) \subseteq PF(X, Y)$.

If f is a function from X into Y, then the set X is called the *domain* of f (it is denoted by Dom f) and Y is called the *codomain* of T (it is denoted by Codom f). The *range* Rg f of the function f is the set of all elements from Y assigned by f to, at least, one element from X, or formally, Rg $f = \{y;\ \exists\, x \in X\ (f(x) = y)\}$. The *domain of definition* also called the *definability domain*, DDom f, of the function f is the set of all elements from X that related by f to, at least, one element from Y is or formally, DDom $f = \{x;\ \exists\, y \in Y\ (f(x) = y)\}$. Thus, for a partial function f, its domain of definition DDom f is the set of all elements for which $f(x)$ is defined.

Taking two mappings (functions) $f: X \to Y$ and $g: Y \to Z$, it is possible to build a new mapping (function) $gf: X \to Z$ that is called the (*sequential*) *composition* or *superposition* of mappings (functions) f and g and defined by the rule $gf(x) = g(f(x))$ for all x from X.

For any set S, $\chi_S(x)$ is its *characteristic function*, also called *set indicator function*, if $\chi_S(x)$ is equal to 1 when $x \in S$ and is equal to 0 when $x \notin S$, and $C_S(x)$ is its partial characteristic function if $C_S(x)$ is equal to 1 when $x \in S$ and is undefined when $x \notin S$.

If $f: X \to Y$ is a function and $Z \subseteq X$, then the restriction $f|_Z$ of f on Z is the function defined only for elements from Z and $f|_Z(z) = f(z)$ for all elements z from Z.

Sequential composition or superposition of binary relations defines sequential composition or superposition of functions, i.e., if and are functions, then.

Sequential composition or superposition of binary relations defines sequential composition or superposition of functions, i.e., if f: $X \to Y$ and g: $Y \to Z$ are functions (mappings), then the mapping (function) $g \circ f$: $X \to Z$ is called the (*sequential*) *composition* or *superposition* of functions (mappings) f and g and defined by the rule $(g \circ f)(x) = g(f(x))$ for all x from X.

A real function f: $\boldsymbol{R} \to \boldsymbol{R}$ is *bounded* if there is a real number c such that $|f(x)| < c$ for all elements x from $\boldsymbol{R}$.

For real functions, there are also *arithmetical compositions*:

$$(g+f)(x) = f(x) + g(x)$$

$$(g \times f)(x) = f(x) \times g(x)$$

A real function f is called *Lipschitz continuous* or has the *Lipschitz property* if there exists a real number $K \geq 0$, such that for all real numbers x and y, we have

$$|f(x) - f(y)| \leq K|x - y|$$

A real function f is called a *contraction* or *contraction function* if there exists a real number k, such that $0 \leq k \leq 1$ and for all real numbers x and y, we have

$$|f(x) - f(y)| < k|x - y|$$

i.e., a contraction is Lipschitz continuous for $K < 1$.

A real function f is called *periodic* if there exists a real number $k > 0$, such that for all real numbers x, we have $f(x) = f(x + k)$.

The *ceiling function* $\lceil x \rceil$ returns the smallest integer that is greater than or equal to x.

The *floor function* $\lfloor x \rfloor$ returns the largest integer that is less than or equal to x.

The *round function* $[x]$ returns the nearest integer to x, while when there are two such integers the larger of them is taken as the value of the function.

A function f from a partially ordered set X to a partially ordered set X is called *monotone* (*antitone*) if $x \leq y$ implies $f(x) \leq f(y)$ ($f(y) \leq f(x)$) for any elements x and y from X.

If U is a correspondence of a set X to a set Y (a binary relation between X and Y), i.e., $U \subseteq X \times Y$, then $U(x) = \{y \in Y;\ (x, y) \in U\}$ and $U^{-1}(y) = \{x \in X;\ (x, y) \in U\}$.

An n-ary relation R in a set X is a subset of the nth power of X, i.e., $R \subseteq X^n$. If $(a_1, a_2, \ldots, a_n) \in R$, then one says that the elements $a_1, a_2, \ldots, a_n$ from X are in relation R.

2. Logical Concepts and Structures

If P and Q are two statements, then P $\rightarrow$ Q means that P implies Q and P $\leftrightarrow$ Q means that P is equivalent to Q.

Logical operations:

- *negation* is denoted by $\neg$ or by $\sim$,
- *conjunction* also called logical "and" is denoted by $\wedge$ or by & or by $\cdot$,
- *disjunction* also called logical "or" is denoted by $\vee$,
- *implication* is denoted by $\rightarrow$ or by $\Rightarrow$ or by $\supset$,
- *equivalence* is denoted by $\leftrightarrow$ or by $\equiv$ or by $\Leftrightarrow$.

The logical symbol $\forall$ is called the *universal quantifier* and means "for any."

The logical symbol $\exists$ is called the *existential quantifier* and means "there exists."

Logical formulas and operations allow mathematicians and logicians to work both with finite and infinite systems. However, to do this in a more constructive form when the number of elements in the system is very big or infinite, mathematicians use the Principle of Induction.

Descriptive Principle of Induction. Given an infinite (very big) system $R = \{x_1, x_2, \ldots, x_n, \ldots\}$ of elements enumerated by natural numbers and a predicate P defined for these elements, if it is proved that $P(x_1)$ is true and assuming for an arbitrary number n, that all $P(x_1), P(x_2), \ldots, P(x_{n-1})$ are true, it is also proved that $P(x_n)$ is true, then $P(x)$ is true for all elements from R.

Constructive Principle of Induction. If an infinite (very big) system $R = \{x_1, x_2, \ldots, x_n, \ldots\}$ of elements enumerated by natural numbers is described by some property represented by a predicate P defined for elements of R, then if there is an algorithm (constructive method) A that builds x_1 and assuming that for an arbitrary number n, all $x_1, x_2, \ldots, x_{n-1}$ are built, A can also build x_n, then can (potentially) build the whole R, i.e., R exists potentially.

3. Topological Concepts and Structures

A *topology* in a set X is a system O(X) of subsets of X that are called *open subsets* and satisfy the following axioms:

T1. $X \in \mathrm{O}(X)$ and $\varnothing \in \mathrm{O}(X)$.

T2. For all subsets A and B of X, if $A, B \in \mathrm{O}(X)$, then $A \cap B \in \mathrm{O}(X)$.

T3. For all subsets A_i of X ($i \in I$), if all $A_i \in \mathrm{O}(X)$, then $\bigcup_{i \in I} A_i \in \mathrm{O}(X)$.

A set X with a topology in it is called a *topological space*.

The complement of an open set is called a *closed set*.

If A is a subset of a topological space, then its *closure* CA is the least closed set that contains A.

If A is a subset of a topological space, then its *interior* Int$A$ is the union of all open subsets of A.

A set $\boldsymbol{B}$ of open sets in X is case a *base*, or a *topological basis*, of a topology τ if any open set in τ is the union of sets from $\boldsymbol{B}$.

Topology in a set can be also defined by a system of neighborhoods of points from this set. In this case, a set is *open* in this topology if it contains a standard neighborhood of each of its points. For instance, if a is a real number and $t \in \boldsymbol{R}^{++}$, then an open interval $\mathrm{O}_t a = \{ x \in \boldsymbol{R}; a - t < x < a + t \}$ is a standard neighborhood of a.

To define a topology, a system of sets has to satisfy the following *neighborhood axioms* (Kuratowski, 1966).

NB1. Any neighborhood of a point $x \in X$ contains this point.

NB2. For any two neighborhoods $\mathrm{O}_1 x$ and $\mathrm{O}_2 x$ of a point $x \in X$, there is a neighborhood Ox of x that is a subset of the intersection $\mathrm{O}_1 x \cap \mathrm{O}_2 x$.

NB3. For any neighborhood Ox of a point $x \in X$ and a point $y \in \mathrm{O}x$, there is a neighborhood Oy of y that is a subset of Ox.

A *local base* at a point x from L is a set $\boldsymbol{LB}$ of neighborhoods of x such that any neighborhood of x contains an element from $\boldsymbol{LB}$.

One more way to define topology in a set is to use the *closure operation* (Kuratowski, 1966).

A topology σ in a set X is *stronger* than topology τ in the same set X if any open in the topology τ set is also open in the topology σ.

If X is a subset of a topological space, then Cl(X) denotes the *closure* of the set X.

A *cluster point* of a set A in a topological space X is a point any neighborhood of which contains infinitely many elements from A.

A system $R = \{X_i; i \in I\}$ of (open) sets X_i in a topological space X is called an (*open*) *cover* of a subset K of X if the union of these sets contains K, i.e., $K \subseteq \cup_{i \in I} X_i$.

A subset K of a topological space X is called *compact* if any open cover of K contains a finite subset, which is also an open covering of K.

A topological space X can satisfy the following axioms (Kelly, 1957):

$\mathbf{T_0}$ (the *Kolmogorov Axiom*). $\forall x, y \in X\, (\exists Ox\, (y \notin Ox) \vee \exists\, Oy\, (x \notin Oy))$.

In other words, for every pair of points a and b there exists an open set U in O such that at least one of the following statements is true: (1) a belongs to

U and *b* does not belong to *U*, and (2) *b* belongs to *U* and *a* does not belongs to *U*.

$\mathbf{T}_1$ (the *Alexandroff Axiom*). $\forall x, y \in X \exists\ Ox\ \exists Oy\ (x \notin Oy\ \&\ y \notin Ox)$.

In other words, for every pair of points *a* and *b* there exists an open set U such that U contains *a* but not *b*. To say that a space is T_1 is equivalent to saying that sets consisting of a single point are closed.

$\mathbf{T}_2$ (the *Hausdorff Axiom*). $\forall x, y \in X \exists Ox\ \exists Oy\ (Ox \cap Oy = \varnothing)$.

In other words, for every pair of points *a* and *b* there exist disjoint open sets which separately contain *a* and *b*. In this case, open sets *separate* points.

Here *Ox*, *Oy* are some neighborhoods of *x* and *y*, respectively.

A topological space, which satisfies the axiom $\mathbf{T}_i$, is called a $\mathbf{T}_i$-*space*. Each axiom $\mathbf{T}_{i+1}$ is stronger than axiom $\mathbf{T}_i$. $\mathbf{T}_0$– *spaces* are also called the *Kolmogorov spaces*. $\mathbf{T}_1$ – *spaces* are also called the *Fréchet spaces*. $\mathbf{T}_2$ – spaces are also called the *Hausdorff spaces* (Kelly, 1957).

There are also $\mathbf{T}_3$ – *spaces* or *regular spaces*, in which for every point *a* and closed set B there exist disjoint open sets which separately contain *a* and B. That is, points and closed sets are separated. Many authors require that $\mathbf{T}_3$-spaces also be $\mathbf{T}_0$ – spaces, since with this added condition, they are also $\mathbf{T}_2$ – spaces (Alexandroff, 1961).

There are also $\mathbf{T}_4$ – *spaces* or *normal spaces*, in which for every pair of closed sets A and B there exist disjoint open sets which separately contain A and B. That is, points and closed sets are separated. Many authors require that $\mathbf{T}_4$ – spaces also satisfy Axiom $\mathbf{T}_1$ (Alexandroff, 1961).

A mapping $f: X \to Y$ is called *continuous* if the inverse image of any open set in the topological space *Y* is an open set in the topological space *X*.

A mapping $f: X \to Y$ is called *sequentially continuous* if the image of any converging sequence in *X* is a converging sequence in *Y*.

A sequence $\{a_i;\ i = 1, 2, 3, \dots\}$ of real (complex) numbers is a Cauchy sequence if for any $\varepsilon \in \boldsymbol{R}^{++}$, there is $n \in \boldsymbol{N}$ such that for any $x \in X$ and any i, $j \geq n$, we have $|\, a_j - a_i \,| < \varepsilon$.

A sequence $\{a_i;\ i = 1, 2, 3, \dots\}$ of real (complex) numbers converges to a number *a* if

$$\lim_{i\to\infty} |\, a_i - a \,| = 0$$

4. Algebraic Concepts and Structures

An *algebraic system* is a structure $A = (X, \Omega, R)$, which consists of: a non-empty set *X* called the *carrier* or the *underlying set* of *A* and elements of which are called the elements of *A*: a family Ω of algebraic operations, which are mappings $\omega_i: X^{ni} \to X\ (i \in I)$; and a family *R* of relations $r_j \subseteq X^{mj}$ (*j*

$\in J$) defined on X. The non-negative integers n_i and m_j are called the *arities* of the respective operations ω_i and relations r_j. When ω is an operation from Ω with arity n, then the image $\omega(a_1, a_2, \ldots, a_n)$ of the element $(a_1, a_2, \ldots, a_n)$ from X^n under the mapping $\omega: X^n \to X$ is called the *value* of the operation ω for elements $a_1, a_2, \ldots, a_n$.

Operations in Ω are usually of two types – inner operations and outer operations.

An *algebraic system* in which there are only relations is called a *model*.

An *algebraic system* in which there are only operations is called a *universal algebra*.

Below definitions of basic universal algebras are given.

A *semigroup* S is a set with an associative binary operation.

A *monoid* S is a semigroup with an identity (unit) element.

A monoid in which all elements have inverse elements is called a *group*. It is possible to consider a group as a universal algebra with one binary operation usually called multiplication, one unary operation, which assigns the inverse to each element, and one nulary operation, which specifies the identity (unit) element.

A *monoid* S is a semigroup with an identity (unit) element.

An *abelian group* A is a group in which the binary operation is commutative.

A *semiring* K is a set with two operations:

addition: $K \times K \to K$ denoted by $x + y$ where x and y belong to K;

multiplication: $K \times K \to K$ denoted by xy where x and y belong to K.

These operations satisfy the following axioms:

1. *Addition is associative*:
   For all x, y, z from K, we have $x + (y + z) = (x + y) + z$.
2. *Addition is commutative*:
   For all x, y from K, we have $x + y = y + x$.
3. *Addition has an identity element*:
   There exists an element 0 from K, called the *zero*, such that $x + 0 = x$ for all x from K.
4. *Multiplication is distributive over addition*:
   For all elements x, y, z from K, we have

$$x(y + z) = xy + xz$$

and

$$(x + y)z = xz + yz$$

The set $\boldsymbol{R}^{+}$ of all nonnegative real numbers and the set $\boldsymbol{W}$ of all whole numbers are semirings.

A universal algebra H that has all properties of a semiring except zero is called an S-ring.

The set $\boldsymbol{R}^{++}$ of all positive real numbers and the set $\boldsymbol{N}$ of all natural numbers are S-rings.

A (left) semimodule M over a semiring K has two operations:

addition: $M\times M \to M$ denoted by $x + y$ where x and y belong to M;

multiplication: $K\times M \to M$ denoted by ax where x belongs to M and a belongs to K.

These operations satisfy the following axioms:

1. *Addition is associative*:
   For all x, y, z from M, we have $x + (y + z) = (x + y) + z$.
2. *Addition is commutative*:
   For all x, y from M, we have $x + y = y + x$.
3. *Addition has an identity element*:
   There exists an element 0 from M, called the *zero*, such that $x + 0 = x$ for all x from M.
4. *Multiplication is distributive over addition* in M:
   For all elements x, from K and y, z from M, we have

$$x(y + z) = xy + xz$$

5. *Multiplication is distributive over addition* in K:
   For all elements x, from M and y, z from K, we have

$$(z + y)x = zx + yx$$

In a *right semimodule M*, elements from M are multiplied from the right by elements from K.

The set $\boldsymbol{R}^{+}$ of all nonnegative real numbers is a left (right) semimodule over the semiring of all whole numbers $\boldsymbol{W}$.

A universal algebra H that has all properties of a semimodule over an S-ring except zero is called an S-module.

The set $\boldsymbol{R}^{++}$ of all positive real numbers is a left (right) S-module over the S-ring of all natural numbers $\boldsymbol{N}$.

A *ring K* is a set with two operations:

addition: $K\times K \to K$ denoted by $x + y$ where x and y belong to K;

multiplication: $K \times K \to K$ denoted by xy where x and y belong to K.

These operations satisfy the following axioms:

5. *Addition is associative*:
   For all x, y, z from K, we have $x + (y + z) = (x + y) + z$.
6. *Addition is commutative*:
   For all x, y from K, we have $x + y = y + x$.
7. *Addition has an identity element*:
   There exists an element 0 from K, called the *zero*, such that $x + 0 = x$ for all x from K.
8. *Addition has inverse element*:
   For any x from K, there exists an element z from K, called the *additive inverse* of x, such that $x + z = 0$.
9. *Multiplication is distributive over addition*:
   For all elements x, y, z from K, we have

$$x(y + z) = xy + xz$$

and

$$(x + y)z = xz + yz$$

The sets of functions $F(\boldsymbol{R})$, $C(\boldsymbol{R})$, $F[a, b]$ and $C[a, b]$ are rings.

A *(left) module* M over a ring K has two operations:

addition: $M \times M \to M$ denoted by $x + y$ where x and y belong to M;

multiplication: $K \times M \to M$ denoted by ax where x belongs to M and a belongs to K.

These operations satisfy the following axioms:

6. *Addition is associative*:
   For all x, y, z from M, we have $x + (y + z) = (x + y) + z$.
7. *Addition is commutative*:
   For all x, y from M, we have $x + y = y + x$.
8. *Addition has an identity element*:
   There exists an element 0 from M, called the *zero*, such that $x + 0 = x$ for all x from M.
9. *Addition has inverse element*:
   For any x from M, there exists an element z from M, called the *additive inverse* of x, such that $x + z = 0$.

10. *Multiplication is distributive over addition* in M:
    For all elements x, from K and y, z from M, we have

$$x(y + z) = xy + xz$$

11. *Multiplication is distributive over addition* in K:
    For all elements x, from M and y, z from K, we have

$$(z + y)x = zx + yx$$

In a *right module M*, elements from M are multiplied from the right by elements from K.

A *field* $\boldsymbol{F}$ is a ring in which:

1. *Multiplication is associative*:
   For all x, y, z from $\boldsymbol{F}$, we have $x(yz) = (xy)z$.
2. *Multiplication is commutative*:
   For all x, y from $\boldsymbol{F}$, we have $xy = yx$.
3. *Multiplication has an identity element*:
   There exists an element e from $\boldsymbol{F}$, such that $xe = ex = x$ for all x from $\boldsymbol{F}$.
4. *Multiplication has inverse element*:
   For any x from K, there exists an element z from $\boldsymbol{F}$, called the *multiplicative inverse* of x, such that $xz = zx = e$.

Sets $F(\boldsymbol{R})$ and $F[a, b]$ are modules over rings $C(\boldsymbol{R})$ and $C[a, b]$, respectively.

A *linear space* or a *vector space* or a *linear vector space* L over a field $\boldsymbol{F}$ has two operations:

addition: $L \times L \to L$ denoted by $x + y$ where x and y belong to L;

scalar multiplication: $\boldsymbol{F} \times L \to L$ denoted by ax where $a \in \boldsymbol{F}$ and $x \in L$.

These operations satisfy the following axioms:

1. Addition is associative:
   For all x, y, z from L, we have $x + (y + z) = (x + y) + z$.
2. Addition is commutative:
   For all x, y from L, we have $x + y = y + x$.
3. Addition has an identity element:

There exists an element 0 from L, called the *zero vector*, such that $x + 0 = x$ for all x from L.

4. Addition has an inverse element:
For any x from L, there exists an element z from L, called the *additive inverse* of x, such that $x + z = 0$.
5. Scalar multiplication is distributive over addition in L:
For all elements a from $\boldsymbol{F}$ and vectors y, z from L, we have

$$a(y + z) = ay + az$$

6. Scalar multiplication is distributive over addition in $\boldsymbol{F}$:
For all element elements a, b from $\boldsymbol{F}$ and any vector $\boldsymbol{y}$ from L, we have

$$(a + b)y = ay + by$$

7. Scalar multiplication is compatible with multiplication in $\boldsymbol{F}$:
For all elements a, b from $\boldsymbol{F}$ and any vector y from L, we have

$$a(by) = (ab)y.$$

8. The identity element 1 from the field $\boldsymbol{F}$ also is an identity element for scalar multiplication:
For all vectors $\boldsymbol{x}$ from L, we have $1\boldsymbol{x} = \boldsymbol{x}$.

The sets of functions $F(\boldsymbol{R})$, $C(\boldsymbol{R})$, $F[a, b]$ and $C[a, b]$ are vector spaces over the field $\boldsymbol{R}$. The sets of functions $F(\boldsymbol{C})$, $C(\boldsymbol{C})$ and $F([a, b], \boldsymbol{C})$ are vector spaces over the field $\boldsymbol{C}$.

Vectors $\boldsymbol{x}_1, \boldsymbol{x}_2, \ldots, \boldsymbol{x}_n$ from L are called *linearly dependent* in L if any there is an equality $\Sigma_{i=1}{}^n a_i \boldsymbol{x}_i = 0$ where a_i are elements from $\boldsymbol{F}$ and not all of them are equal to 0. When there are no such an equality, vectors $\boldsymbol{x}_1, \boldsymbol{x}_2, \ldots, \boldsymbol{x}_n$ are called *linearly independent*.

A system B of linearly independent vectors from L is called a *basis* or *Hamel basis* of the vector space L if any element $\boldsymbol{x}$ from L is equal to a sum $\Sigma_{i=1}{}^n a_i \boldsymbol{x}_i$ where n is some natural number, $\boldsymbol{x}_i$ are elements from B and a_i are elements from $\boldsymbol{F}$.

The number of elements in a basis is called the *dimension* of the vector space L. It is proved that all bases of the same space have the same number of elements. The number of elements in a basis is called the *dimension* of the space L. It is proved that for each vector space L, its dimension is defined in a unique way.

A subset M of a vector space L is called a *linear* (*vector*) *subspace* of L if M is closed with respect to addition and scalar multiplication.

If X are Y are vector spaces over $\boldsymbol{R}$, then a mapping $f: X \to Y$ is called *linear* if $f(c \cdot u + d \cdot v) = c \cdot f(u) + d \cdot f(v)$ for any elements u and v from X and any elements c and d from $\boldsymbol{F}$.

$L(X, Y)$ is the set of all linear mappings from X into Y. $PL(X, Y)$ is the set of all partial linear mappings from X into Y. Naturally, $L(X, Y) \subseteq PL(X, Y)$.

If X is a space of functions and Y is equal to $\boldsymbol{R}$, then a mapping $f: X \to \boldsymbol{R}$ is called a *real functional*.

If X is a space of functions and Y is equal to $\boldsymbol{C}$, then a mapping $f: X \to \boldsymbol{C}$ is called a *complex functional*.

If X and Y are spaces of functions, then a mapping $f: X \to Y$ is called an *operator*.

The space $\boldsymbol{R}$ is a one-dimensional vector (linear) space over itself. The space $\boldsymbol{R}^n$ is an n-dimensional vector (linear) space over $\boldsymbol{R}$.

A linear space A over $\boldsymbol{R}$ is called a *linear algebra* over $\boldsymbol{R}$ if a binary operation called multiplication is also defined in A and this operation satisfies the following additional axioms:

1. *Multiplication is distributive over addition* in A:
   For all elements x, y and z from A, we have

$$x(y + z) = xy + xz$$

2. *Multiplication is distributive over addition* in $\boldsymbol{R}$:
   For all elements x from M and y, z from $\boldsymbol{R}$, we have

$$(z + y)x = zx + yx$$

The sets of functions $F(\boldsymbol{R})$, $C(\boldsymbol{R})$, $F[a, b]$ and $C[a, b]$ are linear algebras over the field $\boldsymbol{R}$.

Note that any linear algebra is also a ring and thus, it is possible to consider modules over linear algebras.

5. Notations from the Theory of Hypernumbers and Extrafunctions

$\boldsymbol{R}^\omega$ is the set of all sequences of real numbers.

If X and Y are topological spaces, then $F(X, Y)$ is the set of all and $C(X, Y)$ is the set of all continuous mappings from X into Y.

If $\boldsymbol{a} = (a_i)_{i \in \omega}$ is a sequence of real numbers, then $\alpha = \mathrm{Hn}(a_i)_{i \in \omega}$ is the real hypernumber determined by $\boldsymbol{a}$.

$\boldsymbol{R}_\omega$ is the set of all real hypernumbers.

$\boldsymbol{R}_\omega^+$ is the set of all real hypernumbers that are larger than or equal to zero.

$\boldsymbol{C}_\omega$ is the set of all complex hypernumbers.

$\boldsymbol{C}_\omega^+$ is the set of all complex hypernumbers that are larger than or equal to zero.

$F(\boldsymbol{R}_{\omega,} \boldsymbol{R}_\omega)$ is the set of all (general) real pointwise extrafunctions.

$C(\boldsymbol{R}_{\omega}, \boldsymbol{R}_\omega)$ of all continuously represented (general) real pointwise extrafunctions.

If $\{f_i; i\in\omega\}$ is a sequence of real functions, then $f = \text{Ep}\{f_i; i\in\omega\}$ is the real restricted pointwise extrafunction determined by the sequence $\{f_i; i\in\omega\}$.

$F(\boldsymbol{R}, \boldsymbol{R}_\omega)$ is the set of all restricted real pointwise extrafunctions.

$C(\boldsymbol{R}, \boldsymbol{R}_\omega)$ of all continuously represented restricted complex pointwise extrafunctions.

$F(\boldsymbol{C}, \boldsymbol{C}_\omega)$ is the set of all restricted complex pointwise extrafunctions.

$C(\boldsymbol{C}, \boldsymbol{C}_\omega)$ is the set of all continuously represented restricted complex pointwise extrafunctions.

If $\mathbf{F}$ is a class of functions, then $\mathbf{F}^\omega$ is the set of all sequences of functions from $\mathbf{F}$.

If $\mathbf{F}$ is a class of functions, then $\mathbf{E}^{\mathbf{F}}{}_{\omega Q}$ is the set of all represented in $\mathbf{F}$ Q-based extrafunctions.

If $\{f_i; i\in\omega\}$ is a sequence of functions from a class of functions $\mathbf{F}$, then $f = \text{EF}_Q(f_i)_{i\in\omega}$ is the represented in $\mathbf{F}$ Q-based extrafunction determined by the sequence $\{f_i; i\in\omega\}$;

$\mathbf{E}^{F(\boldsymbol{R})}{}_\omega Q_{\text{pt}}$ is the set of all represented in $\mathbf{F}$ Q_{pt}-based real extrafunctions.

$\mathbf{E}^{\mathbf{F}}{}_\omega Q_{\text{comp}}$ is the set of all represented in $\mathbf{F}$ Q_{comp}-based real extrafunctions.

$\mathbf{E}^{BI(\boldsymbol{R})}{}_\omega Q_{\text{cp}}$ is the set of all represented in $\mathbf{F}$ Q_{cp}-based real extrafunctions.

$\mathbf{D}_{\mathbf{K}\omega}$ is the class of all real extended distributions with respect to $\mathbf{K}$. It is also denoted by $\mathbf{E}^{\mathbf{F}}{}_\omega Q_{\mathbf{K}}$.

If $\{f_i; i\in\omega\}$ is a sequence of real functions, then $f = \text{Ec}\{f_i; i\in\omega\}$ is the real compactwise extrafunction determined by the sequence $\{f_i; i\in\omega\}$.

$\text{Comp}(\boldsymbol{R}, \boldsymbol{R}_\omega)$ is the set of all real compactwise extrafunctions.

If $\mathbf{K}$ is a class of operators (functionals), then $\mathbf{K}^\omega$ is the set of all sequences of operators (functionals) from $\mathbf{K}$.

If $\mathbf{K}$ is a class of operators (functionals), then $\mathbf{E}^{\mathbf{K}}{}_{\omega Q}$ is the set of all represented in $\mathbf{K}$ Q-based hyperoperators (hyperfunctionals).

If $\{f_i; i\in\omega\}$ is a sequence of operators (functionals) from a class of operators (functionals) $\mathbf{K}$, then $f = \text{EF}_Q(f_i)_{i\in\omega}$ is the real represented in $\mathbf{K}$ Q-based hyperoperators (hyperfunctionals) determined by the sequence $\{f_i; i\in\omega\}$.

REFERENCES

1. Abello, J. M., Pardalos, P. M., & Resende, M. G. C. (Eds.). *Handbook of Massive Data Sets*, Springer, Berlin/New York, 2002.
2. Albert, A. A. (1947). Absolute valued real algebras, *Ann. of Math.*, *48*, 495–501.
3. Albert, A. A. (1949). Absolute valued algebraic algebras, *Bull. Amer. Math. Soc.*, *55*, 763–768.
4. Albeverio, S., Hoegh-Krohn, R., & Mazzucchi, S. *Mathematical Theory of Feynman Path Integral*, Lecture Notes in Mathematics, 523, Springer-Verlag, Berlin/New York, 2008.
5. Albeverio, S., Khrennikov, A. Yu., & Shelkovichs, V. M. *The Theory of P-adic Distributions: Linear and Nonlinear Models*, Cambridge University Press, New York/Cambridge, UK, 2010.
6. Albiac, F., & Kalton, N. J. (2009). Lipschitz structure of quasi-Banach spaces, *Israel J. Math.*, *170*, 317–335.
7. Albiac, F., & Leránoz, C. (2010). Drops in quasi-Banach spaces, *J. Geom. Anal.*, *20*, 525–537.
8. Albiac, F., & Leránoz, C. (2010a). Uniqueness of unconditional basis in quasi-Banach spaces which are not sufficiently Euclidean, *Positivity*, *14*, 579–584.
9. Alexandroff, P. *Elementary Concepts of Topology*, Dover Publications, New York, 1961.
10. Ali, M. K. *Differential Calculus for Locally Convex Topological Vector Spaces*, University of Oklahoma, Norman, 1965.
11. Alsina, C., Schweizer, B., & Sklar, A. (1993). On the definition of probabilistic normed space, *Aequationes Math.*, *46*, 91–98.
12. Alsina, C., Schweizer, B., Sempi, C., & Sklar, A. (1997). On the definition of a probabilistic inner product space, *Rendiconti di Mathematica*, *17*, 115–127.
13. Anderson, R. A. (1986). Almost implies near, *Trans. Amer. Math. Soc.*, *296*, 229–237.
14. Antosik, P., Mikulinski, J., & Sikorski, R. *Theory of Distributions*, Elsevier, Amsterdam, 1973.
15. Aoki, T. (1942). Locally bounded linear topological spaces, *Proc. Imp. Acad.*, 18, No. *10*, 588–594.
16. Arens, R. (1946). The space L_ω and convex topological rings, *Bull. Amer, Math. Soc.*, *52*, 931–935.
17. Arens, R. (1952). A generalization of normed rings, *Pacific J. Math.*, 2, 455–471.
18. Attouch, H., Baillon, J.-B., & Théra, M. (1994). Variational sum of monotone operators, *J. Convex Anal.*, *1*, 1–29.
19. Bachman, G., & Narici, L. *Functional Analysis*, Academic Press Inc., San Diego, CA, 1966.
20. Baeck, T., Fogel, D. B., & Z. Michalewicz, Z. *Evolutionary Computation* 1: *Basic Algorithms and Operators*, CRC Press, New York, 2000.
21. Bag, T., & Samanta, S. K. (2003). Finite dimensional fuzzy-normed linear spaces, *The Journal of Fuzzy Mathematics*, *3*, 687–705.

22. Bagchi, B., & Misra, G. (2001). Homogeneous operators and projective representations of the Möbius group: A survey, *Proceedings of the Indian Academy of Sciences – Mathematical Sciences*, *111*(4), 415–437.
23. Ball, J. A., Bolotnikov, V., Helton, J. W., & Rodman, L. (Eds.). *Topics in Operator Theory* (Operator Theory: Advances and Applications), Birkhäuser Verlag, Basel, Switzerland, 2010.
24. Banach, S. (1922). Sur les opérations dans les ensembles abstraits et leur application aux équations intégrals, *Fundamenta Mathematicae*, *3*, 133–181.
25. Banach, S. (1922a). Sur les fonctions dérivées des fonctions mesurables, *Fundamenta Mathematicae 3*, 128–132.
26. Banach, S. (1929). Sur les fonctionnelles linéaires, *Studia Mathematica*, *1*, 211–216.
27. Banach, S. (1929a). Sur les fonctionnelles linéaires II, *Studia Mathematica*, *1*, 223–239.
28. Banach, S. (1932). *Théory de opérations linéares*, Monografie Matematyczne, 1, Warszawa.
29. Bartle, R. G., & Sherbert, D. R. (1992). *Introduction to Real Analysis*, John Wiley & Sons, Inc., New York.
30. Beck, A., & Teboulle, M. (2009). A fast iterative shrinkage-thresholding algorithm for linear inverse problems. *SIAM J. Imaging Science*, *2*, 183–202.
31. Beckenstein, E., Narici, L., & Suffel, C. (1977). *Topological algebras,* North-Holland, Amsterdam, New York and Oxford.
32. Beer, G., & Levi, S. (2009). Total boundedness and bornologies, *Topology and Its Applications*, *156*(7), 1271–1288.
33. Bellert, S. (1963). On the continuation of the idea of Heaviside of the operational calculus, *Journ Francl. Inst.*, *276*(5), 411–440.
34. Berg, L. (1962). Einfürung in die Operatorenrechnung, Berlin.
35. Berg, L. (1962a). Asymptotische Auffassung der Operatorenrechnung, *Studia Math.*, *21*, 215–229.
36. Berg, L. (1978). Construction of distribution algebras, *Math. Nachr. 82*, 255–262.
37. Beurling, A. (1961). *Quasi-Analyticity and General Distributions*, A. M. S. Summer Institute, Stanford.
38. Björck, G. (1966). Linear partial differential operators and generalized distributions, *Ark. Mat.*, *6*, 351–407.
39. Blackadar, B. (2005). Operator Algebras: Theory of C*-Algebras and von Neumann Algebras, *Encyclopedia of Mathematical Sciences*, Springer-Verlag, Berlin/New York.
40. Blank, J., Exner, P., & Havliček, M. *Hilbert Space Operators in Quantum Physics*, American Institute of Physics, 1994.
41. Bolzano, B. (1851). *Paradoxien des Unendlichen*, C. H. Reclam (translation: Paradoxes of the Infinite, in: Ewald, W. B., (ed.). From Kant to Hilbert: A Source Book in the Foundations of Mathematics, 2 volumes, Oxford University Press, 1996).
42. Bollini, C. G., & Rocca, M. C. (2007). Convolution of Ultradistributions, Field Theory, Lorentz Invariance and Resonances, *Int. J. Theor. Phys.*, *46*, 3030–3042.
43. Bollini, C. G., Escobar, T., & Rocca, M. C. (1999). Convolution of Ultradistributions and Field Theory, *Int. J. of Theor. Phys.*, *38*, 2315–2327.
44. Bonsal, F. F., & Duncan, J. *Complete normed algebras*, Springer-Verlag, Berlin, 1973.
45. Boole, G. *Treatise on the Calculus of Finite Differences*, Dover Publications, New York, 2003.
46. Borel, E. *Le∩ons sur la Théorie des Fonctions,* Gauthier-Villars, Paris, 1927.
47. Borel, E. *Les Nombre Inaccessible*, Gauthier-Villiars, Paris, 1952.

48. Borelli, C., & Forti, G. L. (1995). On a general Hyers–Ulam stability result, *Internat. J. Math. Math. Sci.*, *18*, 229–236.
49. Bourbaki, N. *Espaces Vectoriels Topologiques*, Hermann, Paris, 1953–1955 (Bourbaki, N. *Topological vector spaces*, Springer-Verlag, Berlin/New York, 1987).
50. Bourgin, D. G. (1941). Some Properties of Real Linear Topological Spaces, *Proc. Natl. Acad. Sci. USA*, *27*(11), 539–544.
51. Bourgin, D. G. (1943). Linear topological spaces, *Amer. J. Math.*, *65*, 637–659.
52. Bourgin, D. G. (1951). Classes of transformations and bordering transformations, *Bull. Amer. Math. Soc.*, *57*, 223–237.
53. Bremermann, J. H. *Distributions, complex variables, and Fourier transforms*, Addison-Wesley, Reading, Massachusetts, 1965.
54. Brenner, J., & Burgin, M. (2011). Information as a Natural and Social Operator, *Information: Theories & Applications*, *18*(1), 33–49.
55. Brown, A., & Pearcy, C. *Introduction to Operator Theory* I: *Elements of Functional Analysis*, Springer-Verlag, New York/Heidelberg/Berlin, 1977.
56. Burgin, M. (1976). Recursion Operator and Representability of Functions in the Block-Scheme Language, *Programming and Computer Software*, *2*(4), 13–23.
57. Burgin, M. S. (1980). Functional equivalence of operators and parallel computations, *Programming and Computer Software*, *6*(6), 283–294.
58. Burgin, M. (1982). Products of operators of multidimensional structured model of systems, *Mathematical Social Sciences, 2*, 335–343.
59. Burgin, M. (1993). Differential calculus for extrafunctions, *Notices of the National Academy of Sciences of Ukraine*, *11*, 7–11.
60. Burgin, M. (1993a). Fuzzy continuous functions in object classification, 1st *European Congress on fuzzy and intelligent technologies*, Proceedings, Aachen, 1189–1195.
61. Burgin, M. (1995). Integral calculus for extrafunctions, *Notices of the National Academy of Sciences of Ukraine*, *11*, 14–17.
62. Burgin, M. (1995a). Neoclassical Analysis: Fuzzy Continuity and Convergence, *Fuzzy Sets and Systems*, *75*, 291–299.
63. Burgin, M. (1999). General Approach to Continuity Measures, *Fuzzy Sets and Systems*, *105*(2), 225–231.
64. Burgin, M. (2000). Theory of Fuzzy Limits, *Fuzzy Sets and Systems*, *115*(3), 433–443.
65. Burgin, M. (2002). Theory of Hypernumbers and Extrafunctions: Functional Spaces and Differentiation, *Discrete Dynamics in Nature and Society*, *7*, 201–212.
66. Burgin, M. (2004). Discontinuity Structures in Topological Spaces, *International Journal of Pure and Applied Mathematics*, *16*(4), 485–513.
67. Burgin, M. (2004a). Hyperfunctionals and generalized distributions, In: Krinik, A. C., Swift, R. J. (eds.) *Stochastic Processes and Functional Analysis*, A Dekker Series of Lecture Notes in Pure and Applied Mathematics, *238*, 81–119.
68. Burgin, M. *Fuzzy Continuity in Scalable Topology*, Preprint in Mathematics, math/0512627 (subjects: math.GN; math-ph), 2005, 30 p. (electronic edition: http://arXiv.org).
69. Burgin, M. (2005a). Topology in nonlinear extensions of hypernumbers, *Discrete Dyn. Nat. Soc.* *10*(2), 145–170.
70. Burgin, M. (2005b). Hypermeasures in General Spaces, *International Journal of Pure and Applied Mathematics*, *24*(3), 299–323.
71. Burgin, M. (2006). Scalable Topological Spaces, 5th *Annual International Conference on Statistics, Mathematics and Related Fields*, 2006 Conference Proceedings, Honolulu, Hawaii, 1865–1896.

72. Burgin, M. *Fuzzy Limits of Functions*, Preprint in Mathematics, math. CA/0612676, 2006a, 21 p. (electronic edition: http://arXiv.org).
73. Burgin, M. *Neoclassical Analysis: Calculus closer to the Real World*, Nova Science Publishers, New York, 2008.
74. Burgin, M. (2008). Inequalities in series and summation in hypernumbers, in *Advances in Inequalities for Series*, Nova Science Publishers, New York, 89–120.
75. Burgin, M. (2008a). Hyperintegration approach to the Feynman integral, *Integration: Mathematical Theory and Applications*, *1*, 59–104.
76. Burgin, M. *Theory of Information: Fundamentality, Diversity and Unification*, World Scientific, New York/London/Singapore, 2010.
77. Burgin, M. (2010a). Integration in Bundles with a Hyperspace Base: Indefinite Integration, *Integration*, *2*(4), 39–79.
78. Burgin, M. (2010b). Information Operators in Categorical Information Spaces, *Information*, *1*(1), 119–152.
79. Burgin, M. (2010c). Nonlinear Partial Differential Equations in Extrafunctions, *Integration: Mathematical Theory and Applications*, 2(1), 17–50.
80. Burgin, M. (2011). *Theory of Named Sets*, Nova Science Publishers, New York.
81. Burgin, M. (2011a). Epistemic Information in Stratified M-Spaces, *Information*, 2(2), 697–726.
82. Burgin, M. (2011b). Information Dynamics in a Categorical Setting, *in Information and Computation, World Scientific, New York/London/Singapore*, 35–78.
83. Burgin, M. (2011c). *Differentiation in Bundles with a Hyperspace Base*, Preprint in Mathematics, math.CA/1112.3421, 27 p. (electronic edition: http://arXiv.org).
84. Burgin, M. *Hypernumbers and Extrafunctions: Extending the Classical Calculus*, Springer, New York, 2012.
85. Burgin, M. (2012a). Integration in Bundles with a Hyperspace Base: Definite Integration, *Integration: Mathematical Theory and Applications*, *3*(1), 1–54.
86. Burgin, M. (2012b). Functional integrals, path integrals and the theory of hyperintegration, *Integration: Mathematical Theory and Applications*, 3, 269–304.
87. Burgin, M. (2012c). Fuzzy Continuous Functions in Discrete Spaces, *Annals of Fuzzy Sets, Fuzzy Logic and Fuzzy Systems*, *1*(4), 231–252.
88. Burgin, M. (2013). Semitopological Vector Spaces and Hyperseminorms, *Theory and Applications of Mathematics and Computer Science*, *3*(2), 1–35.
89. Burgin, M. (2013a). Fuzzy Continuity of Almost Linear Operators, *International Journal of Fuzzy System Applications* (IJFSA), *3*(1), 140–150.
90. Burgin, M. (2013b). Nonlinear Cauchy-Kowalewski Theorem in Extrafunctions, in *Topics in Integration Research*, Nova Science Publishers, New York, 2013, 167–202.
91. Burgin, M. (2014). Weighted E-Spaces and Epistemic Information Operators, *Information*, *5*(3), 357–388.
92. Burgin, M. (2015). Operations with Extrafunctions and Integration in Bundles with a Hyperspace Base, in *Functional Analysis and Probability*, Nova Science Publishers, New York, 3–76.
93. Burgin, M., Dantsker, A. M. (1995). A method of solving operator equations of mechanics with theory of Hypernumbers, *Notices of the National Academy of Sciences of Ukraine*, *8*, 27–30 (in Russian).
94. Burgin, M., & Dantsker, A. (2015). Real-Time Inverse Modeling of Control Systems Using Hypernumbers, in *Functional Analysis and Probability*, Nova Science Publishers, New York, 439–456.

95. Burgin, M., Dantsker, A., & Esterhuysen, K. Lithium Battery Temperature Prediction, *Integration*: *Mathematical Theory and Applications*, *3*(4), 2012, 319–331.
96. Burgin, M., & Duman, O. (2011). Approximations by Linear Operators in Spaces of Fuzzy Continuous Functions, *Positivity*, *15*(1), 57–72.
97. Burgin, M., & Glushchenko, V. (1997). Superposition of Fuzzy Continuous Functions, in *Methodological and Theoretical Problems of Mathematics and Information and Computer Sciences*, Ukrainian Academy of Information Sciences, Kiev, 45–51 (in Russian).
98. Burgin, M., & Glushchenko, V. (1998). Spaces of the Fuzzy Continuous Functions, in *On the Nature and Essence of Mathematics*, Appendix, Ukrainian Academy of Information Sciences, Kiev, 113–121 (in Russian).
99. Burgin, M., & Kalina, M. (2005). Fuzzy Conditional Convergence and Nearness Relations, *Fuzzy Sets and Systems*, *149*(3), 383–398.
100. Burgin, M., & Karasik, A. (1976). Operators of multidimensional structured model of parallel computations, *Automation and Remote Control*, *37*(8), 1295–1300.
101. Burgin, M., & Krinik, A. C. (2009). Probabilities and hyperprobabilities, 8th *Annual International Conference on Statistics, Mathematics and Related Fields*, Conference Proceedings, Honolulu, Hawaii, 351–367.
102. Burgin, M., & Krinik, A. C. (2010). Introduction to Conditional Hyperprobabilities, *Integration*: *Mathematical Theory and Applications*, *2*, 285–304.
103. Burgin, M., & Krinik, A. C. (2012). Hyperexpectations of random variables without expectations, *Integration*: *Mathematical Theory and Applications*, *3*, 245–267.
104. Burgin, M., & Krinik, A. C. (2013). *Properties of Conditional Hyperprobabilities*, in Topics in Integration Research, Chapter 15, Nova Science Publishers, New York, 265–288.
105. Burgin, M., & Krinik, A. C. (2015). Hyperexpectation in Axiomatic and Constructive Settings, in *Functional Analysis and Probability*, Nova Science Publishers, New York, 259–288.
106. Burgin, M., & Ralston, J. (2004). PDE and Extrafunctions, *Rocky Mountain Journal of Mathematics*, *34*, 849–867.
107. Byron, F. W., & Fuller, R. W. (1992). *Mathematics of classical and quantum physics*, Courier Dover Publications, New York.
108. Cantor, G. *Grundlagen einer allgemeinen Mannigfaltigkeitslehre*: *Ein mathematisch-philosophischer Versuch in der Lehre des Unendlichen*, Teubner, Leipzig, 1883.
109. Cantor, G. (1886). Über die verschiedenen Ansichten in Bezug auf die actualendlichen Zahlen, in *Bihang Till Koniglen Svenska Vetenskaps Akademens Handligar*, *11*(19), 1–10.
110. Cantor, G. *Gesammelte Abhandlungen Mathematischen und Philosophischen Inhalts*, Springer, Berlin, 1932.
111. Cantrell, C. D. *Modern Mathematical Methods for Physicists and Engineers*, Cambridge University Press, 2000.
112. Carfi, D. (2004). S-linear operators in quantum mechanics and in economics, *Applied Sciences*, *6*, 7–20.
113. Chang, S. C., & Mordeson, J. N. (1994). Fuzzy linear operators and fuzzy-normed linear space, *Bull. Cal. Math. Soc.*, *86*, 429–436.
114. Choi, M. D. (1988). Almost commuting matrices need not be nearly commuting, *Proc. Amer. Math. Soc.*, *102*(3), 529–533.
115. Choquet, G. *Lectures on Analysis*, W. A. Benjamin, Inc., New York/Amsterdam, 1969.

116. Chow, C. C., & Buice, M. A. (2012). *Path Integral Methods for Stochastic Differential Equations*, Preprint in Nonlinear Sciences, nlin.AO (arXiv:1009.5966).
117. Christensen, O. *Functions, Spaces, and Expansions: Mathematical Tools in Physics and Engineering*, Applied and Numerical Harmonic Analysis, Springer, Berlin/New York, 2010.
118. Church, A. *Introduction to Mathematical Logic*, Princeton University Press, Princeton, 1956.
119. Chwistek, L. *The Limits of Science: Outline of Logic and Methodology of Science*, Ksiaznica-Atlas, Lwöw-Warszawa, 1935.
120. Cobzas, S. *Functional Analysis in Asymmetric Normed Spaces*, Springer, Berlin/New York, 2012.
121. Codd, E. F. (1970). A Relational Model of Data for Large Shared Data Banks, *Communications of the ACM, 13*(6), 377–387.
122. Collins, J. C. *Renormalization*, Cambridge University Press, Cambridge, 1984.
123. Colombeau, J.-F. *A mathematical analysis adapted to the multiplication of distributions*, Springer-Verlag, New York/Heidelberg/Berlin, 1986.
124. Combettes, P. L., & Pesquet, J.-C. (2009). Proximal Splitting Methods in Signal Processing, in *Fixed-Point Algorithms for Inverse Problems in Science and Engineering*, 49, series Springer Optimization and Its Applications, 185–212.
125. Connes, A. *Noncommutative geometry*, Academic Press Inc., San Diego, CA, 1994.
126. Conway, J. H. *On Numbers and Games*, Academic Press, London/New York, 1976.
127. Conway, J. B. (1990). *A Course in Functional Analysis*, Springer, Berlin/New York.
128. Courbage, M. (2001). Time operator in quantum mechanics and some stochastic processes with long memory, *Cybernetics and Systems*, 32, No. 3–4, 385–392.
129. Dacorogna, B. *Direct Methods in the Calculus of Variations*, Springer-Verlag, New York, 1989.
130. Dales, H. G., & Polyakov, M. E. *Multi-Normed Spaces*, Preprint in Functional Analysis (math.FA) (arXiv:1112.5148).
131. Date, C. J., Darwen, H., & Lorentzos, N. *Temporal Data & the Relational Model,* Morgan Kaufmann, San Mateo, CA, 2002.
132. Davidson, K. (1985). Almost commuting hermitian matrices, *Math. Scand., 56*, 222–240.
133. Davies, E. B., & Lewis, J. T. (1970). An Operational Approach to Quantum Probability, *Commun. Math. Phys., 17*, 239–260.
134. Davis, T. *The summation of series*, Principia Press of Trinity University, San Antonio, 1962.
135. Dikranjan, D., & Tholen, W. *Categorical Structure of Closure Operators*, Kluwer Academic Publishers, 1995.
136. Ditkin, V. A., & Prudnikov, A. P. *Operational calculus*, Vysshaya Shkola, Moscow, 1966 (in Russian).
137. Dominowski, R. L., & Dallob, P. (1995). Insight and Problem Solving, in *The Nature of Insight*, MIT Press, Harvard, MA, USA, 33–62.
138. Doran, R. S., & Belfi, V. A. *Characterizations of C*-algebras*, Monographs and Textbooks in Pure and Applied Mathematics, 101, Marcel Dekker Inc., New York, 1986.
139. Dosi, A. (2011). Local operator algebras, fractional positivity and the quantum moment problem, *Transactions of the American Mathematical Society, 363*(2), 801–856.
140. Dosiev, A. A. (2008). Local operator spaces, unbounded operators and multinormed C∗-algebras, *J. Funct. Anal.*, 255, 1724–1760.

141. Draper, N. R., & Smith, H. *Applied Regression Analysis*, Wiley Series in Probability and Statistics, Wiley-Interscience, New York, 1998.
142. Duffin, R. J., & L. A. Karlovitz, (1968). Formulation of linear programs in analysis. I. Approximation theory, *SIAM J.Appl. Math.*, *16*, 662–675.
143. Dunford, N., & Schwartz, J. *Linear Operators*, Interscience Publishers, New York, 1958.
144. Edwards, R. E. *Functional Analysis*, Reinehard and Winston, New York, 1965.
145. Edwards, J. R., & Wayment, S. G. (1970). A *v*-integral representation for linear operators on spaces of continuous functions with values in topological vector spaces, *Pacific J. Math.*, *35*(2), 327–330.
146. Effros, E. G., & Ruan, Z.-J., *Operator spaces*, London Math. Soc., Clarendon Press, Oxford (2000).
147. Egorov Yu. V. (1990). A contribution to the theory of generalized functions, *Russian Math. Surveys* 45, 1–49.
148. Ehresmann, Ch. *Introduction à la théorie des structures infinitésimales et des pseudo-groupes de Lie*, Colloque de Topologie et Géométrie Différentielle, No. 11, Strasbourg, 1952.
149. Ehresmann, Ch. *Introduction à la théorie des structures infinitésimales et des pseudo-groupes de Lie, Géométrie Différentielle*, Colloques Internationaux du Centre National de la Recherche Scientifique, Centre National de la Recherche Scientifique, Paris, 1953.
150. Ehrlich, P. (1994). All numbers great and small, in *Real Numbers, Generalizations of the Reals, and Theories of Continua*, Kluwer Academic Publishers, 239–258.
151. Ehrlich, P. (2001). Number systems with simplicity hierarchies: A generalization of Conway's theory of surreal numbers, *The Journal of Symbolic Logic*, *66*, 1231–1258.
152. Estrada, R., & Kanwal, R. P. *Asymptotic Analysis: A Distributional Approach*, Birkhäuser, Boston, 1994.
153. Estrada, R., & Kanwal, R. P. *A Distributional Approach to Asymptotics: Theory and Applications*, Birkhäuser Advanced Texts, Boston, 2002.
154. Eschrig, H. *Topology and Geometry for Physics*, Springer-Verlag, Berlin/Heidelberg, 2011.
155. Evans M., & Harrell, *A Short History of Operator Theory*, 2004 http://www.mathphysics.com/opthy/OpHistory.html.
156. Exel, R., & Loring, T. (1989). Almost commuting unitary matrices, *Proc. Amer. Math. Soc.*, *106*, 913–915.
157. Exner, P., & Havliček, M. *Hilbert Space Operators in Quantum Physics*, Springer, New York, 2008.
158. Felbin, C. (1992). Finite dimensional fuzzy-normed linear space, *Fuzzy Sets and Systems*, *48*, 239–248.
159. Felt, J. E. (1974). ε-Continuity and Shape, *Proc. Amer. Math. Soc.*, *46*(3), 426–430.
160. Ferrer, J., Gregori, V., & Alegre, C. (1993). Quasi-uniform structures in linear lattices, *Rocky Mountain J. Math.*, *23*, 877–884.
161. Fillmore, P. A. *A user's guide to operator algebras*, Canadian Mathematical Society Series of Monographs and Advanced Texts, John Wiley & Sons Inc., New York, 1996.
162. Fisher, B. (1969). Products of generalized functions, *Studia Math.*, *33*, 227–230.
163. Flood, J. *Free topological vector spaces*, Państwowe Wydawn. Nauk., Warszawa, 1984.
164. Forti, G. L. (1987). The stability of homomorphisms and amenability with applications to functional equations, *Abh. Math. Sem. Univ. Hamburg*, *57*, 215–226.

165. Forti, G. L. (1995). Hyers–Ulam stability of functional equations in several variables, *Aequat. Math.*, 50, 143–190.
166. Fox, J. *Applied Regression Analysis and Generalized Linear Models*, Sage, Thousand Oaks, CA, 2016.
167. Fraigniaud, P., Lebhar, E., & Viennot, L. (2008). The Inframetric Model for the Internet, in Proc. *The 27th Conference on Computer Communications*, (*IEEE INFOCOM 2008).* 1085–1093.
168. Frieden, R. B. *Physics from Fisher Information*, Cambridge: Cambridge University Press, 1998.
169. Fuchs, L. *Partially Ordered Algebraic Systems*, Pergamon Press, Oxford/London/New York, 1963.
170. Gamelin, T. W. *Uniform Algebras*, Prentice-Hall, Englewood Cliffs, NJ, 1969.
171. García-Raffi, L. M., Romaguera, S., & Sánchez-Pérez, E. (2002). A. Sequence spaces and asymmetric norms in the theory of computational complexity, *Math. Comput. Model., 36*, 1–11.
172. García-Raffi, L. M., Romaguera, S., & Sánchez-Pérez, E. (2003). On Hausdorff asymmetric normed linear spaces, *Houston J. Math.*, 29, 717–728.
173. Gajda, Z. (1991). On stability of additive mappings, *Internat. J. Math. Math. Sci., 14*, 431–434.
174. Găvruta, P. (1994) A generalization of the Hyers–Ulam–Rassias stability of approximately additive mappings, *J. Math. Anal. Appl., 184*, 431–436.
175. Gelfand, I. M. (1941). Normierte Ringe, *Mat. Sbornik, 9*, 3–24.
176. Gelfand, I. M., & Naimark, M. A. (1943). On the embedding of normed rings into the ring of operators in Hilbert space, *Mat. Sbornik, 12*, 197–213.
177. Gel'fand, I. M., Raikov, D., & Shilov, G. E. *Commutative Normed Rings*, Chelsea P. C., New York, 1964.
178. Gelfand, I. M., & Vilenkin, N. J. *Generalized Functions*, 4: Some Applications of Harmonic Analysis. Rigged Hilbert Spaces, Academic Press, New York, 1964.
179. Geroch, R. *Mathematical Physics*, Chicago Lectures in Physics, Chicago, 1985.
180. Gierz, G. *Bundles of topological vector spaces and their duality*, Springer-Verlag, Berlin/New York, 1982.
181. Giles, J. R. (2000). *Introduction to the Analysis of Normed Linear Spaces*, Cambridge University Press, Cambridge.
182. Giraldo, A., Morón, M. A., Ruiz del Portal, F. R., & Sanjurjo, J. M. R. (2001). Finite approximations to Čech homology, *Journal of Pure and Applied Algebra*, v. *163*(1), 81–92.
183. Goldstein, E. B. *Cognitive Psychology: Connecting Mind, Research, and Everyday Experience*, Thomson Wadsworth, Belmont, 2005.
184. Gonshor, H. *An Introduction to the Theory of Surreal Numbers*, Cambridge University Press, Cambridge, England, 1986.
185. Gordon, Y., & Kalton, N. J. (1994). Local structure theory for quasi-normed spaces, *Bull. Sci. Math., 118*, 441–453.
186. Gordon, Y., & Lewis, D. R. (1991). Dvoretzky's theorem for quasi-normed spaces, *Illinois J. Math., 35*, 250–259.
187. Gromov, M. *Partial differential relations*, Springer-Verlag, Berlin, 1986.
188. Grothendieck, A. *Topological Vector Spaces*, Gordon and Breach, New York, 1992.
189. Grothendieck, A. (1955). *Produits tensoriels topologiques et espaces nucléaires*, Mem. Amer. Math. Soc., *16*, Providence, NY.

190. Gruber, P. M. (1978). Stability of isometries, *Trans. Amer. Math. Soc., 245*, 263–277.
191. Guillen, B., Lallena, J., & Sempi, C. (1999). A study of boundedness in probabilistic normedspaces, *J. Math. Anal. Appl., 232*, 183–196.
192. Guillen, B., Lallena, J., & Sempi, C. (1998). Probabilistic norms for linear operators, *J. Math.Anal. Appl., 220*, 462–476.
193. Guillen, B., Lallena, J., & Sempi, C. (1997). Some classes of probabilistic normed spaces, *Rendicontidi Mathematica, 17*(7), 237–252.
194. Hadamard, J. *Le∩ons sur la propagation des ondes et les équations de l'hydrodynamique*, Hermann, Paris, 1903.
195. Hadamard, *Le∩ons sur le calcul des variations*, Librairie Scientifique, A. Hermann et Fils, Paris, 1910.
196. Halmos, P. (1976/1977). Some unsolved problems of unknown depth about operators on Hilbert space, *Proc. Roy. Soc. Edinburgh Sect.* A, *76*(1), 67–76.
197. Hardy, G. H., Littlewood, J. E., & Pólya, G. *Inequalities*, Cambridge Mathematical Library, Cambridge University Press, Cambridge, 1952.
198. Harland, W. B., Armstrong, R. L., Cox, A. V., Craig, L. E., Smith, A. G., & Smith, D. G. (1990). *A Geologic Time Scale*, Cambridge University Press, Cambridge.
199. Hasler, M. F. (2006). Generalized functions as sequence spaces with ultranorms, *Integral Transforms and Special Functions, 17*(2), 149–156.
200. Hastings, M. B. (2009). Making Almost Commuting Matrices Commute, *Commun. Math. Phys., 291*, 321.
201. Hastings, M., & T. Loring, T. Almost commuting matrices, localized Wannier functions, and the quantum Hall effect, *J. Math. Phys., 51*, 015214.
202. Heaviside, O. (1893). On Operators in Physical Mathematics, I, *Proc. Roy. Soc.*, London, *52*, 504–529.
203. Heaviside, O. (1894). On Operators in Physical Mathematics, II, *Proc. Roy. Soc.*, London, *54*, 105–143.
204. Heaviside, O. *Electromagnetic theory*, London, 1899.
205. Helemski, A. Ya. *The Homology of Banach and Topological Algebras*, Kluwer Academic Publishers, Dordrecht, 1989.
206. Helemski, A. Y. *Quantum Functional Analysis.* AMS Publishers, Providence, R. I., 2010, p. 468.
207. Henle, J. M. (1999). Non-nonstandard analysis: Real infinitesimals, *The Mathematical Intelligencer, 21*(1), 67–73.
208. Henle, J. M., & Kleinberg, E. M. *Infinitesimal Calculus*, M.I.T. Press, Cambridge, MA, 1979.
209. Hocking, J. G., & Young, G. S. *Topology*, Dover Publications, Inc., New York, 1961.
210. Hogbe-Nlend, H. (1970). Complétion, tenseurs et nucléarité en bornologie, *J. Math. Pures Appl., 49*(9), 193–288.
211. Hogbe-Nlend, H. *Bornologies and Functional Analysis*, North-Holland Publishing Co., Amsterdam, 1977.
212. Horváth, J. *Topological Vector Spaces and Distributions*, Addison-Wesley Pub. Co., Reading, Mass., 1966.
213. Hoskins, R. F., & Pinto, J. S. *Theories of Generalised Functions: Distributions, Ultradistributions and Other Generalized Functions*, Horwood Publishing, New York, 2003.
214. House, J. (Ed.) *Inorganic Chemistry*, Elsevier, Amsterdam, 2013.
215. Hauser, J. (2002). A projection operator approach to the optimization of trajectory functionals, Proceedings of the 15th *IFAC World Congress*, 15, pt. 1, Barcelona, Spain, 310–315.

216. Hu, T. C., Klee, V., & Larman, D. (1989). Optimization of Globally Convex Functions, *Coll. Math.*, *27*, 1026–1047.
217. *Hui-Yun, P.,* & Zu Sen, Z. (*1989).* On the validity of the mass–velocity operator in quantum *chemistry, International Journal of Quantum Chemistry*, *36*(1), 15–18.
218. Huang, K. *Quantum Field Theory*: *From Operators to Path Integrals*, John Wiley, New York, 1998.
219. Husain, T. *The Open Mapping and Closed Graph Theorems in Topological Vector Spaces*, Clarendon Press, Oxford, 1965.
220. Hyers, D. H. (1939). Locally Bounded Linear Topological Spaces, *Revista de Ciencias Lima*, *41*, 558–574.
221. Hyers, D. H. (1941). On the stability of the linear functional equation, *Proc. Nat. Acad. Sci.*, *27*, 222–224.
222. Hyers, D. H. (1945). Linear topological spaces, *Bull. Amer. Math. Soc.*, *51*, 1–21.
223. Hyers, D. H. (1983). The stability of homomorphisms and related topics, in *Global Analysis–Analysis on Manifolds*, Texte zur Mathematik, vol. 57, Teubner, 140–153.
224. Hyers, D. H., & Rassias, T. M. (1992). Approximate homomorphisms, *Aequationes Mathematicae*, *44*(2–3), 125–153.
225. Ioku, N., Metafune, G., Sobajima, M., & Spina, C. (2015). L^p–L^q estimates for homogeneous operators, *Commun. Contemp. Math.*, *17*(6), pp.
226. Iqbal, H. J., & Radhi, I. M. A. (2003). Bounded Linear Operators in Probabilistic Normed Space, *Journal of Inequalities in Pure and Applied Mathematics*, *4*(1), 1–7.
227. Isac, G., & Rassias, Th.M. (1993). On the Hyers–Ulam stability of ψ-additive mappings, *J. Approx. Th.*, *72*, 131–137.
228. Isham, C. J., Salam, A., & Strathdee, J. (1972). Infinity Suppression Gravity Modified Quantum Electrodynamics, *Phys. Rev. D5*, 2548.
229. Jacobson, N. *Lectures in Abstract Algebra*, D. Van Nostrand Company, Princeton/New York, 1951.
230. Jacobson, N. (1958). Composition algebras and their automorphisms, *Rendiconti del Circolo Mathematico di Palermo, 7,* 55–80.
231. Jacobson, N. *Basic Algebra*, Dover Publications, New York, 2009.
232. James, I. M. *Topological and Uniform Spaces*, Springer-Verlag, Berlin/New York, 1987.
233. Jarvis, P. D., Bashford, J. D., & Sumner, J. G. (2004). *Path Integral Formulation and Feynman Rules for Phylogenetic Branching Models*, q-bio.PE (arXiv:q-bio/0411047).
234. Jolley, L. B. W. *Summation of Series*, Dover Publications, New York, 1961.
235. Jordan, T. F. *Linear Operators for Quantum Mechanics*, John Wiley and Sons, Inc., 1969.
236. Jordan, C. *Calculus of Finite Differences*, Chelsea, New York, 1965.
237. Jung, S.-M. (1996). On the Hyers–Ulam–Rassias stability of approximately additive mappings, *J. Math. Anal. Appl.*, *204*, 221–226.
238. Jung, S.-M. (1997). Hyers–Ulam–Rassias stability of functional equations, *Dynamic Syst. Appl.*, *6*, 541–566.
239. Jung, S.-M. (1998). Hyers–Ulam–Rassias stability of Jensen's equation and its application. *Proc. Amer. Math. Soc.*, *126*, 3137–3143.
240. Jung, S.-M. (1998a). On the Hyers–Ulam stability of the functional equations that have the quadratic property, *J. Math. Anal. Appl.*, *222*, 126–137.
241. Kac, V., & Cheung, P. *Quantum Calculus*, Springer, Berlin, 2002.

242. Kachkovskiy, I., & Safarov, Y. (2016). Distance to normal elements in C*-algebras of real rank zero, *J. Amer. Math. Soc.*, *29*(1), 61–80.
243. Kalton, N. (2003). Quasi-Banach spaces, in *Handbook of Geometry of Banach Spaces*, 2, (W. B. Johnson and J. Lindenstrauss, Eds), Elsevier, Amsterdam, 1099–1130.
244. Kalton, N., & Sik-Chung Tam, (1993). Factorization theorems for quasi-normed spaces, *Houston J. Math.*, *19*, 301–317.
245. Kanovei, V., & Reeken, M. (2000). On Ulam's problem concerning the stability of approximate homomorphisms, *Tr. Mat. Inst. Steklova*, 231, Dyn. Syst., Avtom. i Beskon. Gruppy, 249–283 (translation from Russian in *Proc. Steklov Inst. Math.* No. 4 (231), 238–270).
246. Kaplansky, I. (1949). Normed algebras, *Duke Math. J.*, *16*, 399–418.
247. Kato, T. *Perturbation Theory for Linear Operators*, Springer, Berlin, Heidelberg, New York, 1980.
248. Katsaras, A. K. (1984). Fuzzy topological vector spaces, *Fuzzy Sets and Systems*, *12*, 143–154.
249. Kelly, J. L. *General Topology*, Van Nostrand Co., Princeton/New York, 1955.
250. Khaleelulla, S. M. *Counterexamples in Topological Vector Spaces*, Lecture Notes in Mathematics, 936, Springer-Verlag, Berlin/Heidelberg, 1982.
251. Kikina, L., Kikina, K., & Gjino, K. (2012). A New Fixed Point Theorem on Generalized Quasimetric Spaces, *International Scholarly Research Notices*, *ISRN Mathematical Analysis*. Article ID 457846, 1–9 pp.
252. Klee, V. L. (1961). Stability of the Fix-Point Property, *Colloquium Math.*, *8*, 43–46.
253. Klee, V. L., & Yandl, A. (1974). Some proximate concepts in topology, *Symposia Math. Publ. Inst. Naz. di Alta Matematica*, Academic Press, *16*, 21–39.
254. Kleinert, H. *Path Integrals in Quantum Mechanics, Statistics, Polymer Physics, and Financial Markets*, World Scientific Publishing Co., New York/London/Singapore, 2006.
255. Knuth, D. E. *Surreal Numbers*: *How Two Ex-Students Turned on to Pure Mathematics and Found Total Happiness*, Reading, Addison-Wesley, Massachusetts, 1974.
256. Kohn, J. J., & Nirenberg, L. (1965). An algebra of pseudo-differential operators, *Comm. Pure Appl. Math.*, *18*(1/2), 269–305.
257. Kolchin, E. R. *Differential Algebra and Algebraic Groups*, Academic Press, New York, 1973.
258. Kolmogoroff, A. (1934). Zur Normierbarkeit eines allgemeinen topologischen linearen Raumes, *Studia Mathematica*, *5*, 29–33.
259. Kolmogorov, A. N., & Fomin, S. V. *Elements of the Theory of Functions and Functional Analysis*, Dover Publications, New York, 1999.
260. Komatsu, H. (1973). Uliradistributions. I: Structure theorems and a characterization, *J. Fac. Sei. Univ. Tokyo*, Sect. IA Math. *20*, 25–105.
261. König, H. (1953). Neue Begrundung der Theorie der Distributionen von L. Schwartz, *Math.Nachr.*, *9*, 129—148.
262. Korányia, A., & Misra, G. (2008). Homogeneous operators on Hilbert spaces of holomorphic functions, *Journal of Functional Analysis*, *254*(9), 2419–2436.
263. Korevaar, J. (1955). Distributions defined by fundamental sequences, *Nederl. Akad. Wetensch. Proc.*, Ser. A, 58 (*Indag. Math.*, 17), I, 368–378; II, 379–389; III, 483–493; IV, 494–503; V, 663–674.
264. Korevaar, J. (1959). Pansions and the theory of Fourier transforms, *Trans. Amer. Math. Soc.*, 91, 53–101.

265. Köthe, G. *Topological vector spaces*, Grundlehren der mathematischen Wissenschaften, v.159: Springer-Verlag, Berlin/New York, 1969.
266. Krasil'shchik, S., Lychagin, V. V., & Vinogradov, A. M. *Geometry of jet spaces and nonlinear partial differential equations*, Gordon and Breach, New York, 1986.
267. Kraus, K. *States, Effects, and Operations*. Springer-Verlag, Berlin, 1983.
268. Krause, E. F. *Taxicab Geometry*, Dover Publications, New York, 1987.
269. Krein, M. G., & Nudel'man, A. A. *The Markov Moment Problem and Extremum Problems*, American Mathematical Society, Providence, RI, 1977.
270. Krishna, S. V., & Sarma, K. K. M. (1994). Separation of fuzzy normed linear space, *Fuzzy Sets and Systems*, *63*, 207–217.
271. Künzi, H.-P. A. (2009). An introduction to quasi-uniform spaces, in *Beyond Topology*, Contemporary Mathematics, 486, American Mathematical Society, Providence, RI, 239–304.
272. Kuratowski, K. (1966). *Topology*, Academic Press, Warszawa, *1*(2), 1968.
273. Kurosh, A. G. *Lectures on General Algebra*, Chelsea P. C., New York, 1963.
274. Kurzweil, J. *Henstock-Kurzweil Integration: Its Relation to Topological Vector Spaces*, World Scientific Publishing Co., 2000.
275. Lael, F., & Nourouzi, K. (2009). Compact Operators Defined on 2-Normed and 2-Probabilistic Normed Spaces, *Mathematical Problems in Engineering*, Article ID 950234, 17 pp.
276. Laugwitz, D., (1960). Anwendungen unendlich kleiner Zahlen I, *J. Reine Angew. Math.*, *207,* 53–60.
277. Laugwitz, D. (1961), Anwendungen unendlich kleiner Zahlen II, *J. Reine Angew. Math.*, *208,* 22–34.
278. Laugwitz, D. *Zahlen und Kontinuum: Eine Einfuhrung in die Infinitesimalmathematik* (Lehrbucher und Monographien zur Didaktik der Mathematik), 1986.
279. Lee, S. (1990). Projection operator Hueckel method and antiferromagnetism, *J. Am. Chem. Soc.*, *112*(19), 6777–6783.
280. Lee, E. B., & Markus, L. *Foundations of Optimal Control Theory*, John Wiley & Sons, New York, 1989.
281. Lemin, A.Yu. (2003). The category of ultrametric spaces is isomorphic to the category of complete, atomic, tree-like, and real graduated lattices LAT*, *Algebra Universalis*, *50,* 35–49.
282. Lemin, A. Yu., & Smirnov, Yu. M. (1986). Groups of isometries of metric and ultrametric spaces and their subgroups, *Russian Math. Surveys*, *41*(6), 213–214.
283. Lerner, V. S. *Information Path Functional and Informational Macrodynamics*, Nova Science Publishers, New York, 2010.
284. Lerner, V. S. (2012). The information path functional approach to solution of a controllable stochastic problem, *Integration*, *3*(1), 55–110.
285. Levine, I. N. *Quantum Chemistry*, Prentice-Hall, Inc., New Jersey, 2000.
286. Lin, H. (1997). Almost commuting self-adjoint matrices and applications, in *Operator algebras and their applications* (Waterloo, ON 1994/1995), Fields Inst. Commun. 13, Amer. Math. Soc., Providence, RI, 193–233.
287. Litvak, A. E. (1998). The Extension of the Finite-Dimensional Version of Krivine's Theorem to Quasi-Normed Spaces, *Convex Geometric Analysis*, MSRI Publications, *34*, 139–148.
288. Litvak, A. E. (2000). Kahane–Khinchin's inequality for the quasi-norms, *Canad. Math. Bull.*, 43, 368–379.

289. Litvak, A. E., Milman, V. D., & G. Schechtman, G. (1998). Averages of norms and quasi-norms, *Math. Ann.*, *312*, 95–124.
290. Liverman, T. P. G. *Generalized Functions and Direct Operational Methods*, Prentice Hall, Englewood Cliffs, New Jersey, 1964.
291. Louden, K. C. *Programming Languages: Principles and Practice*, Cengage Learning, 2003.
292. Lowen, R., Sioen, M., & Ferwulgen, S. (2009). Categorical topology, in *Beyond Topology*, Contemporary Mathematics, 486, American Mathematical Society, Providence, RI, 1–36.
293. Ludwig, G. (1964). Versuch einer axiomatischen Grundlegung der Quantenmechanik und allgemeinerer physikalischer Theorien, *Zeitschrift für Physik*, *181*, 233–260.
294. Lunts, V. A., & Rosenberg, A. L. *Differential calculus in noncommutative algebraic geometry* I. *D-calculus on noncommutative rings*, Max Planck Institute preprint MPI 96–53, Bonn 1996.
295. Lunts, V. A., & Rosenberg, A. L. (1997). Differential operators on noncommutative rings, *Sel. Math.*, N. S., 3, 335–359.
296. Lurye, A. I. *Operational calculus and its application to problems of mechanics*, Gostechizdat, Moscow/Leningrad, 1950 (in Russian).
297. Magaril-Il'yaev, G. G., & Tikhomirov, V. M. *Convex Analysis*: *Theory and Applications*, American Mathematical Society, Providence, RI, 2003.
298. Masani, P. R (1947). Multiplicative Riemann integration in normed rings, *Transactions of the American Mathematical Society*, *61*(1), 147 –192.
299. Maslov, V. P. *Complex Markov Chains and a Feynman Path Integral for Nonlinear Equations*, Nauka, Moscow, 1976 (in Russian).
300. Maslov, V. P. *Operational Methods*, Mir, Moscow, 1976a.
301. Maslov, V. P. *Mathematical Methods in Integral Optics*, MIEM, Moscow, 1983.
302. Mazur, S. (1938). Sur les anneaux lineaires, *C. R. Acad. Sci. Paris*, *207*, 1025–1027.
303. Metzler, R., & Nakano, H. (1966). Quasi-Norm Spaces, *Transactions of the American Mathematical Society*, *123*(1), 1–31.
304. Meyer, R. (2004). Smooth group representations on bornological vector spaces, *Bull. Sci. Math.*, *128*, 127–166.
305. Meyer, R. (2004a). Bornological versus topological analysis in metrizable spaces, *Contemp. Math.*, *363*, 249–278.
306. Mikusinski, J. (1948). Sur la méthode de généralisation de M. Laurent Schwartz et sur la convergence faible, *Fund. Math.*, *35*, 235—239.
307. Mikusinski, J. (1983). Hypernumbers, Part I. Algebra, *Studia Math.*, *77*, 3–16.
308. Milne-Thomson, L. M. *The Calculus of Finite Differences*, Macmillan, London, 1951.
309. Monna, A. F. *Functional Analysis in Historical Perspective*, Wiley, New York and Toronto, 1973.
310. Moore, C., & Russell, A. *Approximate Representations and Approximate Homomorphisms*, Preprint in Mathematics, 2010 (arXiv:1009.6230).
311. Nagumo, M. (1936). Einigean alytische Untersuchungen in linearen metrischen Ringen, *Japan. J. Math.*, *13*, 61–80.
312. Naimark, M. A. *Normed Rings*, P. Noordhoff N. V., Groningen, 1959.
313. Naimark, M. A. *Normed Algebas*, Wolters-Noordhof Publishing Co., Groningen, 1972.
314. Narici, L., & Beckenstein, E. *Topological Vector Spaces*, M. Dekker, New York, 1985.
315. von Neumann, J. (1927). Mathematische Begründung der Quantenmechanik, *Nachricht der Gesellschaft der Wissenschaften zu Göttingen*, 1–57.

316. von Neumann, J. (1929). Zur Algebra der Funktional operatoren und Theorie der normalen Operatoren, *Mathematische Annalen 102,* 370–427.
317. von Neumann, J. (1929a). Proof of the ergodic theorem and the H-theorem in quantum mechanics, *Eur. Phys. J.*, H *57*, 30–70.
318. von Neumann, J. (1932a). Zur Operatoren methode in der klassischen Mechanik, *Ann. Math.*, 33, 587–642.
319. von Neumann, J. (1935). On complete topological spaces, *Trans. Amer. Math. Soc.*, *37*, 1–20.
320. von Neumann, J. *Mathematical Foundations of Quantum Mechanics*, Princeton University Press, Princeton, NJ, 1955 (First published in German in 1932: *Mathematische Grundlagen der Quantenmechank*, Springer, Berlin).
321. von Neumann, J., & Halmos, P. R. (1942). Operator Methods in Classical Mechanics, II, *Ann. Math.*, *43*, 332–350.
322. von Neumann, J., Jordan, P., & Wigner, E. (1934). On an Algebraic Generalization of the Quantum Mechanical formalism, *Ann. Math.*, *35*, 29–64.
323. von Neumann, J., & Murray, F. J. (1936). On Rings of Operators, *Ann. Math.*, *37*, 116–229.
324. von Neumann, J., & Murray, F. J. (1937). On Rings of Operators, II. *AMS Trans.*, *41*, 208–248.
325. Newell, A., & Simon, H. A. *Human Problem Solving*, Prentice Hall, Englewood Cliffs, New Jersey, 1972.
326. Nikolić-Despotović, D., & Pilipović, S. (1988). The quasiasymptotic expansion of tempered distributions of the Stieltjes transform, *Univ. u Novom Sadu Zb. Rad. Prir.-mat. Fak.*, Ser. Mat., *18*(2), 31–44.
327. Nikolić-Despotović, D., & Pilipović, S. (1993). The quasiasymptotic expansion and the moment expansion of tempered distributions, *Publications de L'institut Mathématique*, Nouvelle série, tome *53*(67), 88–94.
328. Nicolis, G., & Prigogine, I. *Self-Organization in Non-Equilibrium Systems*, Willey, New York, 1977.
329. Olver, P. J. *Applications of Lie Groups to Differential Equations*, Springer, New York/Berlin/Heidelberg, 1986.
330. Pantsulaia, G. *Invariant and Quasiinvariant Measures in Infinite-Dimensional Topological Vector Spaces*, Nova Science Publishers, New York, 2007.
331. Partington, J. R. *Linear Operators and Linear Systems: An Analytical Approach to Control Theory*, Cambridge University Press, Cambridge, 2004.
332. Peressini, A. L. *Ordered Topological Vector Spaces*, Harper & Row, New York, 1967.
333. Pietsch, A. *Nukleare Lokalkonvexe Räume*, Akademie-Verlag, Berlin, 1965.
334. Pietsch, A. *History of Banach Spaces and Linear Operators*, Springer, New York/Berlin, 2007.
335. Pilipović, S. (1991). Quasi-asymptotic expansion and the Laplace transformation, *Applicable Anal.*, *35*, 247–261.
336. Pinto, J. S. (1989). Periodic Silva tempered ultradistributions, *Portugaliae mathematica*, *46*(4), 441–452.
337. Pisier, G. *Introduction to Operator Space Theory*, Cambridge University Press, Cambridge, 2003.
338. Poincaré, H. *La Science et l'hypothèse*, Flammarion, Paris, 1902.
339. Poincaré, H. *Scince et Méthode*, Flammarion, Paris, 1908.
340. Poincaré, H. *Dernières Pensées*, Flammarion, Paris, 1913.

341. Prigogine, I. *From Being to Becoming: Time and Complexity in Physical Systems*, Freeman & Co., San Francisco, 1980.
342. Prugovečki, E. (1981). *Quantum mechanics in Hilbert space*, Academic Press, London/New York.
343. Rano, G., & Bag, T. (2015). Bounded linear operators in quasi-normed linear space, *J. Egyptian Math. Soc.*, *23*(2), 303–308.
344. Rassias, T. M. (1978). On the stability of linear mapping in Banach spaces, *Proc. Amer. Math. Soc.*, *72*, 297–300.
345. Rassias, T. M. (2007). Refined Hyers–Ulam approximation of approximately Jensen type mappings, *Bulletin des Sciences Mathématiques*, *131*(1), 89–98.
346. Rassias, T. M., & Rassias, M. J. (2003). On the Ulam stability of Jensen and Jensen type mappings on restricted domains, *J. Math. Anal. Appl.*, *281*, 516–524.
347. Rassias, T. M., & Rassias, M. J. (2005). Asymptotic behavior of alternative Jensen and Jensen type functional equations, *Bull. Sci. Math.*, *129*(7), 545–558.
348. Rhie, G. S., Choi, B. M., & S. K. Dong, (1997). On the completeness of fuzzy-normed linear space, *Math. Japonica*, *45*(1), 33–37.
349. Ribenboim, P. *Functions, Limits, and Continuity*, New York/London/Sydney, 1964.
350. Richardson, C. H. *An Introduction to the Calculus of Finite Differences*, Van Nostrand, New York, 1954.
351. Rickart, C. E. *General Theory of Banach Algebras*, D. van Nostrand, Princeton, N.J. 1960.
352. Riesz, F. (1910). Untersuchungen über Systeme integrierbaren Funktionen, *Math. Ann.*, *69*, 449–497.
353. Riesz, F. (1918). Über lineare Funktional gliehungen, *Acta Math.*, *41*, 71–98.
354. Riesz, F., & Sz.-Nagy, B. *Functional Analysis*, Frederik Ungar P.C., New York, 1955.
355. Robertson, A. P., & Robertson, W. *Topological Vector Spaces*, Cambridge University Press, New York, 1964.
356. Robinson, A. (1961). Non-standard analysis, *Indagationes Mathematicae 23*, 432–440.
357. Robinson, A. *Non-Standard Analysis*, Studies of Logic and Foundations of Mathematics, North-Holland, New York, 1966.
358. Rodríguez-López, J., & Sánchez-Granero, M. A. (2011). Some properties of bornological convergences, *Topology and its Applications*, *158*(1), 101–117.
359. Rolewicz, S. *Metric Linear Spaces*, Mongrafie Matematyczne, 56, PWN, Warsaw, 1972.
360. Rosenthal, P. (1969). Are almost commuting matrices near commuting matrices? *Amer. Math. Monthly*, *76*, 925–926.
361. Ross, K. A. *Elementary Analysis: The Theory of Calculus*, Springer-Verlag, New York/Berlin/Heidelberg, 1996.
362. Roumieu, C. (1960). Sur quelques extensions de la notion de distribution, *Ann. Sei. Ecole Norm. Sup.*, 77(3), 41–121.
363. Rudin, W. *Functional Analysis*, McGraw-Hill, New York, 1991.
364. Sánchez-Álvarez, J. M. (2005). On semi-Lipschitz functions with values in a quasi-normed linear space, *Appl. Gen. Topology*, *6*, 217–228.
365. Sanjurjo, J. M. R. (1989). A Non-continuous Description of the Shape Category of Compacta, *Quart. Journal of Mathematics Oxford*, *40*(2), 351–359.
366. *Sato, M. (1959). Theory of Hyperfunctions, I, Journal of the Faculty of Science, University of Tokyo, Sect. 1, Mathematics, astronomy, physics, chemistry,* *8*(1), 139–193.

367. *Sato, M. (1960). Theory of Hyperfunctions, II, Journal of the Faculty of Science, University of Tokyo, Sect. 1, Mathematics, astronomy, physics, chemistry, 8(2), 387–437.*
368. Sebastiao e Silva, J. (1958). Les fonctions analytiques comme ultradistributions dans le calcul opérationnel, *Math. Ann., 136*, 58–69.
369. Schaefer, H. H., & Wolff, M. P. *Topological Vector Spaces*, Graduate Texts in Mathematics (GTM), 3, Springer-Verlag, Berlin/Heidelberg, 1999.
370. Schmieden, C., & Laugwitz, D. (1958). Eine Erweiterung der Infinitesimalrechnung, *Math. Zeitschr, 69*, 1–39.
371. Schneider, P. *Nonarchimedean Functional Analysis*, Springer, Berlin-Heidelberg-New York, 2001.
372. Schneider, P., & Teitelbaum, J. *Algebras of p-adic distributions and admissible representations*, Preprint in mathematics, math.NT, 2002 (arXiv:math/0206056).
373. Schwartz, L. (1945). Généralisation de la notion de fonction, de dérivation, de transformationde Fourier, et applications mathématiques et physiques, *Annales Univ. Grenoble, 21*, 57–74.
374. Schwartz, L. *Théorie des Distributions*, Vol. I-II, Hermann, Paris, 1950/1951.
375. Schwartz, J. T. *Lectures on the Mathematical Method in Analytical Economics*, Gordon and Breach, New York, 1961.
376. Schwartz, J. T. *Non-Linear Functional Analysis*, Gordon & Breach Science Pub., New York, 1969.
377. Schweizer, B., & Sklar, A. *Probabilistic Metric Space*, Elsevier, North Holland, New York, 1983.
378. Segal, I. E. (1947). Postulates for general quantum mechanics, *Annals of Mathematics, 48*(2), 930–948.
379. Šemrl, P. (1998). Approximately Linear Functionals on Hilbert Spaces, *Journal of Mathematical Analysis and Applications, 226*(2), 466–472.
380. Shuchat, A. H. (1972). Integral representation theorems in topological vector spaces, *Trans. Amer. Math. Soc., 172*, 373–397.
381. Sikorski, R. (1954). A Definition of the Notion of Distribution, *Bull. Acad. Pol. Sci.*, Cl.III, *2*, 207—211.
382. Silva, E. B., Fernandez, D., & Nikolova, L. (2013). Generalized quasi-Banach sequence spaces and measures of noncompactness, *Anais da Academia Brasileira de Ciências, Mathematical Sciences, 85*(2).
383. Smyth, M. B. (1995). Semi-metrics, closure spaces and digital topology, *Theoretical Computer Science, 151*(1), 257–276.
384. Smolyanov, O. G., & S. V. Fomin, (1976). Measures on topological vector spaces, *Uspekhi Mat. Nauk, 31*(4), 3–56.
385. Sobolev S. L. (1936). Méthode nouvelle a résoudre le probléme de Cauchy pour les equations linéaires hyperboliques, *Sbornik, 1*(1), 39–70.
386. Sofo, A. *Computational Techniques for the Summation of Series*, Springer, New York, 2003.
387. Spiegel, M. *Calculus of Finite Differences and Differential Equations*, McGraw-Hill, New York, 1971.
388. Steimann, F. (2001). On the Use and Usefulness of Fuzzy sets in medical AI, *Artificial Intelligence in Medicine, 21*(1–3), 131–137.
389. Stone, M. H. (1930). Linear Transformations in Hilbert Space. III. Operational Methods and Group Theory, *Proceedings of the National Academy of Sciences of the United States of America, 16*(2), 172–175.

390. Stone, M. H. *Linear Transformations in Hilbert Space and their Applications to Analysis*, New York, American Mathematical Society, 1932.
391. Stone, M. H. (1948). The Generalized Weierstrass Approximation Theorem, *Mathematics Magazine*, *21*(4), 167–184; *21*(5), 237–254.
392. Swartz, C. W., & Kurtz, D. S. *Theories of Integration: The Integrals of Riemann, Lebesgue, Henstock-Kurzweil, and McShane*. Series in Real Analysis, 9, World Scientific, New Jersey/London/Singapore, 2004.
393. Synowiec, J. (1983). Distributions: The evolution of a mathematical theory, *Historia Mathematica*, *10*(2), 149–183.
394. Tabor, J. (2004). Stability of the Cauchy functional equation in quasi-Banach spaces, *Ann. Polon. Math.*, *83*, 243–255.
395. Tall, D. O. (1980). The notion of infinite measuring number and its relevance in the intuition of infinity, *Educational Studies in Mathematics*, *11*, 271–284.
396. Temple, G. (1953). Theories and applications of generalized functions, *J. London Math. Soc.*, v. s1–28, Issue 2, 134–148.
397. Temple, G. (1955). The theory of generalized functions, *Proc. Roy. Soc. London*, Math. Ser. A, *228*, 173—190.
398. Todorov, T. D. *Sequential Approach to Colombeau's Theory of Generalized Functions*, Preprint IC/87/26, ICTP, Trieste, 1987.
399. Todorov, T. Pointwise values and fundamental theorem in the algebra of asymptotic functions, in *Non-Linear Theory of Generalized Functions*, Chapman & Hall/CRC Research Notes in Mathematics, *401*, 369–383 (1999).
400. Tomšič, G. (1974). Homogeneous operators, *Studia Mathematica*, *L1*(1), 1–5.
401. Trèves, F. *Topological Vector Spaces, Distributions and Kernels*, Academic Press, Inc., London/New York, 1995.
402. Ulam, S. M. *A Collection of Mathematical Problems,* Interscience Publishers, Inc., New York, 1964.
403. Vestfrid, I. A. (2003). Linear approximation of approximately linear functions, *Aequationes Mathematicae*, *66*(1–2), 37–77.
404. Vlach, M. (1981). Approximation operators in optimization theory, *Zeitschriftfür Operations Research*, *25*(1), 15–23.
405. Volterra, V. (1887). Sopra le funzionei che dependono da altre funzioni, *Memorie della Societa Italiana della Scienze*, *VI*(3), 3–68.
406. Volterra V. (1887a). Sulle equazioni differenziali lineari, *Rendiconti dell'Academia dei Lincei*, *3*, 393–396.
407. Walter, G. G. (1965). Expansions of distributions, *Trans. Amer. Math. Soc.*, *116*, 492–510.
408. Warner, S. *Topological Fields*, North Holland, Mathematics Studies, 157, North-Holland, Amsterdam/London/ New York/Tokyo, 1993.
409. Weierstrass, K. (1885). Über die analytische Darstellbarkeit sogenannter willkürlicher Functionen einer reellen Veränderlichen. *Sitzungsberichte der Königlich Preußischen Akademie der Wissenschaften zu Berlin*, 1885 (II), Erste Mitteilung (part 1) 633–639; Zweite Mitteilung (part 2) 789–805.
410. Weil, A. (1927). Sur calcolo funzionale lineare, *Rendiconti della R. Accademia dei Lincei*, *6*, 773–777.
411. Weil, A. (1927a). Sur les espaces fonctionnels. *C. R. Acad. Sci. Paris*, *184*, 67–69.
412. Weil, A. (1937). Sur les espaces à structure uniforme et sur la topologie générale, *Act. Sci. Ind.*, *551*, Paris.

413. Weinberg, S. *The Quantum Theory of Fields*, Cambridge University Press, Cambridge, 1995.
414. Wilansky, A. *Modern Methods in Topological Vector Spaces*, McGraw-Hill, New York, 1978.
415. Wright, F. B. (1953). Absolute valued algebras, *Proc. Nat. Acad. Sci.*, *39*, 330–332.
416. Xia, Q. (2008). The geodesic problem in nearmetric spaces, *Journal of Geometric Analysis*, *19*(2), 452–479.
417. Xiao, J., & Zhu, X. (2002). On linearly topological structure and property of fuzzy normed linear space, *Fuzzy Sets and Systems*, *125*, 153–161.
418. Yosida, K. (1936). On the groups embedded in the metrical complete ring, *Japan. J. Math.*, *13*, 7–26.
419. Yosida, K. *Functional Analysis*, Springer, New York/Berlin, 1968.
420. Zimmermann, H. J. *Fuzzy Set Theory and Its Applications*, Kluwer Academic Publishers, Boston, MA, 2001.
421. Zlatoš, P., & Špakula, J. (2004). Almost homomorphisms of compact groups, *Illinois J. Math.*, *48*(4), 1183–1189.

INDEX

N

O

W

Z